Applied and Numerical Harmonic Analysis

Applied and Numerical Harmonic Analysis

J.M. Cooper: *Introduction to Partial Differential Equations with MATLAB* (ISBN 0-8176-3967-5)

C.E. D'Attellis and E.M. Fernández-Berdaguer: *Wavelet Theory and Harmonic Analysis in Applied Sciences* (ISBN 0-8176-3953-5)

H.G. Feichtinger and T. Strohmer: *Gabor Analysis and Algorithms* (ISBN 0-8176-3959-4)

T.M. Peters, J.H.T. Bates, G.B. Pike, P. Munger, and J.C. Williams: *Fourier Transforms and Biomedical Engineering* (ISBN 0-8176-3941-1)

A.I. Saichev and W.A. Woyczyński: *Distributions in the Physical and Engineering Sciences* (ISBN 0-8176-3924-1)

R. Tolimieri and M. An: *Time-Frequency Representations* (ISBN 0-8176-3918-7)

G.T. Herman: *Geometry of Digital Spaces* (ISBN 0-8176-3897-0)

A. Procházka, J. Uhlíř, P.J.W. Rayner, and N.G. Kingsbury: *Signal Analysis and Prediction* (ISBN 0-8176-4042-8)

J. Ramanathan: *Methods of Applied Fourier Analysis* (ISBN 0-8176-3963-2)

A. Teolis: *Computational Signal Processing with Wavelets* (ISBN 0-8176-3909-8)

W.O. Bray and Č.V. Stanojević: *Analysis of Divergence* (ISBN 0-8176-4058-4)

G.T Herman and A. Kuba: *Discrete Tomography* (ISBN 0-8176-4101-7)

J.J. Benedetto and P.J.S.G. Ferreira: *Modern Sampling Theory* (ISBN 0-8176-4023-1)

A. Abbate, C.M. DeCusatis, and P.K. Das: *Wavelets and Subbands* (ISBN 0-8176-4136-X)

L. Debnath: *Wavelet Transforms and Time-Frequency Signal Analysis* (ISBN 0-8176-4104-1)

K. Gröchenig: *Foundations of Time-Frequency Analysis* (ISBN 0-8176-4022-3)

D.F. Walnut: *An Introduction to Wavelet Analysis* (ISBN 0-8176-3962-4)

O. Brattelli and P. Jorgensen: *Wavelets through a Looking Glass* (ISBN 0-8176-4280-3)

H.G. Feichtinger and T. Strohmer: *Advances in Gabor Analysis* (ISBN 0-8176-4239-0)

O. Christensen: *An Introduction to Frames and Riesz Bases* (ISBN 0-8176-4295-1)

L. Debnath: *Wavelets and Signal Processing* (ISBN 0-8176-4235-8)

J. Davis: *Methods of Applied Mathematics with a MATLAB Overview* (ISBN 0-8176-4331-1)

G. Bi and Y. Zeng: *Transforms and Fast Algorithms for Signal Analysis and Representations* (ISBN 0-8176-4279-X)

J.J. Benedetto and A. Zayed: *Sampling, Wavelets, and Tomography* (ISBN 0-8176-4304-4)

E. Prestini: *The Evolution of Applied Harmonic Analysis* (ISBN 0-8176-4125-4)

O. Christensen and K.L. Christensen: *Approximation Theory* (ISBN 0-8176-3600-5)

L. Brandolini, L. Colzani, A. Iosevich, and G. Travaglini: *Fourier Analysis and Convexity* (ISBN 0-8176-3263-8)

W. Freeden and V. Michel: *Multiscale Potential Theory* (ISBN 0-8176-4105-X)

O. Calin and D.-C. Chang: *Geometric Mechanics on Riemannian Manifolds* (ISBN 0-8176-4354-0)

J.A. Hogan and J.D. Lakey: *Time-Frequency and Time-Scale Methods* (ISBN 0-8176-4276-5)

C. Heil: *Harmonic Analysis and Applications* (ISBN 0-8176-3778-8)

Harmonic Analysis and Applications

In Honor of John J. Benedetto

Christopher Heil
Editor

Birkhäuser
Boston • Basel • Berlin

Christopher Heil
Georgia Institute of Technology
School of Mathematics
Atlanta, GA 30332-0160
USA

Cover design by Mary Burgess.

Mathematics Subject Classification: 41-02, 42-XX, 42-02, 42B35, 42C15, 42C40, 46-02, 47-02, 94A20

Library of Congress Control Number: 2006926486

ISBN-10 0-8176-3778-8 e-ISBN: 0-8176-4504-7
ISBN-13 978-0-8176-3778-1

Printed on acid-free paper.

Printed in the United States of America. (EB)

9 8 7 6 5 4 3 2 1

www.birkhauser.com

To Papa Benedetto

Mathematical relatives and descendants of John Benedetto at the October 1999 conference (University of Maryland, College Park) in honor of John's 60th birthday

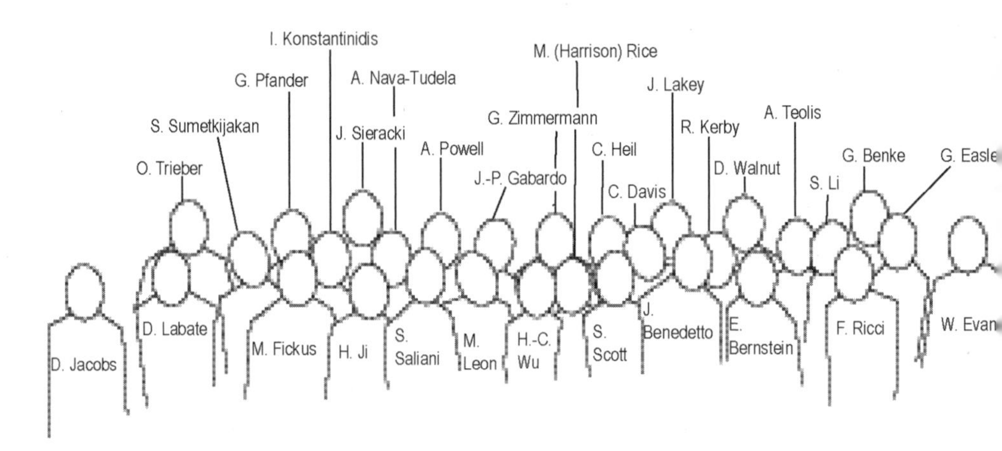

Foreword

This volume is dedicated to John Benedetto. It seems just yesterday that we celebrated his 60th birthday in a memorable conference in College Park. Yet that was October of 1999, and already more than six years have passed. But John is still too young to be fully honored by a single foreword, or even a single volume, that attempts to summarize the impact of his work on harmonic analysis, his students, and his coworkers. Given his continuing high (and even increasing) level of activities, his list of "lifetime achievements" is surely far from complete. Even so, we will make an attempt in this foreword to take a look back, to see the major lines of his work and activities during the past 40 years of his life as a scientist, and to learn from his biography (and bibliography) how the field of harmonic analysis has changed over the years, and in particular to see the vibrant role that John has taken in this process.

John's first paper appeared in 1965, when he was 25 years old, and his first book (the Springer Lecture Notes on *Harmonic Analysis on Totally Disconnected Sets*) when he was 31. By that time he had already published on the subjects of Tauberian algebras, in the theory of generalized functions, and on questions related to spectral synthesis. His work on this latter topic continued through the 1970s, culminating in the insightful volume *Spectral Synthesis* (1975). Only a year later, his text *Real Variables and Integration with Historical Notes* appeared.

In recent years John has published a variety of mathematical papers that include practical applications, yet this is not really a "new" direction for him. Already in 1981 one finds a paper titled "The theory of constructive signal analysis" (Stud. Appl. Math., Vol. 65, pp. 37-80). It is clear from this and his entire bibliography that John has studied the work of Norbert Wiener in great detail, and indeed on many occasions John has confessed that Wiener is his "hero." For example, he connected his earlier work on Tauberian theorems with the uncertainty principle, and in 1989 published (jointly with his former Ph.D. students George Benke and Ward Evans) an n-dimensional version of the Wiener–Plancherel formula, whose 1-dimensional formulation Wiener chose to have on the cover of his autobiography, *I Am a Mathematician.*

In 1990 his first article relating irregular sampling and frame theory appeared, published jointly with another one of his (numerous) students, William Heller. A series of papers on Fourier transform inequalities with weights were written together with his friends, Hans Heinig and Raymond Johnson, between 1982 and 2003.

In the 1990s one finds a substantial number of papers related to wavelet and time-frequency analysis. When the "wavelet-wave" took off around 1990,[1] John was one of the first to realize the relevance of what we nowadays call application-oriented harmonic analysis (more on this later). He encouraged two of his students (Chris Heil and Dave Walnut) to compile, from the material they had collected during their research for their Ph.D. theses, a survey article on wavelets, frames, and time-frequency analysis. This article, published in *SIAM Review* in 1989, was an important source for researchers seeking to enter wavelet theory during its early development. John's book *Wavelets: Mathematics and Applications*, jointly edited with Michael Frazier in 1994, was one of the very first books to contain a nearly complete overview of the most important aspects of wavelet theory, from theory to applications, from the construction of orthonormal wavelet systems to their use in a variety of disciplines, from image compression to turbulence. We do not have space to mention every paper and book of John's, but let us mention his text *Harmonic Analysis and Applications*, which appeared at about the same time (1996). This book provides a detailed account of Fourier analysis, including, for example, the basics of distribution theory and its use in Fourier analysis, classical topics such as summability kernels, and modern topics such as the construction of orthonormal wavelets found by Ingrid Daubechies. John's understanding of the historical development of the field and the impact of history on the directions the field has taken is clearly evident in this volume.

John is not only a dedicated researcher and gifted writer, but is also an engaging lecturer, teacher, and advisor. By any standard in mathematics, both the number of Ph.D. students he has directed and the number he is currently supervising are very high. Yet he devotes to each of them a regular time-slot for the discussion of their work and progress. This provides them invaluable help developing good scientific taste and instincts and the chance to get on their own feet as mathematicians, all on a very individual basis. It is no surprise that a large proportion of his Ph.D. students are engaged in active research positions, both in academia and in industry, within the USA and around the world.

In addition to his research, teaching, and all the standard duties of a mathematician—reviewing papers for journals or proposals for funding agencies, organizing meetings, serving within his department—John has been deeply involved in a number of projects that are establishing harmonic analy-

[1] I happened to spend the academic year 1989/90 in College Park and can therefore report from first-hand experience.

sis not only as an important part of classical mathematics, but also as a vital and highly visible branch of mathematical analysis.

Obviously John's view of harmonic analysis and the role it plays in the modern scientific endeavor are extremely broad. As was the case for his hero, Wiener, there seems for him to be no distinction between "pure" and "applied" mathematics, his work ranges seamlessly across what others perceive as boundaries. Perhaps his many hours spent as an employee of or consultant to such engineering companies as RCA, IBM, and The MITRE Corporation have shown him both how harmonic analysis can be applied and the needs and perceptions of those seeking to solve real-world applications.

Amongst his various activities that have increased the visibility of harmonic analysis I want to first mention that John developed and implemented (together with Wayne Yuhasz at that time at CRC Press) the idea for JFAA, the *Journal of Fourier Analysis and Applications.* The first issue appeared in 1993. I had the great honor to take over from John the chief editorship of this journal in 2000. By this time the journal was well established and had moved to Birkhäuser. It was John's wise decision not to devote the journal only to a single specialized topic, such as wavelets, but rather the entire discipline of Fourier analysis (understood in the widest sense), and at the same time to emphasize the importance of applications, of which we hope to see even more in the years to come.

In a similar direction, John's establishment and editorship of the Birkhäuser book series ANHA, *Applied and Numerical Harmonic Analysis*, again is lending significant visibility to the community. The series contains many of what are now the standard texts in the field. ANHA has become a "first place" for readers and authors to look at, in order to check what is going on in the field, or to find a good series in which to publish.

Most recently John has begun another highly visible initiative by establishing the *Norbert Wiener Center for Harmonic Analysis and Applications* at his home university, the University of Maryland, College Park. The description of the aims and goals of this center shows his views and ambitions, namely, to "provide a national focus for the broad emerging area of Mathematical Engineering." The need for this is clear: while there is more and more need for the development of mathematical tools and algorithms for real-world applications, the mathematical community generally does not provide sufficient "practical" training for its students to address such problems (typically viewed as being "too applied"), while the engineering community likewise does not generally provide sufficient pure mathematical training for its students to address these problems (typically viewed as "too abstract"). But in fact, these two disciplines *are* very close together, and the Norbert Wiener Center seeks to bring them into direct contact, to support each other, and most importantly to bring the questions on each side to the attention of the other side. John has shown in his own work the benefits of such interaction. Let us hope that his spirit, supporting serious mathematical research, ranging through the spectrum of harmonic analysis from "pure mathematics" to "practical applications" will continue on strongly, bringing back new mathematics and new applications. Let us wish John much success in this endeavor and in the years to come.

Vienna, Austria
March 2006

Hans G. Feichtinger

Preface

This volume celebrates the work of John Joseph Benedetto. Mathematician, mentor, father, friend—John has had a profound influence not only on the direction of harmonic analysis and its applications, but also on the entire community of people involved in these endeavors. This volume collects articles from coauthors, students, and colleagues of John, representative of some of the major areas that John has contributed to, including harmonic analysis, frame theory, time-frequency analysis, wavelet theory, and sampling theory and shift-invariant spaces. In recognition of John's own high standards of mathematical exposition, and in order to create a volume of lasting utility, many of the articles in this volume include introductions to or surveys of their representative research directions.

As is clearly illustrated in Hans Feichtinger's foreword to this volume, John's work has covered an enormous breadth of areas, and space limitations have required us to concentrate on the few representative themes mentioned above. Of course, this separation into parts is inherently artificial, as all these directions interrelate in many ways—they are each simply different facets of the vast field that we call harmonic analysis.

We begin in Part I with a tribute to John's early (and continuing) work in classical harmonic analysis. Among the many topics he has worked on in this area are weighted-norm inequalities for the Fourier transform, generalized harmonic analysis, Fourier restriction theorems, uncertainty principles, spectral synthesis, and the interaction between harmonic analysis and number theory. The Gibbs phenomenon (first observed by Wilbraham in 1848) is a fundamental fact in the theory of Fourier series in one dimension. The article by George Benke (who was the first of John's many Ph.D. students) explores Gibbs phenomena in higher dimensions, where the multitude of possible types of discontinuities makes the analysis far more intricate and interesting. Hans Heinig is a longtime coauthor of John, with their first joint paper appearing in 1983. His article uses Fourier inequalities between weighted Lebesgue spaces to derive weighted Sobolev gradient inequalities for a wide range of indices. The final article in this part is by John's 17th Ph.D. student, Georg Zimmermann. It explores semidiscrete multipliers, describing their basic form and deriving representations of such multipliers for a number of important spaces.

In Part II we turn to abstract frame theory. It is not an understatement to say that John's 2003 paper *Finite normalized tight frames* (Adv. Comput. Math., Vol. 18, pp. 357–385), written jointly with his 24th Ph.D. student, Matthew Fickus, revolutionized the theory of frames in finite dimensions. The article by Peter Casazza, Matthew Fickus, Jelena Kovačević, Manuel Leon (John's 20th Ph.D. student), and Janet Tremain surveys and further develops this theory, proving a striking fundamental inequality that is both necessary and sufficient for the existence of finite frames with prescribed norms.

Frame theory, time-frequency analysis, and wavelet theory are all closely intertwined, and John has made fundamental contributions to all of these areas. In Part III we focus on time-frequency analysis (also known as Gabor analysis). John's work on the *Balian–Low Theorems* includes symplectic BLTs in higher dimensions and sharp versions of the BLT. The current state-of-the-art of the Balian–Low Theorems is surveyed in an article by two of his collaborators on this subject, Wojciech Czaja (a former postdoc) and Alexander Powell (John's 29th Ph.D. student). The Zak transform is a fundamental tool in time-frequency analysis and in the analysis and application of Gabor frames. The article by John's 6th Ph.D. student, Jean-Pierre Gabardo, studies and surveys the Zak transform of distributions, and derives necessary and sufficient conditions for the finite linear span of a Gabor system to be dense in the Schwartz space of tempered distributions.

Because of the Balian–Low Theorems, it is *redundant* Gabor frames that must be used in practice. Necessarily, such redundant frames have nonunique dual frames. The article by Eric Hayashi, Shidong Li (John's 15th Ph.D. student), and Tracy Sorrells surveys the existing characterizations of Gabor dual frames, showing the equivalence of each characterization.

Much of the modern theory of quantitative time-frequency analysis, and in particular the introduction, development, and application of the modulation spaces, is due to John's longtime friends Hans Feichtinger and Karlheinz Gröchenig. In his article, Gröchenig surveys the relationship and application of the modulation spaces to the theory of pseudodifferential operators. Today, pseudodifferential operators are finding a wide variety of new applications in wireless communication, medical imaging, and geophysics, and these applications require new tools and techniques, which time-frequency analysis is well suited to deliver.

A basic, simple-to-state, and yet still open question regarding Gabor systems is the following: *are Gabor systems finitely linearly independent*? This is unknown in general even for the case of a collection of four time-frequency shifts of a single L^2 function. How hard can it be to prove that four functions in $L^2(\mathbf{R})$ must be independent? My article surveys this and related problems in Gabor theory.

Part IV is devoted to wavelet theory. In 1997, in an article in the journal *Journal of Fourier Analysis and Applications* (which had been founded by

John only three years before), Xingde Dai, David Larson, and Darrin Speegle surprised the wavelet community by proving the existence of "wavelet sets." The characteristic function of such a set is a single function that generates an orthonormal wavelet basis for $L^2(\mathbf{R}^n)$, whereas wavelet orthonormal basis in higher dimensions generated from multiresolution analyses necessarily require multiple generators when the dilation matrix A satisfies $|\det(A)| > 2$. Together with Manuel Leon (his 20th Ph.D. student), and Songkiat Sumetkijakan (his 26th Ph.D. student), John has written several influential papers on wavelet sets. One of the first wavelet sets constructed by Dai, Larson, and Speegle was the "wedding cake wavelet set." Pictured on the front cover of this volume is one of the wavelet sets constructed by John and Songkiat. This one is John's favorite, and is known as the "wedding night wavelet set."

The first article in Part IV, by David Larson, Eckart Schulz, Darrin Speegle, and Keith Taylor, represents abstract wavelet theory, with an article on explicit cross sections of singly generated group actions. These cross sections are used to characterize those matrices for which there exist minimally supported frequency (MSF) wavelets with infinitely many wavelet generators.

I cannot resist expressing my own personal satisfaction at being able to include the second article in this part, by Kangui Guo, Demetrio Labate, Wang-Q Lim, Guido Weiss, and Edward Wilson. This article surveys a recent development in wavelet constructions in higher dimensions that employs several dilation matrices. This allows the construction of wavelet bases which are well suited for representing the multitude of geometric features that can occur in higher dimensions. Guido has been a friend and mentor to generations of mathematicians, including John and his students in particular, not to mention that he been one of the cornerstone figures of harmonic analysis over the last 50 years. Demetrio is my own very first Ph.D. student, and so is John's "mathematical grandchild." This article represents the continuing impact of stellar mathematicians such as Guido and John on the newest generations of mathematicians.

As noted in Hans Feichtinger's foreword, John's contributions have ranged far across both pure mathematics and applications. Indeed, much of his work shows how naturally harmonic analysis belongs with applications. His Wavelet Auditory Model (WAM), developed jointly with his 14th Ph.D. student, Anthony Teolis, incorporates his work on irregular sampling to build a model of the human cochlear response. Their 1993 WAM article was the first paper in the first issue of the then-new and now-major journal *Applied and Computational Harmonic Analysis.* John's work on regular and irregular sampling, and the closely related area of shift-invariant spaces, is honored in the final portion, Part V, of this volume.

The first article in this section is by Jeffrey Hogan and Joseph Lakey (John's 12th Ph.D. student). It discusses the important special cases of sampling sets that are not lattices but that still possess a periodic structure, e.g., $\mathbb{Z} \cup (\mathbb{Z} + \frac{1}{\pi})$. These types of sampling sets arise in many real-world settings. The article by Bjarte Rom and John's 7th Ph.D. student, David Walnut, addresses another kind of irregular but still somewhat structured sampling geometry. In their setting the sample set is not necessarily well-separated, and thus is outside the bounds of most of the sampling literature. Yet the specific structure addressed, which is a union of lattices (such as $\mathbb{Z} \cup \sqrt{2}\mathbb{Z}$), allows many interesting and useful conclusions to be drawn.

In their article, John's longtime friends Akram Aldroubi, Carlos Cabrelli, and Ursula Molter use the setting of shift-invariant spaces to survey recent results relating frame theory and shift-invariant spaces to learning theory. Larry Baggett, another longtime friend, analyzes the fine structure of shift-invariant spaces. His article, which includes an elegant introduction to frame theory, shows how a certain multiplicity function can be used to quantify the notion of redundancy of frequencies.

The final article, by Peter Casazza, Ole Christensen, Alex Lindner, and Shidong Li, addresses the delicate question of characterizing frame properties of sets of complex exponentials based solely on the density of the set of frequencies of those exponentials.

Although our survey of the articles appearing in this volume is now complete, as a tribute to John's work this volume is certainly far from complete. In the "pure" directions we could well have included articles on generalized harmonic analysis, uncertainty principles, spectral synthesis, and number theory, to name only a few. In the "applied" direction we have omitted mention of John's fundamental work on Σ-Δ quantization, spectral estimation, periodicity detection, EEGs, MRIs—and again this list is incomplete. Sadly, space has limited the number of articles and the number of contributors to a small subset of those who would have eagerly contributed their own tribute to John.

This volume is also incomplete in that it focuses on research alone, while John's contributions go far beyond this. In particular, I am honored to be one of the many Ph.D. students whom John has mentored. Listed on the next two pages is John's mathematical genealogy as it stands at this moment, including his Ph.D. students and grandstudents, his Ph.D. advisor Chandler Davis, and Chandler's advisor Garrett Birkhoff. In the backward direction the graph ends there, for though Garrett Birkhoff was a professor at Harvard, he had no formal Ph.D. degree. In the forward direction, the graph is still far from complete even in listing John's students, as he is still actively advising an entire group of current students. John's dedication to all facets of mathematics, including his research, his Ph.D. students, and all the many undergraduate and graduate students who have taken courses from him, were recognized in 1999 when he was named a Distinguished Scholar/Teacher of the University of Maryland, College Park.

I am sure that all of John's many friends will join with me to wish him health, happiness, and many more years of successful and significant mathematics.

Atlanta, Georgia
March 2006

Christopher Heil

The Benedetto Mathematical Family Tree

- Garrett Birkhoff (no Ph.D.)
  - Chandler Davis (Harvard U.), 1950
    - John J. Benedetto (U. Toronto), 1964
      - George Benke (U. Maryland, College Park), 1971
      - Wan-chen Hsieh (U. Maryland, College Park), 1977
      - Fulvio Ricci (U. Maryland, College Park), 1977
        - Giovanna Carcano (U. Milano), 1985
        - Paolo Ciatti (U. Torino), 1996
        - Bianca di Blasio (U. Torino), 1996
        - Mario Dellanegra (U. Torino), 1998
        - Priscilla Gorelli (U. Torino), 1999
        - Silvia Secco (U. Torino), 1999
        - Daniele Debertol (U. Genova), 2003
      - Ward Evans (U. Maryland, College Park), 1980
      - William Joyner (U. Maryland, College Park), 1983
      - Jean-Pierre Gabardo (U. Maryland, College Park), 1987
        - Xiaojiang Yu (McMaster U.), 2005
      - David Walnut (U. Maryland, College Park), 1989
        - Paul Salamonowicz (George Mason U.), 1999
        - Glenn Easley (George Mason U.), 2000
      - Christopher Heil (U. Maryland, College Park), 1990
        - Demetrio Labate (Georgia Tech), 2000
        - Denise Jacobs (Georgia Tech), 2001
        - Kasso Okoudjou (Georgia Tech), 2003
      - Rodney Kerby (U. Maryland, College Park), 1990
      - George Yang (U. Maryland, College Park), 1990
      - William Heller (U. Maryland, College Park), 1991
      - Joseph Lakey (U. Maryland, College Park), 1991
        - Ying Wang (New Mexico State U.), 2001
        - Sofian Obeidat (New Mexico State U.), 2002
        - Christopher Weaver (New Mexico State U.), 2005

- Erica Bernstein (U. Maryland, College Park), 1992
- Anthony Teolis (U. Maryland, College Park), 1993
- Shidong Li (U. Maryland, Baltimore County), 1993
- Sandra Saliani (U. Maryland, College Park), 1993
- Georg Zimmermann (U. Maryland, College Park), 1994
- Melissa (Harrison) Rice (U. Maryland, College Park), 1998
- Hui-Chuan Wu (U. Maryland, College Park), 1998
- Manuel Leon (U. Maryland, College Park), 1999
- Götz Pfander (U. Maryland, College Park), 1999
- Oliver Treiber (U. Maryland, College Park), 1999
- Sherry Scott (U. Maryland, College Park), 2000
- Matthew Fickus (U. Maryland, College Park), 2001
- Ioannis Konstantinidis (U. Maryland, College Park), 2001
- Songkiat Sumetkijakan (U. Maryland, College Park), 2002
- Anwar Saleh (U. Maryland, College Park), 2002
- Jeffrey Sieracki (U. Maryland, College Park), 2002
- Alexander Powell (U. Maryland, College Park), 2003
- Shijun Zheng (U. Maryland, College Park), 2003
- Joseph Kolesar (U. Maryland, College Park), 2004
- Andrew Kebo (U. Maryland, College Park), 2005
- Juan Romero (U. Maryland, College Park), 2005
- Abdelkrim Bourouihiya (U. Maryland, College Park), 2006[2]
- Aram Tangboondouangjit (U. Maryland, College Park), 2006[3]

[2] To be awarded in May 2006.
[3] To be awarded in May 2006.

Publications of John J. Benedetto

A. Books

1. *Generalized Functions*, Institute for Fluid Dynamics and Applied Mathematics, Technical Report BN-431, University of Maryland, College Park, MD, 1966.
2. *Pseudo-Measures and Harmonic Synthesis*, Department of Mathematics Lecture Notes 5, University of Maryland, College Park, MD, 1968.
3. *Harmonic Analysis on Totally Disconnected Sets*, Lecture Notes in Mathematics, Vol. 202, Springer–Verlag, Berlin, 1971.
4. *Spectral Synthesis*, Pure and Applied Mathematics, No. 66, Academic Press, New York, NY, 1975.
5. *Real Variable and Integration with Historical Notes*, Mathematische Leitfäden, B. G. Teubner, Stuttgart, 1976.
6. *A Mathematical Approach to Mathematics Appreĉiation*, Department of Mathematics Lecture Notes 17, University of Maryland, College Park, MD, 1968.
7. *Harmonic Analysis and Applications*, CRC Press, Boca Raton, FL, 1997.
8. *Integration and Modern Analysis: From Classical Real Variables to Distribution Theory* (with W. Czaja), Birkhäuser, Boston, MA, 2006.

B. Edited Volumes

1. J. J. Benedetto, Editor, *Euclidean Harmonic Analysis*, Lecture Notes in Mathematics, Vol. 779, Springer–Verlag, Berlin, 1979.
2. J. J. Benedetto and M. W. Frazier, Editors, *Wavelets: Mathematics and Applications*, CRC Press, Boca Raton, FL, 1994.
3. J. J. Benedetto and M. W. Frazier, Editors, *Wavelets: Mathematics and Applications* (Japanese Translation), Springer–Verlag, Tokyo, 1995.
4. J. J. Benedetto and P. J. S. G. Ferreira, Editors, *Modern Sampling Theory: Mathematics and Applications*, Birkhäuser, Boston, MA, 2001.
5. J. J. Benedetto and A. I. Zayed, Editors, *Sampling, Wavelets, and Tomography*, Birkhäuser, Boston, MA, 2004.

C. Papers

1. Representation theorem for the Fourier transform of integrable functions, *Bull. Soc. Roy. Sci. de Liège*, **34** (1965), pp. 601–604.
2. Onto criterion for adjoint maps, *Bull. Soc. Roy. Sci. de Liège*, **34** (1965), pp. 605–609.
3. The Laplace transform of generalized functions, *Canad. J. Math.*, **18** (1966), pp. 357–374.
4. Tauberian translation algebras, *Ann. Mat. Pura Appl. (4)*, **74** (1966) pp. 255–282.
5. Analytic representation of generalized functions, *Math. Z.*, **97** (1967), pp. 303–319.
6. A strong form of spectral resolution, *Ann. Mat. Pura Appl. (4)*, **86** (1970), pp. 313–324.
7. Sets without true distributions, *Bull. Soc. Roy. Sci. de Liège*, **39** (1970), pp. 434–437.
8. Support preserving measure algebras and spectral synthesis, *Math. Z.*, **118** (1970), pp. 271–280.
9. Trigonometric sums associated with pseudo-measures, *Ann. Scuola Norm. Sup. Pisa (3)*, **25** (1971), pp. 229–248.
10. Sui problemi di sintesi spettrale, *Rend. Sem. Mat. Fis. Milano*, **41** (1971), pp. 55–61.
11. Il Problema degli insiemi Helson-S, *Rend. Sem. Mat. Fis. Milano*, **41** (1971), pp. 63–68.
12. (LF) spaces and distributions on compact groups and spectral synthesis on $R/2\pi Z$, *Math. Ann.*, **194** (1971), pp. 52–67.
13. Ensembles de Helson et synthèse spectrale, *C. R. Acad. Sci. Paris Sér. A-B*, **274** (1972), pp. A169–A170.
14. Construction de fonctionnelles multiplicatives discontinues sur des algèbres métriques, *C. R. Acad. Sci. Paris Sér. A-B*, **274** (1972), pp. A254–A256.
15. A support preserving Hahn–Banach property to determine Helson-S sets, *Invent. Math.*, **16** (1972), pp. 214–228.
16. Idele characters in spectral synthesis on $R/2\pi Z$, *Ann. Inst. Fourier (Grenoble)*, **23** (1973), pp. 45–64.
17. Pseudo-measure energy and spectral synthesis, *Canad. J. Math.*, **26** (1974), pp. 985–1001.
18. Tauberian theorems, Wiener's spectrum, and spectral synthesis, *Rend. Sem. Mat. Fis. Milano*, **44** (1975), pp. 63–73.
19. The Wiener spectrum in spectral synthesis, *Studies in Appl. Math.*, **54** (1975), pp. 91–115.
20. Idelic pseudo-measures and Dirichlet series, in: *Symposia Mathematica*, Vol. XXII (Convegno sull'Analisi Armonica e Spazi di Funzioni su Gruppi Localmente Compatti, INDAM, Rome, 1976), Academic Press, London, 1977, pp. 205–222.

21. Zeta functions for idelic pseudo-measures, *Ann. Scuola Norm. Sup. Pisa Cl. Sci. (4)*, **6** (1979), pp. 367–377.
22. Fourier analysis of Riemann distributions and explicit formulas, *Math. Ann.*, **252** (1979/1980), pp. 141–164.
23. The theory of constructive signal analysis, *Stud. Appl. Math.*, **65** (1981), pp. 37–80.
24. A closure problem for signals in semigroup invariant systems, *SIAM J. Math. Anal.*, **13** (1982), pp. 180–207.
25. Harmonic analysis and spectral estimation, *J. Math. Anal. Appl.*, **91** (1983), pp. 444–509.
26. Weighted Hardy spaces and the Laplace transform (with H. P. Heinig), in: *Harmonic Analysis* (Cortona, 1982), G. Mauceri, F. Ricci, and G. Weiss, eds., Lecture Notes in Math., Vol. 992, Springer, Berlin, 1983, pp. 240–277.
27. Wiener's Tauberian theorem and the uncertainty principle, in: *Topics in Modern Harmonic Analysis*, Vol. I, II (Turin/Milan, 1982), Ist. Naz. Alta Mat. Francesco Severi, Rome, 1983, pp. 863–887.
28. A local uncertainty principle, *SIAM J. Math. Anal.*, **15** (1984), pp. 988–995.
29. An inequality associated with the uncertainty principle, *Rend. Circ. Mat. Palermo (2)*, **34** (1985), pp. 407–421.
30. Some mathematical methods for spectrum estimation, in: *Fourier Techniques and Applications* (Kensington, 1983), J. F. Price, ed., Plenum, New York, NY, 1985, pp. 73–100.
31. Fourier uniqueness criteria and spectrum estimation theorems, in: *Fourier Techniques and Applications* (Kensington, 1983), J. F. Price, ed., Plenum, New York, NY, 1985, pp. 149–172.
32. Inequalities for spectrum estimation, *Linear Algebra and Applications*, **84** (1986), pp. 377–383.
33. Weighted Hardy spaces and the Laplace transform, II. (with H. P. Heinig and R. Johnson), *Math. Nach.*, **132** (1987), pp. 29–55.
34. Fourier inequalities with A_p weights (with H. P. Heinig and R. Johnson), in: *General Inequalities 5* (Oberwolfach, 1986), Internat. Schriftenreihe Numer. Math., Vol. 80, Birkhäuser, Basel, 1987, pp. 217–232.
35. A quantitative maximum entropy theorem for the real line, *Integral Equations Operator Theory*, **10** (1987), pp. 761–779.
36. Gabor representations and wavelets, in: *Commutative Harmonic Analysis* (Canton, NY, 1987), D. Colella, ed., Contemp. Math., Vol. 91, American Mathematical Society, Providence, RI, 1989, pp. 9–27.
37. An n-dimensional Wiener–Plancherel formula (with G. Benke and W. Evans), *Adv. in Appl. Math.*, **10** (1989), pp. 457–487.
38. Uncertainty principle inequalities and spectrum estimation, in: *Recent Advances in Fourier Analysis and its Applications* (Il Ciocco, 1989), J. S. Byrnes and J. L. Byrnes, eds., NATO Adv. Sci. Inst. Ser. C: Math. Phys. Sci., Vol. 315, Kluwer Acad. Pub., Dordrecht, 1990, pp. 143–182.

39. Irregular sampling and the theory of frames, I (with W. Heller), *Note Mat.*, **10** (1990), suppl. 1, pp. 103–125.
40. Support dependent Fourier transform norm inequalities, (with C. Karanikas), *Rend. Mat. Appl. (7)*, **11** (1991), pp. 157–174.
41. The spherical Wiener–Plancherel formula and spectral estimation, *SIAM J. Math. Anal.*, **22** (1991), pp. 1110–1130.
42. Fourier transform inequalities with measure weights, II (with H. P. Heinig), in: *Function Spaces* (Poznań, 1989), J. Musielak, ed., Teubner-Texte Math., Vol. 120, Teubner, Stuttgart, 1991, pp. 140–151.
43. A multidimensional Wiener–Wintner theorem and spectrum estimation, *Trans. Amer. Math. Soc.*, **327** (1991), pp. 833–852.
44. Stationary frames and spectral estimation, in: *Probabilistic and Stochastic Methods in Analysis, with Applications* (Il Ciocco, 1991), J. S. Byrnes, J. L. Byrnes, K. A. Hargreaves, and K. Berry, eds., NATO Adv. Sci. Inst. Ser. C: Math. Phys. Sci., Vol. 372, Kluwer Acad. Pub., Dordrecht, 1992, pp. 117–161.
45. Uncertainty principles for time-frequency operators (with C. Heil and D. Walnut), in: *Continuous and Discrete Fourier Transforms, Extension Problems and Wiener–Hopf Equations*, Oper. Theory Adv. Appl., Vol. 58, I. Gohberg, ed., Birkhäuser, Basel (1992), pp. 1–25.
46. An auditory motivated time-scale signal representation (with A. Teolis), in: *Proceedings of the 1992 IEEE-SP International Symposium on Time-Frequency and Time-Scale Analysis*, IEEE, Piscataway, NJ, pp. 49–52.
47. Fourier transform inequalities with measure weights (with H. Heinig), *Adv. Math.*, **96** (1992), pp. 194–225.
48. Irregular sampling and frames, in: *Wavelets: A Tutorial in Theory and Applications*, C. K. Chui, ed., Academic Press, Boston, MA, 1992, pp. 445–507.
49. Frame decompositions, sampling, and uncertainty principle inequalities, in: *Wavelets: Mathematics and Applications*, J. J. Benedetto and M. W. Frazier, eds., CRC Press, Boca Raton, FL, 1994, pp. 247–304.
50. Gabor frames for L^2 and related spaces (with D. Walnut), in: *Wavelets: Mathematics and Applications*, J. J. Benedetto and M. W. Frazier, eds., CRC Press, Boca Raton, FL, 1994, pp. 97–162.
51. Multiresolution analysis frames with applications (with S. Li), in: *Proceedings of the 1993 IEEE International Conference on Acoustics, Speech, and Signal Processing* (ICASSP-93), Vol. 3, IEEE, Piscataway, NJ, 1993, pp. 304–307.
52. A wavelet auditory model and data compression (with A. Teolis), *Appl. Comput. Harmon. Anal.*, **1** (1993), pp. 3–28.
53. Local frames (with A. Teolis), in: *Mathematical Imaging: Wavelet Applications in Signal and Image Processing* (San Diego, CA, 1993), A. Aldroubi, ed., Proc. SPIE Vol. 2034, SPIE, Bellingham, WA, 1993, pp. 310–321.
54. From a wavelet auditory model to definitions of the Fourier transform, in: *Wavelets and Their Applications* (Il Ciocco, 1992), J. S. Byrnes,

J. L. Byrnes, K. A. Hargreaves, and K. Berry, eds., NATO Adv. Sci. Inst. Ser. C: Math. Phys. Sci., Vol. 442, Kluwer Acad. Pub., Dordrecht, 1994, pp. 1–17.

55. Noise suppression using a wavelet model (with A. Teolis), in: *Proceedings of the 1994 IEEE International Conference on Acoustics, Speech, and Signal Processing* (ICASSP-94), Vol. 1, IEEE, Piscataway, NJ, 1994, pp. I-17–I-20.
56. Subband coding for sigmoidal nonlinear operations (with S. Saliani), in: *Wavelet Applications* (Orlando, FL, 1994), H. H. Szu, ed., Proc. SPIE Vol. 2242, SPIE, Bellingham, WA, 1993, pp. 19–27.
57. Subband coding and noise reduction in multiresolution analysis frames (with S. Li), in: *Wavelet Applications in Signal and Image Processing II*, (San Diego, CA, 1994), A. F. Laine and M. A. Unser, eds., Proc. SPIE Vol. 2303, SPIE, Bellingham, WA, 1994, pp. 154–165.
58. The definition of the Fourier transform for weighted inequalities (with J. Lakey), *J. Funct. Anal.*, **120** (1994), pp. 403–439.
59. Narrow band frame multiresolution analysis with perfect reconstruction (with S. Li), in: *Proceedings of the 1994 IEEE-SP International Symposium on Time-Frequency and Time-Scale Analysis*, IEEE, Piscataway, NJ, pp. 36–39.
60. Pyramidal Riesz products associated with subband coding and self-similarity (with E. Bernstein), in: *Wavelet Applications II* (Orlando, FL, 1995), H. H. Szu, ed., Proc. SPIE Vol. 2491, SPIE, Bellingham, WA, 1994, pp. 212–221.
61. Wavelet-based analysis of electroencephalogram (EEG) signals for detection and localization of epileptic seizures (with G. Benke, M. Bozek-Kuzmicki, D. Colella, and G. M. Jacyna), in: *Wavelet Applications II* (Orlando, FL, 1995), H. H. Szu, ed., Proc. SPIE Vol. 2491, SPIE, Bellingham, WA, 1994, pp. 760–769.
62. Poisson's summation formula in the construction of wavelet bases (with G. Zimmermann), in: *Proceedings ICIAM/GAMM 95*, O. Mahrenholtz and R. Mennicken, eds., *Z. Angew. Math. Mech.*, Special Issue 2: Applied Analysis (1996), pp. 477–478.
63. Differentiation and the Balian–Low theorem (with C. Heil and D. F. Walnut), *J. Fourier Anal. Appl.*, **1** (1995), pp. 355–402.
64. Wavelet analysis of spectrogram seizure chirps (with D. Colella), in: *Wavelet Applications in Signal and Image Processing III*, (San Diego, CA, 1994), A. F. Laine and M. A. Unser, eds., Proc. SPIE Vol. 2569, SPIE, Bellingham, WA, 1994, pp. 512–521.
65. Local frames and noise reduction (with A. Teolis), *Signal Processing*, **45** (1995), pp. 369–387.
66. Frame signal processing applied to bioelectric data, in: *Wavelets in Medicine and Biology*, A. Aldroubi and M. Unser, eds., CRC Press, Boca Raton, FL, 1996, pp. 467–486.

67. Generalized harmonic analysis and Gabor and wavelet systems, in: *Proc. of the Norbert Wiener Centenary Congress, 1994* (East Lansing, MI, 1994), Proc. Sympos. Appl. Math., Vol. 52, American Mathematical Society, Providence, RI, 1997, pp. 85–113.
68. Sampling operators and the Poisson summation formula, (with G. Zimmermann), *J. Fourier Anal. Appl.*, **3** (1997), pp. 505–523.
69. Wavelet detection of periodic behavior in EEG and ECoG data (with G. E. Pfander), in: *Proceedings of 15th IMACS World Congress on Scientific Computation, Modelling, and Applied Mathematics* (Berlin, 1997), Vol. 1, 1997, pp. 75–80.
70. Gabor systems and the Balian–Low Theorem (with C. Heil and D. F. Walnut), in: *Gabor Analysis and Algorithms: Theory and Applications*, H. G. Feichtinger and T. Strohmer, eds., Birkhäuser, Boston, MA, 1998, pp. 85–122.
71. Noise reduction in terms of the theory of frames, in: *Signal and Image Representation in Combined Spaces*, Y. Zeevi and R. Coifman, eds., Academic Press, New York, NY, 1998, pp. 259–284.
72. Self-similar pyramidal structures and signal reconstruction (with M. Leon and S. Saliani), in: *Wavelet Applications V* (Orlando, FL, 1998), H. H. Szu, ed., Proc. SPIE Vol. 3391, SPIE, Bellingham, WA, 1998, pp. 304–314.
73. The theory of multiresolution frames and applications to filter banks (with S. Li), *Appl. Comput. Harmon. Anal.*, **5** (1998), pp. 389–427.
74. Frames, sampling, and seizure prediction, in: *Advances in Wavelets* (Hong Kong, 1997), K.-S. Lau, ed., Springer–Verlag, Singapore (1999), pp. 1–25.
75. Wavelet periodicity detection algorithms (with G. E. Pfander), in: *Wavelet Applications in Signal and Image Processing VI*, (San Diego, CA, 1998), A. F. Laine and M. A. Unser, eds., Proc. SPIE Vol. 3458, SPIE, Bellingham, WA, 1994, pp. 48–55.
76. A multidimensional irregular sampling algorithm and applications (with H.-C. Wu), in: *Proceedings of the 1999 IEEE International Conference on Acoustics, Speech, and Signal Processing* (ICASSP-99), Vol. 4, IEEE, Piscataway, NJ, 1999, pp. 2039–2042.
77. The construction of multiple dyadic minimally supported frequency wavelets on R^d (with M. Leon), in: *The Functional and Harmonic Analysis of Wavelets and Frames* (San Antonio, 1999), L. W. Baggett and D. R. Larson, eds., Contemp. Math., Vol. 247, American Mathematical Society, Providence, RI, 1999, pp. 43–74.
78. A Beurling covering theorem and multidimensional irregular sampling (with H.-C. Wu), in: *Proceedings of the 1999 International Workshop on Sampling Theory and Applications* (Loen, Norway, 1999), Norwegian University of Science and Technology, Trondheim, 1999, pp. 142–148.
79. Sampling theory and wavelets, in: *Signal Processing for Multimedia* (Il Ciocco, 1998), J. S. Byrnes, ed., NATO Adv. Sci. Inst. Ser. F: Computer Systems Sci., Vol. 174, IOS Press, Amsterdam, 1994, pp. 19–34.
80. Ten books on wavelets, *SIAM Review*, **42** (2000), pp. 127–138.

81. Non-uniform sampling theory and spiral MRI reconstruction (with H.-C. Wu), in: *Wavelet Applications in Signal and Image Processing VIII*, (San Diego, CA, 2000), A. Aldroubi, A. F. Laine, and M. A. Unser, eds., Proc. SPIE Vol. 4119, SPIE, Bellingham, WA, 2000, pp. 130–141.
82. The classical sampling theorem, non-uniform sampling and frames (with P. S. J. G. Ferreira), in: *Modern Sampling Theory: Mathematics and Applications*, J. J. Benedetto and P. S. J. G. Ferreira, eds., Birkhäuser, Boston, MA, 2001, pp. 1–26.
83. Frames, irregular sampling, and a wavelet auditory model (with S. Scott), in: *Nonuniform Sampling: Theory and Practice*, F. Marvasti, ed., Kluwer/Plenum, New York, NY, 2001, pp. 585–618.
84. Wavelet frames: Multiresolution analysis and extension principles (with O. M. Treiber), in: *Wavelet Transforms and Time-Frequency Signal Analysis*, L. Debnath, ed., Birkhäuser, Boston, MA, 2001, pp. 3–36.
85. The construction of single wavelets in d-dimensions (with M. Leon), *J. Geom. Anal.*, **11** (2001), pp. 1–15.
86. MRI signal reconstruction by Fourier frames on interleaving spirals (with A. M. Powell and H.-C. Wu), in: *Proceedings of the 2002 IEEE International Symposium on Biomedical Imaging*, IEEE, Piscataway, NJ, 2002, pp. 717–720.
87. A fractal set constructed from a class of wavelet sets (with S. Sumetkijakan), in: *Inverse Problems, Image Analysis, and Medical Imaging*, (New Orleans, LA, 2001), M. Z. Nashed and O. Scherzer, eds., Contemp. Math., Vol. 313, American Mathematical Society, Providence, RI, 2002, pp. 19–35.
88. Periodic wavelet transforms and periodicity detection (with G. E. Pfander), *SIAM J. Appl. Math.*, **62** (2002), pp. 1329–1368.
89. Finite normalized tight frames (with M. Fickus), *Adv. Comput. Math.*, **18** (2003), pp. 357–385.
90. The Balian–Low theorem and regularity of Gabor systems (with W. Czaja, P. Gadziński, and A. M. Powell), *J. Geom. Anal.*, **13** (2003), pp. 239–254.
91. Weighted Fourier inequalities: new proofs and generalizations (with H. P. Heinig), *J. Fourier Anal. Appl.*, **9** (2003), pp. 1–37.
92. The Balian–Low theorem for the symplectic form on $\mathbb{R}^{2d}$, (with W. Czaja and A. Y. Maltsev), *J. Math. Physics*, **44** (2003), pp. 1735–1750.
93. Local sampling for regular wavelet and Gabor expansions (with N. Atreas and C. Karanikas), *Sampl. Theory Signal Image Process.*, **2** (2003), pp. 1–24.
94. A Wiener–Wintner theorem for $1/f$ power spectra (with R. Kerby and S. E. Scott), *J. Math. Anal. Appl.*, **279** (2003), pp. 740–755.
95. Constructive approximation in waveform design, in: *Advances in Constructive Approximation* (Vanderbilt, 2003), M. Neamtu and E. B. Saff, eds., Nashboro Press, Brentwood, TN, 2004, pp. 89–108.
96. A wavelet theory for local fields and related groups (with R. L. Benedetto), *J. Geom. Anal.*, **14** (2004), pp. 423–456.

97. Sigma-Delta quantization and finite frames (with A. M. Powell and Ö. Yilmaz), in: *Proceedings of the 2004 IEEE International Conference on Acoustics, Speech, and Signal Processing* (ICASSP '04), Vol. 3, IEEE, Piscataway, NJ, 2004, pp. 937–940.
98. Multiscale Riesz products and their support properties (with E. Bernstein and I. Konstantinidis), *Acta Appl. Math.*, **88** (2005), pp. 201–227.
99. Analog to digital conversion for finite frames, (with A. M. Powell and Ö. Yilmaz), in: *Wavelets XI* (San Diego, CA, 2005), M. Papadakis, A. F. Laine, and M. A. Unser, eds., Proc. SPIE Vol. 5914, SPIE, Bellingham, WA, 2000, pp. 370–378.
100. Greedy adaptive discrimination: component analysis by simultaneous sparse approximation (with J. Sieracki), in: *Wavelets XI* (San Diego, CA, 2005), M. Papadakis, A. F. Laine, and M. A. Unser, eds., Proc. SPIE Vol. 5914, SPIE, Bellingham, WA, 2005, pp. 602–610.
101. Frame expansions for Gabor multipliers (with G. E. Pfander), *Appl. Comput. Harmon. Anal.*, **20** (2006), pp. 26–40.
102. Second-order Sigma-Delta ($\Sigma\Delta$) quantization of finite frame expansions (with A. M. Powell and Ö. Yilmaz), *Appl. Comput. Harmon. Anal.*, **20** (2006), pp. 126–148.
103. A (p, q) version of Bourgain's theorem (with A. Powell), *Trans. Amer. Math. Soc.*, **358** (2006), pp. 2489–2505.
104. A Doppler statistic for zero autocorrelation waveforms (with J. Donatelli, I. Konstantinidis, and C. Shaw), in: *Proceedings of the Conference on Information Sciences and Systems* (Princeton, NJ, 2006), pp. 1403–1407.
105. An endpoint $(1, \infty)$ Balian–Low theorem (with W. Czaja, A. M. Powell, and J. Sterbenz), *Math. Res. Lett.*, **13** (2006), pp. 467–474.
106. An optimal example for the Balian–Low uncertainty principle, (with W. Czaja and A. M. Powell), *SIAM J. Math. Anal.*, **38** (2006), pp. 333–345.
107. Tight frames and geometric properties of wavelet sets (with S. Sumetkijakan), *Adv. Comput. Math.*, **24** (2006), pp. 35–56.
108. Geometrical properties of Grassmannian frames for $\mathbb{R}^2$ and $\mathbb{R}^3$ (with J. D. Kolesar), *EURASIP J. Appl. Signal Process.*, Special Issue on Frames and Overcomplete Representations in Signal Processing, **2006** (2006), pp. 1–17.
109. Sigma-Delta ($\Sigma\Delta$) quantization and finite frames (with A. M. Powell and Ö. Yilmaz), *IEEE Trans. Inform. Theory*, **52** (2006), pp. 1990–2005.
110. Introduction, in: *Fundamental Papers in Wavelet Theory*, C. Heil and D. F. Walnut, eds., Princeton University Press, Princeton, NJ, 2006.
111. The construction of d-dimensional MRA frames (with J. Romero), *J. Applied Funct. Anal.*, to appear (2006).
112. Zero autocorrelation waveforms: A Doppler statistic and multifunction problems (with J. Donatelli, I. Konstantinidis, and C. Shaw), in: *Proceedings of the 2006 IEEE International Conference on Acoustics, Speech, and Signal Processing* (ICASSP '06), IEEE, Piscataway, NJ, 2006, to appear.

D. Patents

1. U.S. Patent 5,388,182 (Feb. 7, 1995), Nonlinear method and apparatus for coding and decoding acoustic signals with data compression and noise suppression using cochlear filters, wavelet analysis, and irregular sampling reconstruction, with A. Teolis.

Contents

Harmonic Analysis and Applications

Part I

Harmonic Analysis

1

The Gibbs Phenomenon in Higher Dimensions

George Benke

Department of Mathematics, Georgetown University, Washington, DC 20057, USA
benke@math.georgetown.edu

Summary. The concept of star discontinuity is defined for functions of several variables. A star discontinuity in dimension one is simply a jump discontinuity. It is then shown that in arbitrary dimensions the Gibbs phenomenon for square convergence occurs for periodic functions satisfying appropriate hypotheses at star discontinuities.

Dedicated to John Benedetto—mentor, scholar, friend.

1.1 Introduction

The classical Gibbs phenomenon concerns the behavior of the partial sums of the Fourier series of functions at a jump discontinuity. If a real function of bounded variation has a jump discontinuity at x_0, then for any interval $(x_0, x_0 + \delta)$ there exist infinitely many n and points x_n in the interval such that $S_n(f, x_n) > m + (0.508)h$ where m is the midpoint of the jump and h is the height of the jump. In other words, the partial sums overshoot the size of the jump by at least 16%. A more precise statement is that there exists a $\delta > 0$ such that $S_n(f, (x - x_0)/n)$ converges to $\frac{1}{\pi} \int_0^\infty g(x - u) \frac{\sin u}{u}\, du$ where g is the jump function on the real numbers which matches the jump of f at x_0. This means that as we "zoom in" around the point x_0 the re-scaled graph of $S_n(f, x)$ converges to a universal graph determined solely by the size and location of the jump and is otherwise independent of f.

There are many possible avenues for the exploration of the Gibbs phenomenon. For example Weyl [6], [7] studied the behavior of the partial sums of spherical harmonic expansions of functions defined on the sphere and having a jump discontinuity along a smooth curve on the sphere. See Colzani and Vignati [2] for an investigation along the same lines in the setting of multiple Fourier integrals. One can also consider other summability methods such as Bochner–Riesz summability. This is the theme of Golubov's work [3] and the work of Cheng [1]. An extension along a more classical line is carried out by Helmberg [4], [5] for corner discontinuities of functions defined in the plane.

Here we study the Gibbs phenomenon for d-dimensional 2π-periodic functions for the case of square convergence and for discontinuities which are more general than simple extensions of jump discontinuities. We define the concept of a star discontinuity. This concept is general enough to encompass the corner discontinuities studied by Helmberg and the cliff discontinuities studied by Weyl. Roughly speaking, a star discontinuity is a discontinuity in a neighborhood of which the graph of f looks locally like the graph of a function which depends only on angle and not on the distance from the origin. In dimension one star discontinuities are jump discontinuities. We show that under appropriate hypotheses the square partial sums exhibit a Gibbs phenomenon in the neighborhood of a star discontinuity. Moreover as in the one-dimensional case, the graph of the square partial sums converges, under linear rescaling, to a universal graph depending only on the "shape" of the discontinuity and on no other features of f.

The hypotheses of the main theorem have been formulated so as to yield a widely applicable result and yet keep the proof short, thereby clarifying the essential ideas. This makes the main theorem somewhat awkward to apply directly. To illustrate the use of the theorem, we show that, in dimension two, if the "shape" of the discontinuity has uniformly bounded variation and if the variation (in the sense of Hardy and Krause) of the difference between the function and the "shape" of the discontinuity is bounded, then the hypotheses of the main theorem are satisfied and therefore the function exhibits the Gibbs phenomenon at such star discontinuities.

This paper is organized as follows. In Section 1.2 we state the definitions we need and set down the notation to be used. In Section 1.3 we state and prove the main theorem. In Section 1.4 we deal with the applicability of the main theorem. Here we prove a number of propositions which, along with the main theorem, lead to the result that, in dimension two, the Gibbs phenomenon holds for functions satisfying appropriate hypotheses involving bounds on variations.

1.2 Definitions and Notation

Let $\mathbf{R}$ denote the set of real numbers, let $\mathbf{Z}$ denote the set of integers and let $\mathbf{T}$ denote the interval $(-\pi, \pi]$. For $x \in \mathbf{R}^d$ and $k \in \mathbf{Z}^d$, let kx denote the inner product of k and x. The Dirichlet kernel for the square is

$$D_N(x) = \sum_{\|k\|_\infty \leq N} e^{ikx}.$$

It follows that

$$D_N(x) = \prod_{j=1}^{d} \frac{\sin(N+\frac{1}{2})x_j}{\sin x_j/2},$$

where x_j is the jth component of x. Define $\mathrm{sinc}_N(x)$ by

$$\mathrm{sinc}_N(x) = \prod_{j=1}^{d} \frac{\sin Nx_j}{\sin x_j/2}.$$

The Fourier transform on $\mathbf{T}^d$ is given by

$$\hat{f}(k) = \frac{1}{(2\pi)^d} \int_{-\pi}^{\pi} \cdots \int_{-\pi}^{\pi} f(x)\, e^{-ikx}\, dx.$$

Define

$$S_N f(x) = \sum_{\|k\|_\infty \leq N} \hat{f}(k)\, e^{ikx}.$$

Then

$$S_N f(x) = (D_N * f)(x) = \frac{1}{(2\pi)^d} \int_{-\pi}^{\pi} \cdots \int_{-\pi}^{\pi} f(y)\, D_N(x-y)\, dy.$$

Definition 1.1. *A function* $g\colon \mathbf{R}^d \to \mathbf{C}$ *is called* angular *if* $g(ax) = g(x)$ *for all* $a > 0$ *and* $x \in \mathbf{R}^d$.

Definition 1.2. *Let* g *be an angular function. A function* $f\colon \mathbf{R}^d \to \mathbf{C}$ *has a* star discontinuity of type g at x_0 *if* $\lim_{x\to x_0}(f(x) - g(x-x_0)) = 0$.

Definition 1.3. *Let* f *have a star discontinuity of type* g *at* x_0 *and let* $g_{x_0}(x) = g(x - x_0)$. *We say that the star discontinuity at* x_0 *is* regular with respect to the square *if* $S_N(f - g_{x_0})$ *converges uniformly to* 0 *at* x_0*; that is, if for every* $\varepsilon > 0$ *there exists a* $\delta(\varepsilon) > 0$ *and* $N(\varepsilon)$ *such that* $|S_n(f - g_{x_0})(x)| < \varepsilon$ *for all* $|x - x_0| < \delta$ *and* $n \geq N$. *We say that the star discontinuity is* integrable *if* g *is integrable over all compact sets.*

Definition 1.4. *An angular function* $g\colon \mathbf{R}^d \to \mathbf{C}$ *satisfies* Property U *if for every* $\varepsilon > 0$ *there exists* $N(\varepsilon, g)$ *such that* $n > N$ *implies*

(a) $$\left| \int_{-\pi}^{\pi} (\sin ny_j) \left(\cot\left(\frac{y_j}{2}\right) - \frac{2}{y_j} \right) g\left(\frac{x}{n} - y\right) dy_j \right| < \varepsilon\, (\log n)^{-d+1},$$

(b) $$\left| \int_{-\pi}^{\pi} (\cos ny_j)\, g\left(\frac{x}{n} - y\right) dy_j \right| < \varepsilon\, (\log n)^{-d+1},$$

(c) $$\left| \int_{\mathbf{R}^d \setminus Q_d(n)} \left(\prod_{r=1}^{d} \frac{\sin u_r}{u_r} \right) g(x-u)\, du \right| < \varepsilon,$$

for all $j = 1, \ldots, d$ *and all* $x \in \mathbf{R}^d$ *where* $Q(n) = \prod_{r=1}^{d} [-n\pi, n\pi]$.

Definition 1.5. *A function* $f\colon R \subset \mathbf{R}^d \to \mathbf{C}$ *has* uniformly bounded variation on R *if there exists a constant* B *such that* $\sum_{i=1}^{n} |f(x_{i+1}) - f(x_i)| < B$ *where* $\{x_1, \ldots, x_n\}$ *is any partition of any finite line segment in* R *parallel to any coordinate axis.*

1.3 The Main Theorem

In this section we prove our theorem concerning the Gibbs phenomenon in $\mathbf{R}^d$. The hypotheses are not natural but are chosen so as to make the proof as transparent as possible. In the next section we concern ourselves with more natural hypotheses.

Lemma 1.6. *For complex numbers $a_1, \ldots, a_d$ and $b_1, \ldots, b_d$,*

$$\prod_{k=1}^{d} a_k - \prod_{k=1}^{d} b_k = \sum_{j=1}^{d} (a_j - b_j) \left(\prod_{k=1}^{j-1} a_k \right) \left(\prod_{k=j+1}^{d} b_k \right).$$

Proof. The right-hand sum is a telescoping sum which telescopes to the expression on the left-hand side. □

Theorem 1.7. *Suppose f is a complex-valued function on $\mathbf{R}^d$ which is 2π-periodic in each variable and has a star discontinuity at 0 of type g which is regular with respect to the square. Suppose further that g satisfies Property U. Then f exhibits the Gibbs phenomenon at 0. More specifically, given $\varepsilon > 0$, there exists a $\delta > 0$ and an N such that for $|x| < \delta$ and $n > N$,*

$$\left| S_n f\left(\frac{x}{n}\right) - \frac{1}{\pi^d} \int_{\mathbf{R}^d} \left(\prod_{j=1}^{d} \frac{\sin u_j}{u_j} \right) g(x-u)\, du \right| < \varepsilon.$$

Proof. We write

$$\begin{aligned}
(D_n * f)\left(\frac{x}{n}\right) &- \frac{1}{\pi^d} \int_{\mathbf{R}^d} \left(\prod_{j=1}^{d} \frac{\sin u_j}{u_j} \right) g(x-u)\, du \\
&= \left((D_n * f)\left(\frac{x}{n}\right) - (D_n * g)\left(\frac{x}{n}\right) \right) \\
&\quad + \left((D_n * g)\left(\frac{x}{n}\right) - (\operatorname{sinc}_n * g)\left(\frac{x}{n}\right) \right) \\
&\quad + \left((\operatorname{sinc}_n * g)\left(\frac{x}{n}\right) - \frac{1}{\pi^d} \int_{\mathbf{R}^d} \left(\prod_{j=1}^{d} \frac{\sin u_j}{u_j} \right) g(x-u)\, du \right).
\end{aligned}$$

We will estimate each of these three differences on the right-hand side in turn. Take any $\varepsilon > 0$. First, regularity with respect to the square of the discontinuity at 0 gives immediately the existence of an N_1 and a $\delta > 0$ such that $n > N_1$ and $|x| < \delta$ implies

$$\left| (D_n * f)\left(\frac{x}{n}\right) - (D_n * g)\left(\frac{x}{n}\right) \right| < \varepsilon.$$

Second,

$$(D_n - \mathrm{sinc}_n) * g\left(\frac{x}{n}\right)$$
$$= \frac{1}{2\pi^d} \int_{\mathbf{T}^d} \left[\prod_{k=1}^{d} \left(\frac{\sin(n+\frac{1}{2})y_k}{\sin y_k/2} \right) - \prod_{k=1}^{d} \left(\frac{\sin n y_k}{\sin y_k/2} \right) \right] g\left(\frac{x}{n} - u\right) du. \tag{1.1}$$

By Lemma 1.6 we have

$$\prod_{k=1}^{d} \left(\frac{\sin(n+\frac{1}{2})y_k}{\sin y_k/2} \right) - \prod_{k=1}^{d} \left(\frac{\sin n y_k}{\sin y_k/2} \right) \tag{1.2}$$
$$= \sum_{j=1}^{d} \left(\frac{\sin(n+\frac{1}{2})y_j}{\sin y_j/2} - \frac{\sin n y_j}{\sin y_j/2} \right) \left(\prod_{k=1}^{j-1} \frac{\sin(n+\frac{1}{2})y_k}{\sin y_k/2} \right) \left(\prod_{k=j+1}^{d} \frac{\sin n y_k}{y_k/2} \right).$$

We write

$$A_j(y) = \prod_{k=1}^{j} \frac{\sin(n+\frac{1}{2})y_k}{\sin y_k/2} \quad \text{and} \quad B_j(y) = \prod_{k=j+1}^{d} \frac{\sin n y_k}{y_k/2}.$$

Note that

$$\frac{\sin(n+\frac{1}{2})\theta}{\sin \theta/2} - \frac{\sin n\theta}{\theta/2} = \sin\left(\cot \frac{\theta}{2} - \frac{2}{\theta} \right) + \cos n\theta.$$

Using this observation in (1.2) and then substituting in (1.1) gives

$$(D_n - \mathrm{sinc}_n) * g\left(\frac{x}{n}\right)$$
$$= \frac{1}{2\pi^d} \sum_{j=1}^{d} \int_{\mathbf{T}^d} \sin n y_j \left(\cot \frac{y_j}{2} - \frac{2}{y_j} \right) A_j(j)\, B_j(y)\, g\left(\frac{x}{n} - y\right)$$
$$+ \cos n y_j\, A_j(j)\, B_j(y)\, g\left(\frac{x}{n} - y\right) dy. \tag{1.3}$$

Therefore

$$\left| (D_n - \mathrm{sinc}_n) * g\left(\frac{x}{n}\right) \right|$$
$$\leq \frac{1}{2\pi^d} \sum_{j=1}^{d} \int_{\mathbf{T}^{d-1}} \left| \int_{-\pi}^{\pi} \sin n y_j \left(\cot \frac{y_j}{2} - \frac{2}{y_j} \right) g\left(\frac{x}{n} - y\right) dy_j \right| |A_j(j)\, B_j(y)|\, d\hat{y}_j$$
$$+ \frac{1}{2\pi^d} \sum_{j=1}^{d} \int_{\mathbf{T}^{d-1}} \left| \int_{-\pi}^{\pi} \cos n y_j\, g\left(\frac{x}{n} - y\right) dy_j \right| |A_j(j) B_j(y)|\, d\hat{y}_j, \tag{1.4}$$

where $d\hat{y}_j = dy_1 \cdots dy_{j-1}\, dy_{j+1} \cdots dy_d$. Since g satisfies Property U, there exists an $N_2(\varepsilon, g)$ such that

$$\left| \int_{-\pi}^{\pi} \sin ny_j \left(\cot \frac{y_j}{2} - \frac{2}{y_j} \right) g\Big(\frac{x}{n} - y\Big)\, dy_j \right| < \varepsilon\, (\log n)^{-d+1}$$

and

$$\left| \int_{-\pi}^{\pi} \cos ny_j \, g\Big(\frac{x}{n} - y\Big)\, dy_j \right| < \varepsilon\, (\log n)^{-d+1}$$

for all x and $n \geq N$. Hence the right-hand side of (1.4) is less than

$$\frac{2\varepsilon(\log n)^{-d+1}}{(2\pi)^d} \sum_{j=1}^{d} \Big(\prod_{k=1}^{j-1} \int_{-\pi}^{\pi} \left| \frac{\sin\left(n + \frac{1}{2}\right) y_k}{\sin y_k/2} \right| dy_k \Big) \Big(\prod_{k=j+1}^{d} \int_{-\pi}^{\pi} \left| \frac{\sin ny_k}{y_k/2} \right| dy_k \Big)$$
$$< \frac{2\varepsilon(\log n)^{-d+1}}{(2\pi)^d} \sum_{j=1}^{d} (A \log n)^{j-1}\, (B \log n)^{d-j} < C\varepsilon,$$

where C depends only on d.

Third, we estimate

$$\left| (\mathrm{sinc}_n * g)\left(\frac{x}{n}\right) - \frac{1}{\pi^d} \int_{\mathbf{R}^d} \Big(\prod_{j=1}^{d} \frac{\sin u_j}{u_j} \Big) g(x-u)\, du \right|. \tag{1.5}$$

We have

$$(\mathrm{sinc}_n * g)\left(\frac{x}{n}\right) = \frac{1}{(2\pi)^d} \int_{\mathbf{T}^d} \Big(\prod_{j=1}^{d} \frac{\sin ny_j}{y_j/2} \Big) g\Big(\frac{x}{n} - y\Big)\, dy. \tag{1.6}$$

Since $g(z)$ depends only on the direction of z it follows that $g\left(\frac{x}{n} - y\right) = g(x - ny)$. Making the substitution $u = ny$ in (1.6) gives

$$(\mathrm{sinc}_n * g)\left(\frac{x}{n}\right) = \frac{1}{\pi^d} \int_{-n\pi}^{n\pi} \cdots \int_{-n\pi}^{n\pi} \left(\frac{\sin u_1}{u_1} \cdots \frac{\sin u_d}{u_d} \right) g(x-u)\, du.$$

Therefore (1.5) becomes

$$\left| \frac{1}{\pi^d} \int_{|u_1| > n\pi} \cdots \int_{|u_d| > n\pi} \left(\frac{\sin u_1}{u_1} \cdots \frac{\sin u_d}{u_d} \right) g(x-u)\, du \right|,$$

and by Property U of g, if $n > N_2(\varepsilon, g)$, then

$$\left| (\mathrm{sinc}_n * g)\left(\frac{x}{n}\right) - \frac{1}{\pi^d} \int_{\mathbf{R}^d} \Big(\prod_{j=1}^{d} \frac{\sin u_j}{u_j} \Big) g(x-u)\, du \right| < \varepsilon. \tag{1.7}$$

Combining our three estimates finishes the proof. □

1.4 Auxiliary Results

The aim of this section is to prove a number of auxiliary results which allow us to prove a theorem that asserts the existence of a Gibbs phenomenon in higher dimensions under more natural hypotheses. The ingredients of this involve the notions of bounded variation, integration by parts, increasing and decreasing functions, and second mean value theorems—all in a higher-dimensional context. Integration by parts and the second mean value theorems become very complicated in higher dimensions. Therefore, to keep the exposition under control we will deal with these only in dimension two, and therefore prove only the two-dimensional version of the theorem.

Proposition 1.8. *Let f have uniformly bounded variation in $\mathbf{R}^d$. Then there exists a constant C depending only on f such that for all $m > n$*

$$\left| \int_{Q_d(m)\backslash Q_d(n)} f(x)\left(\prod_{j=1}^{d} \frac{\sin x_j}{x_j}\right) dx \right| \leq C\frac{(\log n)^{d-1}}{n},$$

where $Q_d(r) = \{x \in \mathbf{R}^d \mid -r \leq x_j \leq r, j = 1, \ldots, d\}$, and C depends only on the variation bound of f and on $\|f\|_\infty$.

Proof. The proof proceeds by induction on the dimension. Let $d = 1$, then, as is well known,

$$\left| \int_n^m f(x)\,\frac{\sin x}{x}\,dx \right| \leq C(f)\frac{1}{n}.$$

Suppose the proposition holds for dimension $d - 1$. For $x \in \mathbf{R}^d$ and $1 \leq k \leq d$, let $\hat{x}_k = (x_1, \ldots, x_{k-1}, x_{k+1}, \ldots, x_d)$ and let $Q_d(n)$ denote the cube $\{x \in \mathbf{R}^d \mid -n \leq x_j \leq n, j = 1, \ldots, d\}$. Note that

$$Q_d(m)\backslash Q_d(n) = \bigcup_{k=1}^{d} \left(\{x \,|\, n < |x_k| \leq m\} \cap Q_d(m)\right)$$

and

$$\begin{aligned} &\{x \,|\, n < |x_k| \leq m\} \cap Q_d(m) \\ &\quad = \{x \,|\, n < |x_k| \leq m,\ \hat{x}_k \in Q_{d-1}(n)\} \\ &\quad\quad \bigcup \{x \,|\, n < |x_k| \leq m,\ \hat{x}_k \in Q_{d-1}(m)\backslash Q_{d-1}(n)\}. \end{aligned} \tag{1.8}$$

Denote the sets on the right by $R_k(n,m)$ and $T_k(n,m)$ respectively. Then

$$\begin{aligned} &\int_{R_k(n,m)} f(x)\left(\prod_{j=1}^{d} \frac{\sin x_j}{x_j}\right) dx \\ &\quad = \int_{-n}^{n} \cdots \int_{-n}^{n} \left(\int_n^m f(x)\frac{\sin x_k}{x_k}\,dx_k\right) \prod_{j\neq k} \frac{\sin x_j}{x_j}\,d\hat{x}_k. \end{aligned} \tag{1.9}$$

As noted before,

$$\left| \int_n^m f(x) \frac{\sin x_k}{x_k} \, dx_k \right| \leq C_1 \frac{1}{n},$$

where C_1 depends only on f. Therefore the integral in (1.9) is bounded by $C_2(\log n)^{d-1}/n$ where C_2 depends only on f.

Next consider

$$\int_{T_k(n,m)} f(x) \left(\prod_{j=1}^{d} \frac{\sin x_j}{x_j} \right) dx$$
$$= \int_n^m \left(\int_{Q_{d-1}(m) \setminus Q_{d-1}(n)} f(x) \prod_{j \neq k} \frac{\sin x_j}{x_j} \, d\hat{x}_k \right) \frac{\sin x_k}{x_k} \, dx_k. \quad (1.10)$$

Using the induction hypothesis gives

$$\left| \int_{Q_{d-1}(m) \setminus Q_{d-1}(n)} f(x) \prod_{j \neq k} \frac{\sin x_j}{x_j} \, d\hat{x}_k \right| \leq \frac{C_3(\log n)^{d-2}}{n}.$$

Therefore the integral in (1.10) is bounded by $C_4(\log n)^{d-1}/n$ where C_4 depends only on f. From the decomposition of $Q_{d-1}(m) \setminus Q_{d-1}(n)$ given in (1.8) we get

$$\left| \int_{Q_d(m) - Q_d(n)} f(x) \left(\prod_{j=1}^{d} \frac{\sin x_j}{x_j} \right) dx \right| \leq d(C_2 + C_4) \frac{(\log n)^{d-1}}{n},$$

which completes the proof. □

Proposition 1.9. *Suppose an angular function g defined on $\mathbf{R}^d$ has uniformly bounded variation on an open neighborhood of 0 with variation bound B. Then g has uniformly bounded variation on all of $\mathbf{R}^d$ with the same variation bound B.*

Proof. Let g be angular and of uniformly bounded variation on N, a neighborhood of 0, with bound constant B. Let I be any closed line segment in $\mathbf{R}^d$ parallel to a coordinate axis. Let $\{x_1, \ldots, x_M\}$ be a partition of I. There exists an $a > 0$ such that $aI \subset N$. Then

$$\sum_{k=1}^{M-1} |g(x_{i+1}) - g(x_i)| = \sum_{k=1}^{M-1} |g(ax_{i+1}) - g(ax_i)| \leq B. \quad \square$$

Proposition 1.10. *Suppose an angular function g has uniformly bounded variation on a closed cube Q containing 0. Then g satisfies Property U.*

Proof. For fixed x, n, j, and $y_1, \ldots, y_{j-1}, y_{j+1}, \ldots, y_d$, define h by $h(y_j) = g((x/n) - y)$ where y_j is the jth component of $y = (y_1, \ldots, y_d)$. By Proposition 1.9, h has bounded variation on $[-\pi, \pi]$. Since $(\cot(y_j/2) - (2/y_j))$ also has bounded variation on $[-\pi, \pi]$ it follows, as is well known, that

$$\left| \int_{-\pi}^{\pi} \sin ny_j \left(\cot \frac{y_j}{2} - \frac{2}{y_j} \right) h(y_j)\, dy_j \right| \leq \frac{C}{n}$$

and

$$\left| \int_{-\pi}^{\pi} h(y_j) \cos ny_j \, dy_j \right| \leq \frac{C}{n},$$

where C depends only on h which depends only on g. This shows that g satisfies the requirements (a) and (b) of Property U.

By Proposition 1.9, g has uniformly bounded variation on $\mathbf{R}^d$ and therefore by Proposition 1.8, for all $x \in \mathbf{R}^d$,

$$\left| \int_{Q_d(m)-Q_d(n)} \left(\prod_{j=1}^{d} \frac{\sin u_j}{u_j} \right) g(x-u)\, du \right| \leq C \frac{(\log n)^{d-1}}{n},$$

where C depends only on g. This verifies requirement (c) of Property U and finishes the proof. □

Definition 1.11. *Let f be a complex-valued function defined on the rectangle $R = [a,b] \times [c,d]$. Let $x_1, \ldots, x_n$ and $y_1, \ldots, y_m$ be partitions of the intervals $[a,b]$ and $[c,d]$ respectively. We say that f is of bounded variation in the sense of Hardy and Krause, or f is HKBV in $\mathbf{R}^2$, if there exists a constant B such that*

$$\sum_{i=1}^{n-1} \sum_{j=1}^{m-1} |f(x_{i+1}, y_{j+1}) - f(x_{i+1}, y_j) - f(x_i, y_{j+1}) + f(x_i, y_j)| \leq B,$$

$$\sum_{i=1}^{n-1} |f(x_{i+1}, c) - f(x_i, c)| \leq B,$$

$$\sum_{j=1}^{m-1} |f(a, y_{j+1}) - f(a, y_j)| \leq B,$$

for all partitions.

Definition 1.12. *Let f be a real-valued function on the closed rectangle R in $\mathbf{R}^2$. We say that f is non-decreasing on R if $f(x', y') - f(x, y') - f(x', y) + f(x, y) \geq 0$ for every subrectangle $[x, x'] \times [y, y']$ of R.*

As in the one-dimensional case, it is easy to prove the following.

Proposition 1.13. *If f is HKBV on a rectangle R, then f is the difference of two non-decreasing functions on R.*

The following two-dimensional integration by parts formula can be proved by iterating the one-dimensional formula.

Proposition 1.14. *Let f be continuous and let h be HKBV on the rectangle $[x_1, x_2] \times [y_1, y_2]$. Then*

$$\begin{aligned}
&\int_{x_1}^{x_2}\int_{y_1}^{y_2} g(x,y)\, d_x d_y f(x,y) \\
&\quad = \int_{x_1}^{x_2}\int_{y_1}^{y_2} f(x,y)\, d_x d_y g(x,y) - \int_{y_1}^{y_2} f(x_2,y)\, d_y g(x_2,y) \\
&\qquad + \int_{y_1}^{y_2} f(x_1,y)\, d_y g(x_1,y) - \int_{x_1}^{x_2} f(x,y_2)\, d_x g(x,y_2) \\
&\qquad + \int_{x_1}^{x_2} f(x,y_1)\, d_x g(x,y_1) + f(x_2,y_2)\, g(x_2,y_2) \\
&\qquad - f(x_1,y_2)\, g(x_1,y_2) - f(x_2,y_1)\, g(x_2,y_1) + f(x_1,y_1)\, g(x_1,y_1).
\end{aligned}$$

The next proposition states a second mean value theorem in dimension two.

Proposition 1.15. *Let g be real and non-decreasing on the rectangle $R = [x_1, x_2] \times [y_1, y_2]$ and let f be continuous on R. Then there exist $(x^*, y^*) \in R$, $x' \in [x_1, x_2]$, $x'' \in [x_1, x_2]$, $y' \in [y_1, y_2]$, $y'' \in [y_1, y_2]$ such that*

$$\begin{aligned}
&\int_{x_1}^{x_2}\int_{y_1}^{y_2} g(x,y)\, d_x d_y f(x,y) \\
&\quad = g(x_1,y_1)(f(x^*,y^*) - f(x_1,y') - f(x',y_1) + f(x_1,y_1)) \\
&\qquad - g(x_1,y_2)(f(x^*,y^*) - f(x_1,y') - f(x'',y_2) + f(x_1,y_2)) \\
&\qquad - g(x_2,y_1)(f(x^*,y^*) - f(x_2,y'') - f(x',y_1) + f(x_2,y_1)) \\
&\qquad + g(x_2,y_2)(f(x^*,y^*) - f(x_2,y'') - f(x'',y_2) + f(x_2,y_2)).
\end{aligned}$$

Proof. The integration by parts formula from Proposition 1.14 gives

$$\begin{aligned}
&\int_{x_1}^{x_2}\int_{y_1}^{y_2} g(x,y)\, d_x d_y f(x,y) \\
&\quad = \int_{x_1}^{x_2}\int_{y_1}^{y_2} f(x,y)\, d_x d_y g(x,y) - \int_{y_1}^{y_2} f(x_2,y)\, d_y g(x_2,y) \\
&\qquad + \int_{y_1}^{y_2} f(x_1,y)\, d_y g(x_1,y) - \int_{x_1}^{x_2} f(x,y_2)\, d_x g(x,y_2) \\
&\qquad + \int_{x_1}^{x_2} f(x,y_1)\, d_x g(x,y_1) + f(x_2,y_2)\, g(x_2,y_2) \\
&\qquad - f(x_1,y_2)\, g(x_1,y_2) - f(x_2,y_1)\, g(x_2,y_1) + f(x_1,y_1)\, g(x_1,y_1).
\end{aligned}$$

Since g is non-decreasing, the first mean value theorem applies to each of the integrals on the right side of the previous equation, yielding points (x^*, y^*), x', x'', y', y'' so that

$$\begin{aligned}
\int_{x_1}^{x_2}\int_{y_1}^{y_2} & g(x,y)\,d_x d_y f(x,y) \\
&= f(x^*,y^*)(g(x_2,y_2)-g(x_1,y_2)-g(x_2,y_1)+g(x_1,y_1)) \\
&\quad - f(x_2,y'')\,(g(x_2,y_2)-g(x_2,y_1)) \\
&\quad + f(x_1,y')\,(g(x_1,y_2)-g(x_1,y_1)) \\
&\quad - f(x'',y_2)\,(g(x_2,y_2)-g(x_1,y_2)) \\
&\quad + f(x',y_1)\,(g(x_2,y_1)-g(x_1,y_1)) \\
&\quad + f(x_2,y_2)\,g(x_2,y_2)-f(x_1,y_2)\,g(x_1,y_2) \\
&\quad - f(x_2,y_1)\,g(x_2,y_1)+f(x_1,y_1)\,g(x_1,y_1).
\end{aligned}$$

Regrouping the terms in the previous equation yields the result. □

Theorem 1.16. *Suppose f has bounded variation in the sense of Hardy and Krause on $[-\pi,\pi]\times[-\pi,\pi]$ and $\lim_{(x,y)\to(0,0)} f(x,y)=0$. Then for any $\varepsilon>0$ there exist a $\delta>0$ and N, depending only on ε and f such that $|(D_n * f)(x)|<\varepsilon$ for all $n>N$ and $|x|<\delta$.*

Proof. Since f is HKBV on $[-\pi,\pi]\times[-\pi,\pi]$ and therefore also on $[0,\pi]\times[0,\pi]$, $f=f_1-f_2$ where f_1 and f_2 are the non-decreasing positive and negative variations of f respectively. Since $\lim_{(x,y)\to(0,0)} f(x,y)=0$ it follows that $\lim_{(x,y)\to(0^+,0^+)} f_1(x,y)=0$ and $\lim_{(x,y)\to(0^+,0^+)} f_2(x,y)=0$.

It is easy to see that there exists a constant K such that

$$\left|\int_0^y\int_0^x D_n(u,v)\,du\,dv\right|\le K$$

for all $|x|<\pi$ and $|y|<\pi$ and $n\ge 1$. Given $\varepsilon>0$, choose $0<\delta\le\pi$ such that $|f_1(x,y)|<\varepsilon/(128K)$ for all $(x,y)\in[0,\delta]\times[0,\delta]$. Given $(x,y)\in[-\pi,\pi]\times[-\pi,\pi]$ define

$$F_n(x,y)=\int_0^y\int_0^x D_n(u,v)\,du\,dv.$$

Then by Proposition 1.15 there exist (u_0,v_0), u_1, u_2, v_1, v_2 such that

$$\begin{aligned}
\int_0^\delta\int_0^\delta & f_1(u,v)\,D_n(x-u,y-v)\,d_u\,d_v \\
&= \int_0^\delta\int_0^\delta f_1(u,v)\,d_u d_v F_n(x-u,y-v) \\
&= f_1(0,0)(G(u_0,v_0)-G(0,v_1)-G(u_1,0)+G(0,0)) \\
&\quad - f_1(0,\delta)(G(u_0,v_0)-G(0,v_1)-G(u_2,\delta)+G(0,\delta)) \\
&\quad - f_1(\delta,0)(G(u_0,v_0)-G(\delta,v_2)-G(u_1,0)+G(\delta,0)) \\
&\quad + f_1(\delta,\delta)(G(u_0,v_0)-G(\delta,v_2)-G(u_2,\delta)+G(\delta,\delta)),
\end{aligned}$$

where $G(u,v) = F_n(x-u, y-v)$. Therefore

$$\left| \int_0^\delta \int_0^\delta f_1(u,v)\, D_n(x-u, y-v)\, d_u\, d_v \right| < \frac{\varepsilon}{8}. \tag{1.11}$$

Next consider the three integrals

$$\left(\int_0^\delta \int_\delta^\pi + \int_\delta^\pi \int_0^\delta + \int_\delta^\pi \int_\delta^\pi \right) f_1(u,v)\, D_n(x-u, y-v)\, du\, dv. \tag{1.12}$$

Since f_1 is HKBV it has uniformly bounded variation. Therefore, for fixed v and δ

$$\left| \int_\delta^\pi f_1(u,v) \frac{\sin\left(n+\frac{1}{2}\right)(x-u)}{\sin((x-u)/2)}\, du \right| \le \frac{C}{n},$$

where C is independent of v but depends on δ. Therefore, since

$$\int_0^\delta \left| \frac{\sin\left(n+\frac{1}{2}\right)(x-u)}{\sin((x-u)/2)} \right| du \le C' \log n,$$

it follows that

$$\left| \int_0^\delta \int_\delta^\pi f_1(u,v)\, D_n(x-u, y-v)\, du\, dv \right| \le C'' \frac{\log n}{n}.$$

A similar bound holds for the second integral in (1.12).

Since f_1 is HKBV, arguing as in the one-dimensional case, it can be shown that

$$\left| \int_\delta^\pi \int_\delta^\pi f_1(u,v)\, D_n(x-u, y-v)\, du\, dv \right| \le \frac{C'''}{n^2}.$$

Therefore, there exists an N such that $n > N$ implies

$$\left| \iint_{[0,\pi]\times[0,\pi]\setminus[0,\delta]\times[0,\delta]} f_1(u,v)\, D_n(x-u, y-v)\, du\, dv \right| \le \frac{\varepsilon}{8}.$$

Combining this estimate with the estimate (1.11) gives

$$\left| \int_0^\pi \int_0^\pi f_1(u,v)\, D_n(x-u, y-v)\, d_u\, d_v \right| < \frac{\varepsilon}{4}.$$

We now relabel the δ and N by δ_1 and N_1 respectively.

Next for some $\delta > 0$ and $|x| < \delta$ and $|y| < \delta$ consider

$$\left| \int_{-\pi}^0 \int_0^\pi f(u,v)\, D_n(x-u, y-v)\, du\, dv \right|. \tag{1.13}$$

Using the change of variables $(u,v) \to (u,-v)$ and using the symmetry of D_n, (1.13) becomes

$$\left| \int_0^\pi \int_0^\pi f(u,-v)\, D_n(x-u,-y-v)\, du\, dv \right|,$$

which, letting $g(u,v) = f(u,-v)$ and $(x',y') = (x,-y)$, can be written as

$$\left| \int_0^\pi \int_0^\pi g(u,v)\, D_n(x'-u,y'-v)\, du\, dv \right|,$$

where $|x'| < \delta$ and $|y'| < \delta$.

Therefore the earlier arguments apply and there exist $\delta_2 > 0$ and N_2 such that

$$\left| \int_{-\pi}^0 \int_0^\pi f(u,v)\, D_n(x-u,y-v)\, du\, dv \right| < \frac{\varepsilon}{4}$$

for $|x| < \delta_2$, $|y| < \delta_2$, and $n > N_2$. In a similar way we find δ_3, δ_4, N_3, and N_4 such that

$$\left| \int_{-\pi}^0 \int_{-\pi}^0 f(u,v)\, D_n(x-u,y-v)\, du\, dv \right| < \frac{\varepsilon}{4}$$

for $|x| < \delta_3$, $|y| < \delta_3$, and $n > N_3$ and

$$\left| \int_{-\pi}^0 \int_{\pi}^0 f(u,v)\, D_n(x-u,y-v)\, du\, dv \right| < \frac{\varepsilon}{4}$$

for $|x| < \delta_4$, $|y| < \delta_4$, and $n > N_4$. Letting $\delta = \min\{\delta_1, \delta_2, \delta_3, \delta_4\}$ and $N = \max\{N_1, N_2, N_3, N_3\}$ we have

$$\left| \int_{-\pi}^\pi \int_{-\pi}^\pi f(u,v)\, D_n(x-u,y-v)\, du\, dv \right| < \varepsilon$$

for $|x| < \delta$, $|y| < \delta$ and $n > N$. This completes the proof. □

We are now in a position to state and prove the main theorem of this section.

Theorem 1.17. *Let $f\colon \mathbf{R}^2 \to \mathbf{C}$ be 2π-periodic in each variable. Suppose that f has a star discontinuity of type g at (0,0), and that g has uniformly bounded variation in a neighborhood of (0,0). Suppose further that $f - g$ has bounded variation in the sense of Hardy and Krause. Then f exhibits the Gibbs phenomenon at* $(0,0)$. *More precisely, given $\varepsilon > 0$ there exist a $\delta > 0$ and an N such that for $|(x,y)| < \delta$ and $n > N$*

$$\left| S_n f\left(\frac{(x,y)}{n}\right) - \frac{1}{\pi^2} \int_{-\infty}^\infty \int_{-\infty}^\infty \frac{\sin nx \sin ny}{xy}\, g(x-u,y-v)\, du\, dv \right| < \varepsilon.$$

Proof. By Proposition 1.10, g satisfies Property U. Applying Theorem 1.16 to $f - g$ shows that the star discontinuity at $(0,0)$ is regular with respect to the square. The hypotheses of Theorem 1.7 are satisfied and the conclusion follows. □

References

1. M. T. Cheng, The Gibbs phenomenon and Bochner's summation method. I., *Duke Math J.*, **17** (1950), pp. 83–90; The Gibbs phenomenon and Bochner's summation method. II., *Duke Math J.*, **17** (1950), pp. 477–490.
2. L. Colzani and M. Vignati, The Gibbs phenomenon for multiple Fourier integrals, *J. Approx. Theory*, **80** (1995), pp. 119–131.
3. B. I. Golubov, On the Gibbs phenomenon for Riesz spherical means of multiple Fourier series and Fourier integrals, *Anal. Math.*, **1** (1975), pp. 31–53.
4. G. Helmberg, A corner point Gibbs phenomenon for Fourier series in two dimensions, *J. Approx. Theory*, **100** (1999), pp. 1–43.
5. G. Helmberg, Localization of a corner point Gibbs phenomenon for Fourier series in two dimensions, *J. Fourier Anal. Appl.*, **8** (2002), pp. 29–41.
6. H. Weyl, Die Gibbs'sche Erscheinung in der Theorie der Kugelfunktionen, *Rend. Circ. Mat. Palermo*, **29** (1910), pp. 308–323.
7. H. Weyl, Uber die Gibbs'sche Erscheinung und verwandte Konvergenzphanomene, *Rend. Circ. Mat. Palermo*, **30** (1910), pp. 377–407.

2

Weighted Sobolev Inequalities for Gradients

Hans P. Heinig

Department of Mathematics and Statistics, McMaster University, Hamilton, Ontario L8S 4K1, Canada
heinig@univmail.cis.mcmaster.ca

Summary. We derive from Fourier inequalities between weighted Lebesgue spaces, weighted Sobolev gradient inequalities for a wide range of indices. The weight functions for which these inequalities hold are easily computable, but the norm constants are not optimal.

Dedicated to John J. Benedetto.

2.1 Introduction

The study of weighted norm inequalities for the Fourier transform in Lebesgue spaces is an important part of John Benedetto's many contributions in mathematical analysis (see, e.g., [1], [2], [3] and the literature cited there). In this chapter we apply the most general of these weighted Fourier inequalities and deduce a general weighted Sobolev gradient inequality. That is, we show that for a suitable class of functions f, conditions on weight functions u and w defined on $\mathbb{R}^n$ are given for which the inequality

$$\left(\int_{\mathbb{R}^n} |f(x)|^q\, u(x)\, dx\right)^{1/q} \leq C\left(\int_{\mathbb{R}^n} |(\nabla f)(x)|^p\, w(x)\, dx\right)^{1/p} \tag{2.1}$$

is satisfied for $1 < p,\ q < \infty$. In the case $0 < q \leq p$, $p > 1$ an alternate description of the weight functions is also given for which (2.1) holds. This result is deduced from a theorem of G. Sinnamon [5].

Weight conditions which imply weighted Fourier transform inequalities are given in terms of equimeasurable rearrangements of the weights. For this reason we recall the definition of rearrangement here.

Let (X, μ) be a measure space, where $X \subset \mathbb{R}^n$, and let g be a complex-valued μ-measurable function on X. The distribution function $D_g : [0, \infty) \to [0, \infty)$ of g is defined by

$$D_g(s) = \mu(\{x \in \mathbb{R}^n : |g(x)| > s\}).$$

The equimeasurable decreasing rearrangement of g is $g^*\colon [0,\infty) \to [0,\infty)$, defined by

$$g^*(t) = \inf\{s > 0 : D_g(s) \le t\}.$$

We adhere to the convention $\inf\{\emptyset\} = \infty$, so that if $D_g(s) > t$ for all $s \in [0,\infty)$, then $g^*(t) = \infty$.

In this chapter, the measure μ is always taken to be Lebesgue measure. To illustrate this we have the following.

Example 2.1. If $g(x) = |x|^a$, $a < 0$, $x \in \mathbb{R}^n$, then via polar coordinates in $\mathbb{R}^n$

$$\begin{aligned} D_g(y) &= \mu(\{x \in \mathbb{R}^n : |x|^a > y\}) \\ &= \mu(\{x \in \mathbb{R}^n : |x| < y^{1/a}\}) \\ &= \int_{\Sigma_{n-1}} d\sigma \int_0^{y^{1/a}} t^{n-1}\, dt \\ &= \frac{|\Sigma_{n-1}|}{n}\, y^{n/a}, \end{aligned}$$

where $|\Sigma_{n-1}|$ denotes the surface area of the unit n-sphere in $\mathbb{R}^n$. It follows that $g^*(t) = C_n t^{a/n}$ where $C_n = (n/|\Sigma_{n-1}|)^{a/n}$.

We also recall that the space of rapidly decreasing functions is defined by

$$\mathcal{S}(\mathbb{R}^n) = \{g \in C^\infty(\mathbb{R}^n) : \forall\, \alpha \in \mathbb{N}^n,\ \forall\, \beta \in \mathbb{N}^n,\ \sup |x^\alpha (D^\beta g)(x)| < \infty\}.$$

It is well known that the Fourier transform maps $\mathcal{S}(\mathbb{R}^n)$ onto itself and that the Fourier inversion formula $(\hat{f})\check{} = f$ holds for $f \in \mathcal{S}(\mathbb{R}^n)$.

Throughout, inequalities such as (2.1) are interpreted in the sense that if the right side is finite, so is the left side and the inequality holds. Constants are positive and denoted by C, sometimes with subscripts, and may also be different at different occurrences. As usual, $p' = p/(p-1)$ is the conjugate index of p, similarly for q. Finally, expressions of the form $0 \cdot \infty$ are taken to be zero.

2.2 Results

First we state the weighted Fourier norm inequality in its general form, namely for $1 < p,\ q < \infty$. A new proof of this classical result with good control of the norm constant is given in [1].

Theorem 2.2 ([1, Thm. 1]). *Let u and v be weight functions on $\mathbb{R}^n$ and $1 < p,\ q < \infty$. There is a constant $C > 0$ such that for all $f \in L^p_v(\mathbb{R}^n)$ the inequality*

$$\left(\int_{\widehat{\mathbb{R}}^n} |\hat{f}(\xi)|^q \, u(\xi) \, d\xi\right)^{1/q} \leq C\left(\int_{\mathbb{R}^n} |f(x)|^p \, v(x) \, dx\right)^{1/p} \tag{2.2}$$

is satisfied provided that

(i) *in the case* $1 < p \leq q < \infty$,

$$\sup_{s>0}\left(\int_0^{1/s} u^*(t)\, dt\right)^{1/q}\left(\int_0^s (1/v)^*(t)^{p'-1}\, dt\right)^{1/p'} \equiv C_1 < \infty, \tag{2.3}$$

(ii) *in the case* $1 < q < p$,

$$\left[\int_0^\infty \left(\int_0^{1/s} u^*\right)^{r/q}\left(\int_0^s (1/v)^{*(p'-1)}\right)^{r/q'} (1/v)^*(s)^{p'-1}\, ds\right]^{1/r} \equiv C_2 < \infty, \tag{2.4}$$

where $1/r = 1/q - 1/p$. *Moreover,* C/C_i, $i = 1, 2$ *is bounded above and below by positive constants depending on* p *and* q *only.*

Remark 2.3. (i) Fourier inequalities of the form (2.2) also hold in the range $0 < q < p$, $p > 1$. However the weight conditions are somewhat more complex. See, e.g., [2, Thm. 4.1], where such a result is given in terms of radial measure weights. In one dimension such a result also follows by modifying the proof of [3, Thm. 1].

(ii) Theorem 2.2(i) is sharp in the sense that if u and $1/v$ are radial and as radial functions are decreasing then (2.2) implies (2.3). See, e.g., [3, p. 35], where the one-dimensional case is discussed.

We are now in a position to give the main result.

Theorem 2.4. *Suppose* u *and* w *are weight functions on* $\mathbb{R}^n$ *and* $1 < p$, $q < \infty$. *There is a constant* $C > 0$ *such that for all* $f \in \mathcal{S}(\mathbb{R}^n)$ *and* $|\nabla f| \in L^p_w(\mathbb{R}^n)$ *the inequality* (2.1) *is satisfied, provided that*

(i) *in the case* $1 < p \leq q < \infty$, $n/(n-1) < q$,

$$\left(\int_0^s u^*(t)\, dt\right)^{1/q} \leq C_1 s^{-1/n+1-1/q} \tag{2.5}$$

and

$$\left(\int_0^s (1/w)^*(t)^{p'-1}\, dt\right)^{1/p'} \leq C_2 s^{1/q} \tag{2.6}$$

hold for some constants, C_i, $i = 1, 2$; *and all* $s > 0$;

(ii) *in the case $n/(n-1) < q < p$, the first condition of (2.5) is satisfied and*

$$\left[\int_0^\infty s^{-1-r/q}\left(\int_0^s (1/w)^*(t)^{p'-1}\,dt\right)^{r/p'} ds\right]^{1/r} \equiv C_3 < \infty, \tag{2.7}$$

where $1/r = 1/q - 1/p$.

Proof. (i) Since for $f \in \mathcal{S}(\mathbb{R}^n)$ the Fourier inversion formula holds, we apply inequality (2.2) with $p = q$ and $v(\xi) = |\xi|^q$. Hence

$$\begin{aligned}\left(\int_{\mathbb{R}^n} |f(x)|^q\, u(x)\,dx\right)^{1/q} &\le C\left(\int_{\widehat{\mathbb{R}}^n} |\xi|^q |\hat{f}(\xi)|^q\,d\xi\right)^{1/q}\\ &= C\left(\int_{\widehat{\mathbb{R}}^n} |(\nabla f)\hat{}(\xi)|^q\,d\xi\right)^{1/q},\end{aligned} \tag{2.8}$$

provided

$$\sup_{s>0}\left(\int_0^s u^*(t)\,dt\right)^{1/q}\left(\int_0^{1/s}(1/v)^*(t)^{q'-1}\,dt\right)^{1/q'} \equiv C_1 < \infty.$$

But if $v(\xi) = |\xi|^q$ then by Example 2.1 with $a = -q$, we have $(1/v)^*(t) = C_n t^{-q/n}$. Hence if $n/(n-1) < q$, then

$$\int_0^{1/s}(1/v)^*(t)^{q'-1}\,dt = C\int_0^{1/s} t^{-q(q'-1)/n}\,dt = C(1/s)^{1-q'/n},$$

so that the supremum is finite if the estimate of (2.5) is satisfied. Next

$$\begin{aligned}\left(\int_{\widehat{\mathbb{R}}^n} |(\nabla f)\hat{}(\xi)|^q\,d\xi\right)^{1/q} &= \left(\int_{\widehat{\mathbb{R}}^n}\left(\sum_{j=1}^n\left|\left(\frac{\partial f}{\partial x_j}\right)\hat{}(\xi)\right|^2\right)^{q/2} d\xi\right)^{1/q}\\ &\le \left(\int_{\widehat{\mathbb{R}}^n}\left(\sum_{j=1}^n\left|\left(\frac{\partial f}{\partial x_j}\right)\hat{}(\xi)\right|\right)^{q} d\xi\right)^{1/q}\\ &\le \sum_{j=1}^n\left(\int_{\widehat{\mathbb{R}}^n}\left|\left(\frac{\partial f}{\partial x_j}\right)\hat{}(\xi)\right|^q d\xi\right)^{1/q},\end{aligned}$$

where the first inequality holds trivially and the second is Minkowski's inequality. Applying Theorem 2.2(i) with f replaced by $\frac{\partial f}{\partial x_j}$, $j = 1, 2, \ldots, n$, $u \equiv 1$ and $v = w$, it follows that

$$\begin{aligned}\left(\int_{\widehat{\mathbb{R}}^n}\left|\left(\frac{\partial f}{\partial x_j}\right)\hat{}(\xi)\right|^q d\xi\right)^{1/q} &\le C\left(\int_{\mathbb{R}^n}\left|\frac{\partial f}{\partial x_j}(x)\right|^p w(x)\,dx\right)^{1/p}\\ &\le C\left(\int_{\mathbb{R}^n}\left(\sum_{j=1}^n\left|\frac{\partial f}{\partial x_j}(x)\right|^2\right)^{p/2} w(x)\,dx\right)^{1/p}\\ &= C\left(\int_{\mathbb{R}^n} |(\nabla f)(x)|^p\,w(x)\,dx\right)^{1/p},\end{aligned} \tag{2.9}$$

provided that (2.3) is satisfied with $u = 1$ and $v = w$. But this is clearly the case if the integral estimate of (2.6) holds. This, together with (2.8) proves part (i).

For the proof of part (ii) with $n/(n-1) < q$, we proceed as in the proof of (i) to obtain (2.8) provided the estimate of (2.5) holds. Proceeding as above, we get estimate (2.9) but now with $q < p$ provided that (2.4) is satisfied with $u = 1$ and $v = w$. But then (2.4) becomes

$$\int_0^\infty s^{-r/q} \left(\int_0^s (1/w)^*(t)^{p'-1}\, dt \right)^{r/q'} (1/w)^*(s)^{p'-1}\, ds \equiv C_2^r.$$

On integrating by parts, using the convention $0 \cdot \infty = 0$, we obtain

$$\frac{p'}{q} \int_0^\infty s^{-\frac{r}{q}-1} \left(\int_0^s (1/w)^*(t)^{p'-1}\, dt \right)^{r/p'} ds = C_2^r,$$

which is condition (2.7). This completes the proof. □

Corollary 2.5. *If $f \in \mathcal{S}(\mathbb{R}^n)$, then for $1 < p < n$, the inequality $\|f\|_q \leq C\|\nabla f\|_p$ holds with $q = np/(n-p)$.*

Proof. Applying Theorem 2.4(i) with $u = w = 1$, the result follows if (2.5) and (2.6) are satisfied with these weights. But this is clearly the case if $q = p'$ and $1/q = -1/n + 1/q'$. The condition $n/(n-1) < q$ is then equivalent to $1 < p < n$ and $q = np/(n-p)$. □

This result was proved by G. Talenti [6] with a sharp constant. See also [4, Thm. 8.3].

If $q = p$, $u(x) = |x|^{-p}$, and $w(x) \equiv 1$ in Theorem 2.4(i), then the conditions (2.5) and (2.6) are satisfied only if $p = q = 2$ and $n > 2$. Hence one obtains from (2.1) that

$$\left(\int_{\mathbb{R}^n} |f(x)|^2\, |x|^{-2}\, dx \right)^{1/2} \leq C \left(\int_{\mathbb{R}^n} |(\nabla f)(x)|^2\, dx \right)^{1/2},$$

when $n > 2$.

We conclude with a weight characterization for which a certain gradient inequality with weights holds.

Theorem 2.6 ([5, Thm. 4.1]). *Suppose $1 < p < \infty$, $0 < q < \infty$, and u and v are weight functions on $\mathbb{R}^n$ with $u \not\equiv 0$. The inequality*

$$\left(\int_{\mathbb{R}^n} |f(x)|^q\, u(x)\, dx \right)^{1/q} \leq C \left(\int_{\mathbb{R}^n} |x \cdot (\nabla f)(x)|^p\, v(x)\, dx \right)^{1/p}$$

holds for all $f \in C_0^\infty(\mathbb{R}^n)$ if and only if

(i) *$p = q$ and*

$$\sup_{x\in\mathbb{R}^n}\left(\int_0^1 u(tx)\,t^{n-1}\,dt\right)^{1/p}\left(\int_1^\infty [v(tx)\,t^n]^{1-p'}\,\frac{dt}{t}\right)^{1/p'}<\infty,$$

or

(ii) $q<p$ *and*

$$\left(\int_{\mathbb{R}^n}\left(\int_0^1 u(tx)\,t^{n-1}\,dt\right)^{r/q}\left(\int_1^\infty [v(tx)\,t^n]^{1-p'}\,\frac{dt}{t}\right)^{r/q'} v(x)^{1-p'}\,dx\right)^{1/r}<\infty,$$

where $1/r=1/q-1/p$.

Since $|x\cdot(\nabla f)(x)|\le |x||(\nabla f)(x)|$, then with $w(x)=|x|^p v(x)$ we obtain the following corollary.

Corollary 2.7. *Under the conditions of Theorem 2.6 it follows that*

$$\left(\int_{\mathbb{R}^n}|f(x)|^q\,u(x)\,dx\right)^{1/q}\le C\left(\int_{\mathbb{R}^n}|(\nabla f)(x)|^p\,w(x)\,dx\right)^{1/p},$$

provided that

(i) *for* $p=q$,

$$\sup_{x\in\mathbb{R}^n}|x|\left(\int_0^1 u(tx)\,t^{n-1}\,dt\right)^{1/p}\left(\int_1^\infty [w(xt)\,t^{n-1}]^{1-p'}\,dt\right)^{1/p'}<\infty$$

holds, and

(ii) *for* $q<p$,

$$\left(\int_{\mathbb{R}^n}\left(\int_0^1 u(tx)\,t^{n-1}\,dt\right)^{r/q}\times\left(\int_1^\infty [w(xt)\,t^{n-1}]^{1-p'}\,dt\right)^{r/q'}|x|^r\,w(x)^{1-p'}\,dx\right)^{1/r}<\infty$$

holds.

Observe that in the case $u(x)=|x|^{-p}$, $w(x)=1$, and $q=p$, the supremum in part (i) is satisfied if $n>p$. Hence

$$\left(\int_{\mathbb{R}^n}|f(x)|^p\,|x|^{-p}\,dx\right)^{1/p}\le C\left(\int_{\mathbb{R}^n}|(\nabla f)(x)|^p\,dx\right)^{1/p}$$

is satisfied. But this, as we have seen, cannot be obtained from Theorem 2.4, unless $p=2$.

On the other hand, Corollary 2.5 cannot be obtained from Corollary 2.7 since $q>p$.

Acknowledgments

This research was supported in part by NSERC grant A-4837.

References

1. J. J. Benedetto and H. P. Heinig, Weighted Fourier inequalities: New proofs and generalizations, *J. Fourier Anal. Appl.*, **9** (2003), pp. 1–37.
2. J. J. Benedetto and H. P. Heinig, Fourier transform inequalities with measure weights II, in: *Function Spaces* (Poznań, 1989), Teubner-Texte Math., Vol. 120, Teubner, Stuttgart, 1991, pp. 140–151.
3. J. J. Benedetto, H. P. Heinig, and R. Johnson, Weighted Hardy spaces and the Laplace transform II. *Math. Nachr.*, **132** (1987), pp. 29–55.
4. E. Lieb and M. Loss, *Analysis*, Grad. Studies Math., Vol. 14, American Mathematical Society, Providence, RI, 1997.
5. G. Sinnamon, A weighted gradient inequality, *Proc. Roy. Soc. Edinburgh, Sect. A*, **111** (1989), pp. 329–335.
6. G. Talenti, Best constant in Sobolev inequality, *Ann. Mat. Pura Appl. (4)*, **110** (1976), pp. 353–372.

3

Semidiscrete Multipliers

Georg Zimmermann

Institut für Angewandte Mathematik und Statistik, Universität Hohenheim, 70593 Stuttgart, Germany
Georg.Zimmermann@uni-hohenheim.de

Summary. A semidiscrete multiplier is an operator between a space of functions or distributions on a locally compact Abelian group G on the one hand, and a space of sequences on a discrete subgroup H of G on the other hand, with the property that it commutes with shifts by H. We describe the basic form of such operators and show a number of representation theorems for classical spaces like $\boldsymbol{L}^p$, $\boldsymbol{C}_0$, etc.

We also point out parallels to representation theorems for multipliers.

Dedicated to my academic teacher Prof. John J. Benedetto with many thanks for all the beauty in mathematics he taught me to see.

3.1 Introduction

The term "multiplier" originally stems from its definition as an operator on a Banach algebra which has the property

$$T(\varphi\,\psi) = T(\varphi)\,\psi = \varphi\,T(\psi)\,.$$

The most important and well-studied example is the algebra $\boldsymbol{L}^1(G)$, where G is a locally compact Abelian group. For this case, the following result is classical.

Proposition 3.1 ([9, Thm. 0.1.1]). *Suppose $T : \boldsymbol{L}^1(G) \to \boldsymbol{L}^1(G)$ is a continuous linear transformation. Then the following are equivalent.*

(i) *T commutes with the translation operators, i.e., $T\mathrm{T}_g = \mathrm{T}_g T$ for each $g \in G$.*
(ii) *$T(f_1 * f_2) = T(f_1) * f_2$ for each $f_1,\ f_2 \in \boldsymbol{L}^1(G)$.*
(iii) *There exists a (unique) function k defined on $\widehat{G}$ such that $(Tf)^\wedge = k\,\widehat{f}$ for each $f \in \boldsymbol{L}^1(G)$.*
(iv) *There exists a (unique) measure $\varphi \in \boldsymbol{M}_b(G)$ such that $T(f) = \varphi * f$ for each $f \in \boldsymbol{L}^1(G)$.*

Because of this result, it has become standard to use the term "multiplier" for bounded linear operators which commute with translations. Thus, given a translation-invariant Banach space $\boldsymbol{X}$ of functions or distributions on G, we define the space of *multipliers on* $\boldsymbol{X}(G)$ to be

$$\mathcal{M}\big(\boldsymbol{X}(G)\big) = \big\{T \in \mathcal{L}(\boldsymbol{X}) : \mathrm{T}_g T = T\mathrm{T}_g \ \ \forall g \in G\big\}.$$

The following are probably the most well-known characterizations for multiplier spaces.

Corollary 3.2 ([9, Cor. 0.1.1]). $\mathcal{M}\big(\boldsymbol{L}^1(G)\big)$ *is isometrically isomorphic to* $\boldsymbol{M}_b(G)$ *in the sense that* $T \in \mathcal{M}\big(\boldsymbol{L}^1(G)\big)$ *if and only if there exists* $\varphi \in \boldsymbol{M}_b(G)$ *such that* $T(f) = \varphi * f$, *and in this case* $\|T\|_{\mathcal{L}(\boldsymbol{L}^1)} = \|\varphi\|_{\boldsymbol{M}_b}$.

Proposition 3.3 ([9, Thm. 4.1.1]). $\mathcal{M}\big(\boldsymbol{L}^2(G)\big)$ *is isometrically isomorphic to* $\boldsymbol{L}^\infty(\widehat{G})$ *in the sense that* $T \in \mathcal{M}\big(\boldsymbol{L}^2(G)\big)$ *if and only if there exists* $\varphi \in \big(\boldsymbol{L}^\infty(\widehat{G})\big)^\vee$ *such that* $T(f) = \varphi * f$, *and in this case* $\|T\|_{\mathcal{L}(\boldsymbol{L}^2)} = \|\widehat{\varphi}\|_{\boldsymbol{L}^\infty}$.

We are interested in a related class of operator spaces. In both (discrete) wavelet theory and Gabor analysis, we encounter biorthogonal families of the form $(\mathrm{T}_n\varphi, \mathrm{T}_n\varphi^*)_{n\in\mathbb{Z}}$. For such a family, the coefficient mapping has the form

$$f \mapsto \big(\langle f, \mathrm{T}_n\varphi^*\rangle\big)_{n\in\mathbb{Z}} = \big((f * \widetilde{\varphi^*})(n)\big)_{n\in\mathbb{Z}},$$

where $\widetilde{\varphi^*}(x) = \overline{\varphi^*(-x)}$, and the evaluation map is the semidiscrete convolution product

$$a \mapsto a *' \varphi = \sum_{n\in\mathbb{Z}} a[n]\, \mathrm{T}_n\varphi .$$

Note that both these linear operators commute with integer shifts, but they map functions on the group $\mathbb{R}$ to sequences on the discrete subgroup $\mathbb{Z}$ and vice versa. So they could appropriately be described as *semidiscrete multipliers.*

In what follows, we shall consider a *locally compact Abelian group* G. For our arguments to work, we will also assume that G is first countable, which for a locally compact Abelian group is equivalent to metrizability, and also σ-compactness.

Furthermore, we assume that G has a *discrete, cocompact subgroup* H, so the quotient G/H is compact. Standard examples are $G = \mathbb{R}^d$ and $H = \mathbb{Z}^d$ with $G/H = \mathbb{T}^d$, or $G = \mathbb{T}^d$ and $H = (\frac{1}{N}\mathbb{Z}_N)^d$ with $G/H = \frac{1}{N}\mathbb{T}^d$.

Definition 3.4. *For a translation-invariant space of functions or distributions* $\boldsymbol{X}$ *on* G *and a shift-invariant space of sequences* $\boldsymbol{Y}$ *on* H, *there are two types of* semidiscrete multipliers*:*

(i) *The space of* discretization operators *or* generalized sampling operators

$$\mathcal{M}_H\big(\boldsymbol{X}(G), \boldsymbol{Y}(H)\big) = \big\{T \in \mathcal{L}(\boldsymbol{X}, \boldsymbol{Y}) : \mathrm{T}_h T = T\mathrm{T}_h \ \ \forall h \in H\big\},$$

and

(ii) *the space of* semidiscrete convolution operators

$$\boldsymbol{\mathcal{M}}_H\big(\boldsymbol{Y}(H),\boldsymbol{X}(G)\big)=\big\{T\in\boldsymbol{\mathcal{L}}(\boldsymbol{Y},\boldsymbol{X}):\mathrm{T}_hT=T\mathrm{T}_h\ \ \forall h\in H\big\}\,.$$

We shall study properties of these operator classes in general, and also give characterizations in some special cases. In particular, Theorem 3.22 and Corollary 3.29 will exhibit an interesting parallel to Corollary 3.2 and Proposition 3.3, respectively.

3.1.1 Notation and Conventions

In a slight, fairly common abuse of notation, we shall write

$$\mathrm{e}^{2\pi i\,\gamma g}$$

for the value of a character $\gamma\in\widehat{G}$ evaluated at a point $g\in G$. This should be easier to read than the abstract notation $\gamma(g)$.

As usual, we write

$$(\mathrm{T}_hf)(g):=f(g-h)\quad\text{ and }\quad(\mathrm{M}_\gamma f)(g):=\mathrm{e}^{2\pi i\,\gamma g}\,f(g)$$

for the translation and modulation operators.

It is also very common in harmonic analysis to define

$$\widetilde{\varphi}(g):=\overline{\varphi(-g)}\,,$$

and this implies $\widehat{\widetilde{\varphi}}=\overline{\widehat{\varphi}}$.

Usually, the notation $\langle f_1,f_2\rangle$ is only used for the inner product in a Hilbert space. To avoid having to treat the $\boldsymbol{L}^2$-case separately all the time, we extend this to denote a *sesquilinear* pairing between any locally convex topological vector space $\boldsymbol{X}$ and its dual $\boldsymbol{X}'$. This should not cause too much confusion, but helps to simplify several arguments considerably, since we always have

$$\langle f_1,f_2\rangle=\langle\widehat{f_1},\widehat{f_2}\rangle\,.$$

Strictly speaking, this implies that we consider the dual $\boldsymbol{X}'$ to be the space of continuous *antilinear* functionals on $\boldsymbol{X}$, but this is only a formal difference, since all the spaces we consider are invariant under complex conjugation.

3.2 Generalized Sampling Operators

The standard sampling operator

$$S:f\mapsto\big(\langle f,\mathrm{T}_h\delta\rangle\big)_{h\in H}=(f(h))_{h\in H}$$

only makes sense for functions f which are pointwise well-defined; so we have $S \in \boldsymbol{\mathcal{M}}_H(\boldsymbol{C}_b, \boldsymbol{\ell}^\infty)$ and $S \in \boldsymbol{\mathcal{M}}_H(\boldsymbol{C}_0, \boldsymbol{c}_0)$. In spaces of measurable functions, we have to apply some kind of smoothing operator before we can apply S. In many cases, all generalized sampling operators have this form, as the following result shows.

Proposition 3.5. (i) *Let $\boldsymbol{X}(G)$ and $\boldsymbol{Y}(H)$ be translation-invariant topological vector spaces, and assume that $\delta \in \boldsymbol{Y}'$, i.e., the mapping $\boldsymbol{Y} \to \mathbb{C}$, $a \mapsto a[0]$, is continuous. Then $\boldsymbol{\mathcal{M}}_H(\boldsymbol{X}, \boldsymbol{Y})$ is embedded in $\boldsymbol{X}'$ in the sense that every operator of this type is of the form*

$$S_\varphi : f \mapsto \big(\langle f, \mathrm{T}_h\varphi\rangle\big)_{h\in H}$$

for some $\varphi \in \boldsymbol{X}'$.

(ii) *If in addition, $\boldsymbol{X}$ and $\boldsymbol{Y}$ are Banach spaces, then the embedding is continuous.*

Proof. (i) Let $T \in \boldsymbol{\mathcal{M}}_H(\boldsymbol{X}, \boldsymbol{Y})$. By assumption, the map $f \mapsto (Tf)[0]$ is continuous, and thus $(Tf)[0] = \langle f, \varphi\rangle$ for some $\varphi \in \boldsymbol{X}'$. Furthermore, we have for all $h \in H$ that

$$(Tf)[h] = (\mathrm{T}_{-h}Tf)[0] = (T\mathrm{T}_{-h}f)[0] = \langle \mathrm{T}_{-h}f, \varphi\rangle = \langle f, \mathrm{T}_h\varphi\rangle = (S_\varphi f)[h]\,,$$

so $T = S_\varphi$ as claimed.

(ii) By assumption, we have $|a[0]| \le C\,\|a\|_{\boldsymbol{Y}}$, and thus

$$\begin{aligned}\|\varphi\|_{\boldsymbol{X}'} &= \sup_{\|f\|_{\boldsymbol{X}}=1} \big|\langle f, \varphi\rangle\big| \\ &= \sup_{\|f\|_{\boldsymbol{X}}=1} \big|(S_\varphi f)[0]\big| \\ &\le C \sup_{\|f\|_{\boldsymbol{X}}=1} \|S_\varphi f\|_{\boldsymbol{Y}} = C\,\|S_\varphi\|_{\mathcal{L}(\boldsymbol{X},\boldsymbol{Y})}\,. \square\end{aligned}$$

Corollary 3.6 (Representation Theorem for $\boldsymbol{\mathcal{M}}_H\big(\boldsymbol{X}(G), \boldsymbol{\ell}^\infty(H)\big)$). *Let $\boldsymbol{X}(G)$ be an isometrically translation-invariant Banach space. Then $\boldsymbol{\mathcal{M}}_H\big(\boldsymbol{X}(G), \boldsymbol{\ell}^\infty(H)\big)$ is isometrically isomorphic to $\boldsymbol{X}'(G)$.*

Remark 3.7. (i) Note that in the light of Proposition 3.5, we can read the standard sampling operator as $S = S_\delta \in \boldsymbol{\mathcal{M}}_H(\boldsymbol{C}_0, \boldsymbol{c}_0)$ with $\delta \in \boldsymbol{M}_b = \boldsymbol{C}_0'$.

(ii) If we define the convolution of $f \in \boldsymbol{X}$ with $\widetilde{\varphi} \in \boldsymbol{X}'$ by $(f * \widetilde{\varphi})(g) = \langle f, \mathrm{T}_g\varphi\rangle$, we obtain $S_\varphi(f) = S(f * \widetilde{\varphi}) = (f * \widetilde{\varphi})\big|_H$, i.e., a generalized sampling operator is the composition of an appropriate convolution and standard sampling.

Consequently, we can show the following.

Theorem 3.8 (Representation Theorem for $\mathcal{M}_H(\boldsymbol{C}_0(G), \boldsymbol{c}_0(H))$). *The space $\mathcal{M}_H(\boldsymbol{C}_0(G), \boldsymbol{c}_0(H))$ is isometrically isomorphic to $\boldsymbol{M}_b(G)$. In particular, whenever $S_\varphi \in \mathcal{M}_H(\boldsymbol{C}_0(G), \boldsymbol{\ell}^\infty(H))$, we actually have $S_\varphi \in \mathcal{M}_H(\boldsymbol{C}_0(G), \boldsymbol{c}_0(H))$.*

Proof. By Proposition 3.5, we know that any operator in $\mathcal{M}_H(\boldsymbol{C}_0(G), \boldsymbol{c}_0(H))$ is of the type S_φ for some $\varphi \in \boldsymbol{M}_b(G)$. Conversely, since $\boldsymbol{C}_0 * \boldsymbol{M}_b = \boldsymbol{C}_0$ (e.g., see [9, Thm. 3.3.2]), every such φ induces an operator of the desired type.

Furthermore, by Corollary 3.6, we know that $S_\varphi \in \mathcal{M}_H(\boldsymbol{C}_0, \boldsymbol{\ell}^\infty)$ is equivalent to $\varphi \in \boldsymbol{M}_b$ and thus, by the above, to $S_\varphi \in \mathcal{M}_H(\boldsymbol{C}_0, \boldsymbol{c}_0)$. Since the norms on $\boldsymbol{c}_0$ and $\boldsymbol{\ell}^\infty$ are the same, this also implies the isometry property. □

This decay property of the results of a generalized sampling operator also holds for $\boldsymbol{L}^p(G)$ with $1 < p < \infty$ (but obviously not for $p = 1$ and $p = \infty$).

Corollary 3.9. *For $p \in (1, \infty)$, whenever $S_\varphi \in \mathcal{M}_H(\boldsymbol{L}^p(G), \boldsymbol{\ell}^\infty(H))$, we actually have $S_\varphi \in \mathcal{M}_H(\boldsymbol{L}^p(G), \boldsymbol{c}_0(H))$.*

Proof. By Corollary 3.6, we know that any operator in $\mathcal{M}_H(\boldsymbol{L}^p(G), \boldsymbol{\ell}^\infty(H))$ is of the form $S_\varphi : f \mapsto (f * \widetilde{\varphi})\big|_H$ for some $\varphi \in \boldsymbol{L}^q(G)$, where $\frac{1}{p} + \frac{1}{q} = 1$. Since $\boldsymbol{L}^p * \boldsymbol{L}^q \subseteq \boldsymbol{C}_0$ (e.g., see [7, Thm. 20.19]), this yields the claim. □

However, we should not expect the sampling result to decay faster than the sampled function. So along the same lines as in [9, Thm. 5.2.5], which actually goes back to Hörmander, we can show the following.

Theorem 3.10. *If G is noncompact, we have for $\infty > p_1 > p_2 \geq 1$ that*

$$\mathcal{M}_H(\boldsymbol{L}^{p_1}(G), \boldsymbol{\ell}^{p_2}(H)) = \{0\}\,.$$

This also holds true for $p_1 = \infty$, if we consider (weak,weak)-continuous operators only.*

Proof. For $p_1 < \infty$, we make use of the fact that on a noncompact locally compact Abelian group G,

$$\lim_{g \to \infty} \left\| f + \mathrm{T}_g f \right\|_{\boldsymbol{L}^p} = 2^{1/p} \left\| f \right\|_{\boldsymbol{L}^p}$$

(e.g., see [9, Lem. 3.5.1 (i)]). Let $T \in \mathcal{M}_H(\boldsymbol{L}^{p_1}, \boldsymbol{\ell}^{p_2})$. Then we have for $h \in H$

$$\left\| Tf + \mathrm{T}_h Tf \right\|_{\boldsymbol{\ell}^{p_2}} = \left\| T(f + \mathrm{T}_h f) \right\|_{\boldsymbol{\ell}^{p_2}} \leq \|T\| \left\| f + \mathrm{T}_h f \right\|_{\boldsymbol{L}^{p_1}} .$$

Letting $h \to \infty$, this yields

$$2^{1/p_2} \|Tf\|_{\boldsymbol{\ell}^{p_2}} \leq \|T\|\, 2^{1/p_1} \|f\|_{\boldsymbol{L}^{p_1}} ,$$

i.e.,

$$\|Tf\|_{\boldsymbol{\ell}^{p_2}} \leq 2^{1/p_1 - 1/p_2} \|T\| \, \|f\|_{\boldsymbol{L}^{p_1}}$$

for all $f \in \boldsymbol{L}^{p_1}(G)$. But since $2^{1/p_1 - 1/p_2} < 1$, this contradicts the definition of $\|T\|$, unless $\|T\| = 0$ and thus $T = 0$.

For $p_1 = \infty$, recall that the dual of $\boldsymbol{L}^\infty$ with the weak*-topology is $\boldsymbol{L}^1$, so by Proposition 3.5, $Tf = \big((f * \widetilde{\varphi})(h)\big)_{h \in H}$ for some $\varphi \in \boldsymbol{L}^1(G)$. Now for $\gamma \in \widehat{G}$,

$$T(\mathrm{e}^{2\pi i\, \gamma g}) = \big(\langle \mathrm{e}^{2\pi i\, \gamma g}, \mathrm{T}_h \varphi \rangle\big)_{h \in H} = \big(\overline{\widehat{\varphi}(\gamma)}\, \mathrm{e}^{2\pi i\, \gamma h}\big)_{h \in H},$$

which is in $\boldsymbol{\ell}^{p_2}$ if and only if $\widehat{\varphi}(\gamma) = 0$. Therefore, we must have $\widehat{\varphi} \equiv 0$, i.e., $\varphi \equiv 0$ and thus $T = 0$. □

3.3 Semidiscrete Convolution Operators

Under mild assumptions, we have a characterization similar to Proposition 3.5 for the second type of semidiscrete multipliers. For their description, we need the following concept.

Definition 3.11. *The* semidiscrete convolution *of a function or distribution φ on G with a sequence $a = (a[h])_{h \in H}$ on H is given by*

$$\varphi *' a = a *' \varphi := \sum_{h \in H} a[h]\, \mathrm{T}_h \varphi\,.$$

Remark 3.12. It is worth noting that we can read a semidiscrete convolution as a convolution with a discrete measure in the sense that

$$\varphi *' a = \varphi * \sum_{h \in H} a[h]\, \delta_h\,.$$

In particular, for $a \in \boldsymbol{\ell}^1(H)$, this is simply convolution with a bounded Radon measure.

Proposition 3.13. (i) *Let $\boldsymbol{X}(G)$ and $\boldsymbol{Y}(H)$ be translation-invariant topological vector spaces. Assume that the δ-sequence $\delta = (\delta_{0,h})_{h \in H}$ is an element of $\boldsymbol{Y}$, and that finite sequences are dense in $\boldsymbol{Y}$. Then $\boldsymbol{\mathcal{M}}_H(\boldsymbol{Y}, \boldsymbol{X})$ is embedded in $\boldsymbol{X}$ in the sense that every operator of this type is a* semidiscrete convolution, *i.e., has the form*

$$R_\varphi : a \mapsto a *' \varphi = \sum_{h \in H} a[h]\, \mathrm{T}_h \varphi \tag{3.1}$$

for some $\varphi \in \boldsymbol{X}$.

(ii) *If in addition, $\boldsymbol{X}$ and $\boldsymbol{Y}$ are Banach spaces, then the embedding is continuous.*

Proof. (i) Let $T \in \boldsymbol{\mathcal{M}}_H(\boldsymbol{Y}, \boldsymbol{X})$. Since finite sequences are dense in $\boldsymbol{Y}$, we can express $a \in \boldsymbol{Y}$ as

$$a = \sum_{h \in H} a[h] \, \mathrm{T}_h \delta \,,$$

which yields

$$Ta = \sum_{h \in H} a[h] \, T\mathrm{T}_h \delta = \sum_{h \in H} a[h] \, \mathrm{T}_h T \delta \,.$$

Letting $\varphi := T\delta \in \boldsymbol{X}$, we see that $T = R_\varphi$ as claimed.

(ii) Under the assumptions made, we have

$$\|\varphi\|_{\boldsymbol{X}} = \|\delta *' \varphi\|_{\boldsymbol{X}} = \|R_\varphi \delta\|_{\boldsymbol{X}} \leq \|\delta\|_{\boldsymbol{Y}} \, \|R_\varphi\|_{\mathcal{L}(\boldsymbol{Y},\boldsymbol{X})} \,. \quad \square$$

The two types of semidiscrete multipliers have to be adjoint to each other, of course, and it also comes as no surprise that taking the adjoint preserves φ.

Lemma 3.14. *The operators S_φ and R_φ are adjoint to each other.*

Proof. $\langle S_\varphi f, a \rangle = \sum_{h \in H} \langle f, \mathrm{T}_h \varphi \rangle \, \overline{a[h]} = \left\langle f, \sum_h a[h] \, \mathrm{T}_h \varphi \right\rangle = \langle f, R_\varphi a \rangle \,.$ $\square$

Remark 3.15. The series on the right-hand side of (3.1) is a priori only a formal series. The limit has to be taken in some appropriate sense, e.g., with a summation kernel, if necessary. For details, we refer to the literature on summability theory.

If $\boldsymbol{X}$ is actually isometrically translation-invariant, we can be more specific for the special cases where $\boldsymbol{Y} = \boldsymbol{\ell}^p$.

Proposition 3.16. *Assume that $\boldsymbol{X}$ is an isometrically translation-invariant Banach space, and consider $R_\varphi \in \boldsymbol{\mathcal{M}}_H(\boldsymbol{Y}, \boldsymbol{X})$.*

(i) *If $\boldsymbol{Y} = \boldsymbol{\ell}^1$, then $\sum_h a[h] \, \mathrm{T}_h \varphi$ converges absolutely in $\boldsymbol{X}$.*

(ii) *If $\boldsymbol{Y} = \boldsymbol{\ell}^p$ for some $p \in (1, \infty)$, then the convergence is unconditional in norm in the sense that the finite partial sums form a Cauchy net.*

(iii) *If $\boldsymbol{Y} = \boldsymbol{\ell}^\infty$ and $\boldsymbol{X}$ is a dual space, then for (weak*,weak*)-continuous R_φ, we have unconditional convergence in $\boldsymbol{X}$ with the weak*-topology.*

Proof. (i) The proof follows immediately from the triangle inequality and the isometrical translation-invariance.

(ii) Let $a \in \boldsymbol{Y}$, and $\varepsilon > 0$. There is a finite subset $F \subseteq H$ such that $\left(\sum_{h \in H \setminus F} |a[h]|^p\right)^{1/p} < \varepsilon / \|R_\varphi\|$. Then we have for all finite sets F_1, F_2 containing F that

$$\left\| \sum_{h \in F_1} a[h] \, \mathrm{T}_h \varphi - \sum_{h \in F_2} a[h] \, \mathrm{T}_h \varphi \right\|_{\boldsymbol{X}} = \left\| \sum_{h \in H} b[h] \, \mathrm{T}_h \varphi \right\|_{\boldsymbol{X}} = \|R_\varphi b\|_{\boldsymbol{X}} \,,$$

where

$$b[h] = \begin{cases} a[h]\,, & \text{for } h \in F_1 \setminus F_2, \\ -a[h]\,, & \text{for } h \in F_2 \setminus F_1, \\ 0\,, & \text{otherwise,} \end{cases}$$

so $\|R_\varphi b\|_{\boldsymbol{X}} \le \|R_\varphi\|\, \|b\|_{\boldsymbol{Y}} \le \|R_\varphi\| \left(\sum_{h \in H\setminus F} |a[h]|^p\right)^{1/p} < \varepsilon$.

(iii) The assumptions imply that $R_\varphi^* = S_\varphi$ maps the predual of $\boldsymbol{X}$ into $\boldsymbol{\ell}^1$, hence we have for all $a \in \boldsymbol{\ell}^\infty$ and f in the predual of $\boldsymbol{X}$ that $\langle f, R_\varphi a\rangle = \langle S_\varphi f, a\rangle$ where the series for the latter converges absolutely and hence unconditionally, as claimed. □

Corollary 3.17 (Representation Theorem for $\boldsymbol{\mathcal{M}}_H(\boldsymbol{\ell}^1(H), \boldsymbol{X}(G))$). *Let $\boldsymbol{X}(G)$ be an isometrically translation-invariant Banach space. Then $\boldsymbol{\mathcal{M}}_H(\boldsymbol{\ell}^1(H), \boldsymbol{X}(G))$ is isometrically isomorphic to $\boldsymbol{X}(G)$.*

Corollary 3.18. *If G is noncompact, we have for $\infty > p_1 > p_2 \ge 1$ that*

$$\boldsymbol{\mathcal{M}}_H(\boldsymbol{\ell}^{p_1}(H), \boldsymbol{L}^{p_2}(G)) = \{0\}\,.$$

This also holds true for $p_1 = \infty$ if we consider (weak,weak)-continuous operators only.*

Proof. This follows immediately from Theorem 3.10 and Lemma 3.14. □

3.4 Mixed-Norm Spaces

A. Benedek and R. Panzone have introduced spaces of mixed norms on products of measure spaces in a very general way. We shall see that spaces of this type turn out to be isometric to certain spaces of semidiscrete multipliers.

Definition 3.19 ([1]). *Let (X_i, Σ_i, μ_i) for $i = 1, \ldots, n$ be given σ-finite measure spaces, and $\boldsymbol{P} = (p_1, \ldots, p_n)$ an n-tuple with $p_i \in [1, \infty]$. Then $\boldsymbol{L}^P(X)$ is the Banach space of measurable functions on $(X, \Sigma, \mu) = (\prod X_i, \prod \Sigma_i, \prod \mu_i)$ with norm*

$$\|f\|_{\boldsymbol{L}^P(X)} = \Big\| \cdots \big\| \|f\|_{\boldsymbol{L}^{p_1}(X_1)} \big\|_{\boldsymbol{L}^{p_2}(X_2)} \cdots \Big\|_{\boldsymbol{L}^{p_n}(X_n)}$$
$$= \left(\int_{X_n} \cdots \left(\int_{X_1} \big|f(x_1, \ldots, x_n)\big|^{p_1} \,\mathrm{d}\mu_1 \right)^{p_2/p_1} \cdots \,\mathrm{d}\mu_n \right)^{1/p_n}$$

with the usual conventions for $p_i = \infty$.

If for the quotient G/H there exists a measurable fundamental domain D, we can write $G = H \times D$ as measure spaces and consider the mixed-norm spaces $\boldsymbol{L}^{p_1,p_2}(H \times D)$ of measurable functions on G.

We can avoid making use of D by the following construction. Given a measurable function $f : G \to \mathbb{C}$, define the vector-valued function

$$\vec{f} : G \to \mathbb{C}^H, \quad g \mapsto (f(g+h))_{h\in H}\,.$$

Since we assume G to be σ-compact and H to be discrete, H is countable. So if f is only defined pointwise almost everywhere, the same is still true for $\vec{f}$. This function satisfies

$$\vec{f}(g-h_0) = \big(f((g-h_0)+h)\big)_{h\in H} = \big(f(g+(h-h_0))\big)_{h\in H} = \mathrm{T}_{h_0}(\vec{f}(g))$$

for $h_0 \in H$, i.e., $\|\vec{f}(g)\|_{\boldsymbol{\ell}^{p_1}}$ is well-defined on G/H, and we can introduce the following.

Definition 3.20. *For $p_1,\ p_2 \in [1,\infty]$, we define $\boldsymbol{L}^{p_1,p_2}(H,G/H)$ to be the Banach space of measurable functions on G with norm*

$$\begin{aligned}\|f\|_{\boldsymbol{L}^{p_1,p_2}(H,G/H)} &= \Big\|\|\vec{f}(g)\|_{\boldsymbol{\ell}^{p_1}(H)}\Big\|_{\boldsymbol{L}^{p_2}(G/H)} \\ &= \left(\int_{G/H}\left(\sum_{h\in H}|f(g+h)|^{p_1}\right)^{p_2/p_1} d\mu_{G/H}(g)\right)^{1/p_2},\end{aligned}$$

with the usual convention for $p_i = \infty$.

In other words, $f \in \boldsymbol{L}^{p_1,p_2}(H,G/H)$ if and only if $\vec{f} \in \boldsymbol{L}^{p_2}_{loc}(G,\boldsymbol{\ell}^{p_1}(H))$.

It is not difficult to show that, assuming the existence of a measurable, precompact fundamental domain, this definition agrees with the one above. We should mention the following properties of these spaces.

Lemma 3.21. *Assume that H is discrete and cocompact in G. Then the following hold.*

(i) *For $1 \le p \le \infty$, we have $\boldsymbol{L}^{p,p}(H,G/H) = \boldsymbol{L}^p(G)$.*

(ii) *For $1 \le p_1 < p_2 \le \infty$, we have $\boldsymbol{L}^{p_1,p_2}(H,G/H) \subset \boldsymbol{L}^{p_1} \cap \boldsymbol{L}^{p_2}(G)$.*

(iii) *For $\infty \ge p_1 > p_2 \ge 1$, we have $\boldsymbol{L}^{p_1,p_2}(H,G/H) \supset \boldsymbol{L}^{p_1} + \boldsymbol{L}^{p_2}(G)$.*

Spaces of this type have been considered before in the description of semidiscrete multipliers. In [8], it has been shown that

$$\varphi \in \boldsymbol{L}^{1,q}(\mathbb{Z},\mathbb{T}) \quad\Longrightarrow\quad \begin{array}{l} S_\varphi \in \boldsymbol{\mathcal{M}}_{\mathbb{Z}}(\boldsymbol{L}^p(\mathbb{R}),\boldsymbol{\ell}^p(\mathbb{Z})) \quad \text{and} \\ R_\varphi \in \boldsymbol{\mathcal{M}}_{\mathbb{Z}}(\boldsymbol{\ell}^q(\mathbb{Z}),\boldsymbol{L}^q(\mathbb{R}))\,, \end{array}$$

for $p \in [1,\infty]$, where $\frac{1}{p}+\frac{1}{q} = 1$. We shall see below that this sufficient condition on φ is also necessary in the case $p = 1$. The same holds true for $p = \infty$, if we require S_φ to be (weak*,weak*)-continuous; this is a consequence of Corollary 3.6 and Corollary 3.17, respectively, and the fact that $\boldsymbol{L}^{1,1}(\mathbb{Z},\mathbb{T}) = \boldsymbol{L}^1(\mathbb{R})$. For $p \in (1,\infty)$, the condition on φ is sufficient but not necessary.

Theorem 3.22 (Representation Theorem for $\mathcal{M}_H\big(\boldsymbol{L}^1(G), \boldsymbol{\ell}^p(H)\big)$). *For $p \in [1,\infty]$, the space $\mathcal{M}_H\big(\boldsymbol{L}^1(G), \boldsymbol{\ell}^p(H)\big)$ is isometrically isomorphic to $\boldsymbol{L}^{p,\infty}(H, G/H)$ in the sense that every operator of this type is of the form S_φ for some $\varphi \in \boldsymbol{L}^{p,\infty}(H, G/H)$ with*

$$\|S_\varphi\|_{\mathcal{L}(\boldsymbol{L}^1,\boldsymbol{\ell}^p)} = \|\varphi\|_{\boldsymbol{L}^{p,\infty}}\,.$$

Proof. For $p = \infty$, this is a special case of Corollary 3.6, since we have

$$\boldsymbol{L}^{\infty,\infty}(H, G/H) = \boldsymbol{L}^\infty(G)\,. \tag{3.2}$$

So let $p < \infty$, and assume that $\varphi \in \boldsymbol{L}^{p,\infty}(H, G/H)$. Then we have for $f \in \boldsymbol{L}^1(G)$ that

$$\begin{aligned}
\sum_{h\in H} \big|(f*\widetilde{\varphi})(h)\big|^p &= \sum_{h\in H} \left| \int_G f(g)\,\overline{\varphi(g-h)}\,d\mu_G(g)\right|^p \\
&\le \sum_{h\in H} \left(\int_G |\varphi(g-h)|\, \frac{|f(g)|}{\|f\|_{\boldsymbol{L}^1}}\, d\mu_G(g)\right)^p \|f\|_{\boldsymbol{L}^1}^p \\
&\underset{(*)}{\le} \sum_{h\in H} \int_G |\varphi(g-h)|^p\, \frac{|f(g)|}{\|f\|_{\boldsymbol{L}^1}}\, d\mu_G(g)\, \|f\|_{\boldsymbol{L}^1}^p \\
&= \int_G \left(\sum_{h\in H} |\varphi(g-h)|^p \right) \frac{|f(g)|}{\|f\|_{\boldsymbol{L}^1}}\, d\mu_G(g)\, \|f\|_{\boldsymbol{L}^1}^p \\
&\le \|\varphi\|_{\boldsymbol{L}^{p,\infty}(H,G/H)}^p\, \|f\|_{\boldsymbol{L}^1(G)}^p\,,
\end{aligned}$$

where inequality $(*)$ follows by applying Jensen's inequality on the space

$$\left(G, \frac{|f(g)|}{\|f\|_{\boldsymbol{L}^1}}\, d\mu_G(g)\right).$$

So $S_\varphi \in \mathcal{M}_H\big(\boldsymbol{L}^1(G), \boldsymbol{\ell}^p(H)\big)$ with $\|S_\varphi\|_{\mathcal{L}(\boldsymbol{L}^1,\boldsymbol{\ell}^p)} \le \|\varphi\|_{\boldsymbol{L}^{p,\infty}}$.

Conversely, let $T \in \mathcal{M}_H(\boldsymbol{L}^1(G), \boldsymbol{\ell}^p(H))$. We know by Proposition 3.5 that $T = S_\varphi$ for some $\varphi \in \boldsymbol{L}^\infty(G)$. Consider a neighborhood basis $(V_n)_{n\in\mathbb{N}}$ at $0 \in G$, where all $\overline{V_n}$ are compact and $\overline{V_{n+1}} \subset V_n$. Letting $f_n = (\mu(V_n))^{-1}\,\chi_{V_n}$ yields an approximate identity with $\|f_n\|_{\boldsymbol{L}^1} = 1$, and hence

$$\lim_{n\to\infty} \int_G f_n(g-g_0)\,\overline{\varphi(g)}\,d\mu_G(g) = \overline{\varphi(g_0)}$$

for almost all $g_0 \in G$. Since H is countable, we even have

$$\left| \int_G f_n(g-g_0+h)\,\overline{\varphi(g)}\,d\mu_G(g)\right|^p \to \left|\overline{\varphi(g_0-h)}\right|^p \qquad \text{for all } h \in H$$

for almost all $g_0 \in G$. Applying Fatou's Lemma on H yields

$$\sum_{h\in H} \left|\overline{\varphi(g_0-h)}\right|^p \le \liminf_{n\to\infty} \sum_{h\in H} \left| \int_G f_n(g-g_0+h)\, \overline{\varphi(g)}\, d\mu_G(g) \right|^p$$
$$= \liminf_{n\to\infty} \left\| S_\varphi(\mathrm{T}_{g_0} f_n) \right\|_{\boldsymbol{\ell}^p}^p \le \|S_\varphi\|_{\mathcal{L}(\boldsymbol{L}^1,\boldsymbol{\ell}^p)}^p$$

for almost all $g_0 \in G$, i.e., $\varphi \in \boldsymbol{L}^{p,\infty}(H, G/H)$ with $\|\varphi\|_{\boldsymbol{L}^{p,\infty}} \le \|S_\varphi\|_{\mathcal{L}(\boldsymbol{L}^1,\boldsymbol{\ell}^p)}$.
□

Corollary 3.23 (Representation Theorem for $\mathcal{M}_H(\boldsymbol{\ell}^p(H), \boldsymbol{L}^\infty(G))$). *For $p \in [1,\infty)$, the space $\mathcal{M}_H(\boldsymbol{\ell}^p(H), \boldsymbol{L}^\infty(G))$ is isometrically isomorphic to $\boldsymbol{L}^{q,\infty}(H, G/H)$, where $\frac{1}{p}+\frac{1}{q} = 1$, in the sense that every operator of this type is of the form R_φ for some $\varphi \in \boldsymbol{L}^{q,\infty}(H, G/H)$ with*

$$\|R_\varphi\|_{\mathcal{L}(\boldsymbol{\ell}^p,\boldsymbol{L}^\infty)} = \|\varphi\|_{\boldsymbol{L}^{q,\infty}} .$$

This also holds true for $p = \infty$ if we consider (weak,weak*)-continuous operators only.*

Proof. For $p = 1$, this follows from Corollary 3.17 and (3.2). For $p \in (1,\infty)$, we know that $T \in \mathcal{M}_H(\boldsymbol{\ell}^p(H), \boldsymbol{L}^\infty(G))$ satisfies

$$T = R_\varphi \quad \text{for some } \varphi \in \boldsymbol{L}^\infty(G).$$

Then the restriction of T^* to $\boldsymbol{L}^1(G)$ is $S_\varphi \in \mathcal{M}_H(\boldsymbol{L}^1(G), \boldsymbol{\ell}^q(H))$, so by Theorem 3.22, we have $\varphi \in \boldsymbol{L}^{q,\infty}$. Furthermore, $T = R_\varphi = (S_\varphi)^*$ implies

$$\|T\|_{\mathcal{L}(\boldsymbol{\ell}^p,\boldsymbol{L}^\infty)} = \|S_\varphi\|_{\mathcal{L}(\boldsymbol{L}^1,\boldsymbol{\ell}^q)} = \|\varphi\|_{\boldsymbol{L}^{q,\infty}} .$$

Finally, if $T \in \mathcal{M}_H(\boldsymbol{\ell}^\infty(H), \boldsymbol{L}^\infty(G))$ is (weak*,weak*)-continuous, then we have $T^* \in \mathcal{M}_H(\boldsymbol{L}^1(G), \boldsymbol{\ell}^1(H))$ and the reasoning is the same. □

Remark 3.24. For $G = \mathbb{T}$ with $H = \frac{1}{N}\mathbb{Z}_N = \{\frac{k}{N} : k = 0, \ldots, N-1\}$, and thus $G/H = \frac{1}{N}\mathbb{T}$, the case $p = \infty$ is well known in approximation theory. Given $\varphi \in \boldsymbol{L}^\infty(\mathbb{T})$, the $(\boldsymbol{\ell}^\infty, \boldsymbol{L}^\infty)$-norm of the operator

$$R_\varphi : \mathbb{C}^N \to \boldsymbol{L}^\infty(\mathbb{T}), \quad a \mapsto \sum_k a_k\, \mathrm{T}_{k/N}\varphi ,$$

is the Lebesgue constant of φ, given by

$$\|R_\varphi\|_{\mathcal{L}(\boldsymbol{\ell}^\infty,\boldsymbol{L}^\infty)} = \left\| \sum_k \left|\mathrm{T}_{k/N}\varphi\right| \right\|_{\boldsymbol{L}^\infty(\mathbb{T})} = \|\varphi\|_{\boldsymbol{L}^{1,\infty}(\frac{1}{N}\mathbb{Z}_N, \frac{1}{N}\mathbb{T})} .$$

3.5 Semidiscrete Multipliers under the Fourier Transform

The mixed-norm spaces introduced in the last section play another important role, namely, in the description of the action of semidiscrete multipliers on the Fourier side. The latter turn out to give another reason for the use of the term "multipliers." As usual, we will consider the Fourier transform of a linear operator T to be the operator $\widehat{T} : \widehat{f} \mapsto \widehat{Tf}$, i.e., $\widehat{T} = \mathcal{F} \circ T \circ \mathcal{F}^{-1}$. In particular, we have for example $\widehat{\mathrm{T}_g} = \mathrm{M}_{-g}$ and $\widehat{\mathrm{M}_\gamma} = \mathrm{T}_\gamma$.

The key ingredient for this section is periodization, which we now define.

Definition 3.25. *The* periodization operator *associated with H is given by*

$$P : f \mapsto f_H^\circ := \sum_{h \in H} \mathrm{T}_h f \,.$$

Note that f_H° is indeed H-periodic, so P can be seen as mapping functions (or distributions) on G to functions (or distributions) on G/H.

It is an elementary exercise to show that P is a bounded linear mapping from $\boldsymbol{L}^1(G)$ onto $\boldsymbol{L}^1(G/H)$ with norm $\|P\| = 1$. What is as easy to show, but much more important, is the following meta-theorem.

The Fourier transform of sampling is periodization.

In symbols, this means $\widehat{S} = P$, and by Fourier inversion, also $\widehat{P} = S$. The latter simply states that the Fourier coefficients of f_H° can be obtained by sampling $\widehat{f}$ on $H^\perp$, which can be verified by straightforward calculation. A discussion of this meta-theorem in connection with the Poisson Summation Formula on a variety of spaces, among them the mixed-norm spaces discussed in Section 3.4, can be found in [3].

Since we saw that S_φ consists of convolution with $\widetilde{\varphi}$ followed by sampling, we expect its Fourier transform to be multiplication with $\widehat{\widetilde{\varphi}} = \overline{\widehat{\varphi}}$ followed by periodization. Analogously, we should expect the Fourier transform of the semidiscrete convolution of a sequence a with a function φ to be multiplication of the function $A = \widehat{a}$ on $\widehat{H} = \widehat{G}/H^\perp$ with the function $\widehat{\varphi}$ on $\widehat{G}$. (For this product to make sense, it suffices to recall that a function on $\widehat{G}/H^\perp$ by definition is an $H^\perp$-periodic function on $\widehat{G}$.) These two properties we now show in detail.

Lemma 3.26. (i) *The action of S_φ on the Fourier side is the operator*

$$P_{\widehat{\varphi}} := \widehat{S_\varphi} : \quad \widehat{f} \mapsto (\widehat{f}\, \overline{\widehat{\varphi}})_{H^\perp}^\circ \,.$$

(ii) *The action of R_φ on the Fourier side is the operator*

$$Q_{\widehat{\varphi}} := \widehat{R_\varphi} : \quad A \mapsto A\, \widehat{\varphi} \,.$$

Proof. (i) For $h \in H$, we have

$$\big((\widehat{f}\,\overline{\widehat{\varphi}})^{\circ}_{H^\perp}\big)^{\vee}[h] = \int_{\widehat{G}/H^\perp} \Big(\sum_{\eta \in H^\perp} (\widehat{f}\,\overline{\widehat{\varphi}})(\gamma+\eta) \Big)\, \mathrm{e}^{2\pi i\,\gamma h}\, d\mu_{\widehat{G}/H^\perp}(\gamma)$$
$$= \int_{\widehat{G}} (\widehat{f}\,\overline{\widehat{\varphi}})(\gamma)\, \mathrm{e}^{2\pi i\,\gamma h}\, d\mu_{\widehat{G}}(\gamma)$$
$$= (f * \widetilde{\varphi})(h)\,.$$

Therefore, by Fourier inversion, $\widehat{S_\varphi f} = (\widehat{f}\,\overline{\widehat{\varphi}})^{\circ}_{H^\perp}$ as claimed.

(ii) $R_\varphi a = \sum_h a[h]\, \mathrm{T}_h \varphi$ yields

$$(\widehat{R_\varphi a})(\gamma) = \sum_{h \in H} a[h]\, \mathrm{e}^{-2\pi i\,\gamma h}\, \widehat{\varphi}(\gamma) = A(\gamma)\, \widehat{\varphi}(\gamma)$$

as claimed. □

Note that in analogy to Propositions 3.5 and 3.13, we can describe the operators of type $P_{\widehat{\varphi}}$ and $Q_{\widehat{\varphi}}$ as the operators commuting with the family of modulations $\{\mathrm{M}_h : h \in H\}$, where $(\mathrm{M}_g f)(\gamma) = \mathrm{e}^{2\pi i\,\gamma g} f(\gamma)$ for $g \in G$ and functions f on $\widehat{G}$. Therefore, we consider the spaces of pointwise multipliers

$$\boldsymbol{\mathcal{M}}^{\cdot}_{H^\perp}\big(\boldsymbol{X}(G), \boldsymbol{Y}(G/H)\big) = \big\{T \in \boldsymbol{\mathcal{L}}(\boldsymbol{X}, \boldsymbol{Y}) : \mathrm{M}_\eta T = T\mathrm{M}_\eta \;\; \forall \eta \in H^\perp\big\}\,,$$
$$\text{and} \quad \boldsymbol{\mathcal{M}}^{\cdot}_{H^\perp}\big(\boldsymbol{Y}(G/H), \boldsymbol{X}(G)\big) = \big\{T \in \boldsymbol{\mathcal{L}}(\boldsymbol{Y}, \boldsymbol{X}) : \mathrm{M}_\eta T = T\mathrm{M}_\eta \;\; \forall \eta \in H^\perp\big\}\,.$$

It turns out that we encounter again the mixed-norm spaces introduced above, although in a completely different role.

Theorem 3.27. (i) *For* $\infty > p_1 \geq p_2 \geq 1$, *the space* $\boldsymbol{\mathcal{M}}^{\cdot}_{H^\perp}\big(\boldsymbol{L}^{p_1}(G/H), \boldsymbol{L}^{p_2}(G)\big)$ *is isometrically isomorphic to* $\boldsymbol{L}^{p_2,r}(H, G/H)$, *where* $\frac{1}{r} + \frac{1}{p_1} = \frac{1}{p_2}$, *in the sense that every operator of this type is of the form* Q_φ *for some* $\varphi \in \boldsymbol{L}^{p_2,r}(H, G/H)$ *with*

$$\|Q_\varphi\|_{\boldsymbol{\mathcal{L}}(\boldsymbol{L}^{p_1}, \boldsymbol{L}^{p_2})} = \|\varphi\|_{\boldsymbol{L}^{p_2,r}}\,.$$

This also holds true for $p_1 = \infty$ *if we consider (weak*,weak*)- resp. (weak*, weak)-continuous operators only.*

(ii) *Assume that* G *is nondiscrete. Then we have for* $1 \leq p_1 < p_2 \leq \infty$ *that* $\boldsymbol{\mathcal{M}}^{\cdot}_{H^\perp}\big(\boldsymbol{L}^{p_1}(G/H), \boldsymbol{L}^{p_2}(G)\big) = \{0\}$.

Proof. (i) Let $T \in \boldsymbol{\mathcal{M}}^{\cdot}_{H^\perp}\big(\boldsymbol{L}^{p_1}(G/H), \boldsymbol{L}^{p_2}(G)\big)$ and set $\varphi = T(1)$. By assumption, we have $T(\mathrm{e}^{2\pi i\,\eta g}) = T(\mathrm{M}_\eta 1) = \mathrm{M}_\eta T(1) = \mathrm{e}^{2\pi i \eta g}\, \varphi$ for all $\eta \in H^\perp$. Since trigonometric polynomials are dense in $\boldsymbol{L}^{p_1}(G/H)$, we have $T = Q_\varphi$.

If $p_1 \geq p_2 < \infty$, we have that

$$
\begin{aligned}
& A\,\varphi \in \boldsymbol{L}^{p_2}(G) \qquad \forall A \in \boldsymbol{L}^{p_1}(G/H) \\
\Longleftrightarrow \quad & |A|^{p_2}\,|\varphi|^{p_2} \in \boldsymbol{L}^1(G) \qquad \forall A \in \boldsymbol{L}^{p_1}(G/H) \\
\Longleftrightarrow \quad & |A|^{p_2}\,\left(|\varphi|^{p_2}\right)^\circ_H \in \boldsymbol{L}^1(G/H) \qquad \forall A \in \boldsymbol{L}^{p_1}(G/H) \\
\Longleftrightarrow \quad & B\,\left(|\varphi|^{p_2}\right)^\circ_H \in \boldsymbol{L}^1(G/H) \qquad \forall B \in \boldsymbol{L}^{p_1/p_2}(G/H) \\
\Longleftrightarrow \quad & \left(|\varphi|^{p_2}\right)^\circ_H \in \boldsymbol{L}^{p_1/(p_1-p_2)}(G/H)\,,
\end{aligned}
$$

so $Q_\varphi \in \boldsymbol{\mathcal{M}}^{\cdot}_{H^\perp}\left(\boldsymbol{L}^{p_1}, \boldsymbol{L}^{p_2}\right)$ if and only if $\varphi \in \boldsymbol{L}^{p_2,r}(H, G/H)$.

Then we can estimate for $A \in \boldsymbol{L}^{p_1}(G/H)$ that

$$
\begin{aligned}
\|Q_\varphi\,A\|^{p_2}_{\boldsymbol{L}^{p_2}} &= \int_G |A\,\varphi|^{p_2}\,d\mu_G \\
&= \int_{G/H} |A|^{p_2}\left(|\varphi|^{p_2}\right)^\circ_H\,d\mu_{G/H} \\
&\underset{(*)}{\leq} \left(\int_{G/H} |A|^{p_1}\,d\mu_{G/H}\right)^{p_2/p_1}\left(\int_{G/H}\left(\left(|\varphi|^{p_2}\right)^\circ_H\right)^{r/p_2}d\mu_{G/H}\right)^{p_2/r} \\
&= \|A\|^{p_2}_{\boldsymbol{L}^{p_1}}\,\|\varphi\|^{p_2}_{\boldsymbol{L}^{p_2,r}}\,,
\end{aligned}
$$

where inequality $(*)$ follows from Hölder's inequality with dual exponents $(p,q) = (p_1/p_2, r/p_2)$. Furthermore, we may choose $A = \left(\left(|\varphi|^{p_2}\right)^\circ_H\right)^{r/p_2p_1} \in \boldsymbol{L}^{p_1}(G/H)$ with equality in $(*)$, which yields the claim.

For $p_1 = p_2 = \infty$, it is obvious that $A\,\varphi \in \boldsymbol{L}^\infty(G)$ for all $A \in \boldsymbol{L}^\infty(G/H)$ if and only if $\varphi \in \boldsymbol{L}^\infty(G) = \boldsymbol{L}^{\infty,\infty}(H, G/H)$, and then $\|Q_\varphi\|_{\boldsymbol{\mathcal{L}}(\boldsymbol{L}^\infty)} = \|\varphi\|_{\boldsymbol{L}^\infty}$.

(ii) For the case $p_1 < p_2$, note that since $\boldsymbol{L}^{p_2}(G) \subseteq \boldsymbol{L}^{p_2}_{loc}(G)$, the function φ has to satisfy $A\,\varphi \in \boldsymbol{L}^{p_2}_{loc}(G)$ for all $A \in \boldsymbol{L}^{p_1}(G/H)$, but if G is nondiscrete, this is only possible for $\varphi \equiv 0$. □

Corollary 3.28. (i) *For $\infty > p_1 \geq p_2 \geq 1$, the space $\boldsymbol{\mathcal{M}}^{\cdot}_{H^\perp}\big(\boldsymbol{L}^{p_1}(G), \boldsymbol{L}^{p_2}(G/H)\big)$ is isometrically isomorphic to $\boldsymbol{L}^{q_1,r}(H, G/H)$, where $\frac{1}{p_1} + \frac{1}{q_1} = 1$ and $\frac{1}{r} + \frac{1}{p_1} = \frac{1}{p_2}$, in the sense that every operator of this type is of the form P_φ for some $\varphi \in \boldsymbol{L}^{q_1,r}(H, G/H)$ with*

$$
\|P_\varphi\|_{\boldsymbol{\mathcal{L}}(\boldsymbol{L}^{p_1}, \boldsymbol{L}^{p_2})} = \|\varphi\|_{\boldsymbol{L}^{q_1,r}}\,.
$$

This also holds true for $p_1 = \infty$, if we consider (weak,weak*)- resp. (weak*, weak)-continuous operators only.*

(ii) *Assume that G is nondiscrete. Then we have for $1 \leq p_1 < p_2 \leq \infty$ that $\boldsymbol{\mathcal{M}}^{\cdot}_{H^\perp}\left(\boldsymbol{L}^{p_1}(G), \boldsymbol{L}^{p_2}(G/H)\right) = \{0\}$.*

Proof. For the proof, it suffices to note that $P_\varphi = Q^*_\varphi$. □

It is worth noting that in both cases, φ acts as an attenuation factor [6]. In Q_φ, it has to supply the periodic function A with the desired global decay rate, while in P_φ, multiplication with φ ensures a sufficient decay rate for the periodization operator to be applicable.

Because of the Parseval–Plancherel Theorem, these results transfer directly to a description of the semidiscrete multipliers between $\boldsymbol{L}^2$ and $\boldsymbol{\ell}^2$. For $(G,H) = (\mathbb{R},\mathbb{Z})$, this can be considered folklore in the wavelet community; essentially, it can already be found in [11].

Corollary 3.29. *The spaces $\boldsymbol{\mathcal{M}}_H(\boldsymbol{L}^2(G),\boldsymbol{\ell}^2(H))$ and $\boldsymbol{\mathcal{M}}_H(\boldsymbol{\ell}^2(H),\boldsymbol{L}^2(G))$ are isometrically isomorphic to the space $\boldsymbol{L}^{2,\infty}(H^\perp,\widehat{G}/H^\perp)$ in the sense that every operator of one of these types is of the form S_φ or R_φ, respectively, for some $\varphi \in \big(\boldsymbol{L}^{2,\infty}(H^\perp,\widehat{G}/H^\perp)\big)^\vee$ with*

$$\|S_\varphi\|_{\boldsymbol{\mathcal{L}}(\boldsymbol{L}^2,\boldsymbol{\ell}^2)} = \|R_\varphi\|_{\boldsymbol{\mathcal{L}}(\boldsymbol{\ell}^2,\boldsymbol{L}^2)} = \|\widehat{\varphi}\|_{\boldsymbol{L}^{2,\infty}}\,.$$

At first glance, the results for $(G,H) = (\mathbb{R},\mathbb{Z})$ suggest we should try to apply interpolation methods, since we have

$$\boldsymbol{\mathcal{M}}_{\mathbb{Z}}\big(\boldsymbol{L}^1(\mathbb{R}),\boldsymbol{\ell}^1(\mathbb{Z})\big) \cong \boldsymbol{L}^{1,\infty}(\mathbb{Z},\mathbb{T}) \quad \text{and} \quad \boldsymbol{\mathcal{M}}_{\mathbb{Z}}\big(\boldsymbol{L}^2(\mathbb{R}),\boldsymbol{\ell}^2(\mathbb{Z})\big) \cong \boldsymbol{L}^{2,\infty}(\mathbb{Z},\mathbb{T})\,. \tag{3.3}$$

But note that in the first case, the condition is on φ, while in the second case, it is on $\widehat{\varphi}$, which makes interpolation impossible. The situation is comparable to the classical case, where $\boldsymbol{\mathcal{M}}\big(\boldsymbol{L}^1(\mathbb{R})\big) \cong \boldsymbol{M}_b(\mathbb{R})$ and $\boldsymbol{\mathcal{M}}\big(\boldsymbol{L}^2(\mathbb{R})\big) \cong \boldsymbol{L}^\infty(\mathbb{R})$, but in the first case, $\varphi \in \boldsymbol{M}_b(\mathbb{R})$, while in the second case, $\widehat{\varphi} \in \boldsymbol{L}^\infty(\mathbb{R})$.

The first statement in (3.3) stems from Theorem 3.22, where we have

$$S_\varphi \in \boldsymbol{\mathcal{M}}_H\big(\boldsymbol{L}^1(G),\boldsymbol{\ell}^p(H)\big) \iff \varphi \in \boldsymbol{L}^{p,\infty}(H,G/H)\,,$$

while the second one belongs to the following family of representation theorems obtained from Theorem 3.27 and Corollary 3.28 via the Fourier transformation.

Corollary 3.30. *The spaces $\boldsymbol{\mathcal{M}}_H\big(\widehat{\boldsymbol{L}^{p_1}}(G),\widehat{\boldsymbol{L}^{p_2}}(H)\big)$ and $\boldsymbol{\mathcal{M}}_H\big(\widehat{\boldsymbol{L}^{q_2}}(H),\widehat{\boldsymbol{L}^{q_1}}(G)\big)$ are isometrically isomorphic to the space $\boldsymbol{L}^{q_1,r}(H^\perp,\widehat{G}/H^\perp)$, where $\frac{1}{p_1}+\frac{1}{q_1}=1$ and $\frac{1}{r}+\frac{1}{p_1}=\frac{1}{p_2}$, in the sense that every operator of one of these types is of the form S_φ or R_φ, respectively, for some $\varphi \in \big(\boldsymbol{L}^{q_1,r}(H^\perp,\widehat{G}/H^\perp)\big)^\vee$ with*

$$\|S_\varphi\|_{\boldsymbol{\mathcal{L}}(\widehat{\boldsymbol{L}^{p_1}},\widehat{\boldsymbol{L}^{p_2}})} = \|R_\varphi\|_{\boldsymbol{\mathcal{L}}(\widehat{\boldsymbol{L}^{q_2}},\widehat{\boldsymbol{L}^{q_1}})} = \|\widehat{\varphi}\|_{\boldsymbol{L}^{q_1,r}}\,.$$

In particular, we have for $p_1 = p_2 =: p$ that

$$S_\varphi \in \boldsymbol{\mathcal{M}}_H\big(\widehat{\boldsymbol{L}^p}(G),\widehat{\boldsymbol{L}^p}(H)\big) \iff \widehat{\varphi} \in \boldsymbol{L}^{q,\infty}(H^\perp,\widehat{G}/H^\perp)\,,$$

which for $p = q = 2$ yields the second statement in (3.3).

The case $p = 1$ is also of special interest, since it implies the following analogy to Theorem 3.8.

Theorem 3.31. *Whenever* $S_\varphi \in \boldsymbol{\mathcal{M}}_H\big(\boldsymbol{A}(G), \boldsymbol{\ell}^\infty(H)\big)$, *we actually have* $S_\varphi \in \boldsymbol{\mathcal{M}}_H\big(\boldsymbol{A}(G), \boldsymbol{A}(H)\big)$.

Proof. By Corollary 3.6, we know that $S_\varphi \in \boldsymbol{\mathcal{M}}_H(\boldsymbol{A}, \boldsymbol{\ell}^\infty)$ is equivalent to $\varphi \in \boldsymbol{A}'$, meaning that $\widehat{\varphi} \in \boldsymbol{L}^\infty = \boldsymbol{L}^{\infty,\infty}$. By Corollary 3.30, this implies $S_\varphi \in \boldsymbol{\mathcal{M}}_H(\boldsymbol{A}, \boldsymbol{A})$. □

3.6 Results for Homogeneous Banach Spaces

Definition 3.32. *A Banach space* $\boldsymbol{X}$ *of functions or distributions on* G *is a* homogeneous Banach space, *if it satisfies*

(i) $\big\|\mathrm{T}_g f\big\|_{\boldsymbol{X}} = \|f\|_{\boldsymbol{X}}$ *for all* $f \in \boldsymbol{X}$ *and* $g \in G$, *and*

(ii) *for each* $f \in \boldsymbol{X}$, *the mapping* $g \mapsto \mathrm{T}_g f$ *is continuous from* G *to* $\boldsymbol{X}$.

In other words, the regular representation of G *acts continuously and isometrically on the elements of* $\boldsymbol{X}$.

Classical examples are $\boldsymbol{L}^p(G)$ for $p \in [1, \infty)$ and $\boldsymbol{C}_0(G)$.

If $\boldsymbol{X}(G)$ is homogeneous, we can characterize generalized sampling operators on $\boldsymbol{X}$ with the aid of the vector-valued functions introduced in Section 3.4. First we need to show a few preliminary properties of these functions in connection with the semidiscrete convolution product. Recall that for $f : G \to \mathbb{C}$, we defined

$$\overrightarrow{f} : G \to \mathbb{C}^H, \quad g \mapsto (f(g+h))_{h \in H}\,.$$

Lemma 3.33. *For a sequence* a *on* H *and a function or distibution* φ *on* G, *we have that*

$$\text{(i)} \quad (a *' \varphi)(g) = \big\langle a\,, \overrightarrow{\widetilde{\varphi}}(-g)\big\rangle\,, \qquad \textit{and}$$

$$\text{(ii)} \quad \overrightarrow{(a *' \varphi)}(g) = a * \big(\overrightarrow{\varphi}(g)\big)\,.$$

Proof. (i) $(a *' \varphi)(g) = \sum_h a[h]\, \varphi(g-h) = \sum_h a[h]\, \overline{\widetilde{\varphi}(h-g)} = \big\langle a\,, \overrightarrow{\widetilde{\varphi}}(-g)\big\rangle$.

(ii) For all $h \in H$, we have

$$\begin{aligned} \Big(\overrightarrow{(a *' \varphi)}(g)\Big)[h] &= (a *' \varphi)(g+h) \\ &= \sum_{k \in H} a[k]\, \varphi(g+h-k) \\ &= \sum_{k \in H} a[k]\, (\overrightarrow{\varphi}(g))[h-k] = \Big(a * \big(\overrightarrow{\varphi}(g)\big)\Big)[h]\,. \end{aligned}$$ □

Proposition 3.34. *Let* $\boldsymbol{X}(G)$ *be a homogeneous Banach space, and assume that* $\boldsymbol{Y}(H)$ *is a translation-invariant Banach space with* $\delta \in \boldsymbol{Y}'$. *Then we have for* $\varphi \in \boldsymbol{X}'$ *that* $S_\varphi \in \boldsymbol{\mathcal{M}}_H(\boldsymbol{X}, \boldsymbol{Y})$ *if and only if*

$$\overrightarrow{f * \widetilde{\varphi}} \in \boldsymbol{C}\big(G, \boldsymbol{Y}(H)\big) \quad \textit{for all } f \in \boldsymbol{X}\,.$$

Proof. Note that

$$\begin{aligned}(\overrightarrow{f * \widetilde{\varphi}})(g) &= \big((f * \widetilde{\varphi})(g+h)\big)_{h\in H} \\ &= \big((\mathrm{T}_{-g}(f * \widetilde{\varphi}))(h)\big)_{h\in H} \\ &= \big(((\mathrm{T}_{-g}f) * \widetilde{\varphi})(h)\big)_{h\in H} = S_\varphi(\mathrm{T}_{-g}f)\,.\end{aligned}$$

By homogeneity of $\boldsymbol{X}$, we know that $g \mapsto \mathrm{T}_{-g}f$ is continuous from G to $\boldsymbol{X}$. So if $S_\varphi \in \boldsymbol{\mathcal{M}}_H(\boldsymbol{X}, \boldsymbol{Y})$, then $\overrightarrow{f * \widetilde{\varphi}} : g \mapsto S_\varphi(\mathrm{T}_{-g}f)$ is continuous.

Conversely, assume that $\overrightarrow{f * \widetilde{\varphi}} \in \boldsymbol{C}\big(G, \boldsymbol{Y}(H)\big)$ for all $f \in \boldsymbol{X}$. Then $S_\varphi : f \mapsto \big(\overrightarrow{f * \widetilde{\varphi}}\big)(0)$ is a linear map from $\boldsymbol{X}(G)$ to $\boldsymbol{Y}(H)$. If $f_n \to f$ in $\boldsymbol{X}$ and $S_\varphi f_n \to a$ in $\boldsymbol{Y}$, then

$$a[h] = \lim_{n\to\infty} (S_\varphi f_n)[h] = \lim_{n\to\infty} \langle f_n, \mathrm{T}_h\varphi\rangle = \langle f, \mathrm{T}_h\varphi\rangle = (S_\varphi f)[h]$$

for all $h \in H$, so $a = S_\varphi f$. By the closed graph theorem, this implies that S_φ is bounded and thus an element of $\boldsymbol{\mathcal{M}}_H(\boldsymbol{X}, \boldsymbol{Y})$. □

With these results, we can characterize the operators reverse to those from Theorem 3.8. As usual, we write $\boldsymbol{\ell}^1_{w^*}$ for the space $\boldsymbol{\ell}^1$ with the weak*-topology.

Theorem 3.35 (Representation Theorem for $\boldsymbol{\mathcal{M}}_H\big(\boldsymbol{c}_0(H), \boldsymbol{C}_0(G)\big)$). *The operator R_φ is in $\boldsymbol{\mathcal{M}}_H(\boldsymbol{c}_0, \boldsymbol{C}_0)$, and thus S_φ is in $\boldsymbol{\mathcal{M}}_H(\boldsymbol{M}_b, \boldsymbol{\ell}^1)$ and (weak*,weak*)-continuous, if and only if $\varphi \in \boldsymbol{C}_0$ with $\overrightarrow{\varphi} \in \boldsymbol{C}\big(G, \boldsymbol{\ell}^1_{w^*}(H)\big)$. And in this case, we have*

$$\big\|R_\varphi\big\|_{\boldsymbol{\mathcal{L}}(\boldsymbol{c}_0,\boldsymbol{C}_0)} = \big\|S_\varphi\big\|_{\boldsymbol{\mathcal{L}}(\boldsymbol{M}_b,\boldsymbol{\ell}^1)} = \|\varphi\|_{\boldsymbol{L}^{1,\infty}(H,G/H)}\,.$$

Proof. Assume first that $R_\varphi \in \boldsymbol{\mathcal{M}}_H(\boldsymbol{c}_0, \boldsymbol{C}_0)$. Since $\delta \in \boldsymbol{c}_0$, we know that $\varphi = R_\varphi(\delta) \in \boldsymbol{C}_0$ (compare Proposition 3.13). Also, note that the mapping $G \to \boldsymbol{M}_b$, $g \mapsto \delta_g$, is continuous with the weak*-topology on $\boldsymbol{M}_b$. Since the hypothesis implies that

$$S_\varphi : \boldsymbol{M}_b \to \boldsymbol{\ell}^1 \quad \text{defined by} \quad \delta_g \mapsto (\delta_g * \widetilde{\varphi})\Big|_H = \big(\widetilde{\varphi}(h-g)\big)_{h\in H} = \overrightarrow{\widetilde{\varphi}}(-g)$$

is (weak*,weak*)-continuous, we conclude that $\overrightarrow{\widetilde{\varphi}}(-\,\boldsymbol{\cdot}\,)$ is in $\boldsymbol{C}\big(G, \boldsymbol{\ell}^1_{w^*}(H)\big)$, and therefore $\overrightarrow{\varphi}$ is as well. This shows that the two conditions on φ are necessary.

Conversely, assume that $\overrightarrow{\varphi}$, and therefore $\overrightarrow{\widetilde{\varphi}}(-\,\boldsymbol{\cdot}\,)$ as well, is an element of $\boldsymbol{C}\big(G, \boldsymbol{\ell}^1_{w^*}(H)\big)$. Then we have for all $a \in \boldsymbol{c}_0$ that the mapping

$$a *' \varphi : g \mapsto \big\langle a, \overrightarrow{\widetilde{\varphi}}(-g)\big\rangle \quad \text{(cf. Lemma 3.33.(i))}$$

is continuous on G. Letting K be a compact subset of G containing a fundamental domain for G/H, we conclude that for all $a \in \boldsymbol{C}_0$, the set

$\left\{\langle a, \overrightarrow{\widetilde{\varphi}}(-g)\rangle \;:\; g \in K\right\}$ is the continuous image of a compact set and hence is bounded. By the uniform boundedness principle, this implies that $\|\overrightarrow{\widetilde{\varphi}}(-g)\|_{\boldsymbol{\ell}^1} \le C$ for all $g \in K$. Since for arbitrary $g \in G$, there is $h_g \in H$ such that $g{+}h_g \in K$, we conclude that $\|\overrightarrow{\widetilde{\varphi}}(-g)\|_{\boldsymbol{\ell}^1} = \|\overrightarrow{\widetilde{\varphi}}(-(g{+}h_g))\|_{\boldsymbol{\ell}^1} \le C$ for all $g \in G$. This means $\widetilde{\varphi}$ and thus φ is contained in $\boldsymbol{L}^{1,\infty}$. By Corollary 3.23, this implies $R_\varphi \in \boldsymbol{\mathcal{M}}_H(\boldsymbol{\ell}^\infty, \boldsymbol{L}^\infty)$ with

$$\|R_\varphi\|_{\boldsymbol{\mathcal{L}}(\boldsymbol{\ell}^\infty,\boldsymbol{L}^\infty)} = \|\varphi\|_{\boldsymbol{L}^{1,\infty}} .$$

It remains to show that $a *' \varphi \in \boldsymbol{C}_0$ for all $a \in \boldsymbol{c}_0$. So let $a \in \boldsymbol{c}_0$, and $\varepsilon > 0$ be given, and let $C := \|R_\varphi\|_{\boldsymbol{\mathcal{L}}(\boldsymbol{\ell}^\infty,\boldsymbol{L}^\infty)}$. There is a finite set $F \subset H$ such that $|a[h]| < \varepsilon/(2C)$ for all $h \in H \setminus F$. Letting $a_F := a \cdot \chi_F$ and $a_{H\setminus F} := a - a_F$ yields $a_F *' \varphi \in \boldsymbol{C}_0$ since F is finite, and $\|a_{H\setminus F}\|_{\boldsymbol{\ell}^\infty} \le \varepsilon/(2C)$. Therefore, there is a compact set $K \subset G$ such that $|(a_F *' \varphi)(g)| < \varepsilon/2$ for all $g \notin K$, and we obtain for all these g that

$$\begin{aligned}
|(a *' \varphi)(g)| &\le |(a_F *' \varphi)(g)| + |(a_{H\setminus F} *' \varphi)(g)| \\
&< \quad \varepsilon/2 \quad + \|a_{H\setminus F} *' \varphi\|_{\boldsymbol{L}^\infty} \\
&\le \quad \varepsilon/2 \quad + C\,\|a_{H\setminus F}\|_{\boldsymbol{\ell}^\infty} \\
&\le \quad \varepsilon/2 \quad + C\,\varepsilon/(2C) \qquad = \varepsilon ,
\end{aligned}$$

which yields the claim. □

Remark 3.36. It is worth noting here that while $\overrightarrow{\varphi} \in \boldsymbol{C}(G, \boldsymbol{\ell}^1)$ already implies $\varphi \in \boldsymbol{C}_0$, this does not hold true for $\overrightarrow{\varphi} \in \boldsymbol{C}(G, \boldsymbol{\ell}^1_{w^*})$. As an example, consider $G = \mathbb{R}$ with $H = \mathbb{Z}$. Letting Δ be the hat function

$$\Delta(x) = \begin{cases} 4\,x\,, & x \in [0, \frac{1}{4}]\,, \\ 2 - 4\,x\,, & x \in [\frac{1}{4}, \frac{1}{2}]\,, \\ 0\,, & \text{otherwise,} \end{cases}$$

define φ_1 to be the function

$$\varphi_1(x) = \sum_{n=0}^{\infty} \Delta\big(2^n (x - (n{+}1{-}2^{-n}))\big)$$

(see Fig. 3.1). For each $x_0 \in \mathbb{R}$, the sequence $\overrightarrow{\varphi_1}(x_0)$ has at most one nonzero coefficient and thus is in $\boldsymbol{\ell}^1(\mathbb{Z})$. At $x_0 \notin \mathbb{Z}$, the function $\overrightarrow{\varphi_1}$ is continuous even with the norm topology on $\boldsymbol{\ell}^1$, since φ_1 is continuous and there is a neighborhood U of x_0 such that the common support of $\overrightarrow{\varphi_1}(x)$ for $x \in U$ is finite. But at the integers, the situation is different. For $x_0 = 0$, say, we see that $\overrightarrow{\varphi_1}(x) = 4\,x\,\delta$ for $x \in [0, \frac{1}{4}]$, so $\overrightarrow{\varphi_1}(x) \to \overrightarrow{\varphi_1}(0) = 0$ in the $\boldsymbol{\ell}^1$-norm as $x \to 0+$. For $x \to 0-$, though, we do have $\overrightarrow{\varphi_1}(x) \to 0$ in the weak*-sense,

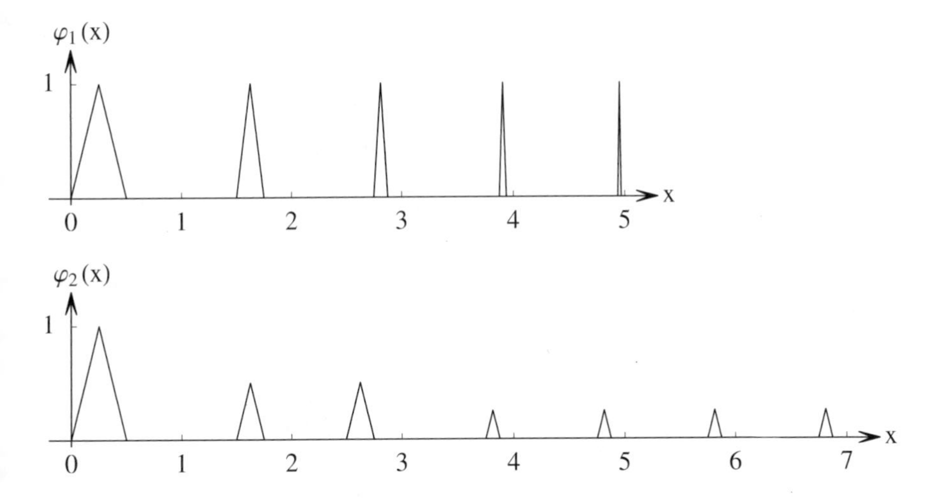

Fig. 3.1. Graphs of the functions φ_1 and φ_2

but not in norm, since $\delta_n \to 0$ $(n \to \infty)$ in $\boldsymbol{\ell}^1_{w^*}$, but not in $\boldsymbol{\ell}^1$. So indeed, $\overrightarrow{\varphi}_1 \notin \boldsymbol{C}(\mathbb{R}, \boldsymbol{\ell}^1)$, but only $\overrightarrow{\varphi}_1 \in \boldsymbol{C}(\mathbb{R}, \boldsymbol{\ell}^1_{w^*})$.

Since we obviously have $\varphi_1 \notin \boldsymbol{C}_0$, this example shows that the condition $\varphi \in \boldsymbol{C}_0$ cannot be omitted in the theorem.

On the other hand, we should note that the two conditions together do not imply $\overrightarrow{\varphi} \in \boldsymbol{C}(\mathbb{R}, \boldsymbol{\ell}^1)$, as the function

$$\varphi_2(x) = \sum_{n=0}^{\infty} \sum_{k=0}^{2^n-1} 2^{-n} \, \Delta\big(2^n(x - (2^n{+}k{-}2^{-n}))\big)$$

shows (again, see Fig. 3.1). Once more, we have $\overrightarrow{\varphi_2} \in \boldsymbol{C}(\mathbb{R}, \boldsymbol{\ell}^1_{w^*})$, but this time also $\varphi_2 \in \boldsymbol{C}_0$. Nevertheless, we have $\left\|\overrightarrow{\varphi_2}(-3/2^{n+1})\right\|_{\boldsymbol{\ell}^1} = 1$ for $n \in \mathbb{N}$ and therefore $\overrightarrow{\varphi_2}(x) \not\to \overrightarrow{\varphi_2}(0) = 0$ in norm as $x \to 0-$, meaning $\overrightarrow{\varphi_2} \notin \boldsymbol{C}(\mathbb{R}, \boldsymbol{\ell}^1)$.

We can also characterize the semidiscrete convolution operators from $\boldsymbol{\ell}^p$ into $\boldsymbol{C}_0$ and, accordingly, the weak*-continuous generalized sampling operators from $\boldsymbol{M}_b$ into $\boldsymbol{\ell}^q$.

For $p = 1$, we know from Corollaries 3.17 and 3.6 that $\boldsymbol{\mathcal{M}}_H(\boldsymbol{\ell}^1, \boldsymbol{C}_0)$ and the (weak*,weak*)-continuous elements of $\boldsymbol{\mathcal{M}}_H(\boldsymbol{M}_b, \boldsymbol{\ell}^\infty)$ are isometrically represented by $\boldsymbol{C}_0$. Also, as in the proof of Theorem 3.8, we can strengthen the second statement somewhat.

Corollary 3.37. *Whenever $S_\varphi \in \boldsymbol{\mathcal{M}}_H(\boldsymbol{M}_b, \boldsymbol{\ell}^\infty)$ is (weak*,weak*)-continuous, we actually have $S_\varphi \in \boldsymbol{\mathcal{M}}_H(\boldsymbol{M}_b, \boldsymbol{c}_0)$.*

Proof. We know from Corollary 3.6 that S_φ is a (weak*,weak*)-continuous element of $\mathcal{M}_H(\boldsymbol{M}_b, \boldsymbol{\ell}^\infty)$ if and only if $\varphi \in \boldsymbol{C}_0$. The claim follows from the fact that $\boldsymbol{M}_b * \boldsymbol{C}_0 = \boldsymbol{C}_0$ (compare the proof of Theorem 3.8). □

Remark 3.38. Note that, e.g., the function $\varphi \equiv 1$ is in $\boldsymbol{M}_b{}' \setminus \boldsymbol{C}_0$, meaning that the linear functional $\mu \mapsto \mu(G)$ is bounded, but not weak*-continuous on $\boldsymbol{M}_b$. The latter can be seen from the fact that for $g \to \infty$, we have $\delta_g \to 0$ in the weak*-sense, but $\langle \delta_g, 1 \rangle = 1 \not\to 0$. Consequently, $S_1 \in \mathcal{M}_H(\boldsymbol{M}_b, \boldsymbol{\ell}^\infty)$ is not (weak*,weak*)-continuous, which, by the above, we may also conclude from the fact that $S_1 \notin \mathcal{M}_H(\boldsymbol{M}_b, \boldsymbol{c}_0)$.

For $p \in (1, \infty)$, we can show a result comparable to Theorem 3.35.

Theorem 3.39 (Representation Theorem for $\mathcal{M}_H\big(\boldsymbol{\ell}^p(H), \boldsymbol{C}_0(G)\big)$). *The operator R_φ is in $\mathcal{M}_H(\boldsymbol{\ell}^p, \boldsymbol{C}_0)$, and thus S_φ is in $\mathcal{M}_H(\boldsymbol{M}_b, \boldsymbol{\ell}^q)$ and is (weak*,weak)-continuous, if and only if $\varphi \in \boldsymbol{C}_0$ with $\vec{\varphi} \in \boldsymbol{C}\big(G, \boldsymbol{\ell}^q_w(H)\big)$. And in this case, we have*

$$\big\|R_\varphi\big\|_{\mathcal{L}(\boldsymbol{\ell}^p, \boldsymbol{C}_0)} = \big\|S_\varphi\big\|_{\mathcal{L}(\boldsymbol{M}_b, \boldsymbol{\ell}^q)} = \|\varphi\|_{\boldsymbol{L}^{q,\infty}(H, G/H)} \,.$$

The proof is—*mutatis mutandis*—almost literally the same as that of Theorem 3.35. Even the function φ_1 from Fig. 3.1 still serves as an example to show that $\vec{\varphi} \in \boldsymbol{C}\big(\mathbb{R}, \boldsymbol{\ell}^q_w(\mathbb{Z})\big)$ does not imply $\varphi \in \boldsymbol{C}_0(\mathbb{R})$. Also, a slight modification of φ_2, namely, the function

$$\varphi_3(x) = \sum_{n=0}^{\infty} \sum_{k=0}^{2^n-1} 2^{-n/q}\, \Delta\big(2^n(x - (2^n + k - 2^{-n}))\big) \,,$$

shows that $\varphi \in \boldsymbol{C}_0$ with $\vec{\varphi} \in \boldsymbol{C}\big(\mathbb{R}, \boldsymbol{\ell}^q_w(\mathbb{Z})\big)$ does not imply $\vec{\varphi} \in \boldsymbol{C}\big(\mathbb{R}, \boldsymbol{\ell}^q(\mathbb{Z})\big)$.

Finally, for $p = \infty$, we can prove a result in the spirit of Corollary 3.18.

Corollary 3.40. *If G is noncompact, then the only element of $\mathcal{M}_H(\boldsymbol{\ell}^\infty, \boldsymbol{C}_0)$ that is (weak*,weak)-continuous is the zero map.*

Proof. Consider $R_\varphi \in \mathcal{M}_H(\boldsymbol{\ell}^\infty, \boldsymbol{C}_0)$. Obviously, $R_\varphi \in \mathcal{M}_H(\boldsymbol{\ell}^\infty, \boldsymbol{L}^\infty)$, so we know from Corollary 3.23 that $\varphi \in \boldsymbol{L}^{1,\infty} \subset \boldsymbol{L}^1$, so $\widehat{\varphi}$ is an element of $\boldsymbol{A} \subset \boldsymbol{C}_0$ and, in particular, is pointwise well-defined. For the constant sequence 1, we have $R_\varphi(1) = \sum_h \varphi(g-h) = \varphi^\circ_H$, which lies in $\boldsymbol{C}_0$ if and only if it vanishes identically. More generally, we can consider the sequence $(\mathrm{e}^{2\pi i\,\gamma h})_{h\in H} \in \boldsymbol{\ell}^\infty$ with

$$R_\varphi\big((\mathrm{e}^{2\pi i\,\gamma h})_{h\in H}\big) = \sum_{h\in H} \mathrm{e}^{2\pi i\,\gamma h}\, \varphi(g-h) \,.$$

This function is an element of $\boldsymbol{C}_0$ if and only if we have

$$\mathrm{e}^{-2\pi i\,\gamma g} \sum_{h\in H} \mathrm{e}^{2\pi i\,\gamma h}\, \varphi(g-h) = \sum_{h\in H} \mathrm{e}^{-2\pi i\,\gamma(g-h)}\, \varphi(g-h) = \big(\mathrm{M}_\gamma \varphi\big)^\circ_H \in \boldsymbol{C}_0 \,,$$

but again, the latter is only possible for $\left(\mathrm{M}_\gamma\varphi\right)^\circ_H \equiv 0$. From this, we may conclude that

$$\widehat{\varphi}(\gamma) = \int_G \varphi(g)\,\mathrm{e}^{-2\pi i\,\gamma g}\,d\mu_G(g) = \int_{G/H} \left(\mathrm{M}_\gamma\varphi\right)^\circ_H(g)\,d\mu_{G/H}(g) = 0$$

for all $\gamma \in \widehat{G}$, so $\widehat{\varphi} \equiv 0$ and thus $\varphi \equiv 0$, which yields the claim. □

Remark 3.41. This result demonstrates the difference between the weak and the weak*-topology on $\boldsymbol{\ell}^1$ in a remarkable way. In Theorem 3.39, it does not matter whether we consider (weak*,weak)- or (weak*,weak*)-continuous operators in $\boldsymbol{\mathcal{M}}_H(\boldsymbol{M}_b, \boldsymbol{\ell}^q)$, since for $q \in (1,\infty)$, the space $\boldsymbol{\ell}^q$ is reflexive. For the case $\boldsymbol{\mathcal{M}}_H(\boldsymbol{M}_b, \boldsymbol{\ell}^1)$, however, Corollary 3.40 implies that the only (weak*,weak)-continuous such operator is the zero map, while the (weak*,weak*)-continuous operators of this type have been described in Theorem 3.35.

We still have not given a complete description of the spaces $\boldsymbol{\mathcal{M}}_H(\boldsymbol{\ell}^{p_1}, \boldsymbol{L}^{p_2})$ and $\boldsymbol{\mathcal{M}}_H(\boldsymbol{L}^{q_2}, \boldsymbol{\ell}^{q_1})$. This is not so surprising, as there is no easy characterization of the spaces $\boldsymbol{\mathcal{M}}(\boldsymbol{L}^{p_1}(G), \boldsymbol{L}^{p_2}(G))$ (e.g., [5], [9]).

Note that the case $p_1 = 1$ and $q_1 = \infty$ is covered by Corollaries 3.17 and 3.6, and the case $p_2 = \infty$ and $q_2 = 1$ by Corollary 3.23 and Theorem 3.22. Furthermore, the cases where $p_1 > p_2$ and $q_2 > q_1$ have been discussed in Corollary 3.18 and Theorem 3.10.

The remaining cases are those where $1 < p_1 \le p_2 < \infty$ and thus $1 < q_2 \le q_1 < \infty$. For these, we present a sufficient condition based on the space of multipliers $\boldsymbol{\mathcal{M}}(\boldsymbol{\ell}^{p_1}, \boldsymbol{\ell}^{p_2}) \cong \boldsymbol{\mathcal{M}}(\boldsymbol{\ell}^{q_2}, \boldsymbol{\ell}^{q_1})$, but we have to leave it as an open question whether this condition is also necessary.

Theorem 3.42. *Let $1 < p_1 \le p_2 < \infty$, and as usual $\frac{1}{p_i} + \frac{1}{q_i} = 1$. If φ is a measurable function on G satisfying*

$$\overrightarrow{\varphi} \in \boldsymbol{L}^{q_1}_{\mathrm{loc}}\Big(G, \boldsymbol{\mathcal{M}}\big(\boldsymbol{\ell}^{p_1}(H), \boldsymbol{\ell}^{p_2}(H)\big)\Big) = \boldsymbol{L}^{q_1}_{\mathrm{loc}}\Big(G, \boldsymbol{\mathcal{M}}\big(\boldsymbol{\ell}^{q_2}(H), \boldsymbol{\ell}^{q_1}(H)\big)\Big),$$

then $R_\varphi \in \boldsymbol{\mathcal{M}}_H\big(\boldsymbol{\ell}^{p_1}(H), \boldsymbol{L}^{p_2}(G)\big)$ and thus $S_\varphi \in \boldsymbol{\mathcal{M}}_H\big(\boldsymbol{L}^{q_2}(G), \boldsymbol{\ell}^{q_1}(H)\big)$.

Proof. Under the assumptions made, we have for all $a \in \boldsymbol{\ell}^{p_1}(H)$ that

$$\begin{aligned}
\left\|a *' \varphi\right\|_{\boldsymbol{L}^{p_2}(G)} &= \Big\|\big\|\overrightarrow{a *' \varphi}\big\|_{\boldsymbol{\ell}^{p_2}(H)}\Big\|_{\boldsymbol{L}^{p_2}(G/H)} \\
&= \Big\|\big\|a * \overrightarrow{\varphi}\big\|_{\boldsymbol{\ell}^{p_2}(H)}\Big\|_{\boldsymbol{L}^{p_2}(G/H)} \\
&\le \Big\|\left\|a\right\|_{\boldsymbol{\ell}^{p_1}(H)} \big\|\overrightarrow{\varphi}\big\|_{\boldsymbol{\mathcal{M}}(\boldsymbol{\ell}^{p_1}, \boldsymbol{\ell}^{p_2})}\Big\|_{\boldsymbol{L}^{p_2}(G/H)} \\
&= \left\|a\right\|_{\boldsymbol{\ell}^{p_1}(H)} \Big\|\big\|\overrightarrow{\varphi}\big\|_{\boldsymbol{\mathcal{M}}(\boldsymbol{\ell}^{p_1}, \boldsymbol{\ell}^{p_2})}\Big\|_{\boldsymbol{L}^{p_2}(G/H)},
\end{aligned}$$

which yields the claim. □

3.7 Pseudo-Interpolation

Consider a function φ on G that is pointwise well-defined and satisfies $\varphi(h) = \delta_{0,h}$ for $h \in H$. Then the operator

$$R_\varphi \circ S_\delta : f \mapsto \sum_{h\in H} f(h)\, \mathrm{T}_h\varphi$$

is called *interpolation* on H with φ, since we have $\big((R_\varphi \circ S_\delta)f\big)(h) = f(h)$ for all $h \in H$.

Consequently, we shall refer to an operator on $\boldsymbol{X}$ of the form $R_\varphi \circ S_{\varphi^*}$ with appropriate $(\varphi, \varphi^*) \in \boldsymbol{X} \times \boldsymbol{X}'$ as *pseudo-interpolation*. In [10], this is called quasi-interpolation, but usually, quasi-interpolation of degree m is defined to be a pseudo-interpolation operator that is *local*, i.e., where φ and φ^* have compact support, and that reproduces polynomials up to degree m (e.g., see [4], [12], [13]).

Theorem 3.10 and Corollary 3.18 imply the following restriction on pseudo-interpolation on $\boldsymbol{L}^p(G)$.

Corollary 3.43. *If G is noncompact, and $R_\varphi \circ S_{\varphi^*}$ is a nontrivial pseudo-interpolation on $\boldsymbol{L}^{p_1}(G)$ factoring through $\boldsymbol{\ell}^{p_2}(H)$ for some p_1, $p_2 \in [1,\infty]$, then necessarily $p_1 = p_2$.*

3.7.1 Shannon's Sampling Theorem

As an application, let us have a closer look at Shannon's Sampling Theorem (see, e.g., [2] for background). For $\Omega > 0$, the *Paley–Wiener space* is defined as

$$\boldsymbol{PW}_\Omega = \big\{ f \in \boldsymbol{L}^2(\mathbb{R}) : \operatorname{supp} \widehat{f} \subseteq [-\tfrac{\Omega}{2}, +\tfrac{\Omega}{2}] \big\}\,.$$

Note that $f \in \boldsymbol{L}^2$ implies $\widehat{f} \in \boldsymbol{L}^2$, so because of the compact support of $\widehat{f}$, we have $\widehat{f} \in \boldsymbol{L}^1$, i.e., $f \in \boldsymbol{A} \subseteq \boldsymbol{C}_0$. As usual, the sinc function is defined via

$$\operatorname{sinc}_\Omega(x) = \frac{\sin \pi\, \Omega\, x}{\pi\, x} \quad \text{with} \quad \widehat{\operatorname{sinc}_\Omega} = \chi_{[-\Omega/2,+\Omega/2]}\,.$$

Proposition 3.44 (Shannon's Sampling Theorem). *For $f \in \boldsymbol{PW}_\Omega$, we have*

$$f(x) = \sum_{n\in\mathbb{Z}} \tfrac{1}{\Omega}\, f(\tfrac{n}{\Omega})\, \operatorname{sinc}_\Omega(x - \tfrac{n}{\Omega})\,, \tag{3.4}$$

where the series converges in $\boldsymbol{L}^2(\mathbb{R})$ and uniformly on $\mathbb{R}$.

If we let $G = \mathbb{R}$ and $H = \frac{1}{\Omega}\mathbb{Z}$ with $\boldsymbol{X}(G) = \boldsymbol{PW}_\Omega$ and $\boldsymbol{Y}(H) = \boldsymbol{\ell}^2(\frac{1}{\Omega}\mathbb{Z})$, and choose $\varphi^* = \delta$ and $\varphi = \operatorname{sinc}_\Omega$, we can express the above as

$$\tfrac{1}{\Omega}\, R_{\operatorname{sinc}_\Omega} \circ S_\delta = \mathrm{id} \quad \text{on } \boldsymbol{PW}_\Omega\,.$$

Since the family $(\frac{1}{\sqrt{\Omega}}\,\mathrm{T}_{n/\Omega}\,\mathrm{sinc}_\Omega)_{n\in\mathbb{Z}}$ is an orthonormal basis for $\boldsymbol{PW}_\Omega$, both maps R_{sinc_Ω} and S_δ are actually multiples of isometries, and to prove the theorem, it suffices to show that $\langle f, \mathrm{T}_{n/\Omega}\,\mathrm{sinc}_\Omega\rangle = f(\frac{n}{\Omega})$ for $f \in \boldsymbol{PW}_\Omega$.

An easier way to understand Shannon's Sampling Theorem is given by the observations we made in Section 3.5. For $G = \mathbb{R}$ and $H = \frac{1}{\Omega}\,\mathbb{Z}$, we have $\widehat{G} = \widehat{\mathbb{R}}$ and $H^\perp = \Omega\,\mathbb{Z}$. Therefore, the concatenation of operators

$$R_{\mathrm{sinc}_\Omega} \circ \tfrac{1}{\Omega}\,S_\delta : \quad f \longmapsto \big(\tfrac{1}{\Omega}\,f(\tfrac{n}{\Omega})\big)_{n\in\mathbb{Z}} \longmapsto \sum_{n\in\mathbb{Z}} \tfrac{1}{\Omega}\,f(\tfrac{n}{\Omega})\,\mathrm{sinc}_\Omega(x - \tfrac{n}{\Omega})$$

on the Fourier side is given by

$$Q_{\chi_{[-\Omega/2,+\Omega/2]}} \circ P_1 : \quad \widehat{f} \longmapsto \widehat{f}^{\,\circ}_{\Omega\,\mathbb{Z}} \longmapsto \widehat{f}^{\,\circ}_{\Omega\,\mathbb{Z}} \cdot \chi_{[-\Omega/2,+\Omega/2]}\,.$$

(The factor $\frac{1}{\Omega}$ vanishes due to the appropriate normalization of the Haar measure.) It is obvious that this operator reproduces $\widehat{f}$ whenever we have $\operatorname{supp}\widehat{f} \subseteq [-\frac{\Omega}{2}, +\frac{\Omega}{2}]$.

How can we increase the space $\boldsymbol{X}$? The sampling operator S_δ is defined on $\boldsymbol{X} = \boldsymbol{C}_b$ with range $\boldsymbol{Y} = \boldsymbol{\ell}^\infty$, but $R_{\mathrm{sinc}_\Omega} \notin \boldsymbol{\mathcal{L}}(\boldsymbol{\ell}^\infty, \boldsymbol{L}^\infty)$ since $\mathrm{sinc}_\Omega \notin \boldsymbol{L}^{1,\infty}$ (cf. Corollary 3.23). For this reason, $\boldsymbol{X} = \boldsymbol{C}_0$ with $\boldsymbol{Y} = \boldsymbol{c}_0$ will not work, either. But it is possible to use $\boldsymbol{X} = \boldsymbol{A}$, i.e., $\widehat{\boldsymbol{X}} = \boldsymbol{L}^1$. Applying Corollary 3.28 with $p_1 = p_2 = \infty$ tells us that $\widehat{\delta} = 1 \in \boldsymbol{L}^\infty = \boldsymbol{L}^{\infty,\infty}$ implies $P_1 \in \boldsymbol{\mathcal{L}}\big(\boldsymbol{L}^1(\widehat{\mathbb{R}}), \boldsymbol{L}^1(\Omega\,\mathbb{T})\big)$ and thus $S_\delta \in \boldsymbol{\mathcal{L}}\big(\boldsymbol{A}(\mathbb{R}), \boldsymbol{A}(\frac{1}{\Omega}\,\mathbb{Z})\big)$. So we obtain $\boldsymbol{Y} = \boldsymbol{A}$. On the other hand, we have $\widehat{\varphi} = \widehat{\mathrm{sinc}_\Omega} = \chi_{[-\Omega/2,+\Omega/2]} \in \boldsymbol{L}^{1,\infty}$ and thus $Q_{\widehat{\varphi}} \in \boldsymbol{\mathcal{L}}\big(\boldsymbol{L}^1(\Omega\,\mathbb{T}), \boldsymbol{L}^1(\widehat{\mathbb{R}})\big)$, i.e., $R_\varphi \in \boldsymbol{\mathcal{L}}\big(\boldsymbol{A}(\frac{1}{\Omega}\,\mathbb{Z}), \boldsymbol{A}(\mathbb{R})\big)$. So $R_{\mathrm{sinc}_\Omega} \circ \frac{1}{\Omega}\,S_\delta \in \boldsymbol{\mathcal{L}}\big(\boldsymbol{A}(\mathbb{R}), \boldsymbol{A}(\mathbb{R})\big)$, and it factors through $\boldsymbol{A}(\frac{1}{\Omega}\,\mathbb{Z})$. By the same argument as before, it is the identity on the space

$$\big\{f \in \boldsymbol{A}(\mathbb{R}) : \operatorname{supp}\widehat{f} \subseteq [-\tfrac{\Omega}{2}, +\tfrac{\Omega}{2}]\big\}\,,$$

and thus actually a projection in $\boldsymbol{A}(\mathbb{R})$ onto this space.

With an appropriate modification, we can also extend $R_{\mathrm{sinc}_\Omega} \circ S_\delta$ to a projection in $\boldsymbol{L}^2(\mathbb{R})$. We stated before that $(\frac{1}{\sqrt{\Omega}}\,\mathrm{T}_{n/\Omega}\,\mathrm{sinc}_\Omega)_{n\in\mathbb{Z}}$ is an orthonormal basis for $\boldsymbol{PW}_\Omega$. Therefore, (3.4) implies that $\delta = \mathrm{sinc}_\Omega$ as elements of $\boldsymbol{PW}'_\Omega$, and thus $S_\delta = S_{\mathrm{sinc}_\Omega}$ on $\boldsymbol{PW}_\Omega$. So the orthogonal projection $\boldsymbol{L}^2 \to \boldsymbol{PW}_\Omega$ can be expressed as $\frac{1}{\Omega}\,R_{\mathrm{sinc}_\Omega} \circ S_{\mathrm{sinc}_\Omega}$. Indeed, Corollary 3.29 confirms that since $\widehat{\mathrm{sinc}_\Omega} = \chi_{[-\Omega/2,+\Omega/2]} \in \boldsymbol{L}^{2,\infty}$, we have $S_{\mathrm{sinc}_\Omega} \in \boldsymbol{\mathcal{L}}(\boldsymbol{L}^2, \boldsymbol{\ell}^2)$ and $R_{\mathrm{sinc}_\Omega} \in \boldsymbol{\mathcal{L}}(\boldsymbol{\ell}^2, \boldsymbol{L}^2)$.

References

1. A. Benedek and R. Panzone, The spaces $\boldsymbol{L}^P$, with mixed norm, *Duke Math. J.*, **28** (1961), pp. 301–324.
2. J. J. Benedetto, Irregular sampling and frames, in: *Wavelets: A Tutorial in Theory and Applications*, C. Chui, ed., Academic Press, Boston, 1992, pp. 445–507.
3. J. J. Benedetto and G. Zimmermann, Sampling multipliers and the Poisson Summation Formula, *J. Fourier Anal. Appl.*, **3** (1997), pp. 505–523.
4. C. de Boor and G. Fix, Spline approximation by quasi-interpolants, *J. Approx. Theory*, **8** (1973), pp. 19–45.
5. R. E. Edwards and G. I. Gaudry, *Littlewood–Paley and Multiplier Theory*, Ergeb. Math. Grenzgeb., **90**, Springer, Berlin, 1977.
6. W. Gautschi, Attenuation factors in practical Fourier analysis, *Numer. Math.*, **18** (1972), pp. 373–400.
7. E. Hewitt and K. A. Ross, *Abstract Harmonic Analysis I*, Grundlehren Math. Wiss., **115**, Springer, Berlin, 1963.
8. R.-Q. Jia and C. A. Micchelli, Using the refinement equations for the construction of pre-wavelets. II: Powers of two, in: *Curves and Surfaces*, P. J. Laurent, A. Le Méhauté, and L. L. Schumaker, eds., Academic Press, Boston, 1991, pp. 209–246.
9. R. Larsen, *An Introduction to the Theory of Multipliers*, Grundlehren Math. Wiss., **175**, Springer, Berlin, 1971.
10. J. Lei, R.-Q. Jia, and E. W. Cheney, Approximation from shift-invariant spaces by integral operators. *SIAM J. Math. Anal.*, **28** (1997), pp. 481–498.
11. S. Mallat, Multiresolution approximations and wavelet orthonormal bases of $\boldsymbol{L}^2(\mathbb{R})$, *Trans. Amer. Math. Soc.*, **315** (1989), pp. 69–87.
12. H. N. Mhaskar, F. J. Narcowich, and J. D. Ward, Quasi-interpolation in shift invariant spaces, *J. Math. Anal. Appl.*, **251** (2000), pp. 356–363.
13. W. Sweldens, Construction and Applications of Wavelets in Numerical Analysis, Ph.D. Thesis, Department of Computer Science, Katholieke Universiteit Leuven, Belgium, 1994.

Part II

Frame Theory

4

A Physical Interpretation of Tight Frames

Peter G. Casazza[1], Matthew Fickus[2], Jelena Kovačević[3], Manuel T. Leon[4], and Janet C. Tremain[5]

[1] Department of Mathematics, University of Missouri, Columbia, MO 65211, USA
pete@math.missouri.edu
[2] Department of Mathematics and Statistics, Air Force Institute of Technology, Wright-Patterson AFB, OH 45433, USA
Matthew.Fickus@afit.edu
[3] Department of Biomedical Engineering, Carnegie Mellon University, Pittsburgh, PA 15213, USA
jelenak@cmu.edu
[4] Department of Mathematics, University of Missouri, Columbia, MO 65211, USA
mleon@math.missouri.edu
[5] Department of Mathematics, University of Missouri, Columbia, MO 65211, USA
janet@math.missouri.edu

Summary. We characterize the existence of finite tight frames whose frame elements are of predetermined length. In particular, we derive a "fundamental inequality" which completely characterizes those sequences which arise as the lengths of a tight frame's elements. Furthermore, using concepts from classical physics, we show that this characterization has an intuitive physical interpretation.

4.1 Introduction

Let $\mathbb{H}_N$ be a finite N-dimensional, real or complex Hilbert space. A finite sequence $\{f_m\}_{m=1}^M$ of vectors is *A-tight* for $\mathbb{H}_N$ if there exists $A \geq 0$ such that

$$A\,\|f\|^2 = \sum_{m=1}^{M} |\langle f, f_m\rangle|^2$$

for all $f \in \mathbb{H}_N$. An *A-tight frame* is an A-tight sequence for which $A > 0$. By polarization, $\{f_m\}_{m=1}^M$ is A-tight for $\mathbb{H}_N$ if and only if

$$Af = \sum_{m=1}^{M} \langle f, f_m\rangle f_m$$

for all $f \in \mathbb{H}_N$. Clearly, any orthonormal basis is a 1-tight frame. However, the converse is false. For example, the tetrahedral vertices

$$f_1 = \frac{1}{2}\begin{bmatrix}1\\1\\1\end{bmatrix}, \quad f_2 = \frac{1}{2}\begin{bmatrix}1\\-1\\-1\end{bmatrix}, \quad f_3 = \frac{1}{2}\begin{bmatrix}-1\\1\\-1\end{bmatrix}, \quad f_4 = \frac{1}{2}\begin{bmatrix}-1\\-1\\1\end{bmatrix}$$

form a 1-tight frame of four elements for $\mathbb{R}^3$. Furthermore, while the elements of an orthonormal basis are of unit length *a priori*, no such assumption is made about the lengths of a tight frame's elements. This raises the question:

> Given positive integers M and N, for what sequences of nonnegative numbers $\{a_m\}_{m=1}^M$ do there exist tight frames $\{f_m\}_{m=1}^M$ for $\mathbb{H}_N$ such that $\|f_m\| = a_m$ for all m?

The answer to this question is the subject of this chapter. In particular, the following result gives a necessary condition upon the lengths.

Proposition 4.1. *If $\{f_m\}_{m=1}^M$ is A-tight for $\mathbb{H}_N$, then*

$$\max_{m=1,\dots,M} \|f_m\|^2 \le A = \frac{1}{N}\sum_{m=1}^M \|f_m\|^2. \tag{4.1}$$

Proof. First note that for any $m = 1, \dots, M$,

$$\|f_m\|^4 = |\langle f_m, f_m\rangle|^2 \le \sum_{m'=1}^M |\langle f_m, f_{m'}\rangle|^2 = A\,\|f_m\|^2.$$

Thus, $\|f_m\|^2 \le A$ for all $m = 1, \dots, M$, yielding the inequality in (4.1). For the equality, let $\{e_n\}_{n=1}^N$ be an orthonormal basis for $\mathbb{H}_N$. By Parseval's identity,

$$\sum_{m=1}^M \|f_m\|^2 = \sum_{m=1}^M\sum_{n=1}^N |\langle f_m, e_n\rangle|^2 = \sum_{n=1}^N\sum_{m=1}^M |\langle e_n, f_m\rangle|^2 = \sum_{n=1}^N A\,\|e_n\|^2 = NA.$$

Dividing by N gives the result. □

Thus, the lengths $\{a_m\}_{m=1}^M$ of a tight frame of M elements for an N-dimensional space must satisfy the *fundamental inequality*

$$\max_{m=1,\dots,M} a_m^2 \le \frac{1}{N}\sum_{m=1}^M a_m^2. \tag{4.2}$$

Remarkably, this easily found necessary condition will prove sufficient as well. That is, we shall show that for any nonzero nonnegative sequence $\{a_m\}_{m=1}^M$ which satisfies (4.2), and for any N-dimensional Hilbert space $\mathbb{H}_N$, there exists a tight frame $\{f_m\}_{m=1}^M$ for $\mathbb{H}_N$ for which $\|f_m\| = a_m$ for all m. Furthermore, when given a sequence $\{a_m\}_{m=1}^M$ which violates (4.2), we shall determine those sequences $\{f_m\}_{m=1}^M$ of norms $\{a_m\}_{m=1}^M$ which are as close as possible to being tight frames, in a natural, intuitive sense.

Of course, this problem does not exist in a vacuum. Frames have been a subject of interest for some time, both in theory and in applications. Finite tight frames, in particular, have garnered attention in the past several years. The theory of frames was first introduced by Duffin and Schaeffer [10] in the 1950s, furthering the study of nonharmonic Fourier series and the time-frequency decompositions of Gabor [13]. Decades later, the subject was reinvigorated following a publication of Daubechies, Grossman, and Meyer [9]. Frames have subsequently become a state-of-the-art signal processing tool.

Frames provide redundant vector space decompositions, which are very attractive from the applied perspective. In particular, frame decompositions are resilient against noise and quantization, and provide numerically stable reconstruction algorithms [6], [8], [15]. Translation-invariant frames in $\ell^2(\mathbb{Z})$, being equivalent to perfect reconstruction oversampled filter banks, have been studied extensively [4], [7], [19], [20]. Frame decompositions have the potential to reveal hidden signal characteristics, and have therefore been used to solve problems of detection [1], [3], [22]. Frames have also been used to design unitary space-time constellations for multiple-antenna wireless systems [16].

Specific types of finite tight frames have been studied to solve problems in communications [14], [17], [21], [25]. In addition, many techniques for constructing finite tight frames have been discovered, several of which involve group theory [5], [23]. Researchers have also been interested in tight frames whose elements are restricted to spheres and ellipsoids [2], [11], as well as the manifold structures of spaces of all such frames [12].

Much of our work is inspired by the characterization of unit-norm tight frames in [2]. Subsequent to the completion of our work, we learned that some of our results were independently obtained by Viswanath and Anantharam [24] in the context of wireless communications. At the conclusion of this chapter, we compare and contrast the two approaches.

In the following section, we motivate our main results by introducing a physical interpretation of frame theory, extending the frame-equivalent notions of force and potential energy first introduced in [2]. Section 4.3 contains several results concerning the minimization of this generalized frame potential, highlighting the connection between optimal energy and tightness. Finally, in Section 4.4, we characterize the lengths of a tight frame's elements in terms of the fundamental inequality, and discuss this characterization from the physical perspective.

4.2 The Physical Theory

Given any nonnegative sequence $\{a_m\}_{m=1}^M$ which satisfies the fundamental inequality (4.2), our goal is to prove the existence of a tight frame $\{f_m\}_{m=1}^M$ for $\mathbb{H}_N$ such that $\|f_m\| = a_m$ for all m. We begin by briefly discussing the special case of this problem when $a_m = 1$ for all m.

A *unit-norm tight frame* (UNTF) is a tight frame whose frame elements are normalized. Recently, several independent proofs have been discovered which show that such frames always exist, that is, for any N-dimensional Hilbert space $\mathbb{H}_N$, and any $M \geq N$, there exists a tight frame of M elements $\{f_m\}_{m=1}^M$ for $\mathbb{H}_N$ such that $\|f_m\| = 1$ for all m. One proof involves an explicit construction of such frames using orthogonal projections of finite Fourier bases [14], [26]. However, this construction does not generalize to the case of nonuniform lengths. Another proof is given in [2], where unit-norm tight frames are referred to as *normalized tight frames*. In that paper, the existence of UNTFs is shown to be a generalization of the classical problem of equally distributing points on spheres. In particular, UNTFs are characterized as the minimizers of a potential energy function, thus guaranteeing their existence.

In this section, we show how the notions of "frame force" and "frame potential," first introduced in [2], may now be extended to the nonuniform setting. Inspired by the Coulomb force, consider the *frame force* $\mathrm{FF}(f_m, f_{m'})$ of $f_{m'} \in \mathbb{R}^N$ upon $f_m \in \mathbb{R}^N$:

$$\mathrm{FF}(f_m, f_{m'}) = 2\langle f_m, f_{m'}\rangle (f_m - f_{m'}) \in \mathbb{R}^N. \tag{4.3}$$

This is essentially the same definition that appears in [2], but without the restriction of f_m and $f_{m'}$ to the unit sphere. Note that, as in [2], we have momentarily restricted ourselves to the context of real spaces, as this will simplify our derivations below. Nevertheless, in the following section, we will generalize the key aspects of the theory to arbitrary real or complex finite-dimensional Hilbert spaces.

The frame force behaves bizarrely compared to the forces of the natural world. For example, this force is not translation-invariant. Also, the force between coincidental points is well-defined, being zero. Most importantly, the frame force "encourages orthogonality"—the force is repulsive when the angle between two vectors is acute, attractive when this angle is obtuse, and zero when the vectors are perpendicular.

Now imagine a physical system of M points $\{f_m\}_{m=1}^M \subset \mathbb{R}^N$ in which each point pushes against all other points according to the frame force, and in which the movement of each f_m is restricted to a sphere of radius a_m about the origin. In particular, imagine how such a physical system would evolve over time. That is, consider a dynamical system of points moving on spheres in which each point is trying to force all other points to be perpendicular to itself. Though a rigorous mathematical analysis of such a system is beyond the scope of this work, we may nevertheless use our intuitive understanding of the physical world to make a reasonable guess as to the asymptotic behavior of the points over time.

In particular, it is reasonable to believe that such a system would asymptotically approach an arrangement of points which is in equilibrium under the frame force, provided there was a mechanism, akin to friction, to dissipate excess energy. The result would be a sequence of vectors $\{f_m\}_{m=1}^M$, with $\|f_m\| = a_m$ for all m, which are "as close to being mutually orthogonal as possible." This is particularly interesting when the number of nonzero radii a_m exceeds the dimension of the space N, since in this case, it is impossible for $\{f_m\}_{m=1}^M$ to be an orthogonal set.

Formally, such equilibria are characterized as the local minimizers of a corresponding potential energy function. For a system of particles pushing against each other according to the frame force, this energy function is known as the "frame potential." To be precise, for $a \geq 0$, let $\mathrm{S}(a)$ be the hypersphere $\{f \in \mathbb{R}^N : \|f\| = a\}$. Similarly, for any nonnegative sequence $\{a_m\}_{m=1}^M$, let

$$\mathrm{S}(\{a_m\}_{m=1}^M) = \mathrm{S}(a_1) \times \cdots \times \mathrm{S}(a_M).$$

The *frame potential* $\mathrm{FP}\colon \mathrm{S}(\{a_m\}_{m=1}^M) \to \mathbb{R}$ is a function for which, given any $\{f_m\}_{m=1}^M, \{g_m\}_{m=1}^M \in \mathrm{S}(\{a_m\}_{m=1}^M)$, the number $\mathrm{FP}(\{f_m\}_{m=1}^M) - \mathrm{FP}(\{g_m\}_{m=1}^M)$ represents the work required to transform $\{g_m\}_{m=1}^M$ into $\{f_m\}_{m=1}^M$ under the influence of the frame force. As such, the frame potential is only unique up to an additive constant. The following result provides an explicit form of this potential in which the additive constant is chosen for the sake of simplicity.

Proposition 4.2. *For any* $\{f_m\}_{m=1}^M \in \mathrm{S}(\{a_m\}_{m=1}^M)$,

$$\mathrm{FP}(\{f_m\}_{m=1}^M) = \sum_{m=1}^M \sum_{m'=1}^M |\langle f_m, f_{m'}\rangle|^2.$$

Proof. Recall $f_m \in \mathrm{S}(a_m) \subset \mathbb{R}^N$ for all m. Thus, for any $m, m' = 1, \ldots, M$,

$$\|f_m - f_{m'}\|^2 = a_m^2 - 2\,\langle f_m, f_{m'}\rangle + a_{m'}^2. \tag{4.4}$$

Solving for $2\langle f_m, f_{m'}\rangle$ in (4.4), and substituting this expression in (4.3), we see that the frame force may be written completely in terms of $f_m - f_{m'}$:

$$\mathrm{FF}(f_m, f_{m'}) = \left(a_m^2 + a_{m'}^2 - \|f_m - f_{m'}\|^2\right)(f_m - f_{m'}).$$

The potential energy is then computed using an anti-gradient, which is accomplished by first anti-differentiating the "scalar force,"

$$-\int (a_m^2 + a_{m'}^2 - x^2)\, x\, dx = \frac{1}{4}\, x^2 \left[x^2 - 2(a_m^2 + a_{m'}^2)\right],$$

and then evaluating at $x = \|f_m - f_{m'}\|$. Next, we simplify using (4.4):

$$\frac{1}{4}\,\|f_m - f_{m'}\|^2 \left[\|f_m - f_{m'}\|^2 - 2(a_m^2 + a_{m'}^2)\right] = \langle f_m, f_{m'}\rangle^2 - \frac{1}{4}\,(a_m^2 + a_{m'}^2)^2.$$

This quantity represents the frame potential energy at f_m from the field generated by $f_{m'}$. The total energy is the sum of these pairwise potentials:

$$\sum_{m=1}^M \sum_{m' \neq m} \left[\langle f_m, f_{m'}\rangle^2 - \frac{1}{4}\,(a_m^2 + a_{m'}^2)^2\right].$$

As noted above, the potential energy function is only unique up to additive constants. We choose to omit the terms $(a_m^2 + a_{m'}^2)^2/4$, and include the diagonal terms $|\langle f_m, f_m\rangle|^2 = a_m^4$, yielding the result. □

In the following sections, we shall use this expression of the frame potential $\mathrm{FP}\colon \mathrm{S}(\{a_m\}_{m=1}^M) \to \mathbb{R}$ to characterize tight frames of lengths $\{a_m\}_{m=1}^M$. Specifically, we characterize such frames in terms of minimizers of FP, that is, in terms of those "maximally orthogonal" sequences $\{f_m\}_{m=1}^M \in \mathrm{S}(\{a_m\}_{m=1}^M)$.

Before continuing, we pause to consider another physical aspect of the frame force (4.3), namely that the magnitude of a frame force field generated by $f_{m'}$ will increase with $\|f_{m'}\|$. That is, we note that points which are farther away from the origin will apply a stronger push than those which are closer. To determine the explicit dependence, consider the *effective component* of the frame force $\mathrm{FF}(f_m, f_{m'})$, which is defined to be the component of $\mathrm{FF}(f_m, f_{m'})$ which lies parallel to the surface of $\mathrm{S}(a_m)$ at f_m. In essence, the effective component is the only part of the frame force that a point will actually "feel," as the point, restricted to a sphere, is prohibited from moving in the direction of the normal component. Formally, the effective component is

$$\begin{aligned}
\mathrm{EFF}(f_m, f_{m'}) &= \mathrm{FF}(f_m, f_{m'}) - \mathrm{Proj}_{f_m}\mathrm{FF}(f_m, f_{m'}) \\
&= \mathrm{FF}(f_m, f_{m'}) - \frac{\langle \mathrm{FF}(f_m, f_{m'}), f_m\rangle}{\|f_m\|^2} f_m \\
&= \frac{2\,|\langle f_m, f_{m'}\rangle|^2}{\|f_m\|^2} f_m - 2\,\langle f_m, f_{m'}\rangle f_{m'} \qquad (4.5)
\end{aligned}$$

for $f_m \neq 0$. Clearly, $\|\mathrm{EFF}(f_m, f_{m'})\|$ grows in proportion to $\|f_{m'}\|^2$. We therefore refer to a_m^2 as the *power* of f_m. From this perspective, the fundamental inequality (4.2) states that there must be a somewhat uniform distribution of power amongst the elements of a tight frame. Conversely, when the fundamental inequality is violated, then at least one of the points is disproportionately powerful compared to the others. In the final section, we shall determine what such an imbalance of power implies for the minimizers of the frame potential.

4.3 The Physical Interpretation of Frames

We have discussed how the frame force may be used to define an intuitive notion of a "maximally orthogonal" sequence of vectors. We now go beyond intuition, and establish a rigorous link between the theory of frames and the physical theory introduced in the previous section. Our work makes repeated use of the canonical linear operators of frame theory. In particular, given $\{f_m\}_{m=1}^M \subset \mathbb{H}_N$, consider the *analysis operator* $F\colon \mathbb{H}_N \to \mathbb{C}^M$,

$$(Ff)(m) = \langle f, f_m\rangle,$$

whose adjoint is the *synthesis operator* $F^*\colon \mathbb{C}^M \to \mathbb{H}_N$,

$$F^* g = \sum_{m=1}^M g(m)\, f_m.$$

Composing these operators gives the *frame operator* $F^*F\colon \mathbb{H}_N \to \mathbb{H}_N$,

$$F^*Ff = \sum_{m=1}^{M} \langle f, f_m \rangle f_m,$$

and the *Gram matrix* $FF^*\colon \mathbb{C}^M \to \mathbb{C}^M$, whose (m, m')th matrix entry is $\langle f_{m'}, f_m \rangle$. We note that $\{f_m\}_{m=1}^M$ is A-tight for $\mathbb{H}_N$ if and only if the corresponding analysis operator satisfies $\|Ff\|^2 = A\,\|f\|^2$ for all $f \in \mathbb{H}_N$, which is in turn equivalent to the frame operator satisfying $F^*F = AI$. In the following result, we use this equivalence to characterize tight frames in terms of the frame force.

Proposition 4.3. *$\{f_m\}_{m=1}^M$ is tight for $\mathbb{R}^N$ if and only if the total effective force field generated by $\{f_m\}_{m=1}^M$ vanishes everywhere, that is, if and only if*

$$\sum_{m=1}^{M} \mathrm{EFF}(f, f_m) = 0$$

for all $f \in \mathbb{R}^N$, $f \neq 0$.

Proof. Let F be the analysis operator of $\{f_m\}_{m=1}^M$. Using the definition of the effective force given in (4.5), we therefore have

$$0 = \sum_{m=1}^{M} \mathrm{EFF}(f, f_m) = \sum_{m=1}^{M} \left[\frac{2\,|\langle f, f_m \rangle|^2}{\|f\|^2} f - 2\,\langle f, f_m \rangle f_m \right]$$

for all $f \in \mathbb{R}^N$, $f \neq 0$, if and only if

$$F^*Ff = \sum_{m=1}^{M} \langle f, f_m \rangle f_m = \sum_{m=1}^{M} \frac{|\langle f, f_m \rangle|^2}{\|f\|^2} f = \frac{\|Ff\|^2}{\|f\|^2} f \tag{4.6}$$

for all $f \in \mathbb{R}^N$, $f \neq 0$. If $\{f_m\}_{m=1}^M$ is tight, then $F^*Ff = Af$ and $\|Ff\|^2 = A\,\|f\|^2$, and so (4.6) holds. For the converse, note that if (4.6) holds, then every $f \neq 0 \in \mathbb{R}^N$ is an eigenvector of F^*F. Thus, $F^*F = AI$ for some $A \geq 0$. □

In light of this result, it is natural to ask whether every real tight frame is in equilibrium with respect to the frame force. Below, we answer this question in the affirmative by showing that every tight frame for $\mathbb{R}^N$ is a global minimizer of the frame potential.

Before continuing, we note that one may "minimize the frame potential" in a more general setting. That is, despite being derived in the context of real Euclidean spaces, the formula for the frame potential makes sense in a general Hilbert space. To be precise, given a possibly complex finite-dimensional Hilbert space $\mathbb{H}_N$, consider the *frame potential* $\mathrm{FP}\colon \mathbb{H}_N^M \to \mathbb{R}$,

$$\mathrm{FP}(\{f_m\}_{m=1}^M) = \sum_{m=1}^M \sum_{m'=1}^M |\langle f_m, f_{m'}\rangle|^2.$$

Often, we shall restrict the domain of FP to the Cartesian product $\mathrm{S}(\{a_m\}_{m=1}^M)$ of the M hyperspheres $\mathrm{S}(a_m) = \{f \in \mathbb{H}_N : \|f\| = a_m\}$. In particular, the following results characterize tight frames of lengths $\{a_m\}_{m=1}^M$ as the global minimizers of this restricted potential, provided that at least one such tight frame exists in the first place. We begin by showing that the frame potential of a sequence is equal to the square of the Hilbert–Schmidt (Frobenius) norm of the corresponding frame operator.

Lemma 4.4. *Let F be the analysis operator of $\{f_m\}_{m=1}^M \subset \mathbb{H}_N$. Then,*

$$\mathrm{FP}(\{f_m\}_{m=1}^M) = \mathrm{Tr}\Big[(F^*F)^2\Big].$$

Proof. Let $\{e_n\}_{n=1}^N$ be an orthonormal basis of $\mathbb{H}_N$. Then

$$\begin{aligned}
\mathrm{FP}(\{f_m\}_{m=1}^M) &= \sum_{m=1}^M \sum_{m'=1}^M \langle f_{m'}, f_m\rangle \langle f_m, f_{m'}\rangle \\
&= \sum_{m=1}^M \sum_{m'=1}^M \Big\langle \sum_{n=1}^N \langle f_{m'}, e_n\rangle e_n, f_m\Big\rangle \langle f_m, f_{m'}\rangle \\
&= \sum_{m=1}^M \sum_{m'=1}^M \sum_{n=1}^N \langle f_{m'}, e_n\rangle \langle e_n, f_m\rangle \langle f_m, f_{m'}\rangle \\
&= \sum_{n=1}^N \sum_{m=1}^M \sum_{m'=1}^M \Big\langle \langle e_n, f_m\rangle f_m, \langle e_n, f_{m'}\rangle f_{m'}\Big\rangle \\
&= \sum_{n=1}^N \langle F^*Fe_n, F^*Fe_n\rangle \\
&= \sum_{n=1}^N \langle (F^*F)^2 e_n, e_n\rangle \\
&= \mathrm{Tr}\Big[(F^*F)^2\Big]. \quad \square
\end{aligned}$$

Proposition 4.5. *For any N-dimensional Hilbert space $\mathbb{H}_N$ and any nonnegative sequence $\{a_m\}_{m=1}^M$, the frame potential* $\mathrm{FP}\colon \mathrm{S}(\{a_m\}_{m=1}^M) \to \mathbb{R}$ *satisfies*

$$\frac{1}{N}\left[\sum_{m=1}^M a_m^2\right]^2 \leq \mathrm{FP}(\{f_m\}_{m=1}^M).$$

Furthermore, this lower bound is achieved if and only if $\{f_m\}_{m=1}^M$ is tight for $\mathbb{H}_N$ and $\|f_m\| = a_m$ for all m.

Proof. Let F be the analysis operator of $\{f_m\}_{m=1}^M \in \mathrm{S}(\{a_m\}_{m=1}^M)$, and let $\{\lambda_n\}_{n=1}^N$ be the eigenvalues of F^*F, counting multiplicities. By Lemma 4.4,

$$\mathrm{FP}(\{f_m\}_{m=1}^M) = \mathrm{Tr}\Big[(F^*F)^2\Big] = \sum_{n=1}^N \lambda_n^2.$$

Meanwhile, letting $\{e_n\}_{n=1}^N$ be any orthonormal basis for $\mathbb{H}_N$, the trace of the frame operator satisfies

$$\sum_{n=1}^N \lambda_n = \mathrm{Tr}(F^*F) = \sum_{n=1}^N \langle F^*Fe_n, e_n\rangle = \sum_{n=1}^N \Big\langle \sum_{m=1}^M \langle e_n, f_m\rangle\, f_m,\, e_n \Big\rangle.$$

Changing the order of summation gives

$$\sum_{n=1}^N \lambda_n = \sum_{m=1}^M \sum_{n=1}^N |\langle f_m, e_n\rangle|^2 = \sum_{m=1}^M \|f_m\|^2 = \sum_{m=1}^M a_m^2.$$

The lower bound is therefore found by solving the constrained minimization problem:

$$\min\left\{ \sum_{n=1}^N \lambda_n^2 : \sum_{n=1}^N \lambda_n = \sum_{m=1}^M a_m^2 \right\}.$$

Using Lagrange multipliers, the minimum is found to occur precisely when

$$\lambda_1 = \cdots = \lambda_N = \frac{1}{N}\sum_{m=1}^M a_m^2.$$

Thus, for any $\{f_m\}_{m=1}^M \in \mathrm{S}(\{a_m\}_{m=1}^M)$,

$$\mathrm{FP}(\{f_m\}_{m=1}^M) = \sum_{n=1}^N \lambda_n^2 \geq \sum_{n=1}^N \left[\frac{1}{N}\sum_{m=1}^M a_m^2\right]^2 = \frac{1}{N}\left[\sum_{m=1}^M a_m^2\right]^2.$$

Furthermore, this lower bound is achieved precisely when all the eigenvalues of F^*F are equal, which can be shown to be equivalent to having $F^*F = AI$ for some $A \geq 0$. Thus, the lower bound is achieved precisely when $\{f_m\}_{m=1}^M$ is A-tight for $\mathbb{H}_N$. □

Note that the previous result does not imply that this lower bound on the frame potential is optimal. Rather, this bound is only achieved when there exists a tight frame $\{f_m\}_{m=1}^M$ for $\mathbb{H}_N$ with $\|f_m\| = a_m$ for all m. We emphasize that this result does not show that such frames actually exist. Indeed, such frames cannot exist when the requisite lengths $\{a_m\}_{m=1}^M$ violate the fundamental inequality.

These ambiguities will be resolved by the main results of the following section. In particular, we show that there always exists a tight frame of lengths

$\{a_m\}_{m=1}^M$, provided $\{a_m\}_{m=1}^M$ satisfies the fundamental inequality, and, in the case when the inequality is violated, we determine the minimum value of the frame potential.

We conclude this section by briefly discussing the minimization of the frame potential in another context. In a finite-dimensional Hilbert space $\mathbb{H}_N$, two sequences $\{f_m\}_{m=1}^M, \{g_m\}_{m=1}^M \subset \mathbb{H}_N$ are *dual frames* if their analysis operators F and G satisfy $G^*F = I$, that is,

$$f = \sum_{m=1}^{M} \langle f, f_m \rangle \, g_m,$$

for all $f \in \mathbb{H}_N$. Any spanning set $\{f_m\}_{m=1}^M \subset \mathbb{H}_N$ has at least one dual frame, namely the *canonical dual* $\{\widetilde{f}_m\}_{m=1}^M \equiv \{(F^*F)^{-1} f_m\}_{m=1}^M$, whose synthesis operator $\widetilde{F}^* = (F^*F)^{-1}F$ is the pseudoinverse of F. Furthermore, when $M > N$, a sequence $\{f_m\}_{m=1}^M \subset \mathbb{H}_N$ will have an infinite number of dual frames. Nevertheless, the canonical dual has been found to be the "optimal" dual in certain applications [15]. The following result shows that the canonical dual is also the optimal dual from the point of view of the frame potential.

Proposition 4.6. *Let $\{f_m\}_{m=1}^M \subset \mathbb{H}_N$ be a spanning set. Then the canonical dual of $\{f_m\}_{m=1}^M$ is the unique dual having minimal frame potential.*

Proof. Let F, $\widetilde{F}$, and G denote the analysis operators of $\{f_m\}_{m=1}^M$, its canonical dual, and any arbitrary dual frame, respectively. As noted above, the synthesis operator of the canonical dual is $\widetilde{F}^* = (F^*F)^{-1}F$, which clearly implies $\widetilde{F}^*F = I$. At the same time, the definition of the arbitrary dual implies $G^*F = I$. Subtracting these two expressions gives $0 = (G - \widetilde{F})^*F$, which, when multiplied by $(F^*F)^{-1}$, yields

$$0 = (G - \widetilde{F})^* F (F^*F)^{-1} = (G - \widetilde{F})^* \widetilde{F}. \tag{4.7}$$

Using (4.7), G^*G may be expanded in a form in which the middle terms vanish:

$$\begin{aligned} G^*G &= \left[\widetilde{F} + (G - \widetilde{F})\right]^* \left[\widetilde{F} + (G - \widetilde{F})\right] \\ &= \widetilde{F}^*\widetilde{F} + \widetilde{F}^*(G - \widetilde{F}) + (G - \widetilde{F})^*\widetilde{F} + (G - \widetilde{F})^*(G - \widetilde{F}) \\ &= \widetilde{F}^*\widetilde{F} + (G - \widetilde{F})^*(G - \widetilde{F}). \end{aligned}$$

Lemma 4.4 then gives

$$\mathrm{FP}(\{g_m\}_{m=1}^M) = \mathrm{Tr}\Big[(G^*G)^2\Big] = \mathrm{Tr}\Big\{\Big[\widetilde{F}^*\widetilde{F} + (G - \widetilde{F})^*(G - \widetilde{F})\Big]^2\Big\}. \tag{4.8}$$

To continue, we expand the square on the right-hand side of (4.8), using Lemma 4.4 and properties of the trace:

$$\begin{aligned}\mathrm{FP}(\{g_m\}_{m=1}^M) &= \mathrm{Tr}\Big[(\widetilde{F}^*\widetilde{F})^2\Big] + \mathrm{Tr}\Big[\widetilde{F}^*\widetilde{F}(G-\widetilde{F})^*(G-\widetilde{F})\Big]\\ &\quad + \mathrm{Tr}\Big[(G-\widetilde{F})^*(G-\widetilde{F})\widetilde{F}^*\widetilde{F}\Big] + \mathrm{Tr}\Big\{\Big[(G-\widetilde{F})^*(G-\widetilde{F})\Big]^2\Big\}\\ &= \mathrm{FP}(\{\widetilde{f}_m\}_{m=1}^M) + 2\,\mathrm{Tr}\Big\{\Big[(G-\widetilde{F})\widetilde{F}^*\Big]^*(G-\widetilde{F})\widetilde{F}^*\Big\}\\ &\quad + \mathrm{Tr}\Big\{\Big[(G-\widetilde{F})^*(G-\widetilde{F})\Big]^2\Big\}.\end{aligned}$$

Since $\big[(G-\widetilde{F})\widetilde{F}^*\big]^*(G-\widetilde{F})\widetilde{F}^*$ and $\big[(G-\widetilde{F})^*(G-\widetilde{F})\big]^2$ are positive semidefinite, their traces are nonnegative, implying

$$\mathrm{FP}(\{g_m\}_{m=1}^M) \geq \mathrm{FP}(\{\widetilde{f}_m\}_{m=1}^M),$$

with equality if and only if $G-\widetilde{F}=0$. Thus, the canonical dual is the unique dual of minimum potential. □

Having found the unique dual of minimal potential, we next characterize those sequences for which the sum of their potential and their dual's potential is minimal.

Proposition 4.7. *A spanning set* $\{f_m\}_{m=1}^M \subset \mathbb{H}_N$ *is a minimizer of*

$$\mathrm{FP}(\{f_m\}_{m=1}^M) + \mathrm{FP}(\{\widetilde{f}_m\}_{m=1}^M)$$

if and only if $\{f_m\}_{m=1}^M$ *is a 1-tight frame for* $\mathbb{H}_N$.

Proof. Let $\{f_m\}_{m=1}^M \subset \mathbb{H}_N$ be a spanning set, and let $\{\lambda_n\}_{n=1}^N$ be the eigenvalues of the corresponding frame operator F^*F, counting multiplicities. As the frame operator of the canonical dual is $(F^*F)^{-1}$, Lemma 4.4 gives

$$\begin{aligned}\mathrm{FP}(\{f_m\}_{m=1}^M) + \mathrm{FP}(\{\widetilde{f}_m\}_{m=1}^M) &= \mathrm{Tr}\Big[(F^*F)^2\Big] + \mathrm{Tr}\Big[(F^*F)^{-2}\Big]\\ &= \sum_{n=1}^N (\lambda_n^2 + \lambda_n^{-2}).\end{aligned} \tag{4.9}$$

For any n, $\lambda_n^2 + 1/\lambda_n^2$ is minimized by letting $\lambda_n = 1$. Thus, (4.9) is bounded below by $2N$, and this lower bound is achieved if and only if $F^*F = I$, that is, if and only if $\{f_m\}_{m=1}^M \subset \mathbb{H}_N$ is a 1-tight frame for $\mathbb{H}_N$. By letting $\{f_m\}_{m=1}^M$ be the union of orthonormal bases with the required number of zero vectors, we see that such frames exist. Thus, the lower bound is indeed a minimum. □

4.4 The Fundamental Inequality

In the first section, we showed that if there exists a tight frame $\{f_m\}_{m=1}^M$ for $\mathbb{H}_N$ with $\|f_m\| = a_m$ for all m, then the fundamental inequality,

$$\max_{m=1,\dots,M} a_m^2 \le \frac{1}{N} \sum_{m=1}^{M} a_m^2,$$

is satisfied. Furthermore, in the previous section, we showed that if such frames exist, then they are minimizers of the frame potential $\mathrm{FP}\colon \mathrm{S}(\{a_m\}_{m=1}^M) \to \mathbb{R}$. In this section, we prove that the converse of the first result is true, and demonstrate a partial converse of the second.

To begin, recall that the fundamental inequality may be interpreted as requiring the powers $\{a_m^2\}_{m=1}^M$ of a tight frame $\{f_m\}_{m=1}^M \in \mathrm{S}(\{a_m\}_{m=1}^M)$ to be somewhat uniform in distribution. For a more precise definition of a "uniform distribution" in this context, consider the following result.

Lemma 4.8. *For any sequence $\{c_m\}_{m=1}^M \subset \mathbb{R}$ with $c_1 \ge \cdots \ge c_M \ge 0$, and for any positive integer N, there is a unique index N_0 with $1 \le N_0 \le N$ such that the inequality*

$$(N-n)c_n > \sum_{m=n+1}^{M} c_m$$

holds for $1 \le n < N_0$, while the opposite inequality

$$(N-n)c_n \le \sum_{m=n+1}^{M} c_m \tag{4.10}$$

holds for $N_0 \le n \le N$.

Proof. We begin by pointing out an implicit assumption of this result, namely that if $M < N$, any summation over an empty set of indices is regarded as zero. Let $\mathcal{I}$ be the set of indices such that (4.10) holds. Note that $\mathcal{I}$ is nonempty since $N \in \mathcal{I}$. Also, if $n \in \mathcal{I}$ then

$$\begin{aligned}
[N-(n+1)]c_{n+1} &= -c_{n+1} + (N-n)c_{n+1} \\
&\le -c_{n+1} + (N-n)c_n \\
&\le -c_{n+1} + \sum_{m=n+1}^{N} c_m \\
&= \sum_{m=n+2}^{N} c_m,
\end{aligned}$$

and so $n+1 \in \mathcal{I}$. N_0 is therefore uniquely defined as the minimum element of $\mathcal{I}$. $\square$

Thus, for a given positive integer N, the index N_0 is the place in the sequence $\{c_m\}_{m=1}^M$ where the terms cease to be larger than the "average" of the smaller remaining terms. Of course, this is not a true average unless

$M = N$. Nevertheless, one expects the index N_0 to be small if the sequence $\{c_m\}_{m=1}^M$ is somewhat evenly distributed, and large if $\{c_m\}_{m=1}^M$ varies greatly.

In the context of sequences $\{f_m\}_{m=1}^M \in \mathrm{S}(\{a_m\}_{m=1}^M) \subset \mathbb{H}_N^M$, we apply Lemma 4.8 to the sequence obtained by rearranging the powers $\{a_m^2\}_{m=1}^M$ in decreasing order. Let the *irregularity* of $\{a_m\}_{m=1}^M$ be $N_0 - 1$, where N_0 is the unique index obtained in this manner. The next result characterizes the fundamental inequality in terms of the irregularity.

Lemma 4.9. *For a positive integer N, a nonnegative sequence $\{a_m\}_{m=1}^M$ satisfies the fundamental inequality if and only if the irregularity of $\{a_m\}_{m=1}^M$ is zero.*

Proof. Without loss of generality, we assume $\{a_m\}_{m=1}^M$ is arranged in decreasing order. Thus, $\{a_m\}_{m=1}^M$ satisfies the fundamental inequality if and only if

$$(N-1)\,a_1^2 = -a_1^2 + N \max_m a_m^2 \leq -a_1^2 + \frac{N}{N}\sum_{m=1}^M a_m^2 = \sum_{m=2}^M a_m^2,$$

that is, when the index obtained by applying Lemma 4.8 to $\{a_m^2\}_{m=1}^M$ is $N_0 = 1$, which is equivalent to an irregularity of $N_0 - 1 = 0$. □

Thus, for an arbitrary nonnegative sequence $\{a_m\}_{m=1}^M$, the irregularity serves to measure the degree to which the fundamental inequality is violated, and partitions the corresponding points $\{f_m\}_{m=1}^M$ into two camps, one strong and the other weak. This idea plays a key role in the following result, in which we completely characterize those sequences in equilibrium under the frame force, namely the local minimizers of the frame potential.

Theorem 4.10. *Let $\mathbb{H}_N$ be an N-dimensional Hilbert space, let $\{a_m\}_{m=1}^M$ be a nonnegative decreasing sequence, and let N_0 be the minimal index n for which*

$$(N-n)a_n^2 \leq \sum_{m=n+1}^M a_m^2 \tag{4.11}$$

holds. Then, any local minimizer of the frame potential $\mathrm{FP}\colon \mathrm{S}(\{a_m\}_{m=1}^M) \to \mathbb{R}$ *may be partitioned as*

$$\{f_m\}_{m=1}^M = \{f_m\}_{m=1}^{N_0-1} \cup \{f_m\}_{m=N_0}^M,$$

where $\{f_m\}_{m=1}^{N_0-1}$ is an orthogonal sequence and $\{f_m\}_{m=N_0}^M$ is a tight sequence for the orthogonal complement of $\mathrm{span}\{f_m\}_{m=1}^{N_0-1}$. *Moreover, $\{f_m\}_{m=N_0}^M$ is a tight frame for this complement when the number of nonzero elements of $\{a_m\}_{m=1}^M$ is at least N.*

Proof. Let F be the analysis operator of a local minimizer $\{f_m\}_{m=1}^M$ of the frame potential $\mathrm{FP}\colon \mathrm{S}(\{a_m\}_{m=1}^M) \to \mathbb{R}$. Let $\{\lambda_j\}_{j=1}^J$ be the nonnegative decreasing sequence of the distinct eigenvalues of F^*F, and let $\{E_j\}_{j=1}^J$ be the corresponding sequence of mutually orthogonal eigenspaces. Consider the sequence of indexing sets $\{\mathcal{I}_j\}_{j=1}^{J}$ defined by

$$\mathcal{I}_j \equiv \{m : F^*F f_m = \lambda_j f_m\} \equiv \{m : f_m \in E_j\} \subseteq \{1, \ldots, M\}.$$

The remainder of the argument is outlined in the form of seven claims.

1. Each f_m is an eigenvector for F^*F.
2. For any $j = 1, \ldots, J$, $\{f_m\}_{m\in\mathcal{I}_j}$ is λ_j-tight for E_j.
3. For any $j < J$, $\{f_m\}_{m\in\mathcal{I}_j}$ is linearly independent.
4. For any $j < J$, $\{\lambda_j^{-1/2} f_m\}_{m\in\mathcal{I}_j}$ is an orthonormal basis for E_j.
5. $\dim E_J = N - M + |\mathcal{I}_J| > 0$.
6. $\{N_0, \ldots, M\} \subseteq \mathcal{I}_J$.
7. $\{N_0, \ldots, M\} = \mathcal{I}_J$.

Claim 1: Each f_m is an eigenvector for F^*F. The proof of this claim is essentially the same as that of [2, Thm. 7.3]. As such, we only provide a brief sketch of the argument. For any $m = 1, \ldots, M$, consider the function obtained by allowing the mth argument of the frame potential to vary, while holding the others constant at the minimizer $\{f_m\}_{m=1}^{M}$, namely $\mathrm{FP}_m \colon \mathbb{H}_N \to \mathbb{R}$,

$$\mathrm{FP}_m(f) = \|f\|^4 + 2\sum_{m'\neq m} |\langle f, f_{m'}\rangle|^2 + \mathrm{FP}(\{f_{m'}\}_{m'\neq m}).$$

Next, note that since $\{f_m\}_{m=1}^{M}$ is a local minimizer of the multi-input function $\mathrm{FP}\colon S(\{a_m\}_{m=1}^{M}) \to \mathbb{R}$, then f_m is a local minimizer of the single-input function $\mathrm{FP}_m colon \mathrm{S}(a_m) \to \mathbb{R}$. Thus, there exists a scalar c for which the corresponding Lagrange equation,

$$\nabla \mathrm{FP}_m(f) = c\,\nabla\|f\|^2,$$

is satisfied at $f = f_m$. These gradients are easily computed:

$$\nabla \mathrm{FP}_m(f) = 4\Big(\|f\|^2 f + \sum_{m'\neq m} \langle f, f_{m'}\rangle f_{m'}\Big),$$
$$\nabla\|f\|^2 = 2f.$$

Setting these two quantities equal at $f = f_m$ gives $4F^*Ff_m = 2cf_m$, and so f_m is an eigenvector of F^*F, as claimed.

As a consequence of the first claim, the elements of the minimizer $\{f_m\}_{m=1}^{M}$ are partitioned according to the eigenvalues. To be precise, we have

$$\bigcup_{j=1}^{J} \mathcal{I}_j = \{1, \ldots, M\}, \quad \mathcal{I}_j \cap \mathcal{I}_{j'} = \emptyset \quad \forall j \neq j',$$

where, without loss of generality, we regard $m \in \mathcal{I}_J$ if $f_m = 0$.

Claim 2: For any $j = 1,\dots,J$, $\{f_m\}_{m\in\mathcal{I}_j}$ is λ_j-tight for E_j. Fix $j = 1,\dots,J$, and let $F_j\colon E_j \to \mathbb{C}^{|\mathcal{I}_j|}$ be the analysis operator of $\{f_m\}_{m\in\mathcal{I}_j}$. Note that as the distinct eigenspaces of F^*F are mutually orthogonal, then $\langle f, f_m\rangle = 0$ for any $f \in E_j$ and any $m \notin \mathcal{I}_j$. Thus, for any $f \in E_j$,

$$\lambda_j f = F^*Ff = \sum_{m=1}^{M} \langle f, f_m\rangle\, f_m = \sum_{m\in\mathcal{I}_j} \langle f, f_m\rangle\, f_m = F_j^*F_j f,$$

and so $F_j^*F_j\colon E_j \to E_j$ satisfies $F_j^*F_j = \lambda_j I$, yielding the claim.

Claim 3: For any $j < J$, $\{f_m\}_{m\in\mathcal{I}_j}$ is linearly independent. Assume to the contrary that $\{f_m\}_{m\in\mathcal{I}_j}$ is linearly dependent for some j with $1 \le j \le J-1$. We shall find a sequence of continuous parametrized curves $\{g_m\}_{m=1}^M : (-1,1) \to \mathrm{S}(\{a_m\}_{m=1}^M)$, such that $\{g_m(0)\}_{m=1}^M = \{f_m\}_{m=1}^M$, and for which

$$\mathrm{FP}(\{g_m(t)\}_{m=1}^M) < \mathrm{FP}(\{f_m\}_{m=1}^M) \tag{4.12}$$

for all t in a neighborhood of the origin, contradicting the assumption that $\{f_m\}_{m=1}^M$ is a local minimizer of the frame potential. To begin, note that E_J is at least one-dimensional, being an eigenspace of F^*F. Fix $h \in E_J$ with $\|h\| = 1$. Since $\{f_m\}_{m\in\mathcal{I}_j}$ is linearly dependent, there exists a nonzero sequence of complex scalars $\{z_m\}_{m\in\mathcal{I}_j}$ such that $|z_m| \le 1/2$ for all $m \in \mathcal{I}_j$, and for which

$$\sum_{m\in\mathcal{I}_j} \overline{z_m}\, a_m\, f_m = 0.$$

For any $t \in (-1,1)$, consider $\{g_m(t)\}_{m=1}^M \subset \mathbb{H}_N$ given by

$$g_m(t) = \begin{cases} \left(1 - t^2\,|z_m|^2\right)^{1/2} f_m + t z_m a_m h, & m \in \mathcal{I}_j,\\ f_m, & m \notin \mathcal{I}_j. \end{cases}$$

Clearly these curves are continuous, and $\{g_m(0)\}_{m=1}^M = \{f_m\}_{m=1}^M$. To verify that $g_m(t) \in \mathrm{S}(a_m)$ for all $m = 1,\dots,M$ and all $t \in (-1,1)$, note that when $m \notin \mathcal{I}_j$, $\|g_m(t)\| = \|f_m\| = a_m$, as claimed. For the remaining case when $m \in \mathcal{I}_j$, note that since $j < J$, the corresponding eigenspaces E_j and E_J are orthogonal, and so $\langle f_m, h\rangle = 0$. In this case, since $\|h\| = 1$, the Pythagorean Theorem gives

$$\|g_m(t)\|^2 = \left(1 - t^2|z_m|^2\right)\|f_m\|^2 + t^2\,|z_m|^2\,a_m^2\,\|h\|^2 = a_m^2,$$

as desired. To prove the remaining property of $\{g_m(t)\}_{m=1}^M$, namely that (4.12) holds for all t in a neighborhood of the origin, we derive a Taylor approximation of $\mathrm{FP}(\{g_m(\cdot)\}_{m=1}^M)\colon (-1,1) \to \mathbb{R}$ about $t = 0$. By the product rule,

$$\frac{\mathrm{d}}{\mathrm{d}t}\mathrm{FP}(\{g_m(t)\}) = 4\,\mathrm{Re}\sum_{m=1}^{M}\sum_{m'=1}^{M} \langle g_m', g_{m'}\rangle\,\langle g_{m'}, g_m\rangle,$$

$$\frac{\mathrm{d}^2}{\mathrm{d}t^2}\mathrm{FP}(\{g_m(t)\}) = 4\,\mathrm{Re}\sum_{m=1}^{M}\sum_{m'=1}^{M}\Big\{\langle g''_m, g_{m'}\rangle\langle g_{m'}, g_m\rangle + \langle g'_m, g'_{m'}\rangle\langle g_{m'}, g_m\rangle + \langle g'_m, g_{m'}\rangle\langle g'_{m'}, g_m\rangle + \langle g'_m, g_{m'}\rangle\langle g_{m'}, g'_m\rangle\Big\}.$$

Meanwhile, for any $m = 1, \dots, M$,

$$g_m(0) = f_m, \quad g'_m(0) = \begin{cases} z_m a_m h, & m \in \mathcal{I}_j, \\ 0, & m \notin \mathcal{I}_j, \end{cases} \quad g''_m(0) = \begin{cases} -|z_m|^2 f_m, & m \in \mathcal{I}_j, \\ 0, & m \notin \mathcal{I}_j. \end{cases}$$

Thus, the first-order Taylor coefficient is

$$\begin{aligned}\frac{\mathrm{d}}{\mathrm{d}t}\mathrm{FP}(\{g_m(t)\})\Big|_{t=0} &= 4\,\mathrm{Re}\sum_{m\in\mathcal{I}_j}\sum_{m'=1}^{M}\langle z_m a_m h, f_{m'}\rangle\langle f_{m'}, f_m\rangle \\ &= 4\,\mathrm{Re}\sum_{m\in\mathcal{I}_j}\langle F^*F z_m a_m h, f_m\rangle \\ &= 4\,\lambda_J \mathrm{Re}\sum_{m\in\mathcal{I}_j} z_m a_m\,\langle h, f_m\rangle.\end{aligned}$$

We now reuse a fact given earlier in the proof of this claim, namely that $\langle f_m, h\rangle = 0$ for all $m \in \mathcal{I}_j$. Thus, as the first-order Taylor coefficient is zero, we compute the second-order coefficient:

$$\begin{aligned}&\frac{1}{2}\frac{\mathrm{d}^2}{\mathrm{d}t^2}\mathrm{FP}(\{g_m(t)\})\Big|_{t=0} \\ &\quad= 2\,\mathrm{Re}\Big\{\sum_{m\in\mathcal{I}_j}\sum_{m'=1}^{M}\langle -|z_m|^2 f_m, f_{m'}\rangle\langle f_{m'}, f_m\rangle && (4.13)\\ &\qquad + \sum_{m\in\mathcal{I}_j}\sum_{m'\in\mathcal{I}_j}\langle z_m a_m h, z_{m'} a_{m'} h\rangle\langle f_{m'}, f_m\rangle && (4.14)\\ &\qquad + \sum_{m\in\mathcal{I}_j}\sum_{m'\in\mathcal{I}_j}\langle z_m a_m h, f_{m'}\rangle\langle z_{m'} a_{m'} h, f_m\rangle && (4.15)\\ &\qquad + \sum_{m\in\mathcal{I}_j}\sum_{m'=1}^{M}\langle z_m a_m h, f_{m'}\rangle\langle f_{m'}, z_m a_m h\rangle\Big\}. && (4.16)\end{aligned}$$

To simplify this expression, note that the right-hand side of (4.13) becomes

$$\begin{aligned}\sum_{m\in\mathcal{I}_j}\sum_{m'=1}^{M}\langle -|z_m|^2 f_m, f_{m'}\rangle\langle f_{m'}, f_m\rangle &= -\sum_{m\in\mathcal{I}_j}|z_m|^2\langle F^*Ff_m, f_m\rangle\\ &= -\sum_{m\in\mathcal{I}_j}|z_m|^2\lambda_j\langle f_m, f_m\rangle\\ &= -\lambda_j\sum_{m\in\mathcal{I}_j}|z_m|^2 a_m^2.\end{aligned}$$

Next, we obtain an alternative expression for (4.14) by moving the scalars $z_m a_m$ and $z_{m'} a_{m'}$:

$$\sum_{m\in\mathcal{I}_j}\sum_{m'\in\mathcal{I}_j}\langle h, h\rangle\langle a_{m'}\overline{z_{m'}}f_{m'}, a_m\overline{z_m}f_m\rangle = \|h\|^2\left\|\sum_{m\in\mathcal{I}_j}\overline{z_m}a_m f_m\right\|^2.$$

Recalling the linear-dependence relation that defined the scalars $\{z_m\}_{m\in\mathcal{I}_j}$, we therefore have that the expression in (4.14) is zero. To continue, we again use the fact that $\langle f_m, h\rangle = 0$ for all $m\in\mathcal{I}_j$. Thus, the expression in (4.15) is also zero. Finally, (4.16) becomes

$$\begin{aligned}\sum_{m\in\mathcal{I}_j}\sum_{m'=1}^{M}\langle z_m a_m h, f_{m'}\rangle\langle f_{m'}, z_m a_m h\rangle &= \sum_{m\in\mathcal{I}_j}|z_m|^2 a_m^2\langle F^*Fh, h\rangle\\ &= \sum_{m\in\mathcal{I}_j}|z_m|^2 a_m^2\lambda_J\langle h, h\rangle\\ &= \lambda_J\sum_{m\in\mathcal{I}_j}|z_m|^2 a_m^2.\end{aligned}$$

Thus, the second-order Taylor coefficient is

$$\frac{1}{2}\frac{\mathrm{d}^2}{\mathrm{d}t^2}\mathrm{FP}(\{g_m(t)\})\Big|_{t=0} = 2(\lambda_J - \lambda_j)\sum_{m\in\mathcal{I}_j}|z_m|^2 a_m^2.$$

To determine the sign of this coefficient, note that $\lambda_j > \lambda_J$ since $j < J$. Also, the sequence $\{z_m\}_{m=1}^M$ is nonzero, by assumption. Furthermore, since $\mathcal{I}_J$ contains the indices of any zero elements of $\{f_m\}_{m=1}^M$ by decree, then $a_m = \|f_m\| > 0$ for all $m\in\mathcal{I}_j$. Thus, the second-order coefficient is strictly less than zero. An explicit, straightforward computation reveals that the third derivative of $\mathrm{FP}(\{g_m(t)\})$ is bounded near zero. Thus, by Taylor's Theorem, $\mathrm{FP}(\{g_m(t)\}_{m=1}^M) < \mathrm{FP}(\{f_m\}_{m=1}^M)$ for all sufficiently small t, a contradiction.

Claim 4: For any $j < J$, $\{\lambda_j^{-1/2}f_m\}_{m\in\mathcal{I}_j}$ is an orthonormal basis for E_j. First note that since $j < J$, we have $\lambda_j > \lambda_J \geq 0$. This, combined with the previous two claims, gives that $\{f_m\}_{m\in\mathcal{I}_j}$ is a linearly independent λ_j-tight frame for E_j. As a tight frame is necessarily a spanning set, $\{f_m\}_{m\in\mathcal{I}_j}$ is a basis for E_j. Thus, $|\mathcal{I}_j| = \dim E_j$, and so the analysis operator $F_j : E_j \to \mathbb{C}^{|\mathcal{I}_j|}$, which already satisfies $F_j^*F_j = \lambda_j I$, must also satisfy $F_jF_j^* = \lambda_j I$. Letting $\{e_m\}_{m\in\mathcal{I}_J}$ be the standard basis for $\mathbb{C}^{|\mathcal{I}_j|}$, we have that for any $m, m' \in \mathcal{I}_j$,

$$\langle f_m, f_{m'}\rangle = \langle F_j^* e_m, F_j^* e_{m'}\rangle = \langle F_j F_j^* e_m, e_{m'}\rangle = \lambda_j \langle e_m, e_{m'}\rangle.$$

Thus, $\{f_m\}_{m\in\mathcal{I}_j}$ is orthogonal and $a_m^2 = \|f_m\|^2 = \lambda_j$ for all $m \in \mathcal{I}_j$.

Claim 5: $\dim E_J = N - M + |\mathcal{I}_J| > 0$. For any $j < J$, the previous claim gives that $\{f_m\}_{m\in\mathcal{I}_j}$ is a basis for E_j. Thus,

$$M = \sum_{j=1}^{J} |\mathcal{I}_j| = |\mathcal{I}_J| + \sum_{j=1}^{J-1} |\mathcal{I}_j| = |\mathcal{I}_J| + \sum_{j=1}^{J-1} \dim E_j = |\mathcal{I}_J| + N - \dim E_J.$$

Thus, $\dim E_J = N - M + |\mathcal{I}_J|$. Next, note that E_J is at least one-dimensional, being an eigenspace of F^*F. Thus, $N - M + |\mathcal{I}_J| \geq 1 > 0$, as claimed.

Claim 6: $\{N_0, \ldots, M\} \subseteq \mathcal{I}_J$. We prove the complementary inclusion, namely $\mathcal{I}_J^C \subseteq \{1, \ldots, N_0 - 1\}$. Take any $n \in \mathcal{I}_j \subseteq \mathcal{I}_J^C$. We sum the values a_m^2 over all $m \in \mathcal{I}_J \cup \{n+1, \ldots, M\}$ in two different ways, noting that this set may either be partitioned into $\mathcal{I}_J \cap \{1, \ldots, n\}$ and $\{n+1, \ldots, M\}$, or alternatively partitioned into $\mathcal{I}_J$ and $\mathcal{I}_J^C \cap \{n+1, \ldots, M\}$. Thus,

$$\sum_{m\in\mathcal{I}_J\cap\{1,\ldots,n\}} a_m^2 + \sum_{m=n+1}^{M} a_m^2 = \sum_{m\in\mathcal{I}_J} a_m^2 + \sum_{m\in\mathcal{I}_J^C\cap\{n+1,\ldots,M\}} a_m^2. \tag{4.17}$$

To estimate the left-hand side of (4.17) from below, note that since $\{a_m\}_{m=1}^M$ is a nonnegative decreasing sequence, then $a_n^2 \leq a_m^2$ for all $m \in \mathcal{I}_J \cap \{1, \ldots, n\}$. Summing over all such m gives

$$|\mathcal{I}_J \cap \{1, \ldots, n\}|\, a_n^2 \leq \sum_{m\in\mathcal{I}_J\cap\{1,\ldots,n\}} a_m^2. \tag{4.18}$$

To estimate the right-hand side of (4.17) from above, note that $a_m^2 \leq a_n^2$ for all $m \in \mathcal{I}_J^C \cap \{n+1, \ldots, M\}$. Summing over all such m gives

$$\sum_{m\in\mathcal{I}_J^C\cap\{n+1,\ldots,M\}} a_m^2 \leq |\mathcal{I}_J^C \cap \{n+1, \ldots, M\}|\, a_n^2. \tag{4.19}$$

Applying the estimates of (4.18) and (4.19) to (4.17) gives

$$|\mathcal{I}_J \cap \{1, \ldots, n\}|\, a_n^2 + \sum_{m=n+1}^{M} a_m^2 \leq \sum_{m\in\mathcal{I}_J} a_m^2 + |\mathcal{I}_J^C \cap \{n+1, \ldots, M\}|\, a_n^2,$$

which reduces to

$$\sum_{m=n+1}^{M} a_m^2 \leq \sum_{m\in\mathcal{I}_J} a_m^2 + \left(|\mathcal{I}_J^C \cap \{n+1, \ldots, M\}| - |\mathcal{I}_J \cap \{1, \ldots, n\}|\right) a_n^2. \tag{4.20}$$

To simplify the right-hand side of (4.20), first note that the elements $\mathcal{I}_J$ may be partitioned into those indices which lie above n, and those that lie below:

$$|\mathcal{I}_J| = |\mathcal{I}_J \cap \{1, \ldots, n\}| + |\mathcal{I}_J \cap \{n+1, \ldots, M\}|. \tag{4.21}$$

Next, note that the $M - n$ elements of $\{n+1, \ldots, M\}$ may be partitioned into those elements which are also in $\mathcal{I}_J$, and those that are not:

$$M - n = |\mathcal{I}_J \cap \{n+1, \ldots, M\}| + |\mathcal{I}_J^C \cap \{n+1, \ldots, M\}|. \tag{4.22}$$

Substituting the expression for $|\mathcal{I}_J \cap \{n+1, \ldots, M\}|$ from (4.21) into (4.22) gives

$$M - n = |\mathcal{I}_J| - |\mathcal{I}_J \cap \{1, \ldots, n\}| + |\mathcal{I}_J^C \cap \{n+1, \ldots, M\}|,$$

which gives

$$|\mathcal{I}_J^C \cap \{n+1, \ldots, M\}| - |\mathcal{I}_J \cap \{1, \ldots, n\}| = -(n - M + |\mathcal{I}_J|). \tag{4.23}$$

Thus, (4.20) becomes

$$\sum_{m=n+1}^{M} a_m^2 \le \sum_{m \in \mathcal{I}_J} a_m^2 - (n - M + |\mathcal{I}_J|)\, a_n^2. \tag{4.24}$$

To further simplify (4.24), recall from the second claim that $\{f_m\}_{m \in \mathcal{I}_J}$ is λ_J-tight for E_J. Proposition 4.1 and the fifth claim therefore give

$$\sum_{m \in \mathcal{I}_J} a_m^2 = (\dim E_J)\, \lambda_J = (N - M + |\mathcal{I}_J|)\, \lambda_J,$$

which reduces (4.24) to

$$\sum_{m=n+1}^{M} a_m^2 \le (N - M + |\mathcal{I}_J|)\, \lambda_J - (n - M + |\mathcal{I}_J|)\, a_n^2. \tag{4.25}$$

To proceed, recall that $n \in \mathcal{I}_j$ for some $j < J$. The fourth claim consequently gives that the corresponding scaled frame element $\lambda_j^{-1/2} f_n$ is an element of an orthonormal basis for E_j. In particular, $\lambda_j^{-1/2} f_n$ is a normalized vector, and so $a_n^2 \equiv \|f_n\|^2 = \lambda_j > \lambda_J$. Since the fifth claim also gives that $N - M + |\mathcal{I}_J| > 0$, it follows that $(N - M + |\mathcal{I}_J|)\, \lambda_J < (N - M + |\mathcal{I}_J|)\, a_n^2$, providing a simple upper bound to the right-hand side of (4.25):

$$\sum_{m=n+1}^{M} a_m^2 < (N - M + |\mathcal{I}_J|) a_n^2 - (n - M + |\mathcal{I}_J|)\, a_n^2 = (N - n)\, a_n^2.$$

Thus, n does not satisfy the inequality (4.11) given in the statement of the result. However, N_0 is defined to be the minimal index that satisfies (4.11). In light of Lemma 4.8, this implies $n < N_0$, proving the claim.

Claim 7: $\{N_0, \ldots, M\} = \mathcal{I}_J$. To prove this claim, we will apply Lemma 4.8 to the sequence $\{c_m\}_{m=1}^M \equiv \{a_m^2\}_{m \in \mathcal{I}_J}$, taking $N \equiv \dim E_J$. Unfortunately, the conclusion of the lemma is more complicated to state in this case, due to the fact that the indices in $\mathcal{I}_J$ are not necessarily evenly spaced. Therefore, rather than state the entire conclusion, we only state the part that we will need. To be precise, we have that if the inequality

$$\left(\dim E_j - |\{1, \ldots, n\} \cap \mathcal{I}_J|\right) a_n^2 \le \sum_{m \in \mathcal{I}_J,\, m > n} a_m^2 \tag{4.26}$$

holds for a certain $n = n_0 \in \mathcal{I}_J$, then the same inequality also holds for all higher indices, that is, for all $n = n_1 \in \mathcal{I}_J$ such that $n_1 > n_0$.

To use this fact to prove the claim that $\{N_0, \ldots, M\} = \mathcal{I}_J$, first note that by the previous claim, it suffices to show $\{1, \ldots, N_0 - 1\} \cap \mathcal{I}_J = \emptyset$. Proceeding by means of contradiction, assume $\{1, \ldots, N_0 - 1\} \cap \mathcal{I}_J \neq \emptyset$, with minimal index n_0 and maximal index n_1. Next, we show that taking $n = n_0$ satisfies (4.26), guaranteeing that taking $n = n_1$ satisfies (4.26) as well.

To show that n_0 satisfies (4.26), note that by the second claim, $\{f_m\}_{m \in \mathcal{I}_J}$ is λ_J-tight for E_J. Thus, Proposition 4.1 gives

$$a_{n_0}^2 = \max_{m \in \mathcal{I}_J} a_m^2 \le \lambda_J = \frac{1}{\dim E_J} \sum_{m \in \mathcal{I}_J} a_m^2.$$

Multiplying this inequality by $\dim E_J$ and subtracting $a_{n_0}^2$ gives

$$(\dim E_J - 1)\, a_{n_0}^2 \le \sum_{m \in \mathcal{I}_J, m > n_0} a_m^2. \tag{4.27}$$

Next, note that since n_0 is the minimal index of $\{1, \ldots, N_0 - 1\} \cap \mathcal{I}_J$, then n_0 is also the minimal index of $\mathcal{I}_J$. Thus, $|\{1, \ldots, n_0\} \cap \mathcal{I}_J| = 1$, making (4.27):

$$\left(\dim E_j - |\{1, \ldots, n_0\} \cap \mathcal{I}_J|\right) a_{n_0}^2 \le \sum_{m \in \mathcal{I}_J,\, m > n_0} a_m^2,$$

verifying that (4.26) holds when $n = n_0$. Thus, (4.26) also holds when $n = n_1$:

$$\left(\dim E_j - |\{1, \ldots, n_1\} \cap \mathcal{I}_J|\right) a_{n_1}^2 \le \sum_{m \in \mathcal{I}_J,\, m > n_1} a_m^2. \tag{4.28}$$

To simplify (4.28), recall that n_1 is defined to be the maximal element of $\{1, \ldots, N_0 - 1\} \cap \mathcal{I}_J$, and that the sixth claim gives $\{N_0, \ldots, M\} \subseteq \mathcal{I}_J$. Thus, $\mathcal{I}_J$ may be partitioned as

$$\mathcal{I}_J = \left(\{1, \ldots, n_1\} \cap \mathcal{I}_J\right) \cup \{N_0, \ldots, M\}, \tag{4.29}$$

so that the right-hand side of (4.28) becomes

$$\sum_{m\in\mathcal{I}_J,\, m>n_1} a_m^2 = \sum_{m=N_0}^{M} a_m^2.$$

The partition given in (4.29) may be used to simplify the left-hand side of (4.28). Specifically, we have

$$\begin{aligned}|\mathcal{I}_J| &= |\{1,\dots,n_1\}\cap\mathcal{I}_J| + |\{N_0,\dots,M\}|\\ &= |\{1,\dots,n_1\}\cap\mathcal{I}_J| + M - N_0 + 1,\end{aligned}$$

implying $|\{1,\dots,n_1\}\cap\mathcal{I}_J| = |\mathcal{I}_J| - M + N_0 - 1$. As the fifth claim also gives $\dim E_J = N - M + |\mathcal{I}_J|$, we have

$$\begin{aligned}\dim E_j - |\{1,\dots,n_1\}\cap\mathcal{I}_J| &= (N - M + |\mathcal{I}_J|) - (|\mathcal{I}_J| - M + N_0 - 1)\\ &= N - N_0 + 1.\end{aligned}$$

Altogether, (4.28) reduces to

$$(N - N_0 + 1)\, a_{n_1}^2 \le \sum_{m=N_0}^{M} a_m^2. \tag{4.30}$$

Next, since $\{a_m^2\}_{m=1}^M$ is decreasing and $n_1 \le N_0 - 1$, then $a_{N_0-1}^2 \le a_{n_1}^2$. Thus, (4.30) implies

$$\left[N - (N_0 - 1)\right] a_{N_0-1}^2 \le \sum_{m=(N_0-1)+1}^{M} a_m^2,$$

namely that $n = N_0 - 1$ satisfies the inequality (4.11) given in the statement of the theorem. However, N_0 is defined to be the minimum index n which satisfies (4.11), a contradiction. Thus, $\{1,\dots,N_0-1\}\cap\mathcal{I}_J = \emptyset$, as claimed.

To complete the proof we need only summarize our progress. By the first and seventh claims,

$$\{f_m\}_{m=1}^M = \{f_m\}_{m\in\mathcal{I}_J^C} \cup \{f_m\}_{m\in\mathcal{I}_J} = \{f_m\}_{m=1}^{N_0-1} \cup \{f_m\}_{m=N_0}^{M},$$

where, by the fourth claim, $\{f_m\}_{m=1}^{N_0-1}$ is an orthogonal basis for $\operatorname{span}\{E_j\}_{j=1}^{J-1}$, and, by the second claim, $\{f_m\}_{m=N_0}^M$ is tight for E_J, which is the orthogonal complement of $\operatorname{span}\{f_m\}_{m=1}^{N_0-1} = \operatorname{span}\{E_j\}_{j=1}^{J-1}$. Finally, in the case where the number of nonzero elements of $\{a_m\}_{m=1}^M$ is at least N, then at least one element of $\{f_m\}_{m=N_0}^M$ is nonzero. By Proposition 4.1, the corresponding tightness constant is then positive, and so $\{f_m\}_{m=N_0}^M$ is a tight frame. □

We note that for the special case of $a_m = 1$ for all m, this result reduces to the main result of [2]. As a consequence of Theorem 4.10, we next demonstrate the sufficiency of the fundamental inequality, answering the question that began our chapter.

Corollary 4.11. *Let* $\mathbb{H}_N$ *be an* N*-dimensional Hilbert space, and let* $\{a_m\}_{m=1}^M$ *be a nonzero nonnegative sequence of real numbers. Then, there exists a tight frame* $\{f_m\}_{m=1}^M$ *for* $\mathbb{H}_N$ *with* $\|f_m\| = a_m$ *for all* m *if and only if* $\{a_m\}_{m=1}^M$ *satisfies the fundamental inequality.*

Proof. The necessity of the fundamental inequality is given by Proposition 4.1. For the converse, let $\{a_m\}_{m=1}^M$ satisfy the fundamental inequality. By Lemma 4.9, the regularity $N_0 - 1$ of $\{a_m\}_{m=1}^M$ is zero. Next, note that the frame potential is clearly continuous over the compact set $\mathrm{S}(\{a_m\}_{m=1}^M)$, and thus has a global minimizer $\{f_m\}_{m=1}^M \in \mathrm{S}(\{a_m\}_{m=1}^M)$. As global minimizers are necessarily local, Theorem 4.10 gives that $\{f_m\}_{m=1}^M$ is of the form

$$\{f_m\}_{m=1}^M = \{f_m\}_{m=1}^0 \cup \{f_m\}_{m=1}^M$$

where $\{f_m\}_{m=1}^M$ is tight for the orthogonal complement of $\{f_m\}_{m=1}^0 = \emptyset$, namely $\mathbb{H}_N$. By combining Proposition 4.1 with the fact that $\{a_m\}_{m=1}^M$ is nonzero, we conclude that the tightness constant is positive. □

As another corollary of Theorem 4.10, we improve upon Proposition 4.5 by finding the actual minimum value of the frame potential when the fundamental inequality is violated.

Corollary 4.12. *Let* $\mathbb{H}_N$ *be an* N*-dimensional Hilbert space, and let* $\{a_m\}_{m=1}^M$ *be a nonnegative decreasing sequence of irregularity* $N_0 - 1$*. Then, any local minimizer of the frame potential* $\mathrm{FP}\colon \mathrm{S}(\{a_m\}_{m=1}^M) \to \mathbb{R}$ *is also a global minimizer, and the minimum value is*

$$\sum_{m=1}^{N_0-1} a_m^4 + \frac{1}{N - N_0 + 1}\left[\sum_{m=N_0}^{M} a_m^2\right]^2.$$

Proof. First note that the frame potential is continuous over the compact set $\mathrm{S}(\{a_m\}_{m=1}^M)$, and thus has at least one local minimizer. By Theorem 4.10, any local minimizer of the frame potential consists of an orthogonal sequence $\{f_m\}_{m=1}^{N_0-1}$ for whose orthogonal complement the sequence $\{f_m\}_{m=N_0}^M$ is tight. We compute the frame potential of any such sequence. Since $\langle f_m, f_{m'}\rangle = 0$ for any $m' \neq m = 1, \ldots, N_0 - 1$,

$$\begin{aligned}\mathrm{FP}(\{f_m\}_{m=1}^M) &= \sum_{m=1}^{N_0-1} |\langle f_m, f_m\rangle|^2 + \sum_{m=N_0}^{M}\sum_{m'=N_0}^{M} |\langle f_m, f_{m'}\rangle|^2 \\ &= \sum_{m=1}^{N_0-1} a_m^4 + \mathrm{FP}(\{f_m\}_{m=N_0}^M).\end{aligned}$$

Note that the definition of the irregularity N_0 implies that the radii $\{a_m\}_{m=1}^{N_0}$ are strictly positive. Thus, the dimension of the span of $\{f_m\}_{m=1}^{N_0-1}$ is $N_0 - 1$, and so the dimension of the orthogonal complement is $(N - N_0 + 1)$. As

$\{f_m\}_{m=N_0}^M$ is tight for this space, the frame potential of $\{f_m\}_{m=N_0}^M$ is given by Proposition 4.5, yielding

$$\mathrm{FP}(\{f_m\}_{m=1}^M) = \sum_{m=1}^{N_0-1} a_m^4 + \frac{1}{N-N_0+1}\left[\sum_{m=N_0}^M a_m^2\right]^2. \tag{4.31}$$

To summarize, there exists at least one local minimizer of the frame potential, and all local minimizers attain the same value (4.31). Thus, every local minimizer is also a global minimizer, and the minimum value is (4.31). □

The strangeness of the frame force notwithstanding, the physics-based perspective of frames introduced in Section 4.2 provides a simple, intuitive explanation of the results above and the fundamental inequality. In particular, this perspective provides an intuitive reason as to why the minimizers look like they do, that is, why the points of every minimizing arrangement may be partitioned into two groups, according to size: larger points that form an orthogonal collection, and smaller points that form a tight frame for the orthogonal complement of the span of the larger points.

To elaborate, recall that in real Euclidean spaces, the frame potential is the total potential energy contained in a system of M points $\{f_m\}_{m=1}^M$ in which each point pushes against the others according to the frame force, and in which the movement of each point is restricted to a sphere of radius a_m. Further recall that the frame force causes each point to seek orthogonality with all others, and that a point on a sphere of a certain radius will generate a stronger force field than any other point on a smaller sphere. Specifically, recall that the effective strength of the "push" given off by a point on $S(a_m)$ is proportional to its *power* a_m^2.

For an intuitive explanation as to why the minimizers of the frame potential have the form that they do, imagine the above system acting dynamically over time, as discussed in Section 4.2. Though each point "wants" to make the others perpendicular to itself, this is clearly impossible if the number of nonzero radii a_m is greater than the dimension N. At the same time, when a single power $a_{m'}^2$ is disproportionately large, it is conceivable that the corresponding point $f_{m'}$ may be strong enough to force all the other points into an orthogonal hyperplane.

Theorem 4.10 verifies that this phenomenon actually occurs, and provides a quantitative way to determine the degree to which it happens. For example, consider a decreasing sequence $\{a_m\}_{m=1}^M$ whose irregularity $N_0 - 1$ is at least one. Let $\{f_m\}_{m=1}^M \in \mathrm{S}(\{a_m\}_{m=1}^M)$ be a local minimizer of the frame potential. Here, the definition of irregularity gives

$$a_1^2 > \frac{1}{N-1}\sum_{m=2}^M a_m^2.$$

At the same time, Theorem 4.10 guarantees that f_1 is necessarily orthogonal to $\{f_m\}_{m=2}^M$. Thus, as long as f_1 is stronger than the "dimensional average" of the remaining points, it is powerful enough to take an entire dimension for itself, leaving the other points to "fight" over the remaining $N-1$ dimensions.

These points then repeat the above scenario on a smaller scale. In particular, if the irregularity of the original sequence is at least two, then

$$a_2^2 > \frac{1}{N-2} \sum_{m=3}^{M} a_m^2.$$

Here, f_2 takes a one-dimensional subspace of the orthogonal complement of f_1 for itself, and lets $\{f_m\}_{m=3}^M$ fight over the $(N-2)$-dimensional leftovers. This process keeps repeating until eventually, perhaps at $N_0 = N$, we move beyond the point of irregularity. Here, none of the remaining points $\{f_m\}_{m=N_0}^M$ is strong enough to overcome the others. Forced to share the orthogonal complement of $\{f_m\}_{m=1}^{N_0}$, these points eventually settle into a tight equilibrium. In the special case when the fundamental inequality is satisfied, all the points $\{f_m\}_{m=1}^M$ will share power with each other, and a tight frame for the entire space will be asymptotically attained. This, at least, is an intuitive explanation of the statement of Theorem 4.10.

We conclude this chapter by recognizing Viswanath and Anantharam's independent discovery of the fundamental inequality during their investigation of the capacity region in synchronous code-division multiple access (CDMA) systems [24]. In a CDMA system, there are M users who share the available spectrum. The sharing is achieved by "scrambling" M-dimensional *user vectors* into smaller, N-dimensional vectors. In terms of frame theory, this scrambling corresponds to the application of a synthesis operator $S = F^*$ corresponding to M distinct N-dimensional *signature* vectors of length $N^{1/2}$. Noise-corrupted versions of these synthesized vectors arrive at a receiver, where the signature vectors are used to help extract the original user vectors.

Viswanath and Anantharam showed that the design of the optimal signature matrix S depends upon the *powers* $\{p_m\}_{m=1}^M$ of the individual users. In particular, they divided the users into two classes: those that are *oversized* and those that are not, by applying the idea of Lemma 4.8 to $\{p_m\}_{m=1}^M$. While the oversized users are assigned orthogonal channels for their personal use, the remaining users have their signature vectors designed so as to be Welch bound equality (WBE) sequences, namely, sequences which achieve the lower bound of Proposition 4.5, which are necessarily tight frames.

When no user is oversized, that is, when the fundamental inequality is satisfied, Viswanath and Anantharam show that the optimal signature sequences S must satisfy $SDS^* = p_{\text{tot}}I$, where D is a diagonal matrix whose entries are the powers $\{p_m\}_{m=1}^M$, and where $p_{\text{tot}} = \sum_{m=1}^M p_m$. By letting $F = N^{-1/2}D^{1/2}S^*$, this problem reduces to finding an $M \times N$ matrix F whose mth row is of norm $p_m^{1/2}$, and such that $F^*F = (p_{\text{tot}}/N)I$. That is, their problem reduces to finding a tight frame for $\mathbb{H}_N$ of lengths $\{p_m^{1/2}\}_{m=1}^M$. While Viswanath and Anantharam gave one solution to this problem using an explicit construction based upon the theory of majorization [18], we have characterized all solutions to this problem using a physics-related viewpoint of frame theory.

Acknowledgments

The authors thank John Benedetto for his insight, advice, and friendship over the years. This chapter only exists because of John. Two of the authors received their Ph.D.'s under John's tutelage. Another two of the authors began their collaborative work on frame theory after first meeting at a conference in honor of John's 60th birthday. Moreover, it was John who, upon hearing a question poised by Hans Feichtinger following a talk given by Edward Saff in 2000, began the work which would become [2], upon which our work is based. The authors also gratefully acknowledge the support of the National Science Foundation grant DMS-0405376.

References

1. J. J. Benedetto and D. Colella, Wavelet analysis of spectrogram seizure chips, in: *Wavelet Applications in Signal and Image Processing III* (San Diego, CA, 1995), A. F. Laine and M. A. Unser, eds., Proc. SPIE Vol. 2569, SPIE, Bellingham, WA, 2000, pp. 512–521.
2. J. J. Benedetto and M. Fickus, Finite normalized tight frames, *Adv. Comput. Math.*, **18** (2003), pp. 357–385.
3. J. J. Benedetto and G. E. Pfander, Periodic wavelet transforms and periodicity detection, *SIAM J. Appl. Math.*, **62** (2002), pp. 1329–1368.
4. H. Bölcskei, F. Hlawatsch, and H. G. Feichtinger, Frame-theoretic analysis of oversampled filter banks, *IEEE Trans. Signal Process.*, **46** (1998), pp. 3256–3269.
5. P. G. Casazza and J. Kovačević, Equal-norm tight frames with erasures, *Adv. Comput. Math.*, **18** (2003), pp. 387–430.
6. Z. Cvetković, Resilience properties of redundant expansions under additive noise and quantization, *IEEE Trans. Inform. Theory*, **49** (2003), pp. 644–656.
7. Z. Cvetković and M. Vetterli, Oversampled filter banks, *IEEE Trans. Signal Process.*, **46** (1998), pp. 1245–1255.
8. I. Daubechies, *Ten Lectures on Wavelets*, SIAM, Philadelphia, 1992.
9. I. Daubechies, A. Grossman, and Y. Meyer, Painless nonorthogonal expansions, *J. Math. Phys.*, **27** (1986) pp. 1271–1283.
10. R. J. Duffin and A. C. Schaeffer, A class of nonharmonic Fourier series, *Trans. Amer. Math. Soc.*, **72** (1952), pp. 341–366.
11. K. Dykema, D. Freeman, K. Kornelson, D. Larson, M. Ordower, and E. Weber, Ellipsoidal tight frames and projection decomposition of operators, *Illinois J. Math.*, **48** (2004), pp. 477–489.
12. K. Dykema and N. Strawn, Manifold structure of spaces of spherical tight frames, *Int. J. Pure Appl. Math.*, to appear.

13. D. Gabor, Theory of communication, *J. Inst. Elec. Eng. (London)*, **93** (1946), pp. 429–457.
14. V. K. Goyal, J. Kovačević, and J. A. Kelner, Quantized frame expansions with erasures, *Appl. Comput. Harmon. Anal.*, **10** (2001), pp. 203–233.
15. V. K. Goyal, M. Vetterli, and N. T. Thao, Quantized overcomplete expansions in $\mathbb{R}^N$: Analysis, synthesis, and algorithms, *IEEE Trans. Inform. Theory*, **44** (1998), pp. 16–31.
16. B. Hochwald, T. Marzetta, T. Richardson, W. Sweldens, and R. Urbanke, Systematic design of unitary space-time constellations, *IEEE Trans. Inform. Theory*, **46** (2000), pp. 1962–1973.
17. R. Holmes and V. Paulsen, Optimal frames for erasures, *Linear Algebra Appl.*, **377** (2004), pp. 31–51.
18. A. Horn, Doubly stochastic matrices and the diagonal of a rotation matrix, *Amer. J. Math.*, **76** (1954), pp. 620–630.
19. J. Kovačević, P. L. Dragotti, and V. K. Goyal, Filter bank frame expansions with erasures, *IEEE Trans. Inform. Theory*, **48** (2002), pp. 1439–1450.
20. T. Strohmer, Finite and infinite-dimensional models for oversampled filter banks, in: *Modern Sampling Theory: Mathematics and Applications*, J. J. Benedetto and P. J. S. G. Ferreira, eds., Birkhäuser, Boston, 2000, pp. 297–320.
21. T. Strohmer and R. Heath, Jr., Grassmannian frames with applications to coding and communications, *Appl. Comput. Harmon. Anal.*, **14** (2003), pp. 257–275.
22. M. Unser, Texture classification and segmentation using wavelet frames, *IEEE Trans. Image Process.*, **4** (1995), pp. 1549–1560.
23. R. Vale and S. Waldron, Tight frames and their symmetries, *Constr. Approx.*, **21** (2005), pp. 83–112.
24. P. Viswanath and V. Anantharam, Optimal sequences and sum capacity of synchronous CDMA systems, *IEEE Trans. Inform. Theory*, **45** (1999), pp. 1984–1991.
25. S. Waldron, Generalized Welch bound equality sequences are tight frames, *IEEE Trans. Inform. Theory*, **49** (2003), pp. 2307–2309.
26. G. Zimmermann, Normalized tight frames in finite dimensions, in: *Recent Progress in Multivariate Approximation* (Witten-Bommerholz, 2000), K. Jetter, W. Haußmann, and M. Reimer, eds., Birkhäuser, Boston, 2001, pp. 249–252.

Part III

Time-Frequency Analysis

5

Recent Developments in the Balian–Low Theorem

Wojciech Czaja[1] and Alexander M. Powell[2]

[1] Institute of Mathematics, University of Wrocław, pl. Grunwaldzki 2/4, 50-384 Wrocław, Poland and Department of Mathematics, University of Maryland, College Park, MD 20742, USA
czaja@math.uni.wroc.pl, wojtek@math.umd.edu

[2] Vanderbilt University, Department of Mathematics, Nashville, TN 37240, USA
alexander.m.powell@vanderbilt.edu

Summary. The Balian–Low Theorem is one of many manifestations of the uncertainty principle in harmonic analysis. Originally stated as a result on the poor time-frequency localization of generating functions of Gabor orthonormal bases, it has become a synonym for many general and abstract problems in time-frequency analysis. In this chapter we present some of the directions in which the Balian–Low Theorem has been extended in recent years.

5.1 Introduction

The harmonic analysis version of the classical *Heisenberg uncertainty principle* states that a function f and its Fourier transform $\hat{f}$ cannot be too well localized simultaneously. This statement can be written in the form of the following inequality:

$$\forall f \in L^2(\mathbb{R}), \quad \|f\|_2^4 \le 16\pi^2 \left(\int_{\mathbb{R}} |f(x)|^2 |x|^2 \, dx \right) \left(\int_{\widehat{\mathbb{R}}} |\hat{f}(\xi)|^2 |\xi|^2 \, d\xi \right), \qquad (5.1)$$

where $\hat{f}$ denotes the Fourier transform of f, and $\widehat{\mathbb{R}} = \mathbb{R}$. Given $f \in L^2(\mathbb{R}^d)$ we use the Fourier transform formally defined by

$$\forall \xi \in \widehat{\mathbb{R}}^d, \quad \hat{f}(\xi) = \mathcal{F}(f)(\xi) = \int_{\mathbb{R}^d} f(x)\, e^{-2\pi i x \cdot \xi} \, dx.$$

The uncertainty principle has many generalizations. We mention here, for example, those of Benedetto–Heinig, Beurling–Hörmander, and Cowling–Price, which involve more general weight functions and various norms, among other possible extensions. We refer the reader to, e.g., the surveys by Folland and Sitaram [28], Benedetto, Heil, and Walnut [15], and by Gröchenig [33], and the treatise by Havin and Jöricke [35] for extensive treatments of the role of uncertainty principle inequalities in harmonic analysis.

The inequality (5.1) is one particular way of studying the joint localization of f and $\hat{f}$, and in this context one often speaks of the joint *time-frequency* localization of a function f. However, there are other ways to represent the time-frequency content of a function. One particularly important approach is the *continuous Gabor transform*, also called the *short-time Fourier transform*, or the *windowed Fourier transform*. Given $f \in L^2(\mathbb{R}^d)$, the continuous Gabor transform $S_g(f) \in L^2(\mathbb{R}^{2d})$ is defined by

$$S_g(f)(x,\xi) = \int_{\mathbb{R}^d} f(t)\, e^{-2\pi i t\cdot\xi}\, \overline{g(t-x)}\, dt,$$

for some fixed function $g \in L^2(\mathbb{R}^d)$.

Let $\Lambda \subset \mathbb{R}^{2d} = \mathbb{R}^d \times \mathbb{R}^d$ be a countable set. Provided $g \in L^2(\mathbb{R}^d)$, we may define a discrete version of S_g in the form of a transformation that assigns to a function $f \in L^2(\mathbb{R}^d)$ a sequence of numbers

$$f \mapsto \{\langle f, g_{m,n}\rangle : (m,n) \in \Lambda\},$$

where the countable collection of functions $g_{m,n}$ is defined by

$$g_{m,n}(x) = e^{2\pi i x\cdot n} g(x-m),$$

for $(m,n) \in \Lambda \subset \mathbb{R}^{2d}$. This collection of functions is called the *Gabor system* or the *Weyl–Heisenberg system* generated by g and Λ, and we denote it by

$$\mathcal{G}(g,\Lambda) = \{g_{m,n} : (m,n) \in \Lambda\}.$$

The function g is called the *generating function* or *window function* of the Gabor system. When $\mathcal{G}(g,\Lambda)$ is an orthonormal basis for $L^2(\mathbb{R}^d)$, we refer to it as a *Gabor orthonormal basis.* The various theorems in this chapter will involve different sets Λ, ranging from rectangular lattices to relatively irregular sets. We shall always either explicitly define Λ or state the structural properties assumed of it.

Roger Balian [1] and, independently, Francis Low [40], stated the following result which relates Gabor systems to the classical uncertainty principle.

Theorem 5.1 (Balian–Low Theorem (BLT)). *Let $ab = 1$, and let $g \in L^2(\mathbb{R})$ have the property that $\mathcal{G}(g, a\mathbb{Z} \times b\mathbb{Z}) = \{g_{m,n} : (m,n) \in a\mathbb{Z} \times b\mathbb{Z}\}$ is a Gabor orthonormal basis for $L^2(\mathbb{R})$. Then*

$$\left(\int_{\mathbb{R}} |g(x)|^2 |x|^2\, dx\right)\left(\int_{\widehat{\mathbb{R}}} |\hat{g}(\xi)|^2 |\xi|^2\, d\xi\right) = \infty. \qquad (5.2)$$

The proofs of Balian and Low both contained a technical gap, but the result turned out to be correct. A complete proof was given by Battle [6], whose approach was quite different: he used techniques analogous to those that are used to prove the classical uncertainty principle. The gap in the original proofs was filled by Coifman, Daubechies, and Semmes in [21], where the result was also extended to hold for Gabor systems that are *exact frames* (also called *Riesz bases*) for $L^2(\mathbb{R})$, cf. [38]. The assumption that $ab = 1$ is redundant. We refer the reader to [16] for a more detailed analysis of this and related facts.

For more on the historical background of the Balian–Low Theorem and the account of its various proofs, we refer the reader to a beautiful survey by Benedetto, Heil, and Walnut [16], and to the books by Daubechies [22] and Gröchenig [32]. For convenience, we follow an established custom and occasionally refer to the Balian–Low Theorem by the acronym BLT in the ensuing sections. We refer the reader to a footnote in [37] for a historical comment on the origin of this acronym.

In Section 5.2 we investigate various recent extensions of the Balian–Low Theorem. We start by exploring possible generalizations to Gabor systems generated by functions of several variables, associated with both lattices and arbitrary countable subsets of $\mathbb{R}^{2d}$. The symplectic lattices stand out naturally in this context and we consider this particular topic from two different points of view. We close Section 5.2 with a quick overview of the "no-go" theorems for Gabor frames generated by several different functions. A natural question in this context is to determine if the Balian–Low Theorem holds for other pairs of weights besides the classical $|x|^2$ and $|\xi|^2$ weights. In Section 5.3 we study the problem of optimality of the weights used in (5.2) and we show that the exponents of $|x|^2$ and $|\xi|^2$ cannot be decreased by any amount. Finally, in Section 5.4, we present a result that is the analogue of the Balian–Low Theorem for the case of *wavelets*, i.e., systems generated by dilations and translations of a single function. This section shows that there are many similarities between the case of Gabor systems and wavelets.

5.2 Extensions of the Classical Balian–Low Theorem

Let us start by introducing some terminology and notation that will be used throughout this chapter. We say that a collection $\{f_k : k \in \mathbb{Z}\} \subset L^2(\mathbb{R}^d)$ of functions is a *frame* for $L^2(\mathbb{R}^d)$, with *frame bounds* A and B, if

$$\forall\, f \in L^2(\mathbb{R}^d), \quad A\,\|f\|_2^2 \le \sum_{k\in\mathbb{Z}} |\langle f, f_k\rangle|^2 \le B\,\|f\|_2^2.$$

A frame is *tight* if $A = B$, and a frame is *exact* if it is no longer a frame after removal of any of its elements. Given any frame $\{f_k : k \in \mathbb{Z}\}$ for $L^2(\mathbb{R}^d)$, there exists a *dual frame* $\{\tilde{f}_k : k \in \mathbb{Z}\}$ for $L^2(\mathbb{R}^d)$ such that

$$\forall\, f \in L^2(\mathbb{R}^d), \quad f = \sum_{k\in\mathbb{Z}} \langle f, f_k\rangle\, \tilde{f}_k = \sum_{k\in\mathbb{Z}} \langle f, \tilde{f}_k\rangle\, f_k, \tag{5.3}$$

where the series converge in $L^2(\mathbb{R}^d)$. The choice of coefficients for expressing f in terms of $\{f_k : k \in \mathbb{Z}\}$ or $\{\tilde{f}_k : k \in \mathbb{Z}\}$ is not unique, unless the frame is a *basis. A frame is a basis if and only if it is exact*, see, e.g., [16].

Given a frame $\{f_k : k \in \mathbb{Z}\}$ for $L^2(\mathbb{R}^d)$, we define the associated *frame operator* S on $L^2(\mathbb{R}^d)$ by

$$\forall\, f \in L^2(\mathbb{R}^d), \quad S(f) = \sum_{k\in\mathbb{Z}} \langle f, f_k\rangle\, f_k.$$

S is a bounded and invertible map of $L^2(\mathbb{R}^d)$ onto itself. Given a frame $\{f_k : k \in \mathbb{Z}\}$, the *canonical dual frame* $\{\tilde{f}_k : k \in \mathbb{Z}\}$ is defined by $\tilde{f}_k = S^{-1}(f_k)$. If a frame is exact then $\{f_k : k \in \mathbb{Z}\}$ and $\{\tilde{f}_k : k \in \mathbb{Z}\}$ are *biorthogonal*, that is,

$$\forall\, k, l \in \mathbb{Z}, \quad \langle f_k, \tilde{f}_l\rangle = \delta_{k,l},$$

where $\delta_{k,l}$ denotes the *Kronecker delta function*, i.e., $\delta_{k,l} = 1$ if $k = l$, and $\delta_{k,l} = 0$ if $k \neq l$.

Let $\Lambda \subset \mathbb{R}^{2d}$ be a countable set. If $g \in L^2(\mathbb{R}^d)$ and Λ generate a Gabor system which is a frame for $L^2(\mathbb{R}^d)$, then we shall refer to $\mathcal{G}(g, \Lambda) = \{g_{m,n} : (m, n) \in \Lambda\}$ as a *Gabor frame* for $L^2(\mathbb{R}^d)$. In the special case where $ab > 0$, $\Lambda = a\mathbb{Z}^d \times b\mathbb{Z}^d$, and $\mathcal{G}(g, \Lambda) = \{g_{m,n} : (m, n) \in a\mathbb{Z}^d \times b\mathbb{Z}^d\}$ is a Gabor frame for $L^2(\mathbb{R}^d)$, it is elementary to show that $S^{-1}(g_{m,n}) = (S^{-1}(g))_{m,n}$.

5.2.1 The Balian–Low Theorem in $\mathbb{R}^d$

The first generalization of the Balian–Low Theorem to $\mathbb{R}^d$ already appeared in [36], see also [32]. This result extends a version of the BLT called the Amalgam Balian–Low Theorem, which is stated in terms of a *Wiener amalgam space* $W(C_0, l^1)$, defined by

$$W(C_0, l^1) = \left\{ f : f \text{ is continuous and } \sum_{k\in\mathbb{Z}^d} \|f(\cdot + k)\|_{L^\infty([0,1]^d)} < \infty \right\}.$$

The next theorem is Corollary 7.5.3 in [36], see also Theorem 8.4.1 in [32].

Theorem 5.2 (Amalgam Balian–Low Theorem). *Suppose $ab = 1$ and $g \in L^2(\mathbb{R}^d)$. If the Gabor system $\mathcal{G}(g, a\mathbb{Z}^d \times b\mathbb{Z}^d) = \{g_{m,n} : (m, n) \in a\mathbb{Z}^d \times b\mathbb{Z}^d\}$ is an exact frame for $L^2(\mathbb{R}^d)$, then*

$$g \notin W(C_0, l^1) \quad \textit{and} \quad \hat{g} \notin W(C_0, l^1).$$

It was shown in [16] that the classical and amalgam versions of the Balian–Low Theorem are independent from each other, see Examples 3.3 and 3.4 in [16]. We refer the reader to [36], [16], [17], and [32] for extended treatments of the Amalgam Balian–Low Theorem and to [37] for some related open questions.

The simultaneous and independent papers [34] and [12] extend the classical version of the Balian–Low Theorem to the case of d-dimensions. The fundamental results to be compared in these papers are Theorems 8 and 11 in [34], and Theorems 2.1 and 2.5 in [12]. Let us introduce some necessary definitions before discussing these results.

Let $v, w \in \mathbb{R}^d$ be non-zero vectors. We define the following *position* and *momentum* operators, wherever they make sense in $L^2(\mathbb{R}^d)$:

$$P_v(f)(x) = \Big(\sum_{j=1}^{d} v_j x_j\Big) f(x)$$

and

$$M_w(f)(x) = \mathcal{F}^{-1}\Big(\Big(\sum_{j=1}^{d} w_j \xi_j\Big)\hat{f}(\xi)\Big)(x) = \mathcal{F}^{-1}(P_w(\hat{f}))(x),$$

where $v = (v_1, \ldots, v_d) = \sum v_j u_j$, $u_j = (0, \ldots, 0, 1, 0, \ldots, 0)$ with 1 in the jth coordinate, and $v_j \in \mathbb{R}$. These unit vectors u_j define the *standard Euclidean basis* $\{u_j : j = 1, \ldots, d\}$ of $\mathbb{R}^d$. If the vectors v and w in the definitions of P_v and M_w are elements of the standard basis, then we shall use the notation P_j and M_j for the operators induced by the jth basis vector u_j.

The following result [34], [12], gives a "weak" extension of Theorem 5.1 to the case of exact frames in higher dimensions with a localization condition stated for general coordinates. The proof is based on Battle's method [6], which, in turn, is a standard technique to prove uncertainty principle inequalities. The 1-dimensional Balian–Low Theorem first appeared in this form in [16].

Theorem 5.3 (Weak Balian–Low Theorem). *Let $\mathcal{G}(g, \mathbb{Z}^{2d}) = \{g_{m,n} : (m,n) \in \mathbb{Z}^d \times \mathbb{Z}^d\}$ be an exact frame for $L^2(\mathbb{R}^d)$ and let $\mathcal{G}(\tilde{g}, \mathbb{Z}^{2d}) = \{\tilde{g}_{m,n} : (m,n) \in \mathbb{Z}^d \times \mathbb{Z}^d\}$ be the canonical dual frame. If v, $w \in \mathbb{R}^d$ satisfy $v \cdot w \neq 0$, then*

$$\|P_v(g)\|_2 \, \|M_w(g)\|_2 \, \|P_v(\tilde{g})\|_2 \, \|M_w(\tilde{g})\|_2 = \infty. \tag{5.4}$$

Proof. We may assume without loss of generality that $|v| = |w| = 1$, where $|\cdot|$ denotes the Euclidean norm in $\mathbb{R}^d$. We shall proceed with a proof by contradiction, and assume that all four functions in (5.4) are elements of $L^2(\mathbb{R}^d)$. Because of the biorthogonality relations for g and $\tilde{g}$, we compute

$$
\begin{aligned}
\langle P_v(g), \tilde{g}_{m,n} \rangle &= \langle P_v(g), \tilde{g}_{m,n} \rangle - \Big(\sum_{j=1}^{d} v_j m_j \Big) \langle g, \tilde{g}_{m,n} \rangle \\
&= \int_{\mathbb{R}^d} \Big(\sum_{j=1}^{d} v_j (x_j - m_j) \Big) g(x) \, \overline{\tilde{g}(x-m)} \, e^{-2\pi i n \cdot x} \, dx \\
&= e^{-2\pi i m \cdot n} \int_{\mathbb{R}^d} \Big(\sum_{j=1}^{d} v_j x_j \Big) \overline{\tilde{g}(x)} \, g(x+m) \, e^{-2\pi i n \cdot x} \, dx \\
&= e^{-2\pi i m \cdot n} \, \langle g_{-m,-n}, P_v(\tilde{g}) \rangle.
\end{aligned} \tag{5.5}
$$

From our assumption that $M_w(g) \in L^2(\mathbb{R}^d)$, it follows that the distributional partial derivative of g, denoted by $\partial_w g$, belongs to $L^2(\mathbb{R}^d)$. By a standard result about Sobolev spaces, e.g., [41, Thm. 1.1], there exists a function h such that $g = h$ a.e., and h is absolutely continuous on almost all straight lines parallel to the vector w. Thus, the distributional directional derivative of g coincides with the classical directional derivative $D_w(g)$ a.e., and so

$$
M_w(g)(x) = \frac{1}{2\pi i} D_w(g)(x) \text{ a.e.}
$$

Moreover, our assumptions imply that $D_w(g), D_w(\tilde{g}) \in L^2(\mathbb{R}^d)$. Therefore, using integration by parts, an appropriate change of variables, and the biorthogonality relations between g and $\tilde{g}$, we can compute

$$
\begin{aligned}
\langle g_{m,n}, M_w(\tilde{g}) \rangle &= \frac{i}{2\pi} \int_{\mathbb{R}^d} g(x-m) \, e^{2\pi i n \cdot x} \, \overline{D_w(\tilde{g})(x)} \, dx \\
&= \frac{1}{2\pi i} \int_{\mathbb{R}^d} D_w \left(g(x-m) \, e^{2\pi i n \cdot x} \right) \overline{\tilde{g}(x)} \, dx \\
&= \frac{1}{2\pi i} \int_{\mathbb{R}^d} \left(D_w(g)(x-m) \, e^{2\pi i n \cdot x} + (w \cdot n) \, g(x-m) \, e^{2\pi i n \cdot x} \right) \overline{\tilde{g}(x)} \, dx \\
&= \frac{e^{2\pi i m \cdot n}}{2\pi i} \int_{\mathbb{R}^d} \left(D_w(g)(x) + 2\pi i \, (w \cdot n) \, g(x) \right) \, e^{2\pi i n \cdot x} \, \overline{\tilde{g}(x+m)} \, dx \\
&= e^{2\pi i m \cdot n} \left(\langle M_w(g), \tilde{g}_{-m,-n} \rangle + (w \cdot n) \, \delta_{m,0} \delta_{n,0} \right) \\
&= e^{2\pi i m \cdot n} \, \langle M_w(g), \tilde{g}_{-m,-n} \rangle.
\end{aligned} \tag{5.6}
$$

Because of (5.5), (5.6), and the frame representation property (5.3), we have

$$\begin{aligned}\langle P_v(g), M_w(\tilde{g})\rangle &= \sum_{m,n\in\mathbb{Z}^d} \langle P_v(g), \tilde{g}_{m,n}\rangle \langle g_{m,n}, M_w(\tilde{g})\rangle \\ &= \sum_{m,n\in\mathbb{Z}^d} \langle g_{-m,-n}, P_v(\tilde{g})\rangle \langle M_w(g), \tilde{g}_{-m,-n}\rangle \\ &= \sum_{m,n\in\mathbb{Z}^d} \langle M_w(g), \tilde{g}_{m,n}\rangle \langle g_{m,n}, P_v(\tilde{g})\rangle \\ &= \langle M_w(g), P_v(\tilde{g})\rangle.\end{aligned} \tag{5.7}$$

It is not difficult to verify that

$$[P_v, M_w] = -\frac{1}{2\pi i}(v\cdot w)\,\mathrm{Id}, \tag{5.8}$$

where the commutator is $[P_v, M_w] = P_vM_w - M_wP_v$, and where Id denotes the identity operator. For example, see [42], where (5.8) appears for the position and momentum operators associated with the standard basis vectors, cf. [16]. Thus, for functions $g,\ \tilde{g} \in L^2(\mathbb{R}^d)$, such that $P_v(g),\ P_v(\tilde{g}) \in L^2(\mathbb{R}^d)$ and $M_w(g),\ M_w(\tilde{g}) \in L^2(\mathbb{R}^d)$, we have

$$\begin{aligned}\langle P_v(g), M_w(\tilde{g})\rangle &= \langle M_w(g), P_v(\tilde{g})\rangle - \frac{1}{2\pi i}(v\cdot w)\langle g,\tilde{g}\rangle \\ &= \langle M_w(g), P_v(\tilde{g})\rangle - \frac{1}{2\pi i}(v\cdot w).\end{aligned}$$

Since we have assumed that $v\cdot w \neq 0$, we obtain a contradiction with our calculation (5.7). □

Remark 5.4. We note here that Theorem 5.3 is stated in greater generality in [34]. It not only applies to exact Gabor frames for $L^2(\mathbb{R}^d)$, but also applies to Gabor systems which are exact frames for the subspaces given by their spans. Gabardo and Han [29, Thm. 1.2] recently proved that a "strong" form of the BLT also holds in the subspace setting.

Let us also mention that the statement and proof of Theorem 5.3 may be rewritten, without any difficulties, for Gabor systems associated with arbitrary lattices Λ in $\mathbb{R}^{2d}$; see Theorem 5.13.

Although Theorem 5.3 is "weak," it is strong enough to imply the following two natural generalizations of the BLT. In contrast to the preceding remark, it is not yet clear if these corollaries extend to arbitrary lattices in the exact frame case; cf. Theorems 5.8 and 5.11.

Corollary 5.5. *Let $\mathcal{G}(g,\mathbb{Z}^{2d}) = \{g_{m,n} : (m,n)\in\mathbb{Z}^d\times\mathbb{Z}^d\}$ be an exact frame for $L^2(\mathbb{R}^d)$. If $v,\ w\in\mathbb{R}^d$ satisfy $v\cdot w\neq 0$, then*

$$\|P_v(g)\|_2\ \|M_w(g)\|_2 = \infty.$$

Proof. In view of Theorem 5.3, it is enough to show that $P_v(g) \in L^2(\mathbb{R}^d)$ if and only if $P_v(\tilde{g}) \in L^2(\mathbb{R}^d)$, and that $M_w(g) \in L^2(\mathbb{R}^d)$ if and only if $M_w(\tilde{g}) \in L^2(\mathbb{R}^d)$. This, in turn, was proved by Daubechies and Janssen [23] for the position and momentum operators associated with the standard basis vectors (also see [16, Thm. 7.7]). The proof for arbitrary operators P_v and M_w is analogous, and it uses the d-dimensional Sobolev space argument in the proof of Theorem 5.3, instead of 1-dimensional considerations. □

Corollary 5.6. *Let $\omega(x)$ be any positive quadratic form on $\mathbb{R}^d$ and let*

$$\mathcal{G}(g, \mathbb{Z}^{2d}) = \{g_{m,n} : (m,n) \in \mathbb{Z}^d \times \mathbb{Z}^d\}$$

be an exact frame for $L^2(\mathbb{R}^d)$. Then

$$\left(\int_{\mathbb{R}^d} \omega(x)\, |g(x)|^2\, dx\right)\left(\int_{\widehat{\mathbb{R}}^d} \omega(\xi)\, |\hat{g}(\xi)|^2\, d\xi\right) = \infty.$$

Further reformulations of the Balian–Low Theorem are possible for more general groups than $\mathbb{R}^d$. Gröchenig [31] studies extensions of the BLT in the context of *locally compact Abelian groups*, where the problems with non-existence of canonical weights and differential operators are solved by using the short-time Fourier transform; cf. [25].

We close this section by showing that the condition $v \cdot w \neq 0$ is necessary in Theorem 5.3. Indeed, let us consider the space $L^2(\mathbb{R}^2)$ with the orthonormal Gabor basis generated by

$$g(x,y) = \chi_{[0,1]}(x)\mathcal{F}^{-1}(\chi_{[0,1]})(y)$$

and the pair of vectors $v = (1,0)$ and $w = (0,1)$. Then

$$\begin{aligned}\|P_v(g)\|_2^2 &= \int_{\mathbb{R}^2} |x\, g(x,y)|^2\, dx\, dy \\ &= \int_{\mathbb{R}} \left|x\, \chi_{[0,1]}(x)\right|^2 dx \int_{\mathbb{R}} \left|\mathcal{F}^{-1}(\chi_{[0,1]})(y)\right|^2 dy < \infty\end{aligned}$$

and

$$\begin{aligned}\|M_w(g)\|_2^2 &= \int_{\mathbb{R}^2} |\eta\, \hat{g}(\xi,\eta)|^2\, d\xi\, d\eta \\ &= \int_{\mathbb{R}} \left|\mathcal{F}(\chi_{[0,1]})(\xi)\right|^2 d\xi \int_{\mathbb{R}} \left|\eta(\chi_{[0,1]})(\eta)\right|^2 d\eta < \infty.\end{aligned}$$

5.2.2 Non-Uniform Gabor Systems

The study of *non-uniform* Gabor systems, i.e., those Gabor systems which are associated with a set Λ which is not a lattice, has increased in recent years because of applications of such systems to problems in signal processing. Of course, not every Λ generates an orthonormal basis or even a frame. For example, in order for a Gabor system to have good signal representation properties, Λ must satisfy certain density conditions, see, e.g., [45], [20].

Before stating a version of the Balian–Low Theorem for non-uniform Gabor systems, we shall need the following lemma, whose proof is similar to the proof of analogous statements in Theorem 5.3.

Lemma 5.7. *Let $\Lambda \subset \mathbb{R}^{2d}$ be a countable set with the property that $\Lambda = -\Lambda$. Let $g \in L^2(\mathbb{R}^d)$, and let $\mathcal{G}(g, \Lambda) = \{g_{m,n} : (m,n) \in \Lambda\}$ be an orthonormal basis for $L^2(\mathbb{R}^d)$. If $P_j(g)$, $M_j(g) \in L^2(\mathbb{R}^d)$, then*

$$\langle g_{m,n}, P_j(g)\rangle = e^{2\pi i m\cdot n}\, \langle P_j(g), g_{-m,-n}\rangle \tag{5.9}$$

and

$$\langle g_{m,n}, M_j(g)\rangle = e^{2\pi i m\cdot n}\, \langle M_j(g), g_{-m,-n}\rangle. \tag{5.10}$$

The next result is a natural generalization of Theorem 5.1 to the case of several dimensions. It may be thought of as an extension of Corollary 5.5 to irregular sets Λ in the case of orthonormal bases and for standard coordinates. However, it will be proven later (see Theorem 5.11) that the particular choice of coordinates in Theorem 5.8 is not necessary and that analogous statements hold for operators P_v and M_w, for arbitrary v, $w \in \mathbb{R}^d$.

Theorem 5.8. *Let $\Lambda \subset \mathbb{R}^{2d}$ be a countable set with the property that $\Lambda = -\Lambda$. Assume that $\mathcal{G}(g, \Lambda) = \{g_{m,n} : (m,n) \in \Lambda\}$ is an orthonormal basis for $L^2(\mathbb{R}^d)$ for some $g \in L^2(\mathbb{R}^d)$. For any $j = 1, \ldots, d$,*

$$\|P_j(g)\|_2\, \|M_j(g)\|_2 = \infty. \tag{5.11}$$

Proof. We shall proceed by contradiction, and assume that there exists $j \in \{1, \ldots, d\}$ such that $P_j(g)$, $M_j(g) \in L^2(\mathbb{R}^d)$. Because of (5.9), (5.10), the representation property of bases, and the fact that $\Lambda = -\Lambda$, we obtain

$$\begin{aligned}\langle M_j(g), P_j(g)\rangle &= \sum_{(m,n)\in\Lambda} \langle M_j(g), g_{m,n}\rangle\, \langle g_{m,n}, P_j(g)\rangle \\ &= \sum_{(m,n)\in\Lambda} \langle g_{-m,-n}, M_j(g)\rangle\, \langle P_j(g), g_{-m,-n}\rangle \\ &= \langle P_j(g), M_j(g)\rangle.\end{aligned} \tag{5.12}$$

On the other hand, again using the classical result from [41] used in the proof of Theorem 5.3, we note that $M_i(g) \in L^2(\mathbb{R}^d)$ implies that $\partial g/\partial x^i$ exists a.e. Thus, using integration by parts, as in the proof of Theorem 5.3, yields

$$\langle M_j(g), P_j(g)\rangle = \langle P_j(g), M_j(g)\rangle + \frac{1}{2\pi i},$$

which, in turn, leads to a contradiction with the calculation (5.12). □

The stronger assumption that $\{g_{m,n} : (m,n) \in \Lambda\}$ is an orthonormal basis in Theorem 5.8 and not an exact frame, as it was in Corollary 5.5, is the price we have paid for extending the BLT from lattices to non-uniform sets. It is an open question as to whether Theorem 5.8 is valid for exact frames. The reader may think of this as yet another way in which the uncertainty principle manifests itself in mathematics.

5.2.3 Gabor Systems Associated with Symplectic Lattices

The standard *symplectic form* Ω on $\mathbb{R}^{2d}$ is defined as

$$\Omega((x,y),(\xi,\eta)) = x \cdot \eta - y \cdot \xi,$$

for any x, y, ξ, $\eta \in \mathbb{R}^d$. A matrix A is *symplectic* if it preserves the symplectic form Ω, i.e., $\Omega(Av, Aw) = \Omega(v, w)$, for all $v, w \in \mathbb{R}^{2d}$. The collection of all such matrices forms the *symplectic group*, and plays a significant role in different areas of mathematics, e.g., in the study of Hamiltonian systems. Following [34] we say that a lattice $\Lambda \subset \mathbb{R}^{2d}$ is *symplectic* if

$$\Lambda = rA(\mathbb{Z}^{2d})$$

for some $r \in \mathbb{R} \setminus \{0\}$ and A a symplectic matrix.

The class of symplectic lattices is important because it contains many examples of *non-separable* lattices, i.e., lattices that cannot be represented as a product of any other two lattices. Because of this, we need a different set of tools for dealing with Gabor systems associated with such non-separable lattices.

Let $w\colon \mathbb{R}^{2d} \to \mathbb{R}^+$ be a weight function with polynomial growth, and such that $w(x_1+x_2, \xi_1+\xi_2) \le w(x_1,\xi_1)\, w(x_2,\xi_2)$, for (x_1,ξ_1), $(x_2,\xi_2) \in \mathbb{R}^{2d}$. Given a Schwartz function $g \in \mathcal{S}(\mathbb{R}^d)$, and $1 \le p \le \infty$, we define the *modulation space* M^p_w to be the set of all tempered distributions $f \in \mathcal{S}'(\mathbb{R}^d)$ such that

$$\|f\|_{M^p_w} = \left(\int_{\mathbb{R}^{2d}} |S_g(f)(x,\xi)|^p\, w(x,\xi)^p \, dx\, d\xi \right)^{1/p} < \infty.$$

Different functions $g \in \mathcal{S}(\mathbb{R}^d)$ yield equivalent norms for M^p_w, e.g., see [32, Prop. 11.3.2]. For our purposes, modulation spaces associated with the following weights are relevant:

$$\forall\, j = 1, \dots, d, \quad m_j(x,\xi) = (1 + |x_j|^2 + |\xi_j|^2)^{1/2}$$

and

$$m(x,\xi) = (1 + |x|^2 + |\xi|^2)^{1/2}.$$

Indeed, if $g \in L^2(\mathbb{R}^d)$, then

$$g \in M^2_{m_j} \quad \Longleftrightarrow \quad \|P_j(g)\|_2\, \|M_j(g)\|_2 < \infty$$

and

$$g \in M_m^2 \quad \Longleftrightarrow \quad \left(\int_{\mathbb{R}^d} |x|^2 |g(x)|^2 \, dx \right) \left(\int_{\widehat{\mathbb{R}}^d} |\xi|^2 |\hat{g}(\xi)|^2 \, d\xi \right) < \infty,$$

see [34].

Theorem 10 in [34] gives the following generalization of the Balian–Low Theorem to the case of exact Gabor frames associated with a symplectic lattice.

Theorem 5.9. *Let $\Lambda \subset \mathbb{R}^{2d}$ be a symplectic lattice. Assume that $g \in L^2(\mathbb{R}^d)$, and that $\mathcal{G}(g, \Lambda) = \{g_{m,n} : (m,n) \in \Lambda\}$ forms an exact frame for $L^2(\mathbb{R}^d)$. Then,*

$$\exists\, j \in \{1, \ldots, d\} \text{ such that } g \notin M_{m_j}^2,$$

and further,

$$g \notin M_m^2.$$

We note here that, besides the assumption about the structure of the set Λ, the statement of Theorem 5.9 is weaker than the statement of Theorem 5.8 in the sense that it only claims existence of *some* $j \in \{1, \ldots, d\}$ for which $\|P_j(g)\|_2 \, \|M_j(g)\|_2 = \infty$. This brings up the open questions of whether the Balian–Low Theorem for exact frames for $L^2(\mathbb{R}^d)$ extends to sets Λ which are more general than symplectic lattices and whether $\|P_j(g)\|_2 \, \|M_j(g)\|_2 = \infty$ holds for all $j = 1, \ldots, d$.

The following result, different but closely related to Theorem 5.9, is due to Feichtinger and Gröchenig [25]. It also appears as Theorem 12 in [34], where it is shown to be a consequence of the Amalgam BLT.

Theorem 5.10. *Let $\Lambda \subset \mathbb{R}^{2d}$ be a symplectic lattice. If $g \in L^2(\mathbb{R}^d)$, and $\mathcal{G}(g, \Lambda) = \{g_{m,n} : (m,n) \in \Lambda\}$ forms an exact frame for $L^2(\mathbb{R}^d)$, then*

$$g \notin M_1^1.$$

Next, we show that the choice of coordinates in Theorem 5.8 (see also Theorem 5.9) is not canonical, i.e., there is no preference for choosing the directional derivatives ∂_j and multiplications by the standard basis coordinates x_j. This means that one can work in any representation of $\mathbb{R}^d$.

Before stating the next result, let us define the following differential operator associated with a vector $v = (v_1, \ldots, v_{2d}) \in \mathbb{R}^{2d}$, by its action on a function h as follows:

$$Q_v(h)(x) = \frac{1}{2\pi i} \nabla(h)(x) + f(x)h(x), \tag{5.13}$$

where

$$\nabla = \sum_{k=1}^{d} v_{k+d} \frac{\partial}{\partial x_k}$$

and

$$f(x) = \sum_{k=1}^{d} v_k x_k.$$

Theorem 5.11. *Let $\Lambda \subset \mathbb{R}^{2d}$ be a countable set with the property that $\Lambda = -\Lambda$. For a function $g \in L^2(\mathbb{R}^d)$, assume that $\mathcal{G}(g,\Lambda) = \{g_{m,n} : (m,n) \in \Lambda\}$ forms an orthonormal basis for $L^2(\mathbb{R}^d)$. For any two vectors v, $w \in \mathbb{R}^{2d}$ for which the symplectic form is non-vanishing, i.e.,*

$$\Omega(v,w) \neq 0,$$

we have

$$\|Q_v(g)\|_2 \, \|Q_w(g)\|_2 = \infty. \tag{5.14}$$

Remark 5.12. The above result is precisely Theorem 3.6 in [12]. The proof presented there depends on defining a generalized Fourier transform which, in turn, allows one to reduce a rather general and comprehensive problem to the Balian–Low Theorem in the standard coordinates as formulated in Theorem 5.8. The main motivation comes from the fact that for any two vectors v, $w \in \mathbb{R}^{2d}$,

$$[Q_v, Q_w] = \frac{1}{2\pi i} \Omega(v,w) \, \mathrm{Id}.$$

This approach is based on the quantum mechanical point of view and thus is both straightforward and natural.

Here, however, we present another elegant proof that was suggested to us by Professor Tim Steger. It is based on the *metaplectic representation* $\mu(A)$ of symplectic matrices, which is a unitary operator such that

$$\forall\, z \in \mathbb{R}^{2d}, \quad T_{A(z)} = \mu(A) T_z \mu(A)^{-1}.$$

Proof (of Theorem 5.11). Fix $j \in \{1,\dots,d\}$. Let S be a symplectic matrix. It follows from the definition of the metaplectic representation that, if $\{g_{m,n} : (m,n) \in \Lambda\}$ is an orthonormal basis for $L^2(\mathbb{R}^d)$, then so is $\{\mu(S)(g)_{m,n} : (m,n) \in S(\Lambda)\}$; cf. [32, Prop. 9.4.4].

Next, choose a symplectic matrix S such that

$$S(v) = e_j \quad \text{and} \quad S(w) = e_{d+j},$$

where $\{e_j : j = 1,\dots,2d\}$ is the standard basis in $\mathbb{R}^{2d}$. This can be done since $\Omega(v,w) \neq 0$.

Theorem 5.8 implies that

$$\|P_j(\mu(S)(g))\|_2 \, \|M_j(\mu(S)(g))\|_2 = \infty,$$

or, equivalently, because of the unitarity of the metaplectic representation,

$$\|\mu(S^{-1})\,(P_j(\mu(S)(g)))\,\|_2 \, \|\mu(S^{-1})\,(M_j(\mu(S)(g)))\,\|_2 = \infty. \tag{5.15}$$

Finally, we use the following invariance property of the Weyl calculus of pseudodifferential operators:

$$\mathrm{Op}(a \circ S^{-1}) = \mu(S^{-1}) \circ \mathrm{Op}(a) \circ \mu(S),$$

see, e.g., [46, XII.7.5*d*]. This observation asserts that the operators in (5.15) are in fact the operators Q_v and Q_w, respectively. □

In view of the above proof, the reader may think of Theorem 5.11 as a result which states that a change of variables is possible in Theorem 5.8, as long as the symplectic form remains non-zero.

We close this section with a generalization of the Weak BLT (Theorem 5.3) to the case of general lattices. This is Theorem 3.7 in [12].

Theorem 5.13. *Let $\Lambda \subset \mathbb{R}^{2d}$ be an arbitrary lattice. For a function $g \in L^2(\mathbb{R}^d)$, assume that $\mathcal{G}(g, \Lambda) = \{g_{m,n} : (m,n) \in \Lambda\}$ forms an exact frame for $L^2(\mathbb{R}^d)$ and let $\tilde{g}$ be the canonical dual to g. For any two vectors v, $w \in \mathbb{R}^{2d}$ for which the symplectic form is non-vanishing, we have*

$$\|Q_v(g)\|_2 \, \|Q_w(g)\|_2 \, \|Q_v(\tilde{g})\|_2 \, \|Q_w(\tilde{g})\|_2 = \infty.$$

5.2.4 Multiwindow Gabor Systems

Various attempts have been made to overcome the localization obstruction posed by the Balian–Low Theorem [47]. One of them was the question of Zeevi and Zibulski, who asked whether one can construct well-localized Gabor-type frames if several generating functions, instead of one, are used. More precisely, they considered systems of functions of the form

$$g^l_{m,n}(x) = e^{2\pi i x n b}\, g^l(x - ma), \quad m, n \in \mathbb{Z},\ l = 1, \ldots, L, \tag{5.16}$$

for some positive constants $0 < a, b$ and $0 < L < \infty$. The first difficulty is to find the right analogue of the critical density, which in the case of a Gabor system with one generator is 1. It was shown in [47] that, for multiply generated systems, one needs to consider

$$ab = L.$$

When this is the case we have yet another manifestation of the Balian–Low Theorem, see [47], [43].

Theorem 5.14. *Let $g^l \in L^2(\mathbb{R})$, $l = 1, \ldots, L$, and let $ab = L$. If the system $\{g^l_{m,n} : m, n \in \mathbb{Z},\ l = 1, \ldots, L\}$ is a frame for $L^2(\mathbb{R})$, then there exists $l \in \{1, \ldots, L\}$ such that*

$$\left(\int_{\mathbb{R}} |x|^2 |g^l(x)|^2 \, dx\right) \left(\int_{\widehat{\mathbb{R}}} |\xi|^2 |\hat{g}^l(\xi)|^2 \, d\xi\right) = \infty.$$

The proof of Theorem 5.14 is based on the method of Coifman, Daubechies, and Semmes [21].

A different approach to the study of Gabor-type systems associated with several generating functions is due to Balan, see e.g., [2], [3]. He studies frames and Riesz bases for the Hilbert space $L^2(\mathbb{R}, \mathbb{C}^N) = L^2(\mathbb{R}) \oplus \cdots \oplus L^2(\mathbb{R})$ with the inner product defined as

$$\langle\langle f_1 \oplus \cdots \oplus f_N, g_1 \oplus \cdots \oplus g_N \rangle\rangle = \sum_{j=1}^{N} \langle f_j, g_j \rangle,$$

where $\langle \cdot, \cdot \rangle$ is the standard inner product in $L^2(\mathbb{R})$. Gabor systems, also called *super Gabor sets* in this context, are the collections of ordered N-tuples of classical Gabor systems with different generators:

$$\mathcal{G}(g^1, \ldots, g^N; a, b) = \{g^1_{ma_1, nb_1} \oplus \cdots \oplus g^N_{ma_N, nb_N} : m, n \in \mathbb{Z}\}.$$

The notion of a frame or a Riesz basis for $L^2(\mathbb{R}, \mathbb{C}^N)$ is different than the notion of multiwindow Gabor frames of Zeevi and Zibulski. For example, if $\mathcal{G}(g^1, \ldots, g^N; a, b)$ is a frame for $L^2(\mathbb{R}, \mathbb{C}^N)$, then each of the systems $\{g^j_{ma_j, nb_j} : m, n \in \mathbb{Z}\}$, $j = 1, \ldots, N$, is a frame for $L^2(\mathbb{R})$, but the converse implication does not hold. Still, various versions of Balian–Low type theorems hold for Riesz bases in $L^2(\mathbb{R}, \mathbb{C}^N)$, see Lemma 3.1, Theorem 3.2, and Theorem 3.4 in [2].

5.3 Optimality of the Balian–Low Theorem

The results of Section 5.2, e.g., Corollary 5.6, show that various functions can be used as weights in the statement of the Balian–Low Theorem. It is therefore natural to ask if the BLT is sharp. That is, can one replace the weights $|x|^2$ and $|\xi|^2$ in (5.2) with something smaller? We shall show in this section that one cannot expect to have the Balian–Low "phenomenon" with powers of $|x|$ and $|\xi|$ less than 2.

Our construction uses the *Zak transform*, which is formally defined for a function $f \in L^2(\mathbb{R})$ as

$$Z(f)(x, \xi) = \sum_{m \in \mathbb{Z}} f(x + m)\, e^{-2\pi i m \xi} = \mathcal{F} f^x(\xi), \qquad (x, \xi) \in \mathbb{R} \times \mathbb{T},$$

where $f^x(k) = f(x+k)$, for $x \in [0,1]$ and $k \in \mathbb{Z}$, and where $\mathcal{F} f^x$ is the Fourier transform of the sequence $\{f^x(k)\}$. *Z is a unitary operator from $L^2(\mathbb{R})$ to* $L^2([0,1] \times \mathbb{T})$, and it has been used extensively in the study of Gabor systems, e.g., [39], [26], [27].

An elementary consequence of the unitarity of Z is the following criterion for Gabor orthonormal bases, e.g., [9].

Proposition 5.15. *Let $g \in L^2(\mathbb{R})$. The set of functions $\mathcal{G}(g,\mathbb{Z}^2) = \{g_{m,n} : (m,n) \in \mathbb{Z}\times\mathbb{Z}\}$ is an orthonormal basis for $L^2(\mathbb{R})$ if and only if $|Z(g)(x,\xi)| = 1$ a.e., for $(x,\xi) \in [0,1] \times \mathbb{T}$.*

In view of Proposition 5.15, we shall construct a function g so that $Z(g)(x,\xi)$ has the smallest possible *singular support*, i.e., the smallest possible closed set of points where the function is not infinitely differentiable. We expect that if $Z(g)$ only has a single singularity, then this will give rise to good decay properties for g and $\hat{g}$.

For our construction, we need to introduce a few auxiliary functions. Let $\varphi \in \mathcal{C}^\infty(\mathbb{R})$ be a function with the following properties:

$$\begin{aligned} &\varphi(x) = -1, && \text{for } x \in (-\infty, 0], \\ &\varphi(x) = 0, && \text{for } x \in [1, \infty), \\ &\varphi(x) \in [-1, 0], && \text{for all } x \in \mathbb{R}. \end{aligned}$$

Let $\psi(x) = \chi_{[0,\infty)}(x)\, x^a$, where $a > 0$ is a fixed number.

Given $0 < \varepsilon < \frac{1}{8}$, let $\gamma \in \mathcal{C}^\infty(\mathbb{R})$ be a function that satisfies

$$\operatorname{supp}(\gamma) \subseteq [-2\varepsilon, 2\varepsilon], \qquad \gamma(x) = 1 \text{ for } x \in [-\varepsilon, \varepsilon],$$

and

$$\gamma(x) \in [0,1] \text{ for all } x \in \mathbb{R}.$$

The next lemma (Lemma 2.2 in [11]) is the basis for our construction.

Lemma 5.16. *There exists a function $H\colon [-\frac{1}{2}, 1] \times [0,1]$, with the following properties (see Fig. 5.1):*

(1) *$H(x,\xi) = 0$ for $x \in [-\frac{1}{2}, 0]$;*

(2) *$H(x,\xi) = \varphi\big(\frac{\xi}{\psi(x)}\big)$ for $x \in (0, 2\varepsilon]$, where $\varepsilon > 0$ is as in the definition of γ;*

(3) *$H(x,0) = 0$ for $x \in [-\frac{1}{2}, 0]$ and $H(x,0) = -1$ for $x \in (0,1]$;*

(4) *$H(1+x,\xi) = H(x,\xi) + (\xi - 1)$ for $x \in [-\frac{1}{2}, 0]$;*

(5) *the function $e^{2\pi i H(x,\xi)}\colon [-\frac{1}{2}, 1] \times \mathbb{T} \to \mathbb{C}$ is of class $\mathcal{C}^\infty$ away from the points $(0,0)$ and $(1,0)$.*

Note that the function H has two singularities in $[-\frac{1}{2}, 1] \times [0,1]$. However, it is easy to find a unit square in the domain of H with a single point of singularity—the minimal singular support mentioned earlier.

Let $a > 0$ be fixed and define H as in Lemma 5.16, and set

$$h(x,\xi) = e^{2\pi i H(x,\xi)},$$

for $(x,\xi) \in [-\frac{1}{2}, 1] \times \mathbb{T}$. By property 4 of Lemma 5.16, the function h can be extended to $\mathbb{R} \times \mathbb{T}$ by means of the formula

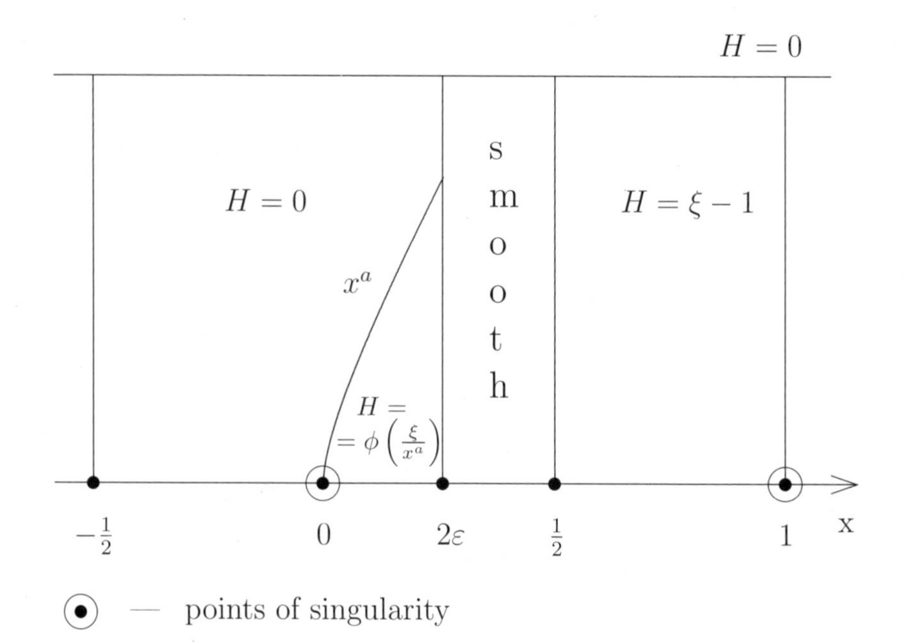

Fig. 5.1. The function H from Lemma 5.16.

$$h(k + x, \xi) = h(x, \xi)\, e^{2\pi i k(\xi - 1)} = h(x, \xi)\, e^{2\pi i k \xi}.$$

Finally, define the function $g = g_a$ on $\mathbb{R}$ by means of the inverse Zak transform of h, i.e.,

$$g(k + x) = Z^{-1}(h)(k + x), \tag{5.17}$$

where $k \in \mathbb{Z}$, $x \in [-\frac{1}{2}, \frac{1}{2})$. We write $g = g_a$ to emphasize the dependence of g on the constant a used above in the definition of ψ.

Notice that the function h, defined as an exponential function with imaginary exponent, is equal to 1 in absolute value. Thus, from Proposition 5.15, it follows that the function g, which is the inverse Zak transform of h, generates a Gabor orthonormal basis for $L^2(\mathbb{R})$. Moreover, it is not difficult to see that the function h extended periodically to $(0,1) \times \mathbb{R}$ is infinitely differentiable. Therefore, the function h extended to $\mathbb{R} \times \mathbb{R}$ is of class $\mathcal{C}^\infty$ away from the set $\mathbb{Z} \times \{0\}$. In view of Proposition 5.15 and the fact that there does not exist a smooth function h on $[0,1] \times \mathbb{T}$ with the property that $Z(h)(k + x, \xi) = Z(h)(x, \xi)\, e^{2\pi i k \xi}$, $k \in \mathbb{Z}$, a set like $\mathbb{Z} \times \{0\}$ is the smallest singular support that one expects for the Zak transform of a generator of a Gabor orthonormal basis.

We now state the main result of this section; cf. Theorem 3.11 in [11].

Theorem 5.17. *If $A = 1 + 1/a$, $B = a + 1$, and $\varepsilon > 0$, then the function $g = g_a \in L^2(\mathbb{R})$ defined in (5.17) has the property that $\mathcal{G}(g, \mathbb{Z}^2) = \{g_{m,n} : (m,n) \in \mathbb{Z} \times \mathbb{Z}\}$ is a Gabor orthonormal basis for $L^2(\mathbb{R})$, with*

$$\int_{\mathbb{R}} |g(x)|^2 \frac{1 + |x|^A}{\log^{1+\varepsilon}(2 + |x|)}\, dx < \infty,$$

and

$$\int_{\widehat{\mathbb{R}}} |\hat{g}(\xi)|^2 \frac{1 + |\xi|^B}{\log^{2+\varepsilon}(2 + |\xi|)}\, d\xi < \infty.$$

In particular, for $a = 1$, we have $A = B = 2$. Thus, the function $g = g_1$ is almost "optimal" from the point of view of Theorem 5.1; that is, both g and $\hat{g}$ are square integrable with respect to all weights $|x|^\alpha$, $\alpha < 2$. This means that one cannot make the Balian–Low Theorem any more restrictive by changing the exponents of the weights.

In addition, we also see that Theorem 5.17 yields the optimality of the multidimensional Balian–Low Theorem. In fact, it is easy to see that the function

$$g(x_1, \ldots, x_d) = g_1(x_1) \cdots g_1(x_d),$$

associated with the lattice $\Lambda = \mathbb{Z}^{2d}$, generates a Gabor orthonormal basis for $L^2(\mathbb{R}^d)$, and has the property that

$$\forall j = 1, \ldots, d,\ \forall \alpha \in (0,2),\ \left(\int_{\mathbb{R}^d} |x_j|^\alpha |g(x)|^2\, dx\right)\left(\int_{\widehat{\mathbb{R}}^d} |\xi_j|^\alpha |\hat{g}(\xi)|^2\, d\xi\right) < \infty.$$

This answers in the affirmative a question posed in [11] about optimality of the BLT in dimensions higher than 1. Additional results concerning optimality in the Balian–Low Theorem can be found in [14].

5.3.1 (p,q) Balian–Low Theorems

The choice of constants A and B in Theorem 5.17 yields

$$\frac{1}{A} + \frac{1}{B} = 1. \tag{5.18}$$

It is natural to ask, in this context, if given a generating function $g \in L^2(\mathbb{R})$ of a Gabor orthonormal basis or exact frame, and constants $1 < A,\ B < \infty$ such that (5.18) holds, is it possible that the time and frequency localization integrals

$$\int_{\mathbb{R}} |x|^A |g(x)|^2\, dx \quad \text{and} \quad \int_{\widehat{\mathbb{R}}} |\xi|^B |\hat{g}(\xi)|^2\, d\xi$$

are both finite? This question remains open in all cases other than the classical one, i.e., $A = B = 2$. We note the following corollary of the work of Feichtinger and Gröchenig in [30] and [25]; see also Theorem 3.14 in [44], and [13].

Theorem 5.18 (**(p,q) Balian–Low Theorem**). *Let $1 < p,\ q < \infty$ be such that $\frac{1}{p}+\frac{1}{q}=1$, and let $\varepsilon > 0$. Let $g \in L^2(\mathbb{R})$ have the property that $\mathcal{G}(g,\mathbb{Z}^2) = \{g_{m,n} : (m,n) \in \mathbb{Z}\times\mathbb{Z}\}$ is a Gabor orthonormal basis for $L^2(\mathbb{R})$. Then*

$$\left(\int_{\mathbb{R}} |g(x)|^2|x|^{(p+\varepsilon)}\,dx\right)\left(\int_{\widehat{\mathbb{R}}} |\hat{g}(\xi)|^2|\xi|^{(q+\varepsilon)}\,d\xi\right) = \infty. \tag{5.19}$$

Because ε is arbitrary, Theorem 5.17 again may be seen as the "optimality" result for the above version of the Balian–Low Theorem, in the sense that the exponents of the weights cannot be decreased below p and q, respectively.

The following extension to the case $(p,q)=(1,\infty)$ is studied in [13].

Theorem 5.19 (**$(1,\infty)$ Balian–Low Theorem**). *Let $g \in L^2(\mathbb{R})$ have the property that $\mathcal{G}(g,\mathbb{Z}^2) = \{g_{m,n} : (m,n) \in \mathbb{Z}\times\mathbb{Z}\}$ is a Gabor orthonormal basis for $L^2(\mathbb{R})$. Then*

$$\left(\int_{\mathbb{R}} |g(x)|^2|x|\,dx\right)\left(\sup_{N>0}\int_{\widehat{\mathbb{R}}} |\hat{g}(\xi)|^2|\xi|^{N}\,d\xi\right) = \infty. \tag{5.20}$$

An optimality result for this $(1,\infty)$ version of the Balian–Low Theorem is obtained by considering the function $g \in L^2(\mathbb{R})$ defined by $\hat{g}(\xi) = \chi_{[0,1]}(\xi)$. Then $\mathcal{G}(g,\mathbb{Z}^2) = \{g_{m,n} : (m,n) \in \mathbb{Z}\times\mathbb{Z}\}$ is a Gabor orthonormal basis for $L^2(\mathbb{R})$, and it is easy to verify that for all $0 \le \varepsilon < 1$,

$$\int_{\mathbb{R}} |x|^{1-\varepsilon}|g(x)|^2 dx < \infty \quad \text{and} \quad \sup_{N>0}\int_{\widehat{\mathbb{R}}.} |\xi|^N|\hat{g}(\xi)|^2 d\xi < \infty.$$

5.4 Wavelets and Uncertainty Principles

Although the Balian–Low Theorem and its generalizations are stated for Gabor systems, there are also similarly motivated results for other classes of orthonormal bases and exact frames. Zeevi and Zibulski [47] considered multiwindow systems of the form

$$\forall\, m,n \in \mathbb{Z} \text{ and } l = 1,\ldots,L, \quad g^l_{m,n,\phi}(x) = g^l(x-n)\,\phi_m(x),$$

where the kernels $\{\phi_m : m \in \mathbb{Z}\}$ form a frame for $L^2([0,1])$ and each function ϕ_m is 1-periodic. Gabor systems with the kernels $e^{2\pi imx}$ are just one example of such systems. The Balian–Low phenomenon still holds in this general case [47]. Another example of the universality of the BLT arises in the work of Balan and Daubechies [5], who studied properties of the optimizers of certain problems in stochastic signal analysis.

Even more general systems can be studied when the localization condition (5.2) is stated for all elements of the basis, not just for the generating function. Given $g \in L^2(\mathbb{R})$ of norm one, and $1 < a < \infty$, we define the *generalized variance* of g as

$$\Delta_a^2(g) = \inf_{x_0 \in \mathbb{R}} \int |x - x_0|^a |g(x)|^2 \, dx.$$

It is straightforward to verify, e.g., [10], that the elements of a Gabor system $\mathcal{G}(g, \mathbb{Z}^2) = \{g_{m,n} : (m,n) \in \mathbb{Z} \times \mathbb{Z}\}$ satisfy

$$\forall\, m, n \in \mathbb{Z}, \quad \Delta_a^2(g_{m,n}) = \Delta_a^2(g) \quad \text{and} \quad \Delta_a^2(\widehat{g_{m,n}}) = \Delta_a^2(\hat{g}). \tag{5.21}$$

In these terms, the Balian–Low Theorem states that if $\{g_{m,n} : (m,n) \in \mathbb{Z} \times \mathbb{Z}\}$ is a Gabor orthonormal basis for $L^2(\mathbb{R})$, then either all $\Delta_2^2(g_{m,n})$ are infinite, or all $\Delta_2^2(\widehat{g_{m,n}})$ are infinite. On the other hand, given $0 < a < 2$, Theorem 5.17 provides an example of a Gabor orthonormal basis $\{g_{m,n} : (m,n) \in \mathbb{Z} \times \mathbb{Z}\}$ such that

$$\sup_{m,n \in \mathbb{Z}} \Delta_a^2(g_{m,n}) < \infty \quad \text{and} \quad \sup_{m,n \in \mathbb{Z}} \Delta_a^2(\widehat{g_{m,n}}) < \infty. \tag{5.22}$$

Bourgain [19] constructed non-Gabor orthonormal bases $\{b_{m,n} : m, n \in \mathbb{Z}\}$ for $L^2(\mathbb{R})$ such that (5.22) holds with $a = 2$; see also [18] for a non-symmetric extension.

But perhaps the most striking similarity to the Balian–Low Theorem as we know it is evident from the work of Battle on localization of wavelets [7], [8]. An *a-adic wavelet* is a system that is generated from a single function $\psi \in L^2(\mathbb{R})$ by a countable family of translations and dilations, i.e.,

$$\forall\, x \in \mathbb{R}, \quad \psi_{j,k}(x) = a^{j/2} \psi(a^j x - k),$$

where j, $k \in \mathbb{Z}$ and a is some positive constant. It is well known that generators of such systems, e.g., *Meyer wavelets*, can possess good time-frequency localization, and thus they do not obey the restrictions posed by the Balian–Low Theorem as they are stated, e.g., in (5.2). However, the following result of Battle [7] shows that there are nonetheless some mild restrictions on the time-frequency localization of wavelets.

Theorem 5.20. *Let $\psi \in L^2(\mathbb{R})$. If the a-adic wavelet $\{\psi_{j,k} : j, k \in \mathbb{Z}\}$ is an orthogonal system, then the generating function ψ cannot have exponential localization in both time and frequency:*

$$\left(\int_{\mathbb{R}} |\psi(x)|^2 e^{|x|} \, dx \right) \left(\int_{\widehat{\mathbb{R}}} |\psi(\xi)|^2 e^{|\xi|} \, d\xi \right) = \infty.$$

The method of proof of Theorem 5.20 is quite different from the known proofs of the Balian–Low Theorem. It is based on a relation between smoothness and vanishing moments of orthogonal wavelets. Battle's result was later extended by Benedetto, Heil, and Walnut [15].

Dziubański and Hernández [24] have shown that this result is sharp. Indeed, they constructed a wavelet $\psi \in L^2(\mathbb{R})$ with subexponential decay, i.e.,

$$\forall\, \varepsilon > 0, \; \exists\, C > 0 \text{ such that } \forall\, x \in \mathbb{R}, \quad |\psi(x)| \leq C e^{-|x|^{1-\varepsilon}},$$

and such that $\hat{\psi}$ has compact support.

Another Balian–Low type result for wavelets is also due to Battle [8]; see also [4] for some extensions. An *a-adic wavelet state* is a wavelet $\{\psi_{j,k} : j,k \in \mathbb{Z}\}$ with the property that the sets $\Psi_j = \{\psi_{j,k} : k \in \mathbb{Z}\}$ are mutually orthogonal to each other. The following result is Theorem 1.1 in [8]. Recall that P and M are the classical position and momentum operators defined in Section 5.2.

Theorem 5.21. *Let $\psi \in L^2(\mathbb{R})$ and let $a > 0$. If $\{\psi_{j,k} : j,k \in \mathbb{Z}\}$ is an a-adic wavelet state, then*

$$\|P(\psi)\|_2 \|M(\psi)\|_2 \geq \frac{1}{4\pi} \frac{1}{1 - \frac{1}{\sqrt{a}}} \|\psi\|_2^2. \tag{5.23}$$

As is suggested by the use of position and momentum operators, Theorem 5.21 is in fact proved with standard uncertainty principle techniques. The key is the addition of the generator R of the group of dilations, in the sense of Stone's theorem:

$$e^{2\pi i a R}(\psi)(x) = e^{a/2} \psi(e^a x).$$

We note that when we let $a \to \infty$, then the constant in (5.23) tends to $1/(4\pi)$, which is the universal constant from the uncertainty principle (5.1).

It is interesting to note that completeness plays no role in the proofs of Battle's results, since only orthogonality is assumed.

Acknowledgments

Some of the results presented in this chapter are the outcome of our collaboration and discussions with John Benedetto, Przemek Gadziński, Andrei Maltsev, Tim Steger, and Jacob Sterbenz. We thank them very much. Of course, special thanks go to Professor John Benedetto for everything, in particular because, despite his well-known love for BLTs, he once joined the first named author for a burrito.

References

1. R. Balian, Un principe d'incertitude fort en théorie du signal ou en mécanique quantique, *C. R. Acad. Sci. Paris*, **292** (1981), pp. 1357–1362.
2. R. Balan, Extensions of no-go theorems to many signal systems, in: *Wavelets, Multiwavelets, and their Applications* (San Diego, 1997), Contemp. Math., Vol. 216, Amer. Math. Soc., Providence, RI, 1998, pp. 3–14.
3. R. Balan, Topological obstructions to localization results, in *Wavelets: Applications in Signal and Image Processing IX*, Proc. SPIE, Vol. 4478, SPIE, Bellingham, WA, 2001, pp. 184–191.

4. R. Balan, An uncertainty inequality for wavelet sets, *Appl. Comput. Harmon. Anal.*, **5** (1998), pp. 106–108.
5. R. Balan and I. Daubechies, Optimal stochastic encoding and approximation schemes using Weyl–Heisenberg sets, in: *Advances in Gabor Analysis*, H. G. Feichtinger and T. Strohmer, eds., Birkhäuser, Boston, 2003, pp. 259–320.
6. G. Battle, Heisenberg proof of the Balian–Low theorem, *Lett. Math. Phys.*, **15** (1988), pp. 175–177.
7. G. Battle, Phase space localization theorem for ondelettes, *J. Math. Phys.*, **30** (1989), pp. 2195–2196.
8. G. Battle, Heisenberg inequalities for wavelet states, *Appl. Comput. Harmon. Anal*, **4** (1997), pp. 119–146.
9. J. J. Benedetto, Gabor representations and wavelets, in: *Commutative Harmonic Analysis* (Canton, NY, 1987), Contemp. Math., Vol. 91, Amer. Math. Soc., Providence, RI, 1989, pp. 9–27.
10. J. J. Benedetto, Frame decompositions, sampling, and uncertainty principles, in: *Wavelets: Mathematics and Applications*, J. J. Benedetto and M. W. Frazier, eds., CRC Press, Boca Raton, FL, 1994, pp. 247–304.
11. J. J. Benedetto, W. Czaja, P. Gadziński, and A. M. Powell, The Balian–Low Theorem and regularity of Gabor systems, *J. Geom. Anal.*, **13** (2003), pp. 217–232.
12. J. J. Benedetto, W. Czaja, and A. Ya. Maltsev, The Balian–Low theorem for the symplectic form on $\mathbb{R}^{2d}$, *J. Math. Phys.*, **44** (2003), pp. 1735–1750.
13. J. J. Benedetto, W. Czaja, A. M. Powell, and J. Sterbenz, A $(1, \infty)$ Balian–Low theorem, *Math. Res. Lett.*, **13** (2006), pp. 467–474.
14. J. J. Benedetto, W. Czaja, and A. M. Powell, An optimal example for the Balian–Low Uncertainty Principle, *SIAM J. Math. Anal.*, **38** (2006), pp. 333–345.
15. J. J. Benedetto, C. Heil, and D. F. Walnut, Uncertainty principles for time-frequency operators, in: *Continuous and Discrete Fourier Transforms, Extension Problems and Wiener–Hopf Equations*, Oper. Theory Adv. Appl., Vol. 58, I. Gohberg, ed., Birkhäuser, Basel, 1992, pp. 1–25.
16. J. J. Benedetto, C. Heil, and D. F. Walnut, Differentiation and the Balian–Low Theorem, *J. Fourier Anal. Appl.*, **1** (1995), pp. 355–402.
17. J. J. Benedetto, C. Heil, and D. F. Walnut, Gabor systems and the Balian–Low Theorem, in *Gabor Analysis and Algorithms*, H. G. Feichtinger and T. Strohmer, eds., Birkhäuser, Boston, MA, 1998, pp. 85–122.
18. J. J. Benedetto and A. M. Powell, A (p, q) version of Bourgain's theorem, *Trans. Amer. Math. Soc.*, **358** (2006), pp. 2489–2505.
19. J. Bourgain, A remark on the uncertainty principle for Hilbertian basis, *J. Func. Anal.*, **79** (1988), pp. 136–143.
20. O. Christensen, B. Deng, and C. Heil, Density of Gabor frames, *Appl. Comput. Harmon. Anal.*, **7** (1999), pp. 292–304.
21. I. Daubechies, The wavelet transform, time-frequency localization and signal analysis, *IEEE Trans. Inform. Theory*, **36** (1990), pp. 961–1005.
22. I. Daubechies, *Ten Lectures on Wavelets*, SIAM, Philadelphia, 1992.
23. I. Daubechies and A. J. E. M. Janssen, Two theorems on lattice expansions, *IEEE Trans. Inform. Theory*, **39** (1993), pp. 3–6.
24. J. Dziubański and E. Hernández, Band-limited wavelets with subexponential decay, *Canad. Math. Bull.*, **41** (1998), pp. 398–403.

25. H. G. Feichtinger and K. Gröchenig, Gabor frames and time-frequency analysis of distributions, *J. Funct. Anal.*, **146** (1997), pp. 464–495.
26. H. G. Feichtinger and T. Strohmer, Eds., *Gabor Analysis and Algorithms: Theory and Applications*, Birkhäuser, Boston, MA, 1998.
27. H. G. Feichtinger and T. Strohmer, Eds., *Advances in Gabor Analysis*, Birkhäuser, Boston, MA, 2003.
28. G. B. Folland and A. Sitaram, The uncertainty principle: A mathematical survey, *J. Fourier Anal. Appl.*, **3** (1997), pp. 207–238.
29. J.-P. Gabardo and D. Han, J.-P. Gabardo and D. Han, Balian–Low phenomenon for subspace Gabor frames, *J. Math. Phys.*, **45** (2004), pp. 3362–3378.
30. K. Gröchenig, An uncertainty principle related to the Poisson summation formula, *Studia. Math.*, **121** (1996), pp. 87–104.
31. K. Gröchenig, Aspects of Gabor analysis on locally compact abelian groups, in: *Gabor Analysis and Algorithms*, H. G. Feichtinger and T. Strohmer, eds., Birkhäuser, Boston, 1998, pp. 211–231.
32. K. Gröchenig, *Foundations of Time-Frequency Analysis*, Birkhäuser, Boston, 2001.
33. K. Gröchenig, Uncertainty principles for time-frequency representations, in: *Advances in Gabor Analysis*, H. G. Feichtinger and T. Strohmer, eds., Birkhäuser, Boston, 2003, pp. 11–30.
34. K. Gröchenig, D. Han, C. Heil, and G. Kutyniok, The Balian–Low Theorem for symplectic lattices in higher dimensions, *Appl. Comput. Harmon. Anal.*, **13** (2002), pp. 169–176.
35. V. Havin and B. Jöricke, *The Uncertainty Principle in Harmonic Analysis*, Springer-Verlag, Berlin, 1994.
36. C. Heil, Wiener Amalgam Spaces in Generalized Harmonic Analysis and Wavelet Theory, Ph.D. Thesis, University of Maryland, College Park, MD, 1990.
37. C. Heil, Linear independence of finite Gabor systems, Chapter 9, this volume (2006).
38. T. Høholdt, H. Jensen, and J. Justesen, Double series representation of bounded signals, *IEEE Trans. Inform. Theory*, **34** (1988), pp. 613–624.
39. A. J. E. M. Janssen, The Zak transform: a signal transform for sampled time-continuous signals, *Philips J. Res.*, **43** (1988), pp. 23–69.
40. F. Low, Complete sets of wave packets, in: *A Passion for Physics—Essays in Honor of Geoffrey Chew*, C. DeTar, J. Finkelstein, and C. I. Tan, eds., World Scientific, Singapore, 1985, pp. 17–22.
41. V. Maz'ja, *Sobolev Spaces*, Springer-Verlag, New York, 1985.
42. A. Messiah, *Quantum Mechanics*, Interscience, New York, 1961.
43. M. Porat, Y. Y. Zeevi, and M. Zibulski, Multi-window Gabor schemes in signal and image representations, in: *Gabor Analysis and Algorithms*, H. G. Feichtinger and T. Strohmer, eds., Birkhäuser, Boston, 1998, pp. 381–408.
44. A. M. Powell, The Uncertainty Principle in Harmonic Analysis and Bourgain's Theorem, Ph.D. Thesis, University of Maryland, College Park, MD, 2003.
45. J. Ramanathan and T. Steger, Incompleteness of sparse coherent states, *Appl. Comput. Harmon. Anal.*, **2** (1995), pp. 148–153.
46. E. M. Stein. *Harmonic Analysis: Real-Variable Methods, Orthogonality, and Oscillatory Integrals*, Princeton University Press, Princeton, NJ, 1993.
47. Y. Y. Zeevi and M. Zibulski, Analysis of multiwindow Gabor-type schemes by frame methods, *Appl. Comput. Harmon. Anal.*, **4** (1997), pp. 188–221.

6

Some Problems Related to the Distributional Zak Transform

Jean-Pierre Gabardo

Department of Mathematics and Statistics, McMaster University, Hamilton, Ontario L8S 4K1, Canada
gabardo@mcmaster.ca

Summary. We define the distributional Zak transform and study some of its properties. We show how the distributional Zak transform can be used as an effective tool in the theory of Gabor systems where the window function belongs to the Schwartz class $\mathcal{S}(\mathbb{R})$ and where the product of the parameters defining the Gabor system is rational. In particular, we obtain a necessary and sufficient condition for the linear span of such a Gabor system to be dense in $\mathcal{S}(\mathbb{R})$ in the topology of $\mathcal{S}(\mathbb{R})$ and, if this is the case, we show that a dual window in the Schwartz class can be constructed. We also characterize when such a Gabor system satisfies the Riesz property.

Dedicated to John Benedetto.

6.1 Introduction

In 1987, I was a Ph.D. student of John J. Benedetto writing my thesis on a topic involving distributions and Fourier transforms. More recently, I became interested in Gabor systems and did some work on them jointly with Deguang Han, mainly in the context of $L^2(\mathbb{R}^d)$. Hence, since this is a paper in honor of John, I thought it would be appropriate for my contribution to include both topics. So the obvious choice was to consider Gabor systems in distributional spaces. Since the Zak transform is such a powerful tool for dealing with Gabor systems in $L^2(\mathbb{R})$, it is natural to want to extend it to spaces of distributions, in particular to the space of tempered distributions on $\mathcal{S}'(\mathbb{R})$. This was actually done by Janssen in 1982 [13] by formally replacing the function $g \in L^2(\mathbb{R})$ in formula (6.1) below by a distribution in $\mathcal{S}'(\mathbb{R})$ and showing that the corresponding series, where each term is viewed as a tensor product of distributions of one variable, converges in $\mathcal{S}'(\mathbb{R}^2)$. Another point of view, using the theory of Boehmians, was also used more recently in [18] to define the Zak transform of a tempered distribution in a more algebraic way. In this chapter, we will follow a somewhat different approach which emphasizes the role that duality plays in the theory. Our goal in this paper is thus first to develop our approach to the distributional Zak transform and secondly to consider certain problems whose solutions require in a natural way the use of the distributional Zak transform.

In Section 6.2, we recall the definition of the Zak transform in $L^2(\mathbb{R})$ and characterize the image of the Schwartz space $\mathcal{S}(\mathbb{R})$. The Zak transform of a distribution in $\mathcal{S}'(\mathbb{R})$ is defined by duality and the image of $\mathcal{S}'(\mathbb{R})$ is also characterized. We will show that, in both cases, the Zak transform is in fact a topological isomorphism between the original space and its image under the Zak transform. In Section 6.3, we consider Gabor systems on $\mathbb{R}$ with associated parameters α, $\beta > 0$, $\alpha\beta \in \mathbb{Q}$, and a window in the Schwartz class $\mathcal{S}(\mathbb{R})$, and we characterize in terms of the Zak transform of the window when the span of the corresponding Gabor system is dense in $\mathcal{S}(\mathbb{R})$ in the topology of $\mathcal{S}(\mathbb{R})$. This can only happen when $\alpha\beta < 1$. In contrast, it is well known (see [1], [5], [16]) that if the window is taken in $L^2(\mathbb{R})$, the fact that the span of the corresponding Gabor system is dense in $L^2(\mathbb{R})$ implies that $\alpha\beta \leq 1$ and the case $\alpha\beta = 1$ can certainly occur.

We also consider the question as to whether or not such a system satisfies the Riesz property. We again characterize whether or not the system satisfies this property in terms of the Zak transform of the window function and show that this property can only be satisfied when $\alpha\beta > 1$. Note that the validity of the Riesz property for a Gabor system in $L^2(\mathbb{R})$ (with the obvious changes in the definition of Riesz property) implies that $\alpha\beta \geq 1$ and the case $\alpha\beta = 1$ can certainly occur (for example, for an orthonormal Gabor system). Finally, we show in Section 6.4, again in the case where $\alpha\beta \in \mathbb{Q}$, that if the linear span of a Gabor system with a window in $\mathcal{S}(\mathbb{R})$ is dense in $\mathcal{S}(\mathbb{R})$, then the system admits a dual window which is also in $\mathcal{S}(\mathbb{R})$ and allows for a reconstruction formula valid for every tempered distribution on $\mathbb{R}$, given in terms of a series converging in the topology of $\mathcal{S}'(\mathbb{R})$. We refer the reader to [4] and [9] for the known results and references on Gabor systems, especially in the irrational case.

We note that we will sometimes use the notation $T(t)$ (resp. $R(t,\nu)$), for a distribution on $\mathbb{R}$ (resp. $\mathbb{R}^2$), which is a slight abuse of notation, but is often very useful, in particular when considering translates of such distributions.

6.2 The Distributional Zak Transform

The Zak transform is an important tool widely used in the theory of Gabor or Weyl–Heisenberg systems (see, for example, [6], [7], [10]) and, since such systems appear typically within the framework of $L^2(\mathbb{R})$, the Zak transform is usually defined only on the space $L^2(\mathbb{R})$. It is perhaps not widely known that the Zak transform can also be defined on the space of tempered distributions on the real line, $\mathcal{S}'(\mathbb{R})$. The definition of the Zak transform on the space $\mathcal{S}'(\mathbb{R})$ is due to Janssen [13].

In this section, we will first recall the definition of the Zak transform on $L^2(\mathbb{R})$, state its main properties and then we will show how this definition and those properties can be extended to the space $\mathcal{S}'(\mathbb{R})$. In doing so, we will follow a different approach than that of Janssen, which perhaps emphasizes more the role of the duality between distributions and test functions.

Definition 6.1. *If $\alpha > 0$ is a parameter and $g \in L^2(\mathbb{R})$, we denote by $\mathcal{Z}_\alpha g$ the Zak transform of g defined by*

$$\mathcal{Z}_\alpha g(t,\nu) = \alpha^{-1/2} \sum_{k\in\mathbb{Z}} g\Big(\frac{t-k}{\alpha}\Big)\, e^{2\pi i k\nu}, \tag{6.1}$$

for $(t,\nu) \in \mathbb{R}^2$.

Note that the series in (6.1) converges in $L^2_{\text{loc}}(\mathbb{R}^2)$. Thus, the Zak transform maps functions on the line to functions on the plane and, in fact, maps $L^2(\mathbb{R})$ functions to a subspace of $L^2_{\text{loc}}(\mathbb{R}^2)$. It is easily shown that $\mathcal{Z}_\alpha$ is a unitary map from $L^2(\mathbb{R})$ to $L^2(I\times I)$, where $I = [0,1]$. We have thus, for any $f, g \in L^2(\mathbb{R})$,

$$\int_{\mathbb{R}} f(x)\,\overline{g(x)}\,dx = \iint_{I^2} \mathcal{Z}_\alpha f(t,\nu)\,\overline{\mathcal{Z}_\alpha g(t,\nu)}\,dt\,d\nu. \tag{6.2}$$

If the Zak transform of some $L^2(\mathbb{R})$-function is known on the unit square $I\times I$, then its values on all of $\mathbb{R}^2$ can be recovered almost everywhere on $\mathbb{R}^2$ using the *quasi-periodicity* relations that the Zak transform of a function must satisfy: if $G(t,\nu) = \mathcal{Z}_\alpha g(t,\nu)$, then

$$G(t+m,\nu) = e^{2\pi i m\nu}\, G(t,\nu), \quad m \in \mathbb{Z}, \tag{6.3}$$

$$G(t,\nu+n) = G(t,\nu), \quad n \in \mathbb{Z}. \tag{6.4}$$

The inverse Zak transform is defined on the space of locally square-integrable functions G on $\mathbb{R}^2$ satisfying the quasi-periodicity relations (6.3) and (6.4) by the formula

$$\big(\mathcal{Z}_\alpha^{-1} G\big)(x) = \alpha^{1/2} \int_I G(\alpha\, x,\nu)\, d\nu, \quad x \in \mathbb{R}. \tag{6.5}$$

Another important property of the Zak transform is the following: for any $g \in L^2(\mathbb{R})$, we have

$$\mathcal{Z}_\alpha\Big[e^{2\pi i m\alpha x}\, g\Big(x - \frac{n}{\alpha}\Big)\Big](t,\nu) = e^{2\pi i m t}\, e^{-2\pi i n\nu}\, \mathcal{Z}_\alpha g(t,\nu), \quad m,n \in \mathbb{Z}. \tag{6.6}$$

Our next goal is to use duality and formula (6.2) to define the Zak transform of a tempered distribution on $\mathbb{R}$ (see [17], [3] for the definition and properties of tempered distributions). In order to do so, we first have to study the Zak transform of the functions in the Schwartz space $\mathcal{S}(\mathbb{R})$. Recall that $\mathcal{S}(\mathbb{R}^d)$ consists of the complex-valued infinitely differentiable functions $\varphi(x)$ on $\mathbb{R}^d$ satisfying

$$||\varphi||_{m,k} := \sup_{\substack{x\in\mathbb{R}^d \\ |\beta|\le k}} |(1+\|x\|^2)^m\, D^\beta \varphi(x)| < \infty, \quad \text{for all } m,k\in\mathbb{N}, \tag{6.7}$$

where $\beta = (\beta_1, \dots, \beta_n)$ is a multi-index and $\|\cdot\|$ denotes the standard Euclidean norm on $\mathbb{R}^d$. The seminorms defined by the right-hand side of (6.7) can be used to make $\mathcal{S}(\mathbb{R}^d)$ a topological vector space (see [17]), in fact a Fréchet space, but in what follows, we will only be concerned with convergence of sequences in $\mathcal{S}(\mathbb{R}^d)$. A sequence of functions $\{\varphi_n\}_{n\ge 0}$ in $\mathcal{S}(\mathbb{R}^d)$ is said to converge in $\mathcal{S}(\mathbb{R}^d)$ to the function $\varphi \in \mathcal{S}(\mathbb{R}^d)$ if $||\varphi_n - \varphi||_{m,k} \to 0$ as $n\to\infty$ for all $(m,k)\in\mathbb{N}^2$. The subspace of $\mathcal{S}(\mathbb{R}^d)$ consisting of all complex-valued infinitely differentiable functions on $\mathbb{R}^d$ having compact support will be denoted by $C_0^\infty(\mathbb{R}^d)$.

Definition 6.2. *We will denote by $\mathcal{A}(\mathbb{R}^2)$ the space of complex-valued infinitely differentiable functions Φ on $\mathbb{R}^2$ satisfying the quasi-periodicity relations* (6.3) *and* (6.4). *A sequence $\{\Phi_n\}_{n\ge 0}$ in $\mathcal{A}(\mathbb{R}^2)$ converges in $\mathcal{A}(\mathbb{R}^2)$ to the function $\Phi \in \mathcal{A}(\mathbb{R}^2)$ if, for all multi-index $\beta\in\mathbb{N}^2$,*

$$D^\beta \Phi_n \to D^\beta \Phi, \quad \text{as} \quad n\to\infty,$$

uniformly on compact subsets of $\mathbb{R}^2$.

Note that, because of the identities (6.3) and (6.4), we can estimate on any compact subset of $\mathbb{R}^2$ the magnitude of a partial derivative of order m of a function in $\mathcal{A}(\mathbb{R}^2)$ in terms of the magnitude of the partial derivatives of order at most m of the same function on $I\times I$. This fact is used in proving the following embedding result. The details are left to the reader.

Proposition 6.3. *The space $\mathcal{A}(\mathbb{R}^2)$ is contained in the space of infinitely differentiable functions on $\mathbb{R}^2$ with polynomial growth, i.e., if $\Phi\in\mathcal{A}(\mathbb{R}^2)$, there exists, for each multi-index $\beta\in\mathbb{N}^2$, an integer $m(\beta)\ge 0$ such that*

$$\sup_{x\in\mathbb{R}^2} \frac{|D^\beta\Phi(x)|}{[1+\|x\|^2]^{m(\beta)}} < \infty.$$

In fact, the following more precise estimate is true: if $\beta = (k,l)\in\mathbb{N}^2$, there exists a constant $C(k)$ such that, for all $(t,\nu)\in\mathbb{R}^2$, we have

$$|D^\beta\Phi(t,\nu)| \le C(k)\,(1+|t|^2)^{k/2} \sup_{\substack{(t',\nu')\in I^2 \\ 0\le j\le k}} \left|\left(\frac{\partial}{\partial t}\right)^j \left(\frac{\partial}{\partial \nu}\right)^l \Phi(t',\nu')\right|. \tag{6.8}$$

The following result originally due to Janssen [13] shows that the image of $\mathcal{S}(\mathbb{R})$ under the Zak transform is precisely the space $\mathcal{A}(\mathbb{R}^2)$.

Theorem 6.4. *The Zak transform $\mathcal{Z}_\alpha$ maps $\mathcal{S}(\mathbb{R})$ onto $\mathcal{A}(\mathbb{R}^2)$. Moreover, the mapping $\mathcal{Z}_\alpha : \mathcal{S}(\mathbb{R}) \to \mathcal{A}(\mathbb{R}^2)$ is a topological isomorphism.*

Proof. Elementary arguments using Definition 6.1 show that the Zak transform of a function in $\mathcal{S}(\mathbb{R})$ has continuous partial derivatives of all orders. Since it satisfies (6.3) and (6.4), it must belong to $\mathcal{A}(\mathbb{R}^2)$. Furthermore, if $\varphi \in \mathcal{S}(\mathbb{R})$ and $\Phi = \mathcal{Z}_\alpha \varphi$, then

$$\left(\frac{\partial}{\partial t}\right)^k \left(\frac{\partial}{\partial \nu}\right)^l \Phi(t,\nu) = \alpha^{-(k+1/2)} \sum_{j\in\mathbb{Z}} D^k \varphi\Big(\frac{t-j}{\alpha}\Big)\,(2\pi i j)^l\, e^{2\pi i j\nu},$$

from which it follows easily that, for each $(k,l) \in \mathbb{N}^2$, there exists a constant $C_{k,l} > 0$ such that

$$\sup_{(t,\nu)\in I\times I} \left|\left(\frac{\partial}{\partial t}\right)^k \left(\frac{\partial}{\partial \nu}\right)^l \Phi(t,\nu)\right| \le C_{k,l}\, ||\varphi||_{k,l+1}.$$

This proves the continuity of the mapping $\mathcal{Z}_\alpha : \mathcal{S}(\mathbb{R}) \to \mathcal{A}(\mathbb{R}^2)$.

Conversely, if $\Phi \in \mathcal{A}(\mathbb{R}^2)$, let $\psi := \mathcal{Z}_\alpha^{-1}\Phi$. Since $\Phi \in C^\infty(\mathbb{R}^2)$, we have

$$\psi^{(k)}(t) = \frac{d}{dt^k}\left\{\alpha^{1/2} \int_I \Phi(\alpha\, t,\nu)\, d\nu\right\} = \alpha^{1/2} \int_I \frac{\partial^k}{\partial t^k}\, [\Phi(\alpha\, t,\nu)]\, d\nu,$$

for all $k \in \mathbb{N}$ and $\psi^{(k)} \in L^2(\mathbb{R})$ for all such k. Furthermore, letting

$$(\mathcal{T}\Phi)(t,\nu) := t\,\Phi(t,\nu) - \frac{1}{2\pi i}\frac{\partial}{\partial \nu}\Phi(t,\nu),$$

it is easily checked that $\mathcal{T}\Phi \in A(\mathbb{R}^2)$. Since, for every $(k,l) \in \mathbb{N}^2$, there exists a constant $D_{k,l} > 0$ such that

$$\left\|\left(\frac{\partial}{\partial t}\right)^k \left(\frac{\partial}{\partial \nu}\right)^l (\mathcal{T}\Phi)\right\|_{L^\infty(I\times I)} \le D_{k,l} \left\|\left(\frac{\partial}{\partial t}\right)^k \left(\frac{\partial}{\partial \nu}\right)^{l+1} \Phi\right\|_{L^\infty(I\times I)},$$

the mapping $\mathcal{T} : A(\mathbb{R}^2) \to A(\mathbb{R}^2)$ is continuous. Also,

$$\begin{aligned}\alpha^{1/2} \int_I (\mathcal{T}\Phi)\,(\alpha\, t,\nu)\, d\nu &= \alpha^{1/2} \left\{\int_I \alpha\, t\, \Phi(\alpha\, t,\nu)\, d\nu - \frac{1}{2\pi i}\Big[\phi(\alpha t,\nu)\Big]_{\nu=0}^{\nu=1}\right\} \\ &= \alpha^{1/2} \int_I \alpha\, t\, \Phi(\alpha\, t,\nu)\, d\nu = \alpha\, t\, \psi(t).\end{aligned}$$

Hence, $t\,\psi(t) \in L^2(\mathbb{R})$, and iterating this procedure shows that, more generally, $t^l\,\psi(t) \in L^2(\mathbb{R})$ for all $l \in \mathbb{N}$ and thus $\psi \in \mathcal{S}(\mathbb{R})$. The previous estimates easily show that the mapping $\mathcal{Z}_\alpha^{-1} : A(\mathbb{R}^2) \to S(\mathbb{R})$ is continuous. □

Suppose now that $G \in L^2_{\text{loc}}(\mathbb{R}^2)$ satisfies the equations (6.3) and (6.4) and let $\Psi \in \mathcal{S}(\mathbb{R}^2)$. Denoting by $\langle\cdot,\cdot\rangle$ the duality between distributions in $\mathcal{S}'(\mathbb{R}^d)$ and test functions in the Schwartz space $\mathcal{S}(\mathbb{R}^d)$, we have

$$\begin{aligned}\langle G,\Psi\rangle &= \iint_{\mathbb{R}^2} G(t,\nu)\,\Psi(t,\nu)\,dt\,d\nu \\ &= \int_{I\times I} G(t,\nu)\sum_{k,l\in\mathbb{Z}^2} e^{2\pi ik\nu}\,\Psi(t+k,\nu+l)\,dt\,d\nu \\ &= \int_{I\times I} G(t,\nu)\,(\mathcal{K}\Psi)(t,\nu)\,dt\,d\nu, \end{aligned} \tag{6.9}$$

where

$$(\mathcal{K}\Psi)(t,\nu) := \sum_{k,l\in\mathbb{Z}^2} e^{2\pi ik\nu}\,\Psi(t+k,\nu+l). \tag{6.10}$$

Proposition 6.5. *If $\Psi\in\mathcal{S}(\mathbb{R}^2)$, then $\overline{\mathcal{K}\Psi}\in\mathcal{A}(\mathbb{R}^2)$, where $\mathcal{K}\Psi$ is defined in equation* (6.10). *Furthermore, the mapping $\mathcal{S}(\mathbb{R}^2)\to\mathcal{A}(\mathbb{R}^2):\Psi\mapsto\overline{\mathcal{K}\Psi}$ is continuous.*

Proof. We first notice that if for some number $R>1$, we have $\|(t,\nu)\|\le R$ and $\|(k,l)\|\ge 2\,R$, then

$$\|(k,l)\|\ge\|(t+k,\nu+l)\|-\|(t,\nu)\|\ge\|(t+k,\nu+l)\|-R\ge\|(t+k,\nu+l)\|-\frac{\|(k,l)\|}{2},$$

and thus $\|(k,l)\|\ge\|(t+k,\nu+l)\|/2$. In particular, if $(t,\nu)\in I^2=I\times I$, we can take $R=\sqrt{2}$ and deduce that

$$\begin{aligned}\sum_{\substack{k,l\in\mathbb{Z}\\ k^2+l^2\ge 8}}\frac{1}{(k^2+l^2)^2} &= \sum_{\substack{k,l\in\mathbb{Z}\\ k^2+l^2\ge 8}}\iint_{I^2}\frac{1}{(k^2+l^2)^2}\,dt\,d\nu \\ &\le \sum_{\substack{k,l\in\mathbb{Z}\\ k^2+l^2\ge 8}}\iint_{I^2}\frac{16}{((k+t)^2+(l+\nu)^2)^2}\,dt\,d\nu \\ &\le \iint_{x^2+y^2\ge 2}\frac{16}{(x^2+y^2)^2}\,dx\,dy<\infty. \end{aligned} \tag{6.11}$$

Note also that, if $\Psi\in\mathcal{S}(\mathbb{R}^2)$, $R>1$ and $m\in\mathbb{N}$, then we have, for $\|(t,\nu)\|\le R$,

$$\begin{aligned}&\sum_{k,l\in\mathbb{Z}}|k|^m\,|\Psi(t+k,\nu+l)|\le\sum_{k,l\in\mathbb{Z}}|k|^m\,\frac{\|\Psi\|_{m+2,0}}{[1+(t+k)^2+(\nu+l)^2]^{m+2}} \\ &\le\|\Psi\|_{m+2,0}\left\{\sum_{\|(k,l)\|\le 2\,R}|k|^m+\sum_{\|(k,l)\|>2\,R}\frac{4\,|k|^m}{[(k)^2+(l)^2]^{m+2}}\right\} \\ &=C(R,m)\,\|\Psi\|_{m+2,0}, \end{aligned} \tag{6.12}$$

where $C(R,m)$ is a finite (because of (6.11)) positive constant. If $p,q\in\mathbb{N}$ and $\beta=(p,q)$, we have

$$D^{\beta}\left(\mathcal{K}\Psi\right)(t,\nu)=\sum_{k,l\in\mathbb{Z}}\sum_{j=0}^{q}\binom{q}{j}(2\pi ik)^{q-j}\,e^{2\pi ik\nu}\,D^{\beta}\Psi(t+k,\nu+l).$$

Thus, using (6.12), it follows that, if $R>1$ and $\|(t,\nu)\|\leq R$,

$$\begin{aligned}|D^{\beta}\left(\mathcal{K}\Psi\right)(t,\nu)| &\leq \sum_{k,l\in\mathbb{Z}}\sum_{j=0}^{q}\binom{q}{j}(2\pi k)^{q-j}\,|D^{\beta}\Psi(t+k,\nu+l)| \\ &\leq \sum_{j=0}^{q}\binom{q}{j}(2\pi)^{q-j}\,C(R,q-j)\,\|D^{\beta}\Psi\|_{q-j+2,0} \\ &\leq D(R,|\beta|)\,\|\Psi\|_{q+2,p+q},\end{aligned} \tag{6.13}$$

where $D(R,|\beta|)$ is a positive constant. This shows that $\mathcal{K}\Psi$ and thus also $\overline{\mathcal{K}\Psi}$ belong to $C^{\infty}(\mathbb{R}^2)$. Since, if $m,n\in\mathbb{N}$,

$$\begin{aligned}\left(\mathcal{K}\Psi\right)(t+m,\nu+n) &= \sum_{k,l\in\mathbb{Z}^2} e^{2\pi ik\nu}\,\Psi(t+m+k,\nu+l+n) \\ &= \sum_{k,l\in\mathbb{Z}^2} e^{2\pi i(k-m)\nu}\,\Psi(t+k,\nu+l) \\ &= e^{-2\pi im\nu}\sum_{k,l\in\mathbb{Z}^2} e^{2\pi ik\nu}\,\Psi(t+k,\nu+l) \\ &= e^{-2\pi im\nu}\,\left(\mathcal{K}\Psi\right)(t,\nu),\end{aligned}$$

it follows that $\overline{\mathcal{K}\Psi}$ also satisfies the equations (6.3) and (6.4) and thus belong to $\mathcal{A}(\mathbb{R}^2)$. The continuity of the mapping $\mathcal{K}$ from $\mathcal{S}(\mathbb{R}^2)$ to $\mathcal{A}(\mathbb{R}^2)$ can be deduced from the estimate (6.13). □

Note now the following fact. If $f\in L^2(\mathbb{R})$ and $\Psi\in\mathcal{S}(\mathbb{R}^2)$, we have, using (6.9) and (6.2),

$$\begin{aligned}\langle\, \mathcal{Z}_{\alpha}f,\Psi\,\rangle &= \iint_{I^2}\mathcal{Z}_{\alpha}f(t,\nu)\,\overline{\overline{\mathcal{K}\Psi(t,\nu)}}\,dt\,d\nu \\ &= \int_{\mathbb{R}} f(t)\,\overline{\left(\mathcal{Z}_{\alpha}^{-1}\left[\overline{\mathcal{K}\Psi}\right]\right)(t)}\,dt \\ &= \int_{\mathbb{R}} f(t)\,\left(\mathcal{H}_{\alpha}\Psi\right)(t)\,dt,\end{aligned} \tag{6.14}$$

where

$$\mathcal{H}_{\alpha}\Psi := \overline{\left(\mathcal{Z}_{\alpha}^{-1}\left[\overline{\mathcal{K}\Psi}\right]\right)}. \tag{6.15}$$

Lemma 6.6. *The mapping $\mathcal{H}_{\alpha}: S(\mathbb{R}^2)\to S(\mathbb{R})$ defined by (6.15) is well-defined and continuous. Moreover, if $\Psi\in\mathcal{S}(\mathbb{R}^2)$, then $\mathcal{H}_{\alpha}\Psi$ is given more explicitly by the formula*

$$(\mathcal{H}_\alpha \Psi)(t) = \alpha^{1/2} \int_{\mathbb{R}} \sum_{k\in\mathbb{Z}} \Psi(\alpha t + k, \nu)\, e^{2\pi i k\nu}\, d\nu, \quad t \in \mathbb{R}. \tag{6.16}$$

Proof. The first statement follows easily from Proposition 6.5 and Theorem 6.4. The formula (6.16) is readily obtained from formulas (6.10) and (6.5). □

Since the identity (6.14) can be formulated as

$$\langle \mathcal{Z}_\alpha f, \Psi \rangle = \langle f, \mathcal{H}_\alpha \Psi \rangle,$$

whenever $f \in L^2(\mathbb{R})$ and $\Psi \in \mathcal{S}(\mathbb{R}^2)$, this suggests that we should use the same formula to define the Zak transform on $S'(\mathbb{R})$.

Definition 6.7. *If $T \in S'(\mathbb{R})$, we define the Zak transform of T, $\mathcal{Z}_\alpha T$, to be the tempered distribution in $S'(\mathbb{R}^2)$ defined by*

$$\langle \mathcal{Z}_\alpha T, \Psi \rangle = \langle T, \mathcal{H}_\alpha \Psi \rangle, \quad \Psi \in \mathcal{S}(\mathbb{R}^2). \tag{6.17}$$

Note that the continuity of the mapping $\mathcal{H}_\alpha$ is used to show that $\mathcal{Z}_\alpha T$ is indeed a distribution in $S'(\mathbb{R}^2)$.

As an illustration, let us compute the Zak transform $\mathcal{Z}_\alpha[\delta_a]$, where δ_a denotes the Dirac mass at the point $a \in \mathbb{R}$. If $\Psi \in \mathcal{S}(\mathbb{R}^2)$, we have

$$\langle \mathcal{Z}_\alpha[\delta_a], \Psi \rangle = \langle \delta_a, \mathcal{H}_\alpha \Psi \rangle = \alpha^{1/2} \int_{\mathbb{R}} \sum_{k\in\mathbb{Z}} \Psi(\alpha a + k, \nu)\, e^{2\pi i k\nu}\, d\nu.$$

We thus obtain the formula

$$\mathcal{Z}_\alpha[\delta_a] = \alpha^{1/2} \sum_{k\in\mathbb{Z}} \delta_{\alpha a + k}(t) \otimes e^{2\pi i k\nu}.$$

Let us next compute $\mathcal{Z}_\alpha\left[e^{2\pi i b x}\right]$, where $b \in \mathbb{R}$:

$$\begin{aligned}
\langle \mathcal{Z}_\alpha\left[e^{2\pi i b x}\right], \Psi \rangle &= \langle e^{2\pi i b x}, \mathcal{H}_\alpha \Psi \rangle \\
&= \alpha^{1/2} \int_{\mathbb{R}} e^{2\pi i b t} \int_{\mathbb{R}} \left\{ \sum_{k\in\mathbb{Z}} \Psi(\alpha t + k, \nu)\, e^{2\pi i k\nu}\, d\nu \right\} dt \\
&= \alpha^{1/2} \sum_{k\in\mathbb{Z}} \iint_{\mathbb{R}^2} e^{2\pi i b t}\, e^{2\pi i k\nu}\, \Psi(\alpha t + k, \nu)\, dt\, d\nu \\
&= \alpha^{-1/2} \sum_{k\in\mathbb{Z}} \iint_{\mathbb{R}^2} e^{2\pi i b(\frac{t-k}{\alpha})}\, e^{2\pi i k\nu}\, \Psi(t, \nu)\, dt\, d\nu.
\end{aligned}$$

This shows that

$$\mathcal{Z}_\alpha\left[e^{2\pi i b x}\right] = \alpha^{-1/2} \sum_{k\in\mathbb{Z}} e^{2\pi i b(\frac{t-k}{\alpha})}\, e^{2\pi i k\nu}.$$

Using the Poisson summation formula (see [3, p. 255]),

$$\sum_{n\in\mathbb{Z}} \delta_n(\nu) = \sum_{k\in\mathbb{Z}} e^{2\pi i k\nu},$$

we have thus

$$\mathcal{Z}_\alpha\left[e^{2\pi i b x}\right] = \alpha^{-1/2}\, e^{2\pi i(\frac{b}{\alpha})t} \sum_{k\in\mathbb{Z}} e^{2\pi i k(\nu-\frac{b}{\alpha})} = \alpha^{-1/2}\, e^{2\pi i(\frac{b}{\alpha})t} \otimes \sum_{n\in\mathbb{Z}} \delta_{n+\frac{b}{\alpha}}(\nu).$$

We show next that the quasi-periodicity properties (6.3) and (6.4) extend to the distributional Zak transform.

Proposition 6.8. *Let $T \in S'(\mathbb{R})$. Then,*

$$\mathcal{Z}_\alpha T(t+m, \nu+n) = e^{2\pi i m\nu}\, \mathcal{Z}_\alpha T(t,\nu), \quad (m,n)\in\mathbb{Z}^2. \tag{6.18}$$

Proof. Note first that, if $\Psi \in \mathcal{S}(\mathbb{R}^2)$ and $(m,n)\in\mathbb{Z}^2$,

$$\begin{aligned}
\mathcal{H}_\alpha[\Psi(\cdot - m, \cdot - n)]\,(t) &= \alpha^{1/2} \int_{\mathbb{R}} \sum_{k\in\mathbb{Z}} \Psi(\alpha t + k - m, \nu - n)\, e^{2\pi i k\nu}\, d\nu \\
&= \alpha^{1/2} \int_{\mathbb{R}} \sum_{k\in\mathbb{Z}} \Psi(\alpha t + k, \nu - n)\, e^{2\pi i (k+m)\nu}\, d\nu \\
&= \alpha^{1/2} \int_{\mathbb{R}} \sum_{k\in\mathbb{Z}} \Psi(\alpha t + k, \nu)\, e^{2\pi i (k+m)\nu}\, d\nu,
\end{aligned}$$

while

$$\begin{aligned}
\mathcal{H}_\alpha\left[e^{2\pi i m\nu}\,\Psi\right](t) &= \alpha^{1/2} \int_{\mathbb{R}} \sum_{k\in\mathbb{Z}} \Psi(\alpha t + k, \nu)\, e^{2\pi i m\nu}\, e^{2\pi i k\nu}\, d\nu \\
&= \alpha^{1/2} \int_{\mathbb{R}} \sum_{k\in\mathbb{Z}} \Psi(\alpha t + k, \nu)\, e^{2\pi i (k+m)\nu}\, d\nu,
\end{aligned}$$

which shows that $\mathcal{H}_\alpha[\Psi(. - m, . - n)] = \mathcal{H}_\alpha\left[e^{2\pi i m\nu}\,\Psi\right]$. We have therefore

$$\begin{aligned}
\langle\, \mathcal{Z}_\alpha T(\cdot + m, \cdot + n), \Psi\,\rangle &= \langle\, \mathcal{Z}_\alpha T, \Psi(\cdot - m, \cdot - n)\,\rangle \\
&= \langle\, T, \mathcal{H}_\alpha\left[\Psi(\cdot - m, \cdot - n)\right]\rangle \\
&= \langle\, T, \mathcal{H}_\alpha\left[e^{2\pi i m\nu}\,\Psi\right]\rangle \\
&= \langle\, \mathcal{Z}_\alpha T, e^{2\pi i m\nu}\,\Psi\,\rangle \\
&= \langle\, e^{2\pi i m\nu}\, \mathcal{Z}_\alpha T, \Psi\,\rangle,
\end{aligned}$$

which proves our claim. □

We will need the following result.

Proposition 6.9. *Let $\Psi_0 \in \mathcal{S}(\mathbb{R}^2)$ satisfy*

$$\sum_{k,l \in \mathbb{Z}^2} \Psi_0(t+k, \nu+l) = 1, \quad (t,\nu) \in \mathbb{R}^2.$$

Then, if $T \in \mathcal{S}'(\mathbb{R})$ and $\varphi \in \mathcal{S}(\mathbb{R})$, we have the identity

$$\langle T, \overline{\varphi} \rangle = \langle \mathcal{Z}_\alpha T, \overline{\mathcal{Z}_\alpha \varphi} \Psi_0 \rangle. \tag{6.19}$$

Proof. First note that, by Proposition 6.3, the function $\overline{\mathcal{Z}_\alpha \varphi} \Psi_0$ belongs to $\mathcal{S}(\mathbb{R}^2)$. By definition,

$$\langle \mathcal{Z}_\alpha T, \overline{\mathcal{Z}_\alpha \varphi} \Psi_0 \rangle = \langle T, \mathcal{H}_\alpha \left[\overline{\mathcal{Z}_\alpha \varphi} \Psi_0 \right] \rangle,$$

so we only need to show that $\overline{\varphi} = \mathcal{H}_\alpha \left[\overline{\mathcal{Z}_\alpha \varphi} \Psi_0 \right]$. This follows from the following computation where we use the equations (6.3), (6.4) and also (6.5) in the last equality:

$$\begin{aligned}
\mathcal{H}_\alpha \left[\overline{\mathcal{Z}_\alpha \varphi} \Psi_0 \right](t) &= \alpha^{1/2} \int_{\mathbb{R}} \sum_{k \in \mathbb{Z}} \overline{\mathcal{Z}_\alpha \varphi(\alpha t + k, \nu)} \, \Psi_0(\alpha t + k, \nu) \, e^{2\pi i k \nu} \, d\nu \\
&= \alpha^{1/2} \int_{\mathbb{R}} \sum_{k \in \mathbb{Z}} e^{-2\pi i k \nu} \, \overline{\mathcal{Z}_\alpha \varphi(\alpha t, \nu)} \, \Psi_0(\alpha t + k, \nu) \, e^{2\pi i k \nu} \, d\nu \\
&= \alpha^{1/2} \int_{\mathbb{R}} \overline{\mathcal{Z}_\alpha \varphi(\alpha t, \nu)} \sum_{k \in \mathbb{Z}} \Psi_0(\alpha t + k, \nu) \, d\nu \\
&= \alpha^{1/2} \int_I \sum_{l \in \mathbb{Z}} \overline{\mathcal{Z}_\alpha \varphi(\alpha t, \nu + l)} \sum_{k \in \mathbb{Z}} \Psi_0(\alpha t + k, \nu + l) \, d\nu \\
&= \alpha^{1/2} \int_I \overline{\mathcal{Z}_\alpha \varphi(\alpha t, \nu)} \sum_{k,l \in \mathbb{Z}} \Psi_0(\alpha t + k, \nu + l) \, d\nu \\
&= \alpha^{1/2} \int_I \overline{\mathcal{Z}_\alpha \varphi(\alpha t, \nu)} \, d\nu = \overline{\varphi(t)}. \qquad \square
\end{aligned}$$

In order to characterize the image in $\mathcal{S}'(\mathbb{R}^2)$ of the Zak transform defined on $\mathcal{S}'(\mathbb{R})$, we need to introduce some new spaces.

Definition 6.10. *We define the spaces $\mathcal{B}(\mathbb{R}^2)$ and $\mathcal{B}'(\mathbb{R}^2)$ as follows.*

(a) *$\mathcal{B}(\mathbb{R}^2)$ consists of all functions of the form $\overline{\Phi(t,\nu)}$, where $\Phi(t,\nu) \in \mathcal{A}(\mathbb{R}^2)$. The topology on $\mathcal{B}(\mathbb{R}^2)$ is the one induced from $\mathcal{A}(\mathbb{R}^2)$, i.e., that of uniform convergence of all partial derivatives on compact subsets of $\mathbb{R}^2$.*

(b) *We define $\mathcal{B}'(\mathbb{R}^2)$ to be the subspace of $\mathcal{S}'(\mathbb{R}^2)$ consisting of the distributions R in $\mathcal{S}'(\mathbb{R}^2)$ satisfying*

$$R(t+m, \nu+n) = e^{2\pi i m \nu} R(t,\nu), \quad (m,n) \in \mathbb{Z}^2. \tag{6.20}$$

The topology on $\mathcal{B}'(\mathbb{R}^2)$ is the one inherited from $\mathcal{S}'(\mathbb{R}^2)$.

We will need the following lemma.

Lemma 6.11. *Let $R \in \mathcal{B}'(\mathbb{R}^2)$ and let Ψ_0 be any function in $C_0^\infty(\mathbb{R}^2)$ satisfying*

$$\sum_{k,l \in \mathbb{Z}^2} \Psi_0(t+k, \nu+l) = 1, \quad (t,\nu) \in \mathbb{R}^2. \tag{6.21}$$

Then, we have

$$\langle R, (\mathcal{K}\Psi)\,\Psi_0 \rangle = \langle R, \Psi \rangle, \quad \Psi \in \mathcal{S}(\mathbb{R}^2). \tag{6.22}$$

Proof. Suppose first that $\Psi \in C_0^\infty(\mathbb{R}^2)$. Since

$$(\mathcal{K}\Psi)(t,\nu)\,\Psi_0(t,\nu) = \sum_{k,l \in \mathbb{Z}^2} e^{2\pi i k \nu}\, \Psi(t+k, \nu+l)\,\Psi_0(t,\nu) \tag{6.23}$$

and the series in (6.23) has only finitely many non-zero terms, as both Ψ and Ψ_0 have compact support, we deduce, using the identity (6.20) that

$$\begin{aligned}
\langle R, (\mathcal{K}\Psi)\,\Psi_0 \rangle &= \sum_{k,l \in \mathbb{Z}^2} \langle R(t,\nu), e^{2\pi i k \nu}\, \Psi(t+k, \nu+l)\,\Psi_0(t,\nu) \rangle \\
&= \sum_{k,l \in \mathbb{Z}^2} \langle e^{2\pi i k \nu}\, R(t-k, \nu-l), \Psi(t,\nu)\,\Psi_0(t-k, \nu-l) \rangle \\
&= \sum_{k,l \in \mathbb{Z}^2} \langle R(t,\nu), \Psi(t,\nu)\,\Psi_0(t-k, \nu-l) \rangle \\
&= \Big\langle R(t,\nu), \sum_{k,l \in \mathbb{Z}^2} \Psi(t,\nu)\,\Psi_0(t-k, \nu-l) \Big\rangle = \langle R, \Psi \rangle.
\end{aligned}$$

Using the density of $C_0^\infty(\mathbb{R}^2)$ in $\mathcal{S}(\mathbb{R}^2)$, the result follows. □

Proposition 6.12. *The topological dual of $\mathcal{B}(\mathbb{R}^2)$ can be identified with the space $\mathcal{B}'(\mathbb{R}^2)$ where the duality between the two spaces is defined by*

$$\langle R, \Phi \rangle_{\mathcal{B}'(\mathbb{R}^2), \mathcal{B}(\mathbb{R}^2)} = \langle R, \Phi\,\Psi_0 \rangle, \quad R \in \mathcal{B}'(\mathbb{R}^2),\ \Phi \in \mathcal{B}(\mathbb{R}^2), \tag{6.24}$$

where Ψ_0 is any function in $C_0^\infty(\mathbb{R}^2)$ satisfying (6.21).

Proof. Note first that the definition given in (6.24) is independent of the choice of the function Ψ_0 in $\mathcal{S}(\mathbb{R}^2)$ satisfying (6.21). Indeed, if Ψ_1 is the difference between two such functions, then it satisfies (6.21) with the right-hand side replaced by 0. This is equivalent to

$$\hat{\Psi}_1(k,l) = 0, \quad (k,l) \in \mathbb{Z}^2, \tag{6.25}$$

where $\hat{\Psi}_1$ denotes the Fourier transform of Ψ_1 defined by

$$\hat{\Psi}_1(\xi) = \mathcal{F}\,[\Psi]\,(\xi) = \iint_{\mathbb{R}^2} e^{-2\pi i \xi . x}\, \Psi_1(x)\, dx, \quad \xi \in \mathbb{R}^2.$$

Furthermore, note that the product $\Phi R \in \mathcal{S}'(\mathbb{R}^2)$ defined by

$$\langle \Phi R, \Psi \rangle = \langle R, \Phi \Psi \rangle, \quad \Psi \in \mathcal{S}(\mathbb{R}^2),$$

is $\mathbb{Z}^2$-periodic. Hence, its distributional Fourier transform is supported on the $\mathbb{Z}^2$-lattice and has the form

$$\mathcal{F}[\Phi R] = \sum_{(k,l)\in\mathbb{Z}^2} c_{k,l}\, \delta_{(k,l)}, \tag{6.26}$$

where $\delta_{(k,l)}$ denotes the Dirac mass at the point (k,l) and where, for some integer $m \geq 0$,

$$\sum_{(k,l)\in\mathbb{Z}^2} \frac{|c_{k,l}|}{[1+k^2+l^2]^m} < \infty.$$

The equations (6.25) and (6.26) easily show that

$$\langle R, \Phi \Psi_1 \rangle = \langle \Phi R, \Psi_1 \rangle = 0,$$

which proves the required independence. The functional on $\mathcal{B}(\mathbb{R}^2)$ defined by the right-hand side of (6.24) is continuous on $\mathcal{B}(\mathbb{R}^2)$. Indeed, it is easily checked using the estimate (6.8) (which is also valid for $\mathcal{B}(\mathbb{R}^2)$) that if $\Phi_n \to 0$ in $\mathcal{B}(\mathbb{R}^2)$, the product $\Phi_n \Psi_0 \to 0$ in $\mathcal{S}(\mathbb{R}^2)$ and thus $\langle R, \Phi_n \Psi_0 \rangle \to 0$. Now let L be a continuous linear functional on the space $\mathcal{B}(\mathbb{R}^2)$. We associate with L the distribution $R \in \mathcal{S}'(\mathbb{R}^2)$ defined by

$$\langle R, \Psi \rangle = L(\mathcal{K}\Psi), \quad \Psi \in \mathcal{S}(\mathbb{R}^2), \tag{6.27}$$

where $\mathcal{K}\Psi$ is defined in (6.10). Proposition 6.5 shows that R is indeed a well-defined element of $\mathcal{S}'(\mathbb{R}^2)$. Let us prove that R satisfies the identity (6.20). Note first that if $\Psi \in \mathcal{S}(\mathbb{R}^2)$ and $(m,n) \in \mathbb{Z}^2$,

$$\mathcal{K}\left[\Psi(\cdot - m, \cdot - n)\right](t,\nu) = e^{2\pi i m \nu}\, \mathcal{K}\left[\Psi\right](t,\nu) = \mathcal{K}\left[e^{2\pi i m \nu}\, \Psi\right](t,\nu).$$

Hence,

$$\begin{aligned}
\langle R(\cdot + m, \cdot + n), \Psi \rangle &= \langle R, \Psi(\cdot - m, \cdot - n) \rangle \\
&= L(\mathcal{K}\left[\Psi(\cdot - m, \cdot - n)\right]) \\
&= L(\mathcal{K}\left[e^{2\pi i m \nu}\, \Psi\right]) \\
&= \langle T, e^{2\pi i m \nu}\, \Psi \rangle \\
&= \langle e^{2\pi i m \nu}\, T, \Psi \rangle,
\end{aligned}$$

showing that R belongs to $\mathcal{B}'(\mathbb{R}^2)$. Next we prove that R satisfies

$$L(\Phi) = \langle R, \Phi \Psi_0 \rangle, \quad \Phi \in \mathcal{B}(\mathbb{R}^2),$$

or, equivalently, that

$$L(\Phi) = L(\mathcal{K}\,[\Phi\,\Psi_0]), \quad \Phi \in \mathcal{B}(\mathbb{R}^2).$$

This will follow if we establish that $\Phi = \mathcal{K}\,[\Phi\,\Psi_0]$, if $\Phi \in \mathcal{B}(\mathbb{R}^2)$. Since, for all $(t,\nu) \in \mathbb{R}^2$,

$$\begin{aligned}
\mathcal{K}\,[\Phi\,\Psi_0]\,(t,\nu) &= \sum_{k,l \in \mathbb{Z}^2} e^{2\pi i k\nu}\, \Phi(t+k,\nu+l)\, \Psi_0(t+k,\nu+l) \\
&= \sum_{k,l \in \mathbb{Z}^2} \Phi(t,\nu)\, \Psi_0(t+k,\nu+l) \\
&= \Phi(t,\nu) \sum_{k,l \in \mathbb{Z}^2} \Psi_0(t+k,\nu+l) = \Phi(t,\nu),
\end{aligned}$$

this identity follows. It remains to show that the continuous linear functional on $\mathcal{B}(\mathbb{R}^2)$ defined by the right-hand side of (6.24) uniquely determines the corresponding distribution $R \in \mathcal{B}'(\mathbb{R}^2)$. This is equivalent to showing that if $R \in \mathcal{B}'(\mathbb{R}^2)$ satisfies

$$\langle\, R, \Phi\,\Psi_0 \,\rangle = 0, \quad \Phi \in \mathcal{B}(\mathbb{R}^2), \tag{6.28}$$

then $R = 0$. This follows immediately by replacing Φ in (6.28) by $\mathcal{K}\Psi$, where Ψ is an arbitrary function in $\mathcal{S}(\mathbb{R}^2)$, and using the identity (6.22) obtained in Lemma 6.11. This completes the proof. □

We now consider the problem of defining the inverse Zak transform on the space $\mathcal{B}'(\mathbb{R}^2)$. To simplify the notation, we introduce the following definition.

Definition 6.13. *If $T \in \mathcal{S}'(\mathbb{R})$, we define $\mathcal{Z}_\alpha^* T$ by*

$$\mathcal{Z}_\alpha^*\,[T(t)] = \overline{\mathcal{Z}_\alpha[\overline{T(t)}]}.$$

The following definition of the inverse Zak transform on $\mathcal{B}'(\mathbb{R}^2)$ is motivated by equation (6.19).

Definition 6.14. *If $R \in \mathcal{B}'(\mathbb{R}^2)$, we define its inverse Zak transform, $\mathcal{Z}_\alpha^{-1} R$, by the formula*

$$\langle\, \mathcal{Z}_\alpha^{-1} R, \varphi \,\rangle = \langle\, R, \mathcal{Z}_\alpha^*\,[\varphi]\,\Psi_0 \,\rangle, \quad \varphi \in \mathcal{S}(\mathbb{R}), \tag{6.29}$$

where Ψ_0 is any function in $C_0^\infty(\mathbb{R}^2)$ satisfying (6.21).

Note that, since the mapping $\mathcal{Z}_\alpha^*$ maps $\mathcal{S}(\mathbb{R})$ onto $\mathcal{B}(\mathbb{R}^2)$, the definition above does not depend on the choice of the function Ψ_0 satisfying (6.21) by the argument given at the beginning of the proof of Proposition 6.12. The fact that the right-hand side of (6.29) yields a well-defined element of $\mathcal{S}'(\mathbb{R}^2)$ follows from the continuity of the mapping $\mathcal{Z}_\alpha^* : \mathcal{S}(\mathbb{R}) \to \mathcal{B}(\mathbb{R}^2)$ (which is a consequence of Theorem 6.4) and that of the mapping $\mathcal{B}(\mathbb{R}^2) \to \mathcal{S}(\mathbb{R}^2) : \Phi \mapsto \Phi\,\Psi_0$ (as can easily be seen from the estimate (6.8)). We will need the following lemma.

Lemma 6.15. *We have the following identities:*

(a) *If $\Psi \in \mathcal{S}(\mathbb{R}^2)$, then*

$$\mathcal{Z}_\alpha^*[\mathcal{H}_\alpha(\Psi)] = \mathcal{K}\Psi. \tag{6.30}$$

(b) *If $\varphi \in \mathcal{S}(\mathbb{R})$, then*

$$\mathcal{H}_\alpha[\mathcal{Z}_\alpha^*[\varphi]\, \Psi_0] = \varphi, \tag{6.31}$$

where Ψ_0 is any function in $C_0^\infty(\mathbb{R}^2)$ satisfying (6.21).

Proof. The identity (6.30) is an immediate consequence of (6.15). To prove (6.31) we note that

$$(\mathcal{Z}_\alpha^*[\varphi]\, \Psi_0)\,(t,\nu) = \alpha^{-1/2} \sum_{k\in\mathbb{Z}} \varphi\Big(\frac{t-k}{\alpha}\Big)\, e^{-2\pi i k\nu}\, \Psi_0(t,\nu).$$

Hence, using the formula (6.16),

$$\begin{aligned}
\mathcal{H}_\alpha[\mathcal{Z}_\alpha^*[\varphi]\, \Psi_0]\,(t) &= \int_{\mathbb{R}} \sum_{l\in\mathbb{Z}} \sum_{k\in\mathbb{Z}} \varphi\left(\frac{\alpha t + l - k}{\alpha}\right) e^{-2\pi i k\nu}\, \Psi_0(\alpha t + l, \nu)\, e^{2\pi i l\nu}\, d\nu \\
&= \int_{\mathbb{R}} \sum_{l\in\mathbb{Z}} \sum_{k\in\mathbb{Z}} \varphi\left(\frac{\alpha t - k}{\alpha}\right) \Psi_0(\alpha t + l, \nu)\, e^{-2\pi i k\nu}\, d\nu \\
&= \sum_{k\in\mathbb{Z}} \varphi\left(\frac{\alpha t - k}{\alpha}\right) \int_{\mathbb{R}} \sum_{l\in\mathbb{Z}} \Psi_0(\alpha t + l, \nu)\, e^{-2\pi i k\nu}\, d\nu \\
&= \sum_{k\in\mathbb{Z}} \varphi\left(\frac{\alpha t - k}{\alpha}\right) \int_{I} \sum_{j\in\mathbb{Z}} \sum_{l\in\mathbb{Z}} \Psi_0(\alpha t + l, \nu + j)\, e^{-2\pi i k\nu}\, d\nu \\
&= \sum_{k\in\mathbb{Z}} \varphi\left(\frac{\alpha t - k}{\alpha}\right) \int_{I} e^{-2\pi i k\nu}\, d\nu = \varphi(t),
\end{aligned}$$

which yields (6.31). □

The following result is essentially due to Janssen [13].

Theorem 6.16. *The mapping $\mathcal{Z}_\alpha : \mathcal{S}'(\mathbb{R}) \to \mathcal{B}'(\mathbb{R}^2)$ is a topological isomorphism with inverse $\mathcal{Z}_\alpha^{-1}$ defined by (6.29).*

Proof. The fact that $\mathcal{Z}_\alpha$ maps $\mathcal{S}'(\mathbb{R})$ to $\mathcal{B}'(\mathbb{R}^2)$ follows from Proposition 6.8. We start by proving that $\mathcal{Z}_\alpha^{-1}[\mathcal{Z}_\alpha T] = T$, for any $T \in \mathcal{S}'(\mathbb{R})$. Indeed, if $\varphi \in \mathcal{S}(\mathbb{R})$, we have, using the definitions of $\mathcal{Z}_\alpha$ and of $\mathcal{Z}_\alpha^{-1}$ and the identity (6.31) that

$$\langle\, \mathcal{Z}_\alpha^{-1}[\mathcal{Z}_\alpha T], \varphi\,\rangle = \langle\, \mathcal{Z}_\alpha T, \mathcal{Z}_\alpha^*[\varphi]\, \Psi_0\,\rangle = \langle T, \mathcal{H}_\alpha[\mathcal{Z}_\alpha^*[\varphi]\, \Psi_0]\,\rangle = \langle T, \varphi\rangle.$$

Next we show that $\mathcal{Z}_\alpha\big[\mathcal{Z}_\alpha^{-1} R\big] = R$, for any $R \in \mathcal{B}'(\mathbb{R})$. If $\Psi \in \mathcal{S}(\mathbb{R}^2)$, we have, using the identity (6.30) and Lemma 6.11, that

$$\begin{aligned}\langle \mathcal{Z}_\alpha\left[\mathcal{Z}_\alpha^{-1}R\right], \Psi \rangle &= \langle \mathcal{Z}_\alpha^{-1}R, \mathcal{H}_\alpha \Psi \rangle \\ &= \langle R, \mathcal{Z}_\alpha^*\left[\mathcal{H}_\alpha \Psi\right] \Psi_0 \rangle \\ &= \langle R, \left[\mathcal{K}\Psi\right] \Psi_0 \rangle \\ &= \langle R, \Psi \rangle,\end{aligned}$$

which proves the required equality. The two identities just proved immediately imply that the mappings $\mathcal{Z}_\alpha : \mathcal{S}'(\mathbb{R}) \to \mathcal{B}'(\mathbb{R}^2)$ and $\mathcal{Z}_\alpha^{-1} : \mathcal{B}'(\mathbb{R}^2) \to \mathcal{S}'(\mathbb{R})$ are surjective and, since their continuity follows immediately from their definition, our claim is proved. □

6.3 Gabor Systems in $\mathcal{S}(\mathbb{R})$ and $\mathcal{S}'(\mathbb{R})$

A Gabor system is a collection of functions generated by a window function $g \in L^2(\mathbb{R})$ and by translations and modulations:

$$\mathbf{G}(g, \alpha, \beta) := \{e^{2\pi i m\alpha x}\, g(x - n\beta) : m, n \in \mathbb{Z}\},$$

where α and β are two positive parameters. To simplify the notation we will write

$$g_{m\alpha, n\beta}(x) = e^{2\pi i m\alpha\, x}\, g(x - n\beta), \quad m, n \in \mathbb{Z}.$$

Gabor systems are usually considered within the framework of Hilbert space theory, i.e., the signals to be expanded in terms of the Gabor system are in $L^2(\mathbb{R})$ and the underlying topology is that of $L^2(\mathbb{R})$. In what follows we consider certain problems where the Gabor systems considered correspond to windows in $\mathcal{S}(\mathbb{R})$ and the convergence is in either $\mathcal{S}(\mathbb{R})$ or $\mathcal{S}'(\mathbb{R})$. These types of systems have been considered by Janssen in the particular case where the window is a Gaussian function in [11] and [12]. For technical reasons, we will always assume that the product of the two parameters associated with the system is rational, i.e.,

$$\alpha\beta = \frac{p}{q}, \quad p, q \in \mathbb{N} \setminus \{0\}, \quad (p, q) = 1. \tag{6.32}$$

Let us start with a simple problem to illustrate how natural and useful the Zak transform is as a tool in this type of situation. The question we consider first is the following: given a function $\varrho \in \mathcal{S}(\mathbb{R})$, when is the linear span of the system $\mathbf{G}(\varrho, \alpha, \beta)$ dense in $\mathcal{S}(\mathbb{R})$? Note that the span of $\mathbf{G}(\varrho, \alpha, \beta)$ is dense in $\mathcal{S}(\mathbb{R})$ if and only if the fact that $T \in \mathcal{S}'(\mathbb{R})$ satisfies

$$\langle T, \varrho_{m\alpha, n\beta} \rangle = 0, \quad m, n \in \mathbb{Z} \tag{6.33}$$

implies that $T = 0$. This question has a very simple answer when $\alpha\beta = 1$.

Theorem 6.17. *Given a function $\varrho \in \mathcal{S}(\mathbb{R})$ and $\alpha > 0$, the linear span of $\mathbf{G}(\varrho, \alpha, \frac{1}{\alpha})$ is never dense in $\mathcal{S}(\mathbb{R})$.*

Proof. Suppose that $T \in \mathcal{S}'(\mathbb{R})$. The product $\mathcal{Z}_\alpha T\, \mathcal{Z}_\alpha^* \varrho$ is a well-defined distribution in $\mathcal{S}'(\mathbb{R}^2)$ which is also $\mathbb{Z}^2$-periodic as can be easily checked using Proposition 6.8. Thus, $\mathcal{Z}_\alpha T\, \mathcal{Z}_\alpha^* \varrho$ can be expanded in $\mathcal{S}'(\mathbb{R})$ as the double Fourier series

$$\mathcal{Z}_\alpha T\, \mathcal{Z}_\alpha^* \varrho = \sum_{m,n \in \mathbb{Z}} c_{m,n}\, e^{2\pi i m t}\, e^{2\pi i n \nu},$$

with Fourier coefficients

$$\begin{aligned} c_{m,n} &= \langle\, \mathcal{Z}_\alpha T(t,\nu)\, \mathcal{Z}_\alpha^* \varrho(t,\nu), e^{-2\pi i m t}\, e^{-2\pi i n \nu}\, \Psi_0(t,\nu)\, \rangle \\ &= \langle\, \mathcal{Z}_\alpha T(t,\nu), \mathcal{Z}_\alpha^* \varrho(t,\nu)\, e^{-2\pi i m t}\, e^{-2\pi i n \nu}\, \Psi_0(t,\nu)\, \rangle, \quad m,n \in \mathbb{N}, \end{aligned}$$

where Ψ_0 is any function in $C_0^\infty(\mathbb{R}^2)$ satisfying (6.21). Using the equation (6.6), we can write

$$c_{m,n} = \langle\, \mathcal{Z}_\alpha T, \mathcal{Z}_\alpha^* \big[\varrho_{-m\alpha, -n\frac{1}{\alpha}}\big]\, \Psi_0\, \rangle, \quad m,n \in \mathbb{Z},$$

or, using the equation (6.19),

$$c_{m,n} = \langle T, \varrho_{-m\alpha, -n\frac{1}{\alpha}}\, \rangle, \quad m,n \in \mathbb{Z}.$$

The fact that T satisfies (6.33) with $\beta = \frac{1}{\alpha}$ is thus equivalent to $c_{m,n} = 0$ for all $m,n \in \mathbb{Z}$ or to $\mathcal{Z}_\alpha T\, \mathcal{Z}_\alpha^* \varrho = 0$. A rather striking property of the Zak transform of a function is that, if it is continuous, it must vanish somewhere on the unit square $I \times I$ ([13]; see also [9, pp. 163–164]). Since $\mathcal{Z}_\alpha^* \varrho$ is continuous on $\mathbb{R}^2$, it must thus have a zero at some point $(t_0, \nu_0) \in I \times I$. Defining

$$R = \sum_{k,l \in \mathbb{Z}} e^{2\pi i k \nu_0}\, \delta_{(t_0+k, \nu_0+l)},$$

where $\delta_{(a,b)}$ denotes the Dirac mass at the point (a,b), it is easily checked that $R \in \mathcal{B}'(\mathbb{R}^2)$ and that $R\, \mathcal{Z}_\alpha^* \varrho = 0$ while $R \neq 0$. Letting $T = \mathcal{Z}_\alpha^{-1} R$, we have thus obtained a non-zero distribution in $\mathcal{S}'(\mathbb{R})$ satisfying (6.33) with $\beta = \frac{1}{\alpha}$ and the result follows. □

We note that the previous result does not have an analogue in the $L^2(\mathbb{R})$ case. In fact, when $\alpha\beta = 1$, the condition for the span of $\mathbf{G}(g, \alpha, \frac{1}{\alpha})$, $g \in L^2(\mathbb{R})$, to be dense in $L^2(\mathbb{R})$ is that the Zak transform of g must be a.e. non-zero on the unit square $I \times I$ (see [10]). This can certainly happen, even under the restriction that the window belongs to $\mathcal{S}(\mathbb{R})$ since it is known that the Zak transform of a Gaussian window has a single zero in $I \times I$ [11].

In order to deal with the case where $\alpha\beta$ is a rational number (i.e., $\alpha\beta = \frac{p}{q}$ as in (6.32)), we need to introduce a matrix version of the Zak transform. The usefulness of this concept to study various properties of Gabor systems in the rational case has been exploited extensively in the work of Zibulski and Zeevi ([19], [20]; see also [5], [8], [15]).

Definition 6.18. *Given a function $\varrho \in \mathcal{S}(\mathbb{R})$, we associate with ϱ the matrix function $\mathcal{G}(t,\nu)$, taking value in $\mathcal{M}_{q,p}$, the space of $q\times p$ matrices with complex entries, defined by*

$$\mathcal{G}_{r,s}(t,\nu) = \mathcal{Z}_\alpha^*\varrho\Big(t - r\,\frac{p}{q}, \nu + \frac{s}{p}\Big), \quad 0 \le r \le q-1,\, 0 \le s \le p-1,$$

for (t,ν) in $\mathbb{R}^2$.

Clearly, the rank of $\mathcal{G}(t,\nu)$ is at most equal to $\min\{p,q\}$ for all (t,ν) in $\mathbb{R}^2$. We need the following lemma.

Lemma 6.19. *If $\varrho \in \mathcal{S}(\mathbb{R})$, if α, p and q are as before and if m, n, r are integers, then we have the identity*

$$\begin{aligned}\mathcal{Z}_\alpha^*\big[\,\varrho_{m\alpha,\frac{1}{\alpha}(nq+r)\frac{p}{q}}\,\big](t,\nu) &= \mathcal{Z}_\alpha^*\big[\,\varrho_{m\alpha,\frac{1}{\alpha}(np+r\frac{p}{q})}\,\big](t,\nu)\\ &= e^{2\pi imt}\,e^{2\pi inp\nu}\,\mathcal{Z}_\alpha^*[\,\varrho\,]\Big(t - r\,\frac{p}{q}, \nu\Big), \quad (t,\nu)\in\mathbb{R}^2.\end{aligned}$$

Proof. We have

$$\begin{aligned}&\mathcal{Z}_\alpha^*\big[\,\varrho_{m\alpha,\frac{1}{\alpha}(np+r\frac{p}{q})}\,\big](t,\nu)\\ &= (\alpha)^{-1/2}\sum_{k\in\mathbb{Z}} e^{2\pi im\alpha\frac{(t-k)}{\alpha}}\,\varrho\Big(\frac{t-k}{\alpha} - \frac{np + r\frac{p}{q}}{\alpha}\Big)\,e^{-2\pi ik\nu}\\ &= (\alpha)^{-1/2}\,e^{2\pi imt}\sum_{k\in\mathbb{Z}}\varrho\Big(\frac{t-k-r\frac{p}{q}}{\alpha}\Big)\,e^{-2\pi i(k-np)\nu}\\ &= e^{2\pi imt}\,e^{2\pi inp\nu}\,\mathcal{Z}_\alpha^*[\,\varrho\,]\Big(t - r\,\frac{p}{q}, \nu\Big),\end{aligned}$$

which proves the lemma. □

We can now answer the question raised earlier about the density of Gabor systems in $\mathcal{S}(\mathbb{R})$. Note that the L^2-version of the result below is known ([5, p. 978]; see also [20], [8]): the same result holds if $\mathcal{S}(\mathbb{R})$ is replaced with $L^2(\mathbb{R})$ and the everywhere equality for the rank is replaced with an a.e. equality.

Theorem 6.20. *Given a function $\varrho \in \mathcal{S}(\mathbb{R})$ and α, $\beta > 0$ with $\alpha\beta = \frac{p}{q}$ as in (6.32), the linear span of $\mathbf{G}(\varrho,\alpha,\beta)$ is dense in $\mathcal{S}(\mathbb{R})$ if and only if*

$$\operatorname{rank}(\mathcal{G}(t,\nu)) = p, \quad \text{for all } (t,\nu)\in\mathbb{R}^2.$$

Proof. Let $T \in \mathcal{S}'(\mathbb{R})$. Note that the condition

$$\langle\, T, \varrho_{m\alpha,n\beta}\,\rangle = 0, \quad m,n\in\mathbb{Z},$$

is equivalent to

$$\langle T, \varrho_{m\alpha,\frac{1}{\alpha}(np+r\frac{p}{q})} \rangle = 0, \quad m,n \in \mathbb{Z},\ r = 0,\ldots,q-1. \tag{6.34}$$

Using the identity (6.19) and Lemma 6.19, we have

$$\begin{aligned}
&\langle T, \varrho_{m\alpha,\frac{1}{\alpha}(np+r\frac{p}{q})} \rangle \\
&= \Big\langle \mathcal{Z}_\alpha T(t,\nu)\,,\, \mathcal{Z}_\alpha^* \varrho\Big(t - r\frac{p}{q},\nu\Big)\, e^{2\pi imt}\, e^{2\pi inp\nu}\, \Psi_0(t,\nu) \Big\rangle \\
&= \Big\langle \mathcal{Z}_\alpha T(t,\nu)\, \mathcal{Z}_\alpha^* \varrho\Big(t - r\frac{p}{q},\nu\Big)\,,\, e^{2\pi imt}\, e^{2\pi inp\nu}\, \Psi_0(t,\nu) \Big\rangle,
\end{aligned} \tag{6.35}$$

where Ψ_0 is any function in $C_0^\infty(\mathbb{R}^2)$ satisfying (6.21). Since the product $\mathcal{Z}_\alpha T(t,\nu)\, \mathcal{Z}_\alpha^* \varrho(t - r\frac{p}{q},\nu)$ belongs to $\mathcal{S}'(\mathbb{R}^2)$ and is $\mathbb{Z}^2$-periodic, it has the Fourier series expansion

$$\mathcal{Z}_\alpha T(t,\nu)\, \mathcal{Z}_\alpha^* \varrho\Big(t - r\frac{p}{q},\nu\Big) = \sum_{m,n\in\mathbb{Z}} c_{m,n}^r\, e^{2\pi imt}\, e^{2\pi in\nu},$$

with Fourier coefficients

$$c_{m,n}^r = \Big\langle \mathcal{Z}_\alpha T(t,\nu)\, \mathcal{Z}_\alpha^* \varrho\Big(t - r\frac{p}{q},\nu\Big)\,,\, e^{-2\pi imt}\, e^{-2\pi in\nu}\, \Psi_0(t,\nu) \Big\rangle, \quad m,n \in \mathbb{N}.$$

Define now, for each $r = 0,\ldots,q-1$, the distribution $R_r \in \mathcal{S}'(\mathbb{R})$, by

$$R_r(t,\nu) = \sum_{l=0}^{p-1} \mathcal{Z}_\alpha T\Big(t,\nu + \frac{l}{p}\Big)\, \mathcal{Z}_\alpha^* \varrho\Big(t - r\frac{p}{q},\nu + \frac{l}{p}\Big).$$

It is clear that each distribution R_r is $\mathbb{Z} \oplus \frac{1}{p}\mathbb{Z}$-periodic. Furthermore, each distribution R_r has the expansion

$$\begin{aligned}
R_r &= \sum_{l=0}^{p-1} \sum_{m,n\in\mathbb{Z}} c_{m,n}^r\, e^{2\pi imt}\, e^{2\pi in(\nu+\frac{l}{p})} \\
&= p \sum_{m,n\in\mathbb{Z}} c_{m,np}^r\, e^{2\pi imt}\, e^{2\pi inp\nu} \\
&= p \sum_{m,n\in\mathbb{Z}} \langle T, \varrho_{-m\alpha,\frac{1}{\alpha}(-np+r\frac{p}{q})} \rangle\, e^{2\pi imt}\, e^{2\pi inp\nu}.
\end{aligned}$$

Hence, the condition (6.34) is equivalent to

$$R_r = 0, \quad r = 0,\ldots,q-1. \tag{6.36}$$

Suppose now that, at some point $(t_0,\nu_0) \in \mathbb{R}^2$, $\mathcal{G}(t_0,\nu_0)$ has rank strictly less than p. We can then find a non-zero vector $(a_0,a_1,\ldots,a_{p-1}) \in \mathbb{C}^p$ satisfying

$$\sum_{l=0}^{p-1} a_l\, \mathcal{Z}_\alpha^* \varrho\Big(t_0 - r\frac{p}{q},\nu_0 + \frac{l}{p}\Big) = 0, \quad r = 0,\ldots,q-1.$$

Defining

$$R = \sum_{l=0}^{p-1} a_l \sum_{k,s\in\mathbb{Z}} e^{2\pi i k(\nu_0+\frac{l}{p})}\, \delta_{(t_0+k,\nu_0+\frac{l}{p}+s)},$$

where $\delta_{(a,b)}$ denotes the Dirac mass at the point (a,b), it is easily checked that $R \in \mathcal{B}'(\mathbb{R}^2)$ and clearly $R \neq 0$. Letting $T = \mathcal{Z}_\alpha^{-1} R$, it follows easily that (6.36) holds. Hence, T is a non-zero distribution in $\mathcal{S}'(\mathbb{R})$ satisfying (6.34) and $\mathbf{G}(\varrho,\alpha,\beta)$ cannot be dense in $\mathcal{S}(\mathbb{R})$. On the other hand, suppose that $\mathcal{G}(t,\nu)$ has rank p for all $(t,\nu) \in \mathbb{R}^2$. In particular, we have $p \leq q$. Fix a point $(t_0,\nu_0) \in \mathbb{R}^2$. Then, we can find p distinct numbers $r_0, r_1, \ldots, r_{p-1} \in \{0,1,\ldots,q-1\}$ such that the $p \times p$ matrix, $\mathcal{Q}(t,\nu)$, with entries

$$\mathcal{Q}(t,\nu)_{i,l} = \mathcal{Z}_\alpha^* \varrho\Big(t - r_i \frac{p}{q}, \nu + \frac{l}{p}\Big), \quad 0 \leq i,l \leq p-1,$$

is invertible when $(t,\nu) = (t_0,\nu_0)$, and thus also on some neighborhood U of that point. Denoting by $\mathcal{Q}^{-1}(t,\nu)$ the inverse of $\mathcal{Q}(t,\nu)$, it is easily seen that all the entries of $\mathcal{Q}^{-1}$ belong to $C^\infty(U)$. If $T \in \mathcal{S}'(\mathbb{R})$ satisfies (6.34) and thus (6.36), we have

$$0 = \sum_{j=0}^{p-1} \mathcal{Q}_{i,j}^{-1}(t,\nu)\, R_{r_j}(t,\nu) = \mathcal{Z}_\alpha T\Big(t, \nu + \frac{i}{p}\Big) \text{ on } U, \quad i = 0,\ldots,p-1,$$

and thus, in particular, $\mathcal{Z}_\alpha T = 0$ on a neighborhood of (t_0,ν_0). Since (t_0,ν_0) was an arbitrary point in $\mathbb{R}^2$, it follows that $\mathcal{Z}_\alpha T = 0$ on all of $\mathbb{R}^2$ and thus that $T = 0$, which concludes the proof. □

The following is an immediate consequence of Theorem 6.17 and Theorem 6.20.

Corollary 6.21. *If a function $\varrho \in \mathcal{S}(\mathbb{R})$ and α, $\beta > 0$ with $\alpha\beta \in \mathbb{Q}$ are such that the linear span of $\mathbf{G}(\varrho,\alpha,\beta)$ is dense in $\mathcal{S}(\mathbb{R})$, then necessarily $\alpha\beta < 1$.*

We now consider a different type of problem and study series obtained as infinite linear combinations of functions in a system $\mathbf{G}(\varrho,\alpha,\beta)$, where $\varrho \in \mathcal{S}(\mathbb{R})$.

Definition 6.22. *A sequence $\{c_{m,n}\}_{(m,n)\in\mathbb{Z}^2}$ of complex numbers is said to have polynomial growth if for some $C > 0$ and some $N \in \mathbb{N}$, we have*

$$|c_{m,n}| \leq C\,(1 + m^2 + n^2)^N, \quad (m,n) \in \mathbb{Z}^2.$$

Lemma 6.23. *Given a function $\varrho \in \mathcal{S}(\mathbb{R})$, α, $\beta > 0$ with $\alpha\beta = \frac{p}{q}$ as in (6.32) and a sequence $\{c_{m,n}\}_{(m,n)\in\mathbb{Z}^2}$ of polynomial growth, the series*

$$\sum_{m,n\in\mathbb{Z}} c_{m,n}\, \varrho_{m\alpha,n\beta} \tag{6.37}$$

converges in $\mathcal{S}'(\mathbb{R})$.

Proof. If $\varphi \in \mathcal{S}(\mathbb{R})$ and Ψ_0 is any function in $C_0^\infty(\mathbb{R}^2)$ satisfying (6.21), we have

$$\sum_{m,n\in\mathbb{Z}} c_{m,n} \left\langle \varrho_{m\alpha,n\beta}, \varphi \right\rangle = \sum_{r=0}^{q-1} \sum_{m,n\in\mathbb{Z}} c_{m,nq+r} \left\langle \varrho_{m\alpha,\frac{1}{\alpha}(np+r\frac{p}{q})}, \varphi \right\rangle$$

$$= \sum_{r=0}^{q-1} \sum_{m,n\in\mathbb{Z}} c_{m,nq+r} \left\langle \mathcal{Z}_\alpha \left[\varrho_{m\alpha,\frac{1}{\alpha}(np+r\frac{p}{q})}\right], (\mathcal{Z}_\alpha^* \varphi)\, \Psi_0 \right\rangle$$

$$= \sum_{r=0}^{q-1} \sum_{m,n\in\mathbb{Z}} c_{m,nq+r} \left\langle e^{2\pi imt}\, e^{-2\pi inp\nu}\, \mathcal{Z}_\alpha \varrho\Big(t - r\frac{p}{q}, \nu\Big), (\mathcal{Z}_\alpha^* \varphi)\, \Psi_0 \right\rangle$$

$$= \sum_{r=0}^{q-1} \sum_{m,n\in\mathbb{Z}} c_{m,nq+r}\, \mathcal{F}\left[\Theta_r\right](-m, n\,p), \tag{6.38}$$

where

$$\Theta_r(t,\nu) = \mathcal{Z}_\alpha \varrho\Big(t - r\frac{p}{q}, \nu\Big)\, \mathcal{Z}_\alpha^* \varphi(t,\nu)\, \Psi_0(t,\nu), \quad r = 0, \ldots, q-1.$$

Since $\Theta_r \in C_0^\infty(\mathbb{R}^2)$, the function $\mathcal{F}\left[\Theta_r\right] \in \mathcal{S}(\mathbb{R}^2)$ for each $r = 0, \ldots, q-1$ and the series in (6.38) is thus absolutely convergent for each $\varphi \in S(\mathbb{R}^2)$, which proves the lemma. □

In the following, we will be interested in the uniqueness of the coefficients in a series as in (6.37). This motivates the following definition.

Definition 6.24. *Given a function $\varrho \in \mathcal{S}(\mathbb{R})$ and α, $\beta > 0$ with $\alpha\beta \in \mathbb{Q}$, the system $\mathbf{G}(\varrho, \alpha, \beta)$ is said to have the Riesz property if, for any sequence $\{c_{m,n}\}_{(m,n)\in\mathbb{Z}^2}$ of polynomial growth, the fact that*

$$\sum_{m,n\in\mathbb{Z}} c_{m,n}\, \varrho_{m\alpha,n\beta} = 0$$

in $\mathcal{S}'(\mathbb{R})$ implies that $c_{m,n} = 0$ for all $(m,n) \in \mathbb{Z}^2$.

The following result can be seen as dual to Theorem 6.20. An L^2-version of this result (with the equality for the rank being replaced with an a.e. equality) can be found in [2] (see also [8]).

Theorem 6.25. *Given a function $\varrho \in \mathcal{S}(\mathbb{R})$ and α, $\beta > 0$ with $\alpha\beta = \frac{p}{q}$ as in (6.32), the system $\mathbf{G}(\varrho, \alpha, \beta)$ has the Riesz property if and only if*

$$\operatorname{rank}(\mathcal{G}(t,\nu)) = q, \quad \textit{for all } (t,\nu) \in \mathbb{R}^2.$$

Proof. If $\{c_{m,n}\}$ is a sequence of polynomial growth, define the sequences $\{c_{m,n}^r\}$ by

$$c^r_{m,n} = c_{m,nq+r}, \quad r = 0, \ldots, q-1.$$

The fact that

$$\sum_{m,n \in \mathbb{Z}} c_{m,n}\, \varrho_{m\alpha,n\beta} = 0 \tag{6.39}$$

in $\mathcal{S}'(\mathbb{R})$ is thus equivalent to

$$\sum_{r=0}^{q-1} \sum_{m,n \in \mathbb{Z}} c^r_{m,n}\, \varrho_{m\alpha,\frac{1}{\alpha}(nq+r)\frac{p}{q}} = 0$$

in $\mathcal{S}'(\mathbb{R})$ or, applying $\mathcal{Z}^*_\alpha$ to both sides of the previous equation and using Lemma 6.19, to the identity

$$\sum_{r=0}^{q-1} \sum_{m,n \in \mathbb{Z}} c^r_{m,n}\, e^{2\pi imt}\, e^{2\pi inp\nu}\, \mathcal{Z}^*_\alpha \varrho\Big(t - r\frac{p}{q}, \nu\Big) = 0$$

in $\mathcal{S}'(\mathbb{R}^2)$. Since each sequence $\{c^r_{m,n}\}$, $r = 0, \ldots, q-1$, has polynomial growth, the Fourier series

$$\sum_{m,n \in \mathbb{Z}} c^r_{m,n}\, e^{2\pi imt}\, e^{2\pi inp\nu}, \quad r = 0, \ldots, q-1,$$

converges in $\mathcal{S}'(\mathbb{R}^2)$ to a distribution S_r which is clearly $\mathbb{Z} \oplus \frac{1}{p}\mathbb{Z}$-periodic and, conversely, any such distribution can be expressed as a Fourier series with coefficients $\{c^r_{m,n}\}$ which have polynomial growth. Thus, the existence of a non-identically zero sequence of polynomial growth satisfying (6.39) is equivalent to the existence of $\mathbb{Z} \oplus \frac{1}{p}\mathbb{Z}$-periodic distributions in $\mathcal{S}'(\mathbb{R}^2)$, S_0, $S_1, \ldots, S_{q-1}$, which are not all zero and satisfy the equation

$$\sum_{r=0}^{q-1} S_r(t,\nu)\, \mathcal{Z}^*_\alpha \varrho\Big(t - r\frac{p}{q}, \nu\Big) = 0, \tag{6.40}$$

or, using the $\mathbb{Z} \oplus \frac{1}{p}\mathbb{Z}$-periodicity of each S_r, the equations

$$\sum_{r=0}^{q-1} S_r(t,\nu)\, \mathcal{Z}^*_\alpha \varrho\Big(t - r\frac{p}{q}, \nu + \frac{l}{p}\Big) = 0, \quad l = 0, \ldots, p-1. \tag{6.41}$$

Suppose now that, at some point $(t_0, \nu_0) \in \mathbb{R}^2$, the rank of $\mathcal{G}(t_0, \nu_0)$ is strictly less than q. There exists then a non-zero vector $(a_0, a_1, \ldots, a_{q-1}) \in \mathbb{C}^q$ satisfying

$$\sum_{r=0}^{q-1} \mathcal{Z}^*_\alpha \varrho\Big(t_0 - r\frac{p}{q}, \nu_0 + \frac{l}{p}\Big)\, a_r = 0, \quad l = 0, \ldots, p-1. \tag{6.42}$$

Defining

$$S_r = a_r \sum_{m,n\in\mathbb{Z}} \delta_{(t_0+m,\nu_0+\frac{n}{p})}, \quad r = 0,\ldots,q-1,$$

it is easily checked that (6.41) holds since, using (6.42), for each $(m,n) \in \mathbb{Z}^2$,

$$\begin{aligned}
&\sum_{r=0}^{q-1} a_r\, \mathcal{Z}_\alpha^* \varrho\Big(t_0 + m - r\,\frac{p}{q}, \nu_0 + n + \frac{l}{p}\Big) \\
&= \sum_{r=0}^{q-1} a_r\, e^{-2\pi i m(\nu_0+\frac{l}{p})}\, \mathcal{Z}_\alpha^* \varrho\Big(t_0 - r\,\frac{p}{q}, \nu_0 + \frac{l}{p}\Big) = 0, \quad l = 0,\ldots,p-1.
\end{aligned}$$

This shows that the system $\mathbf{G}(\varrho,\alpha,\beta)$ does not have the Riesz property. On the other hand, suppose that $\mathcal{G}(t,\nu)$ has rank q for all $(t,\nu) \in \mathbb{R}^2$. In particular, we have $q \le p$. Fix a point $(t_0,\nu_0) \in \mathbb{R}^2$. Then, we can find q distinct numbers $l_0, l_1, \ldots, l_{q-1} \in \{0,1,\ldots,p-1\}$ such that the $q \times q$ matrix, $\mathcal{R}(t,\nu)$, with entries

$$\mathcal{R}(t,\nu)_{r,i} = \mathcal{Z}_\alpha^* \varrho\Big(t - r\,\frac{p}{q}, \nu + \frac{l_i}{p}\Big), \quad 0 \le r, i \le q-1,$$

is invertible when $(t,\nu) = (t_0,\nu_0)$, and thus also on some neighborhood V of that point. Denoting by $\mathcal{R}^{-1}(t,\nu)$ the inverse of $\mathcal{R}(t,\nu)$, it is easily seen that all the entries of $\mathcal{R}^{-1}$ belong to $C^\infty(V)$. If $S_0, S_1, \ldots, S_{q-1}$ are $\mathbb{Z}\oplus\frac{1}{p}\,\mathbb{Z}$-periodic distributions in $\mathcal{S}'(\mathbb{R}^2)$ satisfying the equations (6.41), we have

$$\begin{aligned}
0 &= \sum_{i=0}^{q-1}\sum_{r=0}^{q-1} S_r(t,\nu)\, \mathcal{Z}_\alpha^* \varrho\Big(t - r\,\frac{p}{q}, \nu + \frac{l_i}{p}\Big)\, \mathcal{R}^{-1}(t,\nu)_{i,s} \\
&= \sum_{r=0}^{q-1} S_r(t,\nu) \sum_{i=0}^{q-1} \mathcal{Z}_\alpha^* \varrho\Big(t - r\,\frac{p}{q}, \nu + \frac{l_i}{p}\Big)\, \mathcal{R}^{-1}(t,\nu)_{i,s} \\
&= S_s(t,\nu) \text{ on } V, \quad s = 0,\ldots,q-1.
\end{aligned}$$

Since (t_0,ν_0) was an arbitrary point in $\mathbb{R}^2$, we have $S_s = 0$ on all of $\mathbb{R}^2$, for each $s = 0,\ldots,q-1$, which shows that the system $\mathbf{G}(\varrho,\alpha,\beta)$ has the Riesz property. This proves our claim. □

Corollary 6.26. *If a function $\varrho \in \mathcal{S}(\mathbb{R})$ and α, $\beta > 0$ with $\alpha\beta \in \mathbb{Q}$ are such that the system $\mathbf{G}(\varrho,\alpha,\beta)$ has the Riesz property, then necessarily $\alpha\beta > 1$.*

Proof. When $\alpha\beta \ne 1$, this follows from the previous theorem using a simple rank argument and when $\alpha\beta = 1$ from the fact that the Zak transform of ϱ vanishes somewhere on the unit square $I \times I$. □

6.4 The Dual Window in $\mathcal{S}(\mathbb{R})$

In this last section, we consider the problem of reconstructing a tempered distribution T on $\mathbb{R}$ from the data $\{\langle T, \varrho_{m\alpha,n\beta}\rangle : m,n \in \mathbb{Z}\}$, where $\varrho \in \mathcal{S}(\mathbb{R})$, in the case where $\alpha\beta$ is rational and less than 1. We first prove the following lemma.

Lemma 6.27. *Given a function $\varrho \in \mathcal{S}(\mathbb{R})$, α, $\beta > 0$ with $\alpha\beta = \frac{p}{q}$ as in (6.32) and a distribution $T \in \mathcal{S}'(\mathbb{R})$, the sequence $\{\langle T, \varrho_{m\alpha,n\beta}\rangle\}_{(m,n)\in\mathbb{Z}^2}$ is of polynomial growth.*

Proof. Let Ψ_0 be any function in $C_0^\infty(\mathbb{R}^2)$ satisfying (6.21). Then, repeating the computation in (6.35), we have, for each $r = 0, \ldots, q-1$,

$$\langle T, \varrho_{m\alpha, \frac{1}{\alpha}(np + r\frac{p}{q})}\rangle = \Big\langle \mathcal{Z}_\alpha T(t,\nu)\, \mathcal{Z}^*_\alpha \varrho\Big(t - r\frac{p}{q}, \nu\Big),\, e^{2\pi i m t}\, e^{2\pi i n p \nu}\, \Psi_0(t,\nu)\Big\rangle,$$

and the claim follows easily from the fact that, for each $r = 0, \ldots, q-1$, the distribution $\mathcal{Z}_\alpha T(t,\nu)\, \mathcal{Z}^*_\alpha \varrho\big(t - r\frac{p}{q}, \nu\big)$ has locally finite order (see [17]). □

We conclude this paper with the following result which shows that if $\alpha\beta \in \mathbb{Q}$ and $\varrho \in \mathcal{S}(\mathbb{R})$ is such that every $T \in \mathcal{S}'(\mathbb{R})$ is uniquely determined by the data $\{\langle T, \varrho_{m\alpha,n\beta}\rangle : m, n \in \mathbb{Z}\}$, then there exists a dual window in $\mathcal{S}(\mathbb{R})$ providing a perfect reconstruction formula in terms of a series converging in $\mathcal{S}'(\mathbb{R})$. There is a corresponding result in the L^2-case proved by Janssen ([14]; see also [9] for another approach): if $\varrho \in \mathcal{S}(\mathbb{R})$ is such that the system $\mathbf{G}(\varrho, \alpha, \beta)$ forms a frame for $L^2(\mathbb{R})$, then the dual window (which is used in the L^2-reconstruction formula) also belongs to $\mathcal{S}(\mathbb{R})$. The theorem proved below is perhaps a bit surprising since the L^2-result does not hold under the weaker assumption that the span of the system $\mathbf{G}(\varrho, \alpha, \beta)$ is dense in $L^2(\mathbb{R})$ (for example for a Gaussian window when $\alpha\beta = 1$).

Theorem 6.28. *Given a function $\varrho \in \mathcal{S}(\mathbb{R})$ and α, $\beta > 0$ with $\alpha\beta = \frac{p}{q}$ as in (6.32), suppose that the system $\mathbf{G}(\varrho, \alpha, \beta)$ is dense in $\mathcal{S}(\mathbb{R})$. Then there exists a function $\psi \in \mathcal{S}(\mathbb{R})$ such that*

$$T = \sum_{m,n\in\mathbb{Z}} \langle T, \varrho_{m\alpha,n\beta}\rangle\, \psi_{-m\alpha,n\beta}, \quad T \in \mathcal{S}'(\mathbb{R}), \tag{6.43}$$

where the series in (6.43) converges in $\mathcal{S}'(\mathbb{R})$.

Proof. Note first that if ϱ and ψ are arbitrary functions in $\mathcal{S}(\mathbb{R})$, the corresponding series in (6.43) converges in $\mathcal{S}'(\mathbb{R})$, for all $T \in \mathcal{S}'(\mathbb{R})$, using Lemma 6.27 and Lemma 6.23. In particular, we can define an operator $\mathcal{U} : \mathcal{S}'(\mathbb{R}) \to \mathcal{S}'(\mathbb{R})$ by the formula

$$\begin{aligned}\mathcal{U}(T) &= \sum_{m,n\in\mathbb{Z}} \langle T, \varrho_{m\alpha,n\beta}\rangle\, \overline{\varrho_{m\alpha,n\beta}},\\ &= \sum_{r=0}^{q-1} \sum_{m,n\in\mathbb{Z}} \langle T, \varrho_{m\alpha,\frac{1}{\alpha}(nq+r)\frac{p}{q}}\rangle\, (\overline{\varrho})_{-m\alpha,\frac{1}{\alpha}(nq+r)\frac{p}{q}}, \quad T \in \mathcal{S}'(\mathbb{R}).\end{aligned}$$

Using equation (6.19) and Lemma 6.19, we compute

$$\begin{aligned}
&\mathcal{Z}_\alpha[\mathcal{U}(T)]\,(t,\nu)\\
&\quad=\sum_{r=0}^{q-1}\sum_{m,n\in\mathbb{Z}}\Big\{\Big\langle \mathcal{Z}_\alpha T(t,\nu)\,\mathcal{Z}_\alpha^*\varrho\Big(t-r\,\frac{p}{q},\nu\Big),\,e^{2\pi imt}\,e^{2\pi inp\nu}\,\Psi_0(t,\nu)\Big\rangle\\
&\qquad\qquad\times e^{-2\pi imt}\,e^{-2\pi inp\nu}\,\mathcal{Z}_\alpha[\,\overline{\varrho}\,]\Big(t-r\,\frac{p}{q},\nu\Big)\Big\}.
\end{aligned}$$

Since $\mathcal{Z}_\alpha T(t,\nu)\,\mathcal{Z}_\alpha^*\varrho(t-r\,\frac{p}{q},\nu)$ is $\mathbb{Z}^2$-periodic,

$$\begin{aligned}
&\sum_{m,n\in\mathbb{Z}}\Big\langle \mathcal{Z}_\alpha T(t,\nu)\,\mathcal{Z}_\alpha^*\varrho\Big(t-r\,\frac{p}{q},\nu\Big),\,e^{2\pi imt}\,e^{2\pi inp\nu}\,\Psi_0(t,\nu)\Big\rangle\,e^{-2\pi imt}\,e^{-2\pi inp\nu}\\
&\quad=\frac{1}{p}\sum_{l=0}^{p-1}\mathcal{Z}_\alpha T\Big(t,\nu+\frac{l}{p}\Big)\,\mathcal{Z}_\alpha^*\varrho\Big(t-r\,\frac{p}{q},\nu+\frac{l}{p}\Big).
\end{aligned}$$

This shows that

$$\begin{aligned}
&\mathcal{Z}_\alpha[\mathcal{U}(T)]\,(t,\nu)\\
&\quad=\frac{1}{p}\sum_{r=0}^{q-1}\sum_{l=0}^{p-1}\mathcal{Z}_\alpha T\Big(t,\nu+\frac{l}{p}\Big)\,\mathcal{Z}_\alpha^*\varrho\Big(t-r\,\frac{p}{q},\nu+\frac{l}{p}\Big)\mathcal{Z}_\alpha[\,\overline{\varrho}\,]\Big(t-r\,\frac{p}{q},\nu\Big)\\
&\quad=\frac{1}{p}\sum_{l=0}^{p-1}\sum_{r=0}^{q-1}\mathcal{Z}_\alpha T\Big(t,\nu+\frac{l}{p}\Big)\,\mathcal{Z}_\alpha^*\varrho\Big(t-r\,\frac{p}{q},\nu+\frac{l}{p}\Big)\,\overline{\mathcal{Z}_\alpha^*\varrho}\Big(t-r\,\frac{p}{q},\nu\Big),
\end{aligned}$$

or, equivalently, using the $\mathbb{Z}^2$-periodicity of $\mathcal{Z}_\alpha T\,\mathcal{Z}_\alpha^*\varrho$, that

$$\begin{aligned}
&\mathcal{Z}_\alpha[\mathcal{U}(T)]\,\Big(t,\nu+\frac{l_1}{p}\Big)\\
&\quad=\frac{1}{p}\sum_{l_2=0}^{p-1}\sum_{r=0}^{q-1}\mathcal{Z}_\alpha^*\varrho\Big(t-r\,\frac{p}{q},\nu+\frac{l_2}{p}\Big)\,\overline{\mathcal{Z}_\alpha^*\varrho}\Big(t-r\,\frac{p}{q},\nu+\frac{l_1}{p}\Big)\,\mathcal{Z}_\alpha T\Big(t,\nu+\frac{l_2}{p}\Big)\\
&\quad=\frac{1}{p}\sum_{l_2=0}^{p-1}\,[\mathcal{G}^*(t,\nu)\,\mathcal{G}(t,\nu)]_{l_1,l_2}\,\mathcal{Z}_\alpha T\Big(t,\nu+\frac{l_2}{p}\Big),
\end{aligned}$$

for each $l_1=0,\dots,p-1$, where $\mathcal{G}^*(t,\nu)$ denotes the adjoint of $\mathcal{G}(t,\nu)$. Letting $\mathcal{P}(t,\nu)=\mathcal{G}^*(t,\nu)\,\mathcal{G}(t,\nu)$, for all $(t,\nu)\in\mathbb{R}^2$, it is easily seen that $\mathcal{P}$ is a $p\times p$ matrix-valued function whose entries are all in $C^\infty(\mathbb{R}^2)$ and $\mathbb{Z}^2$-periodic. Furthermore, for any integer $l_1=0,\dots,p-1$,

$$\mathcal{Z}_\alpha[\mathcal{U}(T)]\,\Big(t,\nu+\frac{l_1}{p}\Big)=\frac{1}{p}\sum_{l_2=0}^{p-1}\mathcal{P}(t,\nu)_{l_1,l_2}\,\mathcal{Z}_\alpha T\Big(t,\nu+\frac{l_2}{p}\Big). \tag{6.44}$$

By Theorem 6.20, the fact that the system $\mathbf{G}(\varrho, \alpha, \beta)$ is dense in $\mathcal{S}(\mathbb{R})$ is equivalent to $\operatorname{rank}(G(t,\nu)) = p$ for all $(t,\nu) \in \mathbb{R}^2$, i.e., to $\mathcal{P}(t,\nu)$ being invertible at every such point. Using this and the facts that the entries of $\mathcal{P}^{-1}(t,\nu)$ are also in $C^\infty(\mathbb{R}^2)$ and $\mathbb{Z}^2$-periodic, it follows from equation (6.44) that the mapping $\mathcal{U} : \mathcal{S}'(\mathbb{R}) \to \mathcal{S}'(\mathbb{R})$ is a topological isomorphism with inverse $\mathcal{U}^{-1}$ satisfying for any integer $l_1 = 0, \ldots, p-1$,

$$\mathcal{Z}_\alpha\big[\mathcal{U}^{-1}(T)\big]\left(t, \nu + \frac{l_1}{p}\right) = p \sum_{l_2=0}^{p-1} \mathcal{P}^{-1}(t,\nu)_{l_1,l_2}\, \mathcal{Z}_\alpha T\left(t, \nu + \frac{l_2}{p}\right). \tag{6.45}$$

For the same reason, the mapping $\mathcal{U} : \mathcal{S}(\mathbb{R}) \to \mathcal{S}(\mathbb{R})$ is a topological isomorphism with inverse satisfying the same equation (6.45). We have, in particular, for any distribution $T \in \mathcal{S}'(\mathbb{R})$, that

$$T = \sum_{m,n\in\mathbb{Z}} \langle T, \varrho_{m\alpha,n\beta} \rangle\, \mathcal{U}^{-1}\big(\overline{\varrho_{m\alpha,n\beta}}\big). \tag{6.46}$$

Since, for any $\varphi \in \mathcal{S}(\mathbb{R})$, we have $\mathcal{U}(\varphi_{m\alpha,n\beta}) = (\mathcal{U}\varphi)_{m\alpha,n\beta}$, it follows that $\mathcal{U}^{-1}(\varphi_{m\alpha,n\beta}) = (\mathcal{U}^{-1}\varphi)_{m\alpha,n\beta}$. Therefore,

$$\mathcal{U}^{-1}\big(\overline{\varrho_{m\alpha,n\beta}}\big) = \mathcal{U}^{-1}\big((\overline{\varrho})_{-m\alpha,n\beta}\big) = \big(\mathcal{U}^{-1}(\overline{\varrho})\big)_{-m\alpha,n\beta} \tag{6.47}$$

and (6.43) follows from (6.46) and (6.47) if we define $\psi = \mathcal{U}^{-1}(\overline{\varrho})$, which concludes the proof. □

It seems reasonable to expect the previous result to hold even in the case where $\alpha\beta$ is irrational but, unfortunately, the Zak transform techniques, which are so powerful in the rational case, do not seem to be of any help in the irrational one. We leave this as an open problem for the interested reader.

Acknowledgments

The author was supported by an NSERC grant.

References

1. L. Baggett, Processing a radar signal and representations of the discrete Heisenberg group, *Colloq. Math.*, **60/61** (1990), pp. 195–203.
2. R. Balan, I. Daubechies and V. Vaisshampayan, The analysis and design of windowed Fourier frame based multiple description encoding schemes, *IEEE Trans. Inform. Theory*, **46** (2000), pp. 2491–2536.
3. J. J. Benedetto, *Harmonic analysis and applications*, CRC Press, Boca Raton, FL, 1997.

4. O. Christensen, *An Introduction to Frames and Riesz bases*, Birkhäuser, Boston, 2003.
5. I. Daubechies, The wavelet transform, time-frequency localization and signal analysis, *IEEE Trans. Inform. Theory*, **36** (1990), pp. 961–1005.
6. H. G. Feichtinger and T. Strohmer, eds., *Gabor Analysis and Algorithms: Theory and Applications*, Birkhäuser, Boston, 1998.
7. H. G. Feichtinger and T. Strohmer, eds., *Advances in Gabor Analysis*, Birkhäuser, Boston, 2003.
8. J.-P. Gabardo and D. Han, Balian–Low phenomenon for subspace Gabor frames, *J. Math. Phys.*, **45** (2004), pp. 3362–3378.
9. K. Gröchenig, *Foundations of Time-Frequency Analysis*, Birkhäuser, Boston, 2001.
10. C. E. Heil and D. F. Walnut, Continuous and discrete wavelet transforms, *SIAM Review*, **31** (1989), pp. 628–666.
11. A. J. E. M. Janssen, Gabor representation of generalized functions, *J. Math. Anal. Appl.*, **83** (1981), pp. 377–394.
12. A. J. E. M. Janssen, Weighted Wigner distributions vanishing on lattices, *J. Math. Anal. Appl.*, **80**, (1981), pp. 156–167.
13. A. J. E. M. Janssen, Bargmann transform, Zak transform, and coherent states, *J. Math. Phys.*, **23** (1982), pp. 720–731.
14. A. J. E. M. Janssen, Duality and biorthogonality for Weyl–Heisenberg frames, *J. Fourier. Anal. Appl.*, **1** (1995), pp. 403–436.
15. A. J. E. M. Janssen, The duality condition for Weyl–Heisenberg frames, in: *Advances in Gabor Analysis*, H. G. Feichtinger and T. Strohmer, eds., Birkhäuser, Boston, 2003, pp. 33–84.
16. M. Rieffel, Von Neumann algebras associated with pairs of lattices in Lie groups, *Math. Ann.*, **257** (1981), pp. 403–418.
17. L. Schwartz, *Théorie des Distributions*, Hermann, Paris, 1966.
18. A. I. Zayed and P. Mikusiński, On the extension of the Zak transform, *Methods Appl. Anal.*, **2** (1995), pp. 160–172.
19. M. Zibulski and Y. Y. Zeevi, Oversampling in the Gabor scheme, *IEEE Trans. Signal Proc.*, **41** (1993), pp. 2679–2687.
20. M. Zibulski and Y. Y. Zeevi, Analysis of multiwindow Gabor-type schemes by frame methods, *Appl. Comput. Harmon. Anal.*, **4** (1997), pp. 188–121.

7

Gabor Duality Characterizations

Eric Hayashi[1], Shidong Li[2], and Tracy Sorrells[3]

[1] Department of Mathematics, San Francisco State University, San Francisco, CA 94132, USA
hayashi@math.sfsu.edu

[2] Department of Mathematics, San Francisco State University, San Francisco, CA 94132, USA
shidong@sfsu.edu

[3] Department of Mathematics, San Francisco State University, San Francisco, CA 94132, USA

Summary. Gabor duality studies have resulted in a number of characterizations of dual Gabor frames, among which the Wexler–Raz identity and the operator approach reformulation by Janssen and by Daubechies, Landau, and Landau are well known. A concise overview of existing Gabor duality characterizations is presented. In particular, we demonstrate that the Gabor duality conditions by Wexler and Raz [23] and by Daubechies, Landau, and Landau [6], and the parametric dual Gabor formula of [15] are equivalent.

Dedicated to Professor John Benedetto.

7.1 Introduction

Let $g \in L^2(\mathbf{R})$, and $a, b > 0$ be given. Gabor's seminal article [9] (1946) deals with "regular lattice" systems $\{g_{mb,na} \equiv g(\cdot - na)e^{2\pi imb\cdot}\}_{m,n\in\mathbf{Z}}$, where g is the Gaussian waveform. Gabor proposed to expand a function f by the series $f = \sum_{mn} c_{m,n} g_{mb,na}$ (quote of John Benedetto [2]). In 1952, Duffin and Schaeffer [7] formulated the notion of *frames* for the study of nonharmonic Fourier series. A broader study of Gabor expansions followed in the wake of the recent boom in wavelet and frame analysis. In this context, Gabor expansions are frame expansions: if $\{g_{mb,na}\}_{m,n}$ is a frame for $L^2(\mathbf{R})$ then

$$\forall\, f \in L^2(\mathbf{R}), \quad f = \sum_{mn} \langle f, \gamma(m,n)\rangle\, g_{mb,na},$$

where $\{\gamma(m,n)\}_{m,n}$ is a dual Gabor frame. Note that $\{\gamma(m,n)\}_{m,n}$ is not unique when $\{g_{mb,na}\}_{m,n}$ is inexact.

In most Gabor duality studies, the assumption is made that one is only interested in finding Gabor dual frames that are generated via the Heisenberg group structure by single *Gabor dual functions* γ, i.e., $\gamma(m,n) = \gamma_{mb,na}$. In this context, Wexler and Raz formulated a (pointwise biorthogonal) Gabor duality condition via the Poisson summation formula [23] (1990) (see also Section 7.2.1). At the same time, from the frame-theoretical point of view, it was shown that the *standard dual* Gabor frame necessarily has the Heisenberg group structure, e.g., [3] (1990). That is, $S_{g;b,a}^{-1}(g_{mb,na}) = \left(S_{g;b,a}^{-1}g\right)_{mb,na}$ for $m,\ n \in \mathbf{Z}$. Here, $S_{g;b,a}$ is the frame operator defined by $S_{g;b,a}f = \sum_{mn}\langle f, g_{mb,na}\rangle\, g_{mb,na}$ for $f \in L^2(\mathbf{R})$.

The standard dual frame has the important property that the corresponding Gabor coefficients have the minimum ℓ^2-norm among all the possible Gabor coefficients $\{c_{m,n}\}$ [4], [11]. For this reason, it is also called the *minimum dual* [12], [13], or sometimes the *canonical dual.*

In one of Wexler–Raz's original efforts [23], the dual γ^0 with the least square norm was sought. It was later shown (e.g., [12], [6]) that such a least square-norm solution γ^0 coincides with the standard dual Gabor function $S_{g;b,a}^{-1}g$, i.e., $\gamma^0 = S_{g;b,a}^{-1}g$. A simple examination of this property is also presented in Section 7.4 using the dual Gabor formula described in [15], [17] (one of the equivalent Gabor duality characterizations).

However, the least square property of the standard dual is not necessarily the best choice for applications with specific numerical requirements. There have been several studies of alternative Gabor dual functions, including those by Daubechies, Landau, and Landau [6] (see also [1], [5]) and by Janssen [12], [13]. Also, from the dual frame point of view, a slightly different approach is taken in [15], [17], where a formula for all dual Gabor functions via a parameter function in $L^2(\mathbf{R})$ is derived. This last approach allows for the construction of all dual Gabor functions via a parametric equation. Since the generality of the formula was not explicitly stated in [15] and [17], we include in this survey a proof that these Gabor dual characterizations are equivalent. It could also serve as a concise overview on Gabor duality conditions. Both the Wexler–Raz identity and the parametric dual Gabor formula allow for computation of nonstandard/alternative dual Gabor functions which may be better adapted for specialized applications; see [6], [23], [19], [18], [17], [14].

The organization of this chapter is as follows. Section 7.2 describes the three major Gabor duality conditions, namely the Wexler–Raz identity, the dual lattice identity by Daubechies, Landau, and Landau, and the parametric dual Gabor formula. In Section 7.3, the equivalence of these three Gabor duality conditions is presented. Then in Section 7.4, alternative dual Gabor functions and the optimality of the Gabor duals are discussed.

7.2 Gabor Duality Characterizations

Following the notation in [6], assume that $\{g_{mb,na}\}$ is a Bessel sequence, and define the *analysis operator* $T_{g;b,a} : L^2(\mathbf{R}) \to \ell^2$ by

$$T_{g;b,a} f = \{\langle f, g_{mb,na}\rangle\}_{m,n\in\mathbf{Z}}, \qquad f \in L^2(\mathbf{R}).$$

Then $T_{g;b,a}$ is bounded and its adjoint $T^*_{g;b,a}$ is given by

$$T^*_{g;b,a} c = \sum_{m,n\in\mathbf{Z}} c_{m,n} g_{mb,na}, \qquad c = \{c_{m,n}\} \in \ell^2.$$

In particular, if $\{g_{mb,na}\}$ is a frame and $\{\gamma_{mb,na}\}$ is its dual frame, then $T^*_{g;b,a} T_{\gamma;b,a} = T^*_{\gamma;b,a} T_{g;b,a} = I$.

7.2.1 The Wexler–Raz Identity

The *Wexler–Raz identity,* as rigorously formulated by Janssen [12], states the following.

Proposition 7.1. *Let* $\{g_{mb,na}\}$ *be a Gabor frame for* $L^2(\mathbf{R})$*, and let* $\{\gamma_{mb,na}\}$ *be a Bessel sequence in* $L^2(\mathbf{R})$*. Then* $\{\gamma_{mb,na}\}$ *is a dual Gabor frame for* $L^2(\mathbf{R})$ *if and only if*

$$\langle \gamma, g_{\frac{m}{a},\frac{n}{b}}\rangle = ab\, \delta_{m0}\delta_{n0}, \qquad m, n \in \mathbf{Z}.$$

Remark 7.2. The Wexler–Raz identity transforms an identity on the Gabor lattice $\{mb, na\}$ to an identity on a *dual Gabor lattice* $\{\frac{m}{a}, \frac{n}{b}\}$. It turns out that much can be said about the relationship of the regular Gabor lattice and the dual Gabor lattice in the "Gabor duality principle," independently derived by Janssen [12, Thm. 3.1], Daubechies, Landau, and Landau [6], and Ron and Shen [20] (see also the discussion and references in [10]).

The Gabor duality principle states the following fact.

Proposition 7.3. $\{g_{mb,na}\}$ *is a frame for* $L^2(\mathbf{R})$ *with frame bounds* A *and* B *if and only if* $\{g_{\frac{m}{a},\frac{n}{b}}\}$ *is a Riesz sequence in* $L^2(\mathbf{R})$ *with frame bounds* abA *and* abB*.*

Here a *Riesz sequence* is a sequence that forms a Riesz basis for its closed linear span.

7.2.2 The Dual Lattice Identity

Relationships between the Gabor and dual Gabor lattices have been studied in a number of papers (as noted above). In particular, the following relationship appears in the papers by Janssen [12], [13], and Daubechies, Landau, and Landau [6]; compare also the paper of Tolimieri and Orr [21].

Proposition 7.4. *Let f, g, $h \in L^2(\mathbf{R})$ and a, $b > 0$ be such that $T_{f;b,a}$, $T_{g;b,a}$, and $T_{h;\frac{1}{a},\frac{1}{b}}$ are bounded. Then*

$$T^*_{f;b,a} T_{g;b,a} h = \frac{1}{ab} T^*_{h;\frac{1}{a},\frac{1}{b}} T_{g;\frac{1}{a},\frac{1}{b}} f. \tag{7.1}$$

Note that the boundedness of $T_{g;\frac{1}{a},\frac{1}{b}}$ is implied by the boundedness of $T_{g;b,a}$. This is discussed extensively in [6].

Equation (7.1) is known variously as the *Janssen representation of the frame operator* [10, Sec. 7.2], the *Wexler–Raz identity* [6], the *fundamental identity of Gabor analysis* [13], [8], or (the terminology that we shall use) the *dual lattice identity.*

Wexler–Raz's original identity can now be reformulated as the following operator identity, where we use the notation $e_{m,n} \equiv \{\delta_{mj}\delta_{nk}\}_{j,k\in\mathbf{Z}}$.

Proposition 7.5. *Let $\{g_{mb,na}\}$ be a frame for $L^2(\mathbf{R})$, and assume that $\gamma \in L^2(\mathbf{R})$ is such that $T_{\gamma;b,a}$ is bounded. Then $\{\gamma_{mb,na}\}$ is a dual Gabor frame if and only if*

$$T_{g;\frac{1}{a},\frac{1}{b}} \gamma = ab\, e_{0,0}, \tag{7.2}$$

or, equivalently,

$$\frac{1}{ab} T_{g;\frac{1}{a},\frac{1}{b}} T^*_{\gamma;\frac{1}{a},\frac{1}{b}} = I. \tag{7.3}$$

The Wexler–Raz operator identity in (7.3) on the Gabor dual lattice (derived in [6]) reveals a symmetric expression for the Gabor frame identity $T^*_{g;b,a} T_{\gamma;b,a} = I$ on the regular lattice. To understand (7.3), we can see that (7.2) is equivalent to

$$T_{g;\frac{1}{a},\frac{1}{b}} T^*_{\gamma;\frac{1}{a},\frac{1}{b}} e_{m,n} = ab\, e_{m,n},$$

which gives rise to (7.3).

7.2.3 The Dual Gabor Formula

It was shown in [15] that among all the dual Gabor frames, those that are generated by translations and modulations of (single) Gabor dual functions can be explicitly constructed. The class of all Gabor dual functions can be given in a parametric formula.

Proposition 7.6. *Let $\{g_{mb,na}\}$ be a Gabor frame for $L^2(\mathbf{R})$, and set $\tilde{\gamma} \equiv S^{-1}_{g;b,a} g$. Then γ is a Gabor dual function if and only if*

$$\gamma = \tilde{\gamma} + \xi - \sum_{mn} \langle \tilde{\gamma}, g_{mb,na} \rangle\, \xi_{mb,na}, \tag{7.4}$$

where $\xi \in L^2(\mathbf{R})$ is such that $\{\xi_{mb,na}\}$ is a Bessel sequence in $L^2(\mathbf{R})$.

Remark 7.7. The dual Gabor formula (7.4) was derived from the general dual frame formula [15]. While Gabor duality studies by Wexler and Raz and by others are mainly focused on the study of dual Gabor frames that are generated by one window function γ, i.e., $\gamma(m,n) = \gamma_{mb,na}$, the paper [15] also pointed out constructively that dual Gabor frames may be generated by entirely different waveforms that do not have the Heisenberg group structure.

7.3 Equivalence of Gabor Duality Characterizations

We will show in this section that the aforementioned Gabor duality characterizations are equivalent. We begin with the following elementary lemmas.

Lemma 7.8. *Let $W_{mb,na} = E_{mb}T_{na}$ be the composition of the modulation and translation operators, i.e., $W_{mb,na}f(t) \equiv f(t-na)e^{2\pi imbt} = f_{mb,na}$. If $\{f_{mb,na}\}$ and $\{g_{mb,na}\}$ are Bessel sequences, then $W_{mb,na}$ commutes with $T^*_{f;b,a}T_{g;b,a}$ for every m, $n \in \mathbf{Z}$.*

Proof. The proof of this commutativity is similar to that of the commutativity between $W_{mb,na}$ and the Gabor frame operator $S_{g;b,a}$. □

Lemma 7.9. *If $A : L^2(\mathbf{R}) \to L^2(\mathbf{R})$ is bounded and commutes with $W_{mb,na}$ for every m, $n \in \mathbf{Z}$, then $AT^*_{f;b,a} = T^*_{Af;b,a}$ whenever $T^*_{f;b,a}$ and $T^*_{Af;b,a}$ are well-defined.*

Proof. Let $e_{m,n} = \{\delta_{mj}\delta_{nk}\}_{j,k\in\mathbf{Z}}$. Then the assertion follows from the following computation:

$$\begin{aligned} T^*_{Af;b,a}e_{m,n} &= (Af)_{mb,na} \\ &= W_{mb,na}(Af) \\ &= A(W_{mb,na}f) \\ &= Af_{mb,na} \\ &= AT^*_{f;b,a}e_{m,n}. \quad \square \end{aligned}$$

Lemma 7.10. *Let $c = \{c_{m,n}\}$ be a nontrivial sequence in ℓ^2, and let $f \in L^2(\mathbf{R})$ be any smooth and nontrivial function of compact support such that $|\mathrm{supp}(f)| < \min\{a, \frac{1}{b}\}$. Then $T_{f;\frac{1}{a},\frac{1}{b}}$ and $T_{f;b,a}$ are bounded, and $T^*_{f;b,a}(c) \neq 0$.*

Proof. This can be seen by the fact that the translates of f are disjoint with respect to the translations specified by both the lattice and the dual lattice. □

Now we can prove the desired equivalence.

Theorem 7.11. *Let $\{g_{mb,na}\}$ be a Gabor frame in $L^2(\mathbf{R})$ with standard dual frame $\{\tilde{\gamma}_{mb,na}\}$. Suppose $\gamma \in L^2(\mathbf{R})$. Then the following statements are equivalent.*

(i) $T^*_{\gamma;b,a}$ *is bounded and* $T^*_{\gamma;b,a}T_{g;b,a} = I$.

(ii) $\gamma = \tilde{\gamma} + \xi - \sum_{mn} \langle \tilde{\gamma}, g_{mb,na}\rangle\, \xi_{mb,na}$, *where* $\xi \in L^2(\mathbf{R})$ *is such that* $\{\xi_{mb,na}\}$ *is a Bessel sequence.*

(iii) $T_{g;\frac{1}{a},\frac{1}{b}}\gamma = ab\, e_{0,0}$.

Proof. (i) $\Rightarrow$ (ii). Suppose that $T^*_{\gamma;b,a}T_{g;b,a} = I$, i.e., that $\{\gamma_{mb,na}\}$ is a dual Gabor frame. We shall show that γ has the form given in statement (ii). Let $\xi = \gamma - \tilde{\gamma}$ be the difference between the dual function γ and the standard dual function $\tilde{\gamma}$. Then $\{\xi_{mb,na}\}$ is a Bessel sequence, and $\xi_{mb,na} = \gamma_{mb,na} - \tilde{\gamma}_{mb,na}$. Hence, $T^*_{\xi;b,a} = T^*_{\gamma;b,a} - T^*_{\tilde{\gamma};b,a}$ and is bounded, and

$$T^*_{\xi;b,a}T_{g;b,a} = T^*_{\gamma;b,a}T_{g;b,a} - T^*_{\tilde{\gamma};b,a}T_{g;b,a} = I - I = 0.$$

Therefore,

$$\gamma = \tilde{\gamma} + \xi - 0 = \tilde{\gamma} + \xi - T^*_{\xi;b,a}T_{g;b,a}\tilde{\gamma} = \tilde{\gamma} + \xi - \sum_{m,n} \langle \tilde{\gamma}, g_{mb,na}\rangle\, \xi_{mb,na}.$$

(ii) $\Rightarrow$ (i). Suppose that $\xi \in L^2(\mathbf{R})$ is such that $\{\xi_{mb,na}\}$ is a Bessel sequence. Then $\gamma = \tilde{\gamma} + \xi - \sum_{mn} \langle \tilde{\gamma}, g_{mb,na}\rangle\, \xi_{mb,na}$ is in $L^2(\mathbf{R})$ and $\{\gamma_{mb,na}\}$ is Bessel. Set $A = T^*_{\xi;b,a}T_{g;b,a}$. By Lemmas 7.8 and 7.9 we have $T^*_{A\tilde{\gamma};b,a} = AT^*_{\tilde{\gamma};b,a}$, so

$$T^*_{\gamma;b,a} = T^*_{(\tilde{\gamma}+\xi-A\tilde{\gamma});b,a} = T^*_{\tilde{\gamma};b,a} + T^*_{\xi;b,a} - T^*_{A\tilde{\gamma};b,a} = T^*_{\tilde{\gamma};b,a} + T^*_{\xi;b,a} - AT^*_{\tilde{\gamma};b,a}.$$

Therefore,

$$T^*_{\gamma;b,a}T_{g;b,a} = T^*_{\tilde{\gamma};b,a}T_{g;b,a} + T^*_{\xi;b,a}T_{g;b,a} - AT^*_{\tilde{\gamma};b,a}T_{g;b,a} = I + A - AI = I.$$

The equivalence of (i) and (iii) is established in [6], and we include a proof here for a quick reference.

(i) $\Rightarrow$ (iii). Assume that $T^*_{\gamma;b,a}T_{g;b,a} = I$. For every "nice" function f (smooth and having small compact support as specified in Lemma 7.10), $T_{f;b,a}$ and $T_{f;\frac{1}{a},\frac{1}{b}}$ will be bounded. So by the dual lattice identity (7.1),

$$T^*_{f;\frac{1}{a},\frac{1}{b}}T_{g;\frac{1}{a},\frac{1}{b}}\gamma = ab\, T^*_{\gamma;b,a}T_{g;b,a}f = abf = ab\, T^*_{f;\frac{1}{a},\frac{1}{b}}e_{0,0}.$$

Hence,

$$T^*_{f;\frac{1}{a},\frac{1}{b}}\left(T_{g;\frac{1}{a},\frac{1}{b}}\gamma - ab\, e_{0,0}\right) = 0. \tag{7.5}$$

By Lemma 7.10, (7.5) implies that $T_{g;\frac{1}{a},\frac{1}{b}}\gamma - ab\, e_{0,0} = 0$.

(iii) $\Rightarrow$ (i). Consider any nice $f \in L^2(\mathbf{R})$, say f in the Schwartz class. Then $T_{f;\frac{1}{a},\frac{1}{b}}$ is bounded, and

$$T^*_{f;\frac{1}{a},\frac{1}{b}}T_{g;\frac{1}{a},\frac{1}{b}}\gamma = ab\, T^*_{f;\frac{1}{a},\frac{1}{b}}e_{0,0} = abf.$$

Therefore, by the dual lattice relation (7.1),

$$f = \frac{1}{ab} T^*_{f;\frac{1}{a},\frac{1}{b}} T_{g;\frac{1}{a},\frac{1}{b}} \gamma = T^*_{\gamma;b,a} T_{g;b,a} f.$$

Thus, $T^*_{\gamma;b,a} T_{g;b,a} f = f$ holds for all f in a set whose linear span is dense in $L^2(\mathbf{R})$, implying that $T^*_{\gamma;b,a} T_{g;b,a} = I$. □

Remark 7.12. (a) The dual Gabor function formula (7.4) can be substantially simplified by working out the summation term. A standard calculation of Gabor representations, e.g., [14], shows that

$$\xi - \sum_{m,n} \langle \tilde{\gamma}, g_{mb,na} \rangle \, \xi_{mb,na} = \sum_n H(\cdot, n)\, \xi(\cdot - na),$$

where

$$H(\cdot, n) \equiv \begin{cases} 1 - \sum_k \tilde{\gamma}(\cdot + \frac{k}{b})\, \overline{g(\cdot + \frac{k}{b})}, & n = 0, \\ -\sum_k \tilde{\gamma}(\cdot + \frac{k}{b})\, \overline{g(\cdot + \frac{k}{b} - na)}, & n \neq 0. \end{cases}$$

Therefore, formula (7.4) becomes

$$\gamma = \tilde{\gamma} + \sum_n H(\cdot, n)\, \xi(\cdot - na),$$

which is much more computationally efficient than the original form using complex-valued Gabor sequences.

(b) The dual Gabor formula (7.4) (item (ii) of Theorem 7.11) can also be written as

$$\begin{aligned} \gamma &= \tilde{\gamma} + \xi - \sum_{mn} \langle \tilde{\gamma}, g_{mb,na} \rangle \, \xi_{mb,na} \\ &= \tilde{\gamma} + \xi - \sum_{mn} \langle \xi, g_{\frac{m}{a},\frac{n}{b}} \rangle \, \tilde{\gamma}_{\frac{m}{a},\frac{n}{b}}. \end{aligned}$$

This simply results from the dual lattice identity (7.1), applied to the summation term of (7.4). Therefore, the modification terms (the second and the third terms) in the dual Gabor formula constitute a component in the orthogonal complement of the subspace spanned by the Riesz basic sequence $\{g_{\frac{m}{a},\frac{n}{b}}\}$.

(c) The form of $H(\cdot, n)$ is strikingly reminiscent of the *Walnut representation of the frame operator*, compare [10, Sec. 7.1], [11, Thm. 4.2.1], [22].

7.4 Discussion: Optimality Considerations and Alternative Gabor Duals

In this section we will demonstrate how the dual Gabor formula can be used in optimality considerations and for finding alternative Gabor duals. Considerations of alternative Gabor duals by Daubechies, Landau, and Landau and by Janssen will also be discussed.

7.4.1 The Least Square Property of the Standard Dual via the Dual Gabor Formula

We provide another reasoning in this section showing that if γ is any alternative dual Gabor function for a function g that generates a Gabor frame, then $\|S_{g;b,a}^{-1} g\|_2 \leq \|\gamma\|_2$. First we require the following lemma.

Lemma 7.13. *Let f, g, p, and q be functions in $L^2(\mathbf{R})$ such that $T_{f;b,a}$, $T_{g;b,a}$, $T_{p;b,a}$, and $T_{q;b,a}$ are bounded. Then*

$$\langle T_{p;b,a}^* T_{f;b,a} g,\, q\rangle = \langle p,\, T_{q;b,a}^* T_{g;b,a} f\rangle.$$

Proof.

$$\begin{aligned}
\langle T_{p;b,a}^* T_{f;b,a} g,\, q\rangle &= \sum_{m,n} \langle g,\, f_{mb,na}\rangle \langle p_{mb,na},\, q\rangle \\
&= \sum_{m,n} \langle g,\, f_{mb,na}\rangle \langle p,\, q_{-mb,-na} e^{-2\pi i mnab}\rangle \\
&= \Big\langle p,\, \sum_{m,n} \langle f_{mb,na},\, g\rangle\, q_{-mb,-na} e^{-2\pi i mnab}\Big\rangle \\
&= \Big\langle p,\, \sum_{m,n} \langle f,\, g_{-mb,-na} e^{-2\pi i mnab}\rangle\, q_{-mb,-na} e^{-2\pi i mnab}\Big\rangle \\
&= \Big\langle p,\, \sum_{m,n} \langle f,\, g_{-mb,-na}\rangle\, q_{-mb,-na}\Big\rangle \\
&= \langle p,\, T_{q;b,a}^* T_{g;b,a} f\rangle. \quad \square
\end{aligned}$$

We can now easily establish the following.

Proposition 7.14. *Let $\{g_{mb,na}\}$ be a Gabor frame for $L^2(\mathbf{R})$. Let $\tilde{\gamma} = S_{g;b,a}^{-1} g$, the standard dual Gabor function. Assume that γ is an alternative dual Gabor function. Then*

$$\langle \gamma - \tilde{\gamma}, \tilde{\gamma}\rangle = 0.$$

Proof. By the dual Gabor formula (7.4),

$$\begin{aligned}
\langle \gamma - \tilde{\gamma}, \tilde{\gamma}\rangle &= \langle \xi, \tilde{\gamma}\rangle - \sum_{m,n} \langle \tilde{\gamma}, g_{mb,na}\rangle \langle c_{mb,na}, \tilde{\gamma}\rangle \\
&= \langle \xi, \tilde{\gamma}\rangle - \langle T_{\xi;b,a}^* T_{g;b,a} \tilde{\gamma}, \tilde{\gamma}\rangle \\
&= \langle \xi, \tilde{\gamma}\rangle - \langle \xi, T_{\tilde{\gamma};b,a}^* T_{\tilde{\gamma};b,a} g\rangle \\
&= \langle \xi, \tilde{\gamma}\rangle - \langle \xi, S_{\tilde{\gamma};b,a} g\rangle \\
&= \langle \xi, \tilde{\gamma}\rangle - \langle \xi, S_{g;b,a}^{-1} g\rangle = 0.
\end{aligned}$$

Here in the third equality we have used Lemma 7.13. $\square$

With this observation, the minimum norm property of $\tilde{\gamma}$ follows.

Corollary 7.15. *Let $\{g_{mb,na}\}$ be a Gabor frame. Let $\tilde{\gamma}$ be the standard dual Gabor function, and let γ be an alternative dual Gabor function given by (7.4). Then*

$$\|\gamma\|_2^2 = \|\tilde{\gamma}\|_2^2 + \|\xi - T^*_{\xi;b,a} T_{g;b,a}\tilde{\gamma}\|_2^2.$$

7.4.2 Compactly Supported Dual Gabor Functions via the Dual Gabor Formula

One application of the Gabor dual formula is to find dual Gabor functions with compact support, particularly for discrete finite-dimensional cases. The dual Gabor formula for discrete finite-dimensional spaces is identical to (7.4), except for the change of variable t from time into the index and the range of summations [17]. Assume that L is the dimensionality of the signal space. Assume that parameters a and $1/b$ both divide L and $N \equiv L/a$, $K \equiv Lb$. Then the dual Gabor formula becomes

$$\gamma(t) - \tilde{\gamma}(t) = \sum_{n=0}^{N-1} H(t,n)\,\xi(t - na), \quad t = 0, 1, \ldots, L-1, \tag{7.6}$$

where $H(t,n)$ is given by, for $0 \le t \le L-1$ and $0 \le n \le N-1$,

$$H(t,n) \equiv \begin{cases} 1 - \sum_{k=0}^{K-1} \tilde{\gamma}(t + \frac{k}{b})\,\overline{g(t + \frac{k}{b})}, & n = 0, \\ -\sum_{k=0}^{K-1} \tilde{\gamma}(t + \frac{k}{b})\,\overline{g(t + \frac{k}{b} - na)}, & n \neq 0. \end{cases}$$

Assume now that we are to find a dual Gabor function $\gamma(t)$ that vanishes on the two tails for $t = 0, 1, \ldots, M-1$ and $t = L-M, L-M+1, \ldots, L-1$ for sufficiently small M. Then, from (7.6), we are to solve for $\{\xi(t)\}_{t=0}^{L-1}$ from the system of equations

$$-\tilde{\gamma}(t) = \sum_{n=0}^{N-1} H(t,n)\xi(t - na), \quad 0 \le t \le M-1;\ L-M \le t \le L-1.$$

When $\{g_{mb,na}\}$ is a redundant Gabor frame system, for sufficiently small M, the above system of equations always has solutions.

We mention that compactly supported Gabor duals γ together with compactly supported g can produce a dimension-invariant Gabor frame and dual frame system for sufficiently large L [16]. That is, once L reaches a certain size, there is no need to re-compute the dual Gabor function γ. When L is further enlarged, padding g and γ with zeros to the enlarged dimension would be sufficient. In such a way, a pair of compactly supported Gabor duals obtained in a finite-dimensional space of certain dimension L can be applied to spaces of any larger dimension that is a multiple of a and $1/b$. This is very convenient for discrete-time signal processing.

7.4.3 Other Alternative Constructions

Janssen in [13] also constructed duals of compact support by the Wexler–Raz identity in some special cases. The basic idea is to assume that $ab = 1/q$ for some integer $q \geq 2$, and to assume that g and γ are supported on intervals $[0, ra]$ and $[ua, va]$, respectively, for integers $r > 0$, $u < 0$, and $v > r$. Under such assumptions, the Wexler–Raz identity then becomes a system of equations with a banded system matrix, with which solutions likely exist. We refer the reader to [13] for a further discussion.

Another notable construction of alternative Gabor duals is by Daubechies, Landau, and Landau [6]. The idea is to introduce an operator Λ so as to find an optimal γ such that $\|\Lambda\gamma\|_2$ is minimized. One way to do that is to find the function G with the smallest L^2-norm that satisfies

$$T_{g;\frac{1}{a},\frac{1}{b}} \Lambda^{-1} G = ab\, e_{0,0},$$

and then find γ through $\gamma = \Lambda^{-1}G$. In doing so, the invertibility of an operator Λ is assumed. We refer to [6] for a further discussion.

References

1. R. Balan and I. Daubechies, Optimal stochastic encoding and approximation schemes using Weyl–Heisenberg sets, in: *Advances in Gabor Analysis*, H. G. Feichtinger and T. Strohmer eds., Birkhäuser, Boston, 2003, pp. 259–320.
2. J. J. Benedetto, Irregular sampling and frames, in: *Wavelets: A Tutorial in Theory and Applications*, C. K. Chui, ed., Academic Press, Boston, 1992, pp. 445–507.
3. I. Daubechies, The wavelet transform, time-frequency localization and signal analysis, *IEEE Trans. Inform. Theory*, **39** (1990), pp. 961–1005.
4. I. Daubechies, *Ten Lectures on Wavelets*, SIAM, Philadelphia, 1992.
5. I. Daubechies, Better dual functions for Gabor time-frequency lattices, in: *Approximation theory VIII*, Vol. 2 (College Station, TX, 1995), C. K. Chui and L. L. Schumaker, eds., World Sci. Publishing, River Edge, NJ, 1995, pp. 113–116.
6. I. Daubechies, H. Landau, and Z. Landau, Gabor time-frequency lattices and the Wexler–Raz identity, *J. Fourier Anal. Appl.*, **1** (1995), pp. 437–478.
7. R. J. Duffin and A. C. Schaeffer, A class of nonharmonic Fourier series, *Trans. Amer. Math. Soc.*, **72** (1952), pp. 341–366.
8. H. G. Feichtinger and G. Zimmermann, A Banach space of test functions for Gabor analysis, in: *Gabor Analysis and Algorithms: Theory and Applications*, H. G. Feichtinger and T. Strohmer eds., Birkhäuser, Boston, 1998, pp. 123–170.
9. D. Gabor, Theory of communication, *J. Inst. Elec. Eng. (London)*, **93** (1946), pp. 429–457.
10. K. Gröchenig, *Foundations of Time-Frequency Analysis*, Birkhäuser, Boston, 2001.
11. C. E. Heil and D. F. Walnut, Continuous and discrete wavelet transforms, *SIAM Review*, **31** (1989), pp. 628–666.

12. A. J. E. M. Janssen, Duality and biorthogonality for Weyl–Heisenberg frames, *J. Fourier Anal. Appl.*, **1** (1995), pp. 403–436.
13. A. J. E. M. Janssen, The duality condition for Weyl–Heisenberg frames, in: *Gabor Analysis and Algorithms: Theory and Applications*, H. G. Feichtinger and T. Strohmer eds., Birkhäuser, Boston, 1998, pp. 33–84.
14. S. Li, A fast and parametric algorithm for discrete Gabor expansions and the role of various dual windows, in: *Wavelet Applications II*, H. H. Szu, ed., Proc. SPIE Vol. 2491, 1995, pp. 935–946.
15. S. Li, On general frame decompositions, *Numer. Funct. Anal. Optim.*, **16** (1995), pp. 1181–1191.
16. S. Li, Compactly supported Gabor duals and a dimension invariance property, in preparation.
17. S. Li and D. M. Healy Jr., A parametric class of discrete Gabor expansions, *IEEE Trans. Signal Proc.*, **44** (1996), pp. 1201–211.
18. S. Qian and D. Chen, Discrete Gabor transform, *IEEE Trans. Signal Proc.*, **41** (1993), pp. 2429–2438.
19. S. Qian, K. Chen, and S. Li, Optimal biorthogonal sequences for finite discrete Gabor expansion, *Signal Processing*, **27** (1992), pp. 177–185.
20. A. Ron and Z. Shen, Weyl–Heisenberg frames and Riesz bases in $L_2(\mathbf{R}^d)$, *Duke Math. J.*, **89** (1997), pp. 237–282.
21. R. Tolimieri and R. Orr, Poisson summation, the ambiguity function, and the theory of Weyl–Heisenberg frames, *J. Fourier Anal. Appl.*, **1** (1995), pp. 233–247.
22. D. F. Walnut, Continuity properties of the Gabor frame operator, *J. Math. Anal. Appl.*, **165** (1992), pp. 479–504.
23. J. Wexler and S. Raz, Discrete Gabor expansions, *Signal Processing*, **21** (1990), pp. 207–220.

8

A Pedestrian's Approach to Pseudodifferential Operators

Karlheinz Gröchenig

Institute of Biomathematics and Biometry, GSF - National Research Center for Environment and Health, Ingolstädter Landstraße 1, 85764 Neuherberg, Germany; Current address: Department of Mathematics, University of Vienna, Nordbergstrasse 15, A-1090 Vienna, Austria
karlheinz.groechenig@univie.ac.at

Summary. Pseudodifferential operators are an indispensable tool for the study of partial differential equations and are therefore a branch of classical analysis. In this chapter we offer an approach using time-frequency methods. In this approach time-frequency representations that are standard in signal analysis are used to set up the formalism of pseudodifferential operators, and certain classes of function spaces and symbols, the modulation spaces, arise naturally in the investigation. Although the approach is "pedestrian" and based more on engineering intuition than on "hard" analysis, strong results on boundedness and Schatten class properties are within its scope.

8.1 Motivation

Pseudodifferential operators are a generalization of partial differential operators and play a fundamental role in the analysis and understanding of partial differential equations (PDEs). As such they are often part of "hard analysis."

In this chapter we present a more elementary approach in the language of signal analysis and using methods from time-frequency analysis.

Let $A = \sum_{|\alpha|\le N} \sigma_\alpha(x)\, D^\alpha$ be a partial differential operator. If we use the Fourier inversion formula $f(x) = \int_{\mathbb{R}^d} \hat{f}(\xi)\, e^{2\pi i\xi\cdot x}\, d\xi$ and then differentiate under the integral, the partial derivatives are

$$D^\alpha f(x) = \int_{\mathbb{R}^d} \hat{f}(\xi)\, (2\pi i\xi)^\alpha\, e^{2\pi i\xi\cdot x}\, d\xi,$$

and thus the differential operator can be written informally as

$$Af(x) = \sum_{|\alpha|\le N} \sigma_\alpha(x)\, D^\alpha f(x) = \int_{\mathbb{R}^d} \underbrace{\Bigg(\sum_{|\alpha|\le N} \sigma_\alpha(x)\, (2\pi i\xi)^\alpha \Bigg)}\, \hat{f}(\xi)\, e^{2\pi i x\cdot\xi}\, d\xi\,.$$

Abbreviating the term in parentheses by $\sigma(x,\xi) = \sum_{|\alpha|\leq N} \sigma_\alpha(x)\,(2\pi i\xi)^\alpha$, we obtain

$$Af(t) = \int_{\mathbb{R}^d} \sigma(x,\xi)\,\hat{f}(\xi)\,e^{2\pi i x\cdot\xi}\,d\xi\,.$$

This form suggests introducing a general class of operators by allowing more general functions for σ. We therefore make the following definition.

Definition 8.1. *Given a symbol $\sigma \in \mathcal{S}'(\mathbb{R}^{2d})$, the pseudodifferential operator σ^{KN} is given by*

$$\sigma^{KN} f(x) = \int_{\mathbb{R}^d} \sigma(x,\xi)\,\hat{f}(\xi)\,e^{2\pi i x\cdot\xi}\,d\xi\,. \tag{8.1}$$

To distinguish σ^{KN} from other types of pseudodifferential operators, the mapping $\sigma \to \sigma^{KN}$ is called the Kohn–Nirenberg correspondence, *and σ is the* Kohn–Nirenberg symbol *of the operator.*

In the literature σ^{KN} is often denoted by $\sigma(x,D)$ as a reminder that it is a generalization of a partial differential operator.

Let us massage formula (8.1) a bit more:

$$\begin{aligned}
\sigma^{KN} f(x) &= \int_{\mathbb{R}^d} \sigma(x,\xi)\,\hat{f}(\xi)\,e^{2\pi i x\cdot\xi}\,d\xi \\
&= \iint_{\mathbb{R}^{2d}} \sigma(x,\xi)\,e^{2\pi i (x-y)\cdot\xi}\,f(y)\,d\xi\,dy \\
&= \iint_{\mathbb{R}^{2d}} \hat{\sigma}(\eta, y-x)\,e^{2\pi i \eta\cdot x}\,f(y)\,dy\,d\eta \\
&= \iint_{\mathbb{R}^{2d}} \hat{\sigma}(\eta, u)\,e^{2\pi i \eta\cdot x}\,f(u+x)\,du\,d\eta\,.
\end{aligned} \tag{8.2}$$

This formula stars the two fundamental operations of time-frequency analysis, namely the operators of translation and modulation

$$T_x f(t) = f(t-x) \qquad \text{and} \qquad M_\xi f(t) = e^{2\pi i \xi\cdot t} f(t)\,.$$

Their compositions are the time-frequency shifts $T_x M_\xi$ or $M_\xi T_x$.

We can now formally write the operator σ^{KN} as a superposition of time-frequency shifts in the form

$$\sigma^{KN} = \iint_{\mathbb{R}^{2d}} \hat{\sigma}(\eta, u)\,M_\eta T_{-u}\,du\,d\eta\,. \tag{8.3}$$

This is the *spreading representation* of the Kohn–Nirenberg correspondence. This representation of σ^{KN} suggests that we use the mathematical structure of time-frequency shifts and the corresponding methods of time-frequency analysis for the study of pseudodifferential operators.

This is the objective of this chapter. While in (8.3) we lose the connection to differentiation and PDEs, the spreading representation has a concrete interpretation in signal processing and is used to model communication channels and time-varying systems.

Assume that a signal f is transmitted from a sender to a receiver. For instance, this could be a call with a cellular phone. In the process of transmission the signal will undergo a number of distortions. For one, the signal, an electromagnetic wave, will propagate along different paths and will be reflected by obstacles, before some of the paths arrive at the receiver. As a consequence of different path lengths, the received signal $\tilde{f}$ will be a weighted superposition of time shifts. If the sender or the receiver or both are moving, then in addition f will be modified by the Doppler effect. This effect results in a frequency shift, and since $\widehat{M_\omega f}(\xi) = \hat{f}(\xi - \omega)$, the Doppler effect results in a weighted superposition of modulations. If we neglect other types of distortion, then the received signal $\tilde{f}$ is therefore a weighted superposition of time-frequency shifts, in other words, $\tilde{f}$ is of the form (8.3). In this model of mobile communications there is no reference to derivatives at all, so in this context it makes sense to investigate pseudodifferential operators with time-frequency methods.

In this chapter we offer a pedestrian approach to the analysis of pseudodifferential operators that is geared towards the understanding of pseudodifferential operators in the context and with the language of signal analysis. We will exclusively focus on the Kohn–Nirenberg correspondence. Other forms of pseudodifferential operators have a similar theory and the results discussed in this chapter can be translated easily.

We will first develop the basic concepts of time-frequency analysis in Section 8.2 and then introduce the modulation spaces in Section 8.3. These are a family of function and distribution spaces that occur naturally in time-frequency analysis, and they play an important role in the investigation of pseudodifferential operators. Then we prepare the ground for the analysis of pseudodifferential operators by making rigorous the above motivation and developing the necessary formalism. In Section 8.4 we derive some of the main results on pseudodifferential operators within time-frequency analysis. On the one hand, we describe how a pseudodifferential operator modifies the time-frequency content of a function, in other words, we study the mapping properties between modulation spaces; on the other hand, we use modulation spaces as symbol classes for pseudodifferential operators.

Although this chapter is intended as an elementary introduction to a classical topic from the point of view of time-frequency analysis, there will be several new aspects in our treatment. First, we emphasize the interaction between pseudodifferential operators and well-known time-frequency representations and the connection to signal analysis. Second, we focus on the Kohn–Nirenberg correspondence. The Kohn–Nirenberg correspondence can be extended with only minor notational changes to the analysis of operators on general locally compact Abelian groups. This is important for the treatment of discrete or digitized signals. Third, we give a non-standard definition of modulation spaces and in this way avoid some of the difficulties of the usual definitions. In particular, this approach does not require an extensive setup of (ultra) distribution theory.

8.1.1 Notation

$\mathcal{S}(\mathbb{R}^d)$ denotes the Schwartz class of test functions on $\mathbb{R}^d$, and its dual $\mathcal{S}'(\mathbb{R}^d)$ is the space of tempered distributions equipped with the weak* topology. We always take the duality $\langle\, .\, ,\, .\, \rangle$ as extending the inner product on $L^2(\mathbb{R}^d)$, thus $\langle f, g\rangle$ is conjugate-linear in g. As a consequence, Parseval's formula

$$\langle f, g\rangle = \langle \hat{f}, \hat{g}\rangle \tag{8.4}$$

holds for $(f,g) \in L^2(\mathbb{R}^d) \times L^2(\mathbb{R}^d)$ as well as for $(f,g) \in \mathcal{S}'(\mathbb{R}^d) \times \mathcal{S}(\mathbb{R}^d)$.

8.2 Basic Concepts of Time-Frequency Analysis

8.2.1 Time-Frequency Representations and the Definition of Pseudodifferential Operators

Time-frequency representations are a common tool in engineering, in particular in signal analysis. The idea is to combine information about time *and* frequency of a signal f in a joint representation of f on the time-frequency "plane" $\mathbb{R}^{2d}$. Time-frequency representations also arise naturally in the study of pseudodifferential operators. To see how they arise, let us look at the weak interpretation of the integral (8.1) defining σ^{KN}. Assuming first that $\sigma \in L^1(\mathbb{R}^{2d})$ and $f, g \in \mathcal{S}(\mathbb{R}^d)$, the integral (8.1) converges absolutely, and we may apply Fubini's Theorem to interchange the order of integration:

$$\begin{aligned}\langle \sigma^{KN} f, g\rangle &= \int_{\mathbb{R}^d} \left(\int_{\mathbb{R}^d} \sigma(x,\xi)\, \hat{f}(\xi)\, e^{2\pi i x\cdot\xi}\, d\xi \right) \overline{g(x)}\, dx \\ &= \int_{\mathbb{R}^d} \int_{\mathbb{R}^d} \sigma(x,\xi) \left(\hat{f}(\xi)\, e^{2\pi i x\cdot\xi}\, \overline{g(x)} \right) d\xi\, dx\,.\end{aligned}$$

This motivates the following definition.

Definition 8.2. *Let $f, g \in L^2(\mathbb{R}^d)$. The (cross)* Rihaczek distribution *of f and g is defined to be the function on $\mathbb{R}^{2d}$*

$$R(f,g)(x,\xi) = f(x)\, \overline{\hat{g}(\xi)}\, e^{-2\pi i x\cdot\xi}\,.$$

We can now write the action of σ^{KN} as

$$\langle \sigma^{KN} f, g\rangle = \langle \sigma, R(g,f)\rangle\,. \tag{8.5}$$

Note that the first inner product is for functions on $\mathbb{R}^d$, whereas the second inner product is for functions on $\mathbb{R}^{2d}$. We now choose (8.5) as the rigorous definition of σ^{KN}. The advantage of this weak definition becomes apparent in the next lemma.

Lemma 8.3. *If $\sigma \in \mathcal{S}'(\mathbb{R}^{2d})$, then the sesquilinear form (8.5) defines a continuous operator σ^{KN} from $\mathcal{S}(\mathbb{R}^d)$ to $\mathcal{S}'(\mathbb{R}^d)$. Thus the Kohn–Nirenberg correspondence is defined for all symbols $\sigma \in \mathcal{S}'(\mathbb{R}^{2d})$.*

Proof. If $f, g \in \mathcal{S}(\mathbb{R}^d)$, then $R(g,f)(x,\xi) = \overline{\hat{f}(\xi)}\, g(x)\, e^{-2\pi i x\cdot\xi}$ is in $\mathcal{S}(\mathbb{R}^{2d})$, therefore $\langle \sigma, R(g,f)\rangle$ is well-defined for every $\sigma \in \mathcal{S}'(\mathbb{R}^{2d})$. Fix $f \in \mathcal{S}(\mathbb{R}^d)$, and assume that $g_n \in \mathcal{S}(\mathbb{R}^d)$ converges to $g \in \mathcal{S}(\mathbb{R}^d)$ in the topology of $\mathcal{S}$. Then $R(g_n, f) \to R(g,f)$ in $\mathcal{S}(\mathbb{R}^{2d})$. Thus $g \to \langle \sigma, R(g,f)\rangle$ is a continuous functional on $\mathcal{S}(\mathbb{R}^d)$ and defines a unique tempered distribution $\sigma^{KN} f \in \mathcal{S}'(\mathbb{R}^d)$. If $f_n \in \mathcal{S}(\mathbb{R}^d)$ converges to $f \in \mathcal{S}(\mathbb{R}^d)$ in $\mathcal{S}$, then by the same argument $R(g, f_n) \to R(g,f)$ in $\mathcal{S}(\mathbb{R}^{2d})$ and thus $\langle \sigma, R(g, f_n)\rangle \to \langle \sigma, R(g,f)\rangle$ for all $g \in \mathcal{S}(\mathbb{R}^d)$. This means that $\sigma^{KN} f_n \xrightarrow{w^*} \sigma^{KN} f$. Consequently, the operator σ^{KN} is continuous from $\mathcal{S}(\mathbb{R}^d)$ to $\mathcal{S}'(\mathbb{R}^d)$ with the weak* topology. Clearly, if $\sigma \in L^1(\mathbb{R}^{2d})$, then the weak definition coincides with the integral in (8.1). □

If we use the weak definition of the spreading representation (8.3), we may argue similarly and obtain

$$\langle \sigma^{KN} f, g\rangle = \int_{\mathbb{R}^d} \left(\iint_{\mathbb{R}^{2d}} \hat{\sigma}(\eta, u)\, M_\eta T_{-u} f(t)\, du\, d\eta \right) \overline{g(t)}\, dt$$

$$= \int_{\mathbb{R}^{2d}} \hat{\sigma}(\eta, u) \overline{\int_{\mathbb{R}^d} g(t)\, \overline{M_\eta T_{-u} f(t)}\, dt}\, du\, d\eta\, .$$

Again, these manipulations are well-defined when $\hat{\sigma} \in L^1(\mathbb{R}^{2d})$ and $f, g \in \mathcal{S}(\mathbb{R}^d)$. The expression under the inner integral is one of the most fundamental objects of time-frequency analysis.

Definition 8.4. *Let $f, g \in L^2(\mathbb{R}^d)$. The short-time Fourier transform (STFT) of f with respect to the window g is defined as*

$$V_g f(x,\xi) = \int_{\mathbb{R}^d} f(t)\overline{g(t-x)}\, e^{-2\pi i \xi\cdot t}\, dt = \langle f, M_\xi T_x g\rangle\, . \tag{8.6}$$

The spreading representation can therefore be written as

$$\langle \sigma^{KN} f, g\rangle_{\mathbb{R}^d} = \int_{\mathbb{R}^{2d}} \hat{\sigma}(\eta, u)\, \overline{V_f g(-u, \eta)}\, du\, d\eta\, . \tag{8.7}$$

Remark 8.5. (a) The Rihaczek distribution $R(f,f)$ is a naive attempt to combine a function and its Fourier transform into a single object, simply by taking their (modified) tensor product $f(x)\, \overline{\hat{f}(\xi)}\, e^{-2\pi i x\cdot\xi}$. Like the Rihaczek distribution $R(f,g)$, the STFT is a sesquilinear form in f, g, but usually the window g is thought to be fixed and the STFT is interpreted as a linear mapping $f \to V_g f$. Thinking of g as a compactly supported non-negative smooth function centered at the origin, the product $f \cdot T_x \bar{g}$ can be thought of as a smooth cut-off of f to a neighborhood of $x \in \mathbb{R}^d$. Then $V_g f(x,\xi)$ is the Fourier transform of the local piece of f near x and measures the amplitude of f in the frequency band near $\xi \in \mathbb{R}^d$ at time x.

Both the Rihaczek distribution $R(f,f)$ and the short-time Fourier transform $V_g f$ are examples of time-frequency representations. They provide simultaneous information about time *and* frequency in a single object (at the expense of doubling the number of variables). In this sense the short-time Fourier transform is a mathematical analog to a *musical score.*

(b) Whereas the integral in (8.6) makes sense for f, $g \in L^2(\mathbb{R}^d)$, the inner product $\langle f, M_\xi T_x g\rangle$ can be extended to other dual pairings of spaces that are invariant under time-frequency shifts.

(c) In the following we will always assume that the window is in a "small" space, for instance, $g \in \mathcal{S}(\mathbb{R}^d)$. Then $V_g f$ is defined for $f \in \mathcal{S}'(\mathbb{R}^d)$ and can be used as a tool to analyze the time-frequency properties of distributions.

(d) Similarly to Lemma 8.3, one can show that the sequilinear form defined by the right-hand side of (8.7) defines a continuous operator from $\mathcal{S}(\mathbb{R}^d)$ to $\mathcal{S}'(\mathbb{R}^d)$. Based on the informal calculation in (8.2) and in anticipation of a rigorous argument in Corollary 8.10, we denote this operator by σ^{KN}.

8.2.2 Properties of the Short-Time Fourier Transform

Before we continue the investigation of pseudodifferential operators, we investigate some of the properties of the two time-frequency representations that arise in the definition of the Kohn–Nirenberg correspondence. Clearly, in this approach an understanding of pseudodifferential operators goes hand in hand with an understanding of time-frequency representations.

First we summarize a few formulas about time-frequency shifts. These are easy consequences of the definitions and do not require explicit proof.

Lemma 8.6. *Let $f \in \mathcal{S}'(\mathbb{R}^d)$, $x, \xi \in \mathbb{R}^d$. Then*

(i) $\widehat{M_\xi T_x f} = T_\xi M_{-x} \hat{f}$,

(ii) *Canonical commutation relations (CCR):*

$$T_x M_\xi = e^{-2\pi i x\cdot\xi} M_\xi T_x\,.$$

Next we list several formulas for the STFT, each of which occurs in certain calculations in this chapter.

Lemma 8.7. *If f, $g \in L^2(\mathbb{R}^d)$, then $V_g f$ is uniformly continuous on $\mathbb{R}^{2d}$, and*

$$V_g f(x,\xi) = (f\cdot T_x\bar{g})^\wedge(\xi) = \langle f, M_\xi T_x g\rangle = \langle \hat{f}, T_\xi M_{-x}\hat{g}\rangle = e^{-2\pi i x\cdot\xi} V_{\hat{g}}\hat{f}(\xi,-x)\,.$$

We state explicitly the *fundamental identity of time-frequency analysis*:

$$V_g f(x,\xi) = e^{-2\pi i x\cdot\xi}\, V_{\hat g}\hat f(\xi,-x)\,. \tag{8.8}$$

This identity makes transparent in what sense the STFT carries *simultaneous* information about f and $\hat f$ in a single representation.

It is useful to express the STFT in a different form that emphasizes some structural properties. We introduce the partial Fourier transform $\mathcal{F}_2$ in the second coordinate, an (asymmetric) coordinate transform, and the tensor product notation:

$$\mathcal{F}_2 F(x,\xi) = \int_{\mathbb{R}^d} F(x,t)\, e^{-2\pi i \xi\cdot t}\, dt\,,$$

$$\mathcal{T} F(x,t) = F(t,t-x)\,,$$

$$(f\otimes g)(x,y) = f(x)g(y).$$

Then we have the following.

Lemma 8.8. *For $f,\ g \in L^2(\mathbb{R}^d)$,*

$$V_g f = \mathcal{F}_2\mathcal{T}(f\otimes \bar g)\,. \tag{8.9}$$

In addition, (8.9) makes sense for $f,\ g \in \mathcal{S}'(\mathbb{R}^d)$ and defines the STFT on $\mathcal{S}'(\mathbb{R}^d)$.

Lemma 8.9. *The two-dimensional Fourier transform of the Rihaczek distribution is a rotated short-time Fourier transform. If $f,\ g\in\mathcal{S}(\mathbb{R}^d)$, then*

$$\widehat{R(f,g)}(\eta,u) = V_g f(-u,\eta)\,, \qquad \forall\,(\eta,u)\in\mathbb{R}^{2d}\,. \tag{8.10}$$

If $f,\ g\in\mathcal{S}'(\mathbb{R}^d)$, then (8.10) holds in a distributional sense as $\widehat{R(f,g)} = \widetilde{V_g f}$, where $\widetilde{F}(\eta,u) = F(-u,\eta)$.

Proof. For $f,\ g\in\mathcal{S}(\mathbb{R}^d)$ the following manipulations are rigorous:

$$\begin{aligned}
\widehat{R(f,g)}(\eta,u) &= \int_{\mathbb{R}^d}\int_{\mathbb{R}^d} f(x)\,\overline{\hat g(\xi)}\, e^{-2\pi i x\cdot\xi}\, e^{-2\pi i(\eta\cdot x + u\cdot\xi)}\, dx\, d\xi \\
&= \int_{\mathbb{R}^d} f(x)\,\overline{\left(\int_{\mathbb{R}^d} \hat g(\xi)\, e^{2\pi i\xi\cdot(x+u)}\, d\xi\right)}\, e^{-2\pi i\eta\cdot x}\, dx \\
&= \int_{\mathbb{R}^d} f(x)\,\overline{g(x+u)}\, e^{-2\pi i\eta\cdot x}\, dx \\
&= V_g f(-u,\eta)\,.
\end{aligned}$$

If $f,\ g\in\mathcal{S}'(\mathbb{R}^d)$, then we choose two sequences $f_n,\ g_n\in\mathcal{S}(\mathbb{R}^d)$ such that $f_n \overset{w^*}{\to} f$ and $g_n \overset{w^*}{\to} g$. Then by the weak* continuity of the Fourier transform on $\mathcal{S}'(\mathbb{R}^d)$ and by (8.10) we obtain that $\widehat{R(f,g)} = w^*\text{-}\lim_{n\to\infty} \widehat{R(f_n,g_n)} = w^*\text{-}\lim_{n\to\infty}\widetilde{V_{g_n} f_n} = \widetilde{V_g f}$. $\square$

We can now establish rigorously the equivalence of the Kohn–Nirenberg correspondence and the spreading representation of pseudodifferential operators.

Corollary 8.10. *If $\sigma \in \mathcal{S}'(\mathbb{R}^{2d})$ and $f,\, g, \in \mathcal{S}(\mathbb{R}^d)$, then*

$$\langle \sigma^{KN} f,\, g\rangle = \langle \hat{\sigma},\, \widetilde{V_f g}\rangle\,. \tag{8.11}$$

If $\sigma \in \mathcal{S}'$ is a (measurable) function of polynomial growth, then (8.11) is simply the weak version of the vector-valued integral

$$\sigma^{KN} f = \int_{\mathbb{R}^d}\int_{\mathbb{R}^d} \hat{\sigma}(\eta, u)\, M_\eta T_{-u} f\, du\, d\eta\,.$$

Proof. The statement follows from Parseval's relation (8.4) and Lemma 8.9: if $f,\, g \in \mathcal{S}(\mathbb{R}^d)$, then

$$\langle \sigma, R(g,f)\rangle = \langle \hat{\sigma}, \widehat{R(g,f)}\rangle = \langle \hat{\sigma}, \widetilde{V_f g}\rangle\,.$$

If the distribution σ has polynomial growth, then the "inner product" makes sense as an absolutely convergent integral, and we obtain

$$\begin{aligned}\langle \sigma^{KN} f, g\rangle &= \int_{\mathbb{R}^d}\int_{\mathbb{R}^d} \hat{\sigma}(\eta, u)\, \overline{V_f g(-u,\eta)}\, du\, d\eta \\ &= \int_{\mathbb{R}^d}\int_{\mathbb{R}^d} \hat{\sigma}(\eta, u)\, \langle M_\eta T_{-u} f, g\rangle\, du\, d\eta\,.\end{aligned}$$

This is the weak form of the spreading representation. □

Next we investigate some general properties of the Rihaczek distribution and the short-time Fourier transform that are used frequently.

Lemma 8.11 (Covariance Property). *Whenever $V_g f$ is defined,*

$$V_g(T_u M_\eta f)(x,\xi) = e^{-2\pi i u\cdot\xi}\, V_g f(x-u, \xi-\eta)$$

for $x, u, \xi, \eta \in \mathbb{R}^d$. In particular,

$$|V_g(T_u M_\eta f)(x,\xi)| = |V_g f(x-u,\xi-\eta)|\,.$$

The next statement is crucial and occurs in many calculations. It is usually referred to as the *orthogonality relations.*

Theorem 8.12 (Orthogonality Relations). *If $f_1,\, f_2,\, g_1,\, g_2 \in L^2(\mathbb{R}^d)$, then*

$$\int_{\mathbb{R}^d}\int_{\mathbb{R}^d} R(f_1, g_1)(x,\xi)\, \overline{R(f_2,g_2)(x,\xi)}\, dx\, d\xi = \langle f_1, f_2\rangle\, \overline{\langle g_1, g_2\rangle}\,,$$

and

$$\int_{\mathbb{R}^d}\int_{\mathbb{R}^d} V_{g_1} f_1(x,\xi)\, \overline{V_{g_2} f_2(x,\xi)}\, dx\, d\xi = \langle f_1, f_2\rangle\, \overline{\langle g_1, g_2\rangle}\,. \tag{8.12}$$

In particular, $V_g f \in L^2(\mathbb{R}^{2d})$ for $f,\, g \in L^2(\mathbb{R}^d)$.

Proof. The first formula follows immediately from the definition and from Plancherel's Theorem:

$$
\begin{aligned}
&\int_{\mathbb{R}^d}\int_{\mathbb{R}^d} R(f_1,g_1)(x,\xi)\overline{R(f_2,g_2)(x,\xi)}\,d\xi\,dx \\
&\qquad = \int_{\mathbb{R}^d}\int_{\mathbb{R}^d} f_1(x)\,\overline{\hat{g}_1(\xi)}\,e^{-2\pi i x\cdot\xi}\,\overline{f_2(x)\,\overline{\hat{g}_2(\xi)}\,e^{-2\pi i x\cdot\xi}}\,dx\,d\xi \\
&\qquad = \int_{\mathbb{R}^d} f_1(x)\,\overline{f_2(x)}\,dx \int_{\mathbb{R}^d} \hat{g}_2(\xi)\,\overline{\hat{g}_1(\xi)}\,d\xi \\
&\qquad = \langle f_1, f_2\rangle\,\langle \widehat{g_2}, \widehat{g_1}\rangle \\
&\qquad = \langle f_1, f_2\rangle\,\overline{\langle g_1, g_2\rangle}\,.
\end{aligned}
$$

The second formula then follows from Plancherel's Theorem and Lemma 8.9. Alternatively, we can use the representation of $V_g f$ given in Lemma 8.8. Since the partial Fourier transform and the coordinate transform $\mathcal{T}$ are unitary, we obtain

$$
\begin{aligned}
\big\langle V_{g_1} f_1,\, V_{g_2} f_2\big\rangle_{L^2(\mathbb{R}^{2d})} &= \big\langle \mathcal{F}_2\mathcal{T}(f_1\otimes\overline{g_1}),\, \mathcal{F}_2\mathcal{T}(f_2\otimes\overline{g_2})\big\rangle \\
&= \big\langle f_1\otimes\overline{g_1},\, f_2\otimes\overline{g_2}\big\rangle_{L^2(\mathbb{R}^{2d})} \\
&= \langle f_1, f_2\rangle\,\overline{\langle g_1, g_2\rangle}\,. \qquad \square
\end{aligned}
$$

Choosing $g_1 = g_2 = g$ and $f_1 = f_2 = f$, we obtain the isometry property of the STFT. In a sense, it is equivalent to Plancherel's Theorem for the Fourier transform.

Corollary 8.13. *If $\|g\|_2 = 1$, then the STFT is an isometry from $L^2(\mathbb{R}^d)$ into $L^2(\mathbb{R}^{2d})$, and*

$$\|V_g f\|_2 = \|f\|_2, \qquad \forall\, f \in L^2(\mathbb{R}^d)\,.$$

As always, a Plancherel Theorem leads to an inversion formula. Specifically, a weak interpretation of (8.12) yields the *inversion formula for the STFT.*

Corollary 8.14 (Inversion Formula for the STFT). *Suppose that $g,\gamma \in L^2(\mathbb{R}^d)$ with $\langle g,\gamma\rangle \neq 0$. Then for all $f \in L^2(\mathbb{R}^d)$,*

$$f = \frac{1}{\langle \gamma, g\rangle}\iint_{\mathbb{R}^{2d}} V_g f(x,\xi)\, M_\xi T_x \gamma\, d\xi\, dx\,, \tag{8.13}$$

where the vector-valued integral is to be understood in the weak sense.

Proof. The weak interpretation of (8.13) means that for all $h \in L^2(\mathbb{R}^d)$ we should have

$$\langle f, h\rangle = \frac{1}{\langle \gamma, g\rangle} \iint_{\mathbb{R}^{2d}} V_g f(x,\xi)\, \langle M_\xi T_x \gamma, h\rangle \, d\xi\, dx\,.$$

The right-hand side is just $\frac{1}{\langle \gamma, g\rangle}\langle V_g f, V_\gamma h\rangle$ and by (8.12) it equals $\langle f, h\rangle$, as claimed. □

Remark 8.15. The inversion formula extends to other spaces, e.g., if $g, \gamma \in \mathcal{S}$, then (8.13) holds for all $f \in \mathcal{S}'$ in a weak sense. Since the proofs require only technicalities in distribution theory, but no new ideas, we omit the precise arguments. Equivalent to the extended inversion formula is the Parseval formula for the STFT: *if $g \in \mathcal{S}(\mathbb{R}^d)$ and $\|g\|_2 = 1$, then for all $f \in \mathcal{S}'(\mathbb{R}^d)$ and $\varphi \in \mathcal{S}(\mathbb{R}^d)$ we have*

$$\big\langle V_g f, V_g \varphi \big\rangle = \langle f, \varphi\rangle\,. \tag{8.14}$$

8.2.3 Time-Frequency Analysis of Test Functions and Distributions

As an example of how the formalism of time-frequency representations can be used to analyze functions and distributions, we characterize the Schwartz class by means of the STFT.

Theorem 8.16. *Fix a nonzero $g \in \mathcal{S}(\mathbb{R}^d)$ and write $z = (x,\xi) \in \mathbb{R}^{2d}$. The following are equivalent:*

(i) $f \in \mathcal{S}(\mathbb{R}^d)$,

(ii) $V_g f \in \mathcal{S}(\mathbb{R}^{2d})$,

(iii) $|V_g f(z)| = \mathcal{O}\big((1+|z|)^{-N}\big)$ *for all* $N \geq 0$.

Further, if $f \in \mathcal{S}'(\mathbb{R}^d)$, then $V_g f(z) = \mathcal{O}\big((1+|z|)^N\big)$ for some $N \geq 0$.

Proof (idea). The equivalence of (i) and (ii) follows from Lemma 8.8 and the fact that the partial Fourier transform and linear coordinate transformations are isomorphisms on $\mathcal{S}(\mathbb{R}^{2d})$. Thus if $f, g \in \mathcal{S}(\mathbb{R}^d)$, then $V_g f = \mathcal{F}_2 \mathcal{T}(f \otimes \bar{g}) \in \mathcal{S}(\mathbb{R}^{2d})$. Conversely, if $V_g f \in \mathcal{S}(\mathbb{R}^{2d})$, then $f(x)\,\bar{g}(y) = \mathcal{T}^{-1}\mathcal{F}_2^{-1} V_g f \in \mathcal{S}(\mathbb{R}^{2d})$. By fixing y, we obtain that $f \in \mathcal{S}(\mathbb{R}^d)$.

Clearly, (ii) implies (iii). The subtle part is the implication (iii) $\Rightarrow$ (i), which is based on the inversion formula. Applying the operator $X^\beta D^\alpha$ to both sides of (8.13) and using the triangle inequality, one obtains the estimate

$$\|X^\beta D^\alpha f\|_\infty \leq \int_{\mathbb{R}^{2d}} |V_\varphi f(x,\xi)|\, \|X^\beta D^\alpha (M_\xi T_x \varphi)\|_\infty \, dx\, d\xi\,.$$

Next one shows the polynomial estimate $\|X^\beta D^\alpha (M_\xi T_x \varphi)\|_\infty = \mathcal{O}(|\xi|^{|\alpha|} + |x|^{|\beta|})$. Since the STFT $V_\varphi f$ decays rapidly, the integral and the norm are finite for all multi-indices α, $\beta \geq 0$, and this means that $f \in \mathcal{S}(\mathbb{R}^d)$. For details see [33] and [24, Ch. 11.2]. □

Remark 8.17. Short-time Fourier transforms are functions with remarkable local properties and, in many contexts, behave almost like analytic functions. In the case of Theorem 8.16, the decay of the STFT implies automatically C^∞-smoothness. Other statements where decay of the STFT implies its smoothness can be found in [10], [18], [58].

We summarize the domain of the short-time Fourier transform in the following table.

f	**g**	$\mathbf{V_g f}$
$L^2(\mathbb{R}^d)$	$L^2(\mathbb{R}^d)$	$L^2(\mathbb{R}^{2d})$, continuous
$L^p(\mathbb{R}^d)$	$L^{p'}(\mathbb{R}^d)$	$L^\infty(\mathbb{R}^{2d})$, continuous
$\mathcal{S}'(\mathbb{R}^d)$	$\mathcal{S}(\mathbb{R}^d)$	polynomial growth, continuous
$\mathcal{S}'(\mathbb{R}^d)$	$\mathcal{S}'(\mathbb{R}^d)$	$\mathcal{S}'(\mathbb{R}^{2d})$

8.2.4 Representation of Operators: General Formalism

Recall the *Schwartz Kernel Theorem: If A is continuous from $\mathcal{S}(\mathbb{R}^d)$ into $\mathcal{S}'(\mathbb{R}^d)$ (under the weak* topology), then there is a unique distributional kernel $k \in \mathcal{S}'(\mathbb{R}^{2d})$, such that*

$$\langle Af, g\rangle = \langle k, g \otimes \bar{f}\rangle .$$

If the kernel k is locally integrable, then A can be written as an integral operator, because

$$\langle k, g \otimes \bar{f}\rangle = \iint_{\mathbb{R}^{2d}} k(x,y)\, f(y)\, \overline{g(x)}\, dy\, dx .$$

Thus Schwartz's theorem asserts that every reasonable operator is a "generalized integral operator" [39], [49].

Next we show that every such operator can also be written as a pseudodifferential operator σ^{KN} for some distributional symbol $\sigma \in \mathcal{S}'(\mathbb{R}^{2d})$.

Proposition 8.18. *Assume that $\sigma \in \mathcal{S}'(\mathbb{R}^{2d})$. Write $\mathcal{T}'F(x,y) = F(x, y-x)$, then σ^{KN} has the distributional kernel*

$$k = \mathcal{T}'\mathcal{F}_2\sigma \in \mathcal{S}'(\mathbb{R}^{2d}) .$$

Furthermore

$$\big\langle \sigma^{KN} f,\, g\big\rangle = \big\langle \sigma,\, R(g,f)\big\rangle = \big\langle k,\, g \otimes \bar{f}\big\rangle , \qquad \forall\, f, g \in \mathcal{S}(\mathbb{R}^d) .$$

Proof. We take up the calculation begun in (8.2). If $f \in L^1(\mathbb{R}^d)$ and $\sigma \in \mathcal{S}(\mathbb{R}^{2d})$, then

$$\begin{aligned} \sigma^{KN} f(x) &= \int_{\mathbb{R}^d} \sigma(x,\xi)\, \hat{f}(\xi)\, e^{2\pi i x\cdot\xi}\, d\xi \\ &= \int_{\mathbb{R}^d} \left(\int_{\mathbb{R}^d} \sigma(x,\xi)\, e^{2\pi i (x-y)\cdot \xi}\, d\xi \right) f(y)\, dy , \end{aligned}$$

and so the kernel is

$$k(x,y) = \int_{\mathbb{R}^d} \sigma(x,\xi)\, e^{2\pi i(x-y)\cdot\xi}\, d\xi = \mathcal{F}_2\sigma(x, y-x) = \mathcal{T}'\mathcal{F}_2\sigma(x,y)\,.$$

If $\sigma \in \mathcal{S}'(\mathbb{R}^{2d})$, then by the Schwartz Kernel Theorem there is a distributional kernel $k \in \mathcal{S}'(\mathbb{R}^{2d})$ such that $\langle \sigma^{KN} f, g\rangle = \langle k, g \otimes \bar{f}\rangle$. Now let $\sigma_n \in \mathcal{S}(\mathbb{R}^{2d})$ be a sequence converging weak* to σ. Then

$$\begin{aligned}
\langle \sigma^{KN} f,\, g\rangle &= \langle \sigma,\, R(g,f)\rangle \\
&= \lim_{n\to\infty} \langle \sigma_n,\, R(g,f)\rangle \\
&= \lim_{n\to\infty} \langle (\sigma_n)^{KN} f,\, g\rangle \\
&= \lim_{n\to\infty} \langle \mathcal{T}'\mathcal{F}_2\sigma_n,\, g\otimes \bar{f}\rangle \\
&= \langle \mathcal{T}'\mathcal{F}_2\sigma,\, g\otimes\bar{f}\rangle\,.
\end{aligned}$$

So $k = \mathcal{T}'\mathcal{F}_2\sigma \in \mathcal{S}'(\mathbb{R}^{2d})$. □

We may now summarize the operator formalism as follows.

Theorem 8.19. *Let A be a continuous linear operator mapping $\mathcal{S}(\mathbb{R}^d)$ into $\mathcal{S}'(\mathbb{R}^d)$. Then there exist tempered distributions k, F and $\sigma \in \mathcal{S}'(\mathbb{R}^{2d})$ that yield the following representations of A.*

(a) *Schwartz Kernel Theorem: A is an "extended integral operator"*

$$\langle Af, g\rangle = \langle k, g\otimes\bar{f}\rangle, \qquad f,g \in \mathcal{S}(\mathbb{R}^d).$$

(b) *Kohn–Nirenberg correspondence: A is a pseudodifferential operator*

$$A = \sigma^{KN}.$$

(c) *Spreading representation: A is a superposition of time-frequency shifts*

$$A = \iint_{\mathbb{R}^{2d}} F(x,\xi)\, M_\xi T_x\, dx\, d\xi.$$

The transition formulas are

$$\begin{aligned}
\sigma &= \mathcal{F}_2^{-1}\mathcal{T}'^{\,-1} k, \\
F(x,\xi) &= \hat{\sigma}(\xi, -x).
\end{aligned}$$

Remark 8.20. With equal right we might have studied other calculi for pseudodifferential operators. A general calculus would be the t-pseudodifferential operators, which are defined as

$$\mathrm{Op}_t(\sigma)f(x) = \int\int \sigma((1-t)x+ty,\xi)\, e^{2\pi i\xi\cdot(x-y)}\, f(y)\, d\xi\, dy\,.$$

Thus for $t = 0$ one obtains $\mathrm{Op}_0(\sigma) = \sigma^{KN}$; the case $t = 1/2$ corresponds to the Weyl transform. Although the Weyl transform is often more convenient because it possesses more symmetry properties, we prefer to outline the case of the Kohn–Nirenberg correspondence. Many formulas are simpler to derive and, moreover, the formalism of the Kohn–Nirenberg correspondence can be easily extended to arbitrary locally compact Abelian (LCA) groups [20]. Indeed, all we need is the structure of time-frequency shifts, and these can be defined on any LCA group. By contrast, the mapping $x \to x/2$ that is required in the Weyl calculus makes sense only on a subclass of LCA groups.

8.3 Modulation Spaces

So far we have studied the concept of pseudodifferential operators and the tight interplay between pseudodifferential operators and time-frequency representations. This study leads to several different representations of arbitrary operators. Before we proceed to the investigation of mapping properties and boundedness properties we shall study the appropriate class of function spaces. Once again we abide by the principle that the appropriate function spaces should be closely related to the concepts involved in the definition of pseudodifferential operators. In other words, the function spaces should be related to time-frequency shifts and time-frequency representations.

The goal is to quantify the time-frequency behavior of a function or distribution. Using the metaphor of the STFT as a mathematical score, we want to quantify how much and where in the time-frequency plane a function is concentrated. It is therefore natural to impose a norm on $V_g f$ and thereby define a norm on f.

8.3.1 Definition of Modulation Spaces

Informally we start with a suitable function space Y on $\mathbb{R}^{2d}$, for instance, $Y = L^p(\mathbb{R}^{2d})$, and then we define the *modulation space* $M(Y)$ by the norm

$$\|f\|_{M(Y)} = \|V_g f\|_Y$$

for a suitable, fixed window g.

Since Y should be a measure for the time-frequency concentration of f, the norm of Y should only depend on the modulus of $V_g f$ and Y should be invariant under translations in the time-frequency plane, i.e., $M(Y)$ should be invariant under time-frequency shifts. Obeying further principles from the theory of function spaces, we will consider a whole class of function spaces Y on $\mathbb{R}^{2d}$ and require that this class be invariant under duality and interpolation.

In this way, we arrive naturally at the class of mixed-norm spaces $L^{p,q}_m$ on $\mathbb{R}^{2d}$ as choices for Y. For $p, q < \infty$, these are defined by the norm

$$\|F\|_{L^{p,q}_m} = \left(\int_{\mathbb{R}^d} \left(\int_{\mathbb{R}^d} |F(x,\xi)|^p \, m(x,\xi)^p \, dx \right)^{q/p} d\xi \right)^{1/q},$$

with the usual modifications when either $p = \infty$ or $q = \infty$.

The weight m quantifies the growth or decay of $V_g f$ in the time-frequency plane. For explicitness we may think of m as a function of the form

$$m(z) = e^{a|z|^b} \, (1+|z|)^s \, \big(\log(e+|z|)\big)^t$$

for $a, s, t \in \mathbb{R}$ and $0 \le b \le 1$. We postpone the precise definition of admissible weights and proceed to the formal definition of the modulation spaces.

Definition 8.21. *Let $\varphi(t) = e^{-\pi t \cdot t}$ be the Gaussian and*

$$\mathcal{H}_0 = \left\{ f = \sum_{j=1}^{n} c_j M_{\xi_j} T_{x_j} \varphi : (x_j, \xi_j) \in \mathbb{R}^{2d}, n \in \mathbb{N} \right\}$$

be the linear subspace of all finite linear combinations of time-frequency shifts of the Gaussian. Then $\mathcal{H}_0$ is invariant under the Fourier transform and is a very small subspace of $\mathcal{S}$.

Next consider the norm

$$\begin{aligned} \|f\|_{M^{p,q}_m} &= \|V_\varphi f\|_{L^{p,q}_m} \\ &= \left(\int_{\mathbb{R}^d} \left(\int_{\mathbb{R}^d} |V_\varphi f(x,\xi)|^p \, m(x,\xi)^p \, dx \right)^{q/p} d\xi \right)^{1/q}. \end{aligned} \tag{8.15}$$

We define the modulation space $M^{p,q}_m$ to be the completion of $\mathcal{H}_0$ with respect to the $M^{p,q}_m$-norm, if $0 < p, q < \infty$, and the weak completion if $p = \infty$ or $q = \infty$.*

Furthermore, we write $M^p_m = M^{p,p}_m$ and $M^{p,q} = M^{p,q}_m$ when $m \equiv 1$.

This definition is not the standard definition of the modulation spaces and seems to be rather cumbersome, because we define an element in a modulation space by an abstract process of completion. On the other hand, the Banach space property is evident without further effort.

Theorem 8.22. (a) *The modulation space $M^{p,q}_m$ is a Banach space for $1 \le p, q \le \infty$. If p, $q < \infty$, then $\mathcal{H}_0$ is norm-dense in $M^{p,q}_m$.*

(b) *If the weight m satisfies the property*

$$m(z_1 + z_2) \le C \, v(z_1) \, m(z_2) \qquad \forall \, z_1, z_2 \in \mathbb{R}^{2d},$$

where v is a submultiplicative weight function, then $M^{p,q}_m$ is invariant under time-frequency shifts and

$$\|M_\xi T_x f\|_{M_m^{p,q}} \le C\, v(x,\xi)\, \|f\|_{M_m^{p,q}}\,.$$

(c) *If $m(\xi,-x) \le C\, m(x,\xi)$, then M_m^p is invariant under the Fourier transform; in fact, the Fourier transform is an isomorphism of M_m^p.*

Proof. (a) The Banach space property and the density of $\mathcal{H}_0$ in $M_m^{p,q}$ are part of the construction [48].

(b) To show the invariance properties, we note that for $f \in \mathcal{H}_0$ or even for $f \in L^2(\mathbb{R}^d)$ we have

$$\begin{aligned}
\|M_\xi T_x f\|_{M_m^{p,q}} &= \left(\int_{\mathbb{R}^d}\left(\int_{\mathbb{R}^d} |V_\varphi(M_\xi T_x f)(u,\eta)|^p\, m(u,\eta)^p\, du\right)^{q/p} d\eta\right)^{1/q}\\
&= \left(\int_{\mathbb{R}^d}\left(\int_{\mathbb{R}^d} |V_\varphi f(u-x,\eta-\xi)|^p\, m(u,\eta)^p\, du\right)^{q/p} d\eta\right)^{1/q}\\
&= \left(\int_{\mathbb{R}^d}\left(\int_{\mathbb{R}^d} |V_\varphi f(u,\eta)|^p\, m(u+x,\eta+\xi)^p\, du\right)^{q/p} d\eta\right)^{1/q}\\
&\le C\, v(x,\xi)\left(\int_{\mathbb{R}^d}\left(\int_{\mathbb{R}^d} |V_\varphi f(u,\eta)|^p\, m(u,\eta)^p\, du\right)^{q/p} d\eta\right)^{1/q}.
\end{aligned}$$

By density, this inequality carries over to the closure of $\mathcal{H}_0$, i.e., to $M_m^{p,q}$.

(c) The invariance of the Fourier transform follows by a similar argument from the fundamental formula (8.8), and the calculation

$$\begin{aligned}
\iint_{\mathbb{R}^{2d}} |V_\varphi f(x,\xi)|^p\, m(x,\xi)^p\, dx\, d\xi &= \iint_{\mathbb{R}^{2d}} |V_{\hat\varphi} \hat f(\xi,-x)|^p\, m(x,\xi)^p\, dx\, d\xi\\
&= \iint_{\mathbb{R}^{2d}} |V_\varphi \hat f(x,\xi)|^p\, m(-\xi,x)^p\, dx\, d\xi\\
&\le C\|V_\varphi f\|^p_{M_m^p}\,. \qquad \square
\end{aligned}$$

Remark 8.23. (a) While the elements in $\mathcal{H}_0$ are concrete C^∞-functions with rapid decay, the general elements of a modulation space are by definition abstract elements in some completion of $\mathcal{H}_0$. For decent weights m the modulation spaces are always tempered distributions. Specifically, if $m(z) = \mathcal{O}(|z|^N)$ for some $N > 0$, then $M_m^{p,q}$ is in fact the subspace of tempered distributions $f \in \mathcal{S}'(\mathbb{R}^d)$ for which $\|f\|_{M_m^{p,q}}$ is finite. And if $m \ge 1$ and $1 \le p,q \le 2$, then $M_m^{p,q}$ is actually a subspace of $L^2(\mathbb{R}^d)$. However, if $v(z) = e^{a|z|^b}, b < 1$, then $M_v^1 \subseteq \mathcal{S}(\mathbb{R}^d)$ and $\mathcal{S}'(\mathbb{R}^d) \subseteq M_{1/v}^\infty$, and we would have to use ultradistributions [2] to define $M_m^{p,q}$ as a subspace of "something" [45].

(b) Since, for fixed $\xi \in \mathbb{R}^{2d}$, $V_g f$ is a smoothed version of f and, for fixed $x \in \mathbb{R}^d$, $V_g f$ is a smoothed version of $\hat f$, we may think of $f \in M^{p,q}$ as roughly meaning that $f \in L^p$ and $\hat f \in L^q$. While this is completely wrong as a rigorous statement, it is sufficient as an intuition to explain a number of properties of modulation spaces, for instance, the convolution and pointwise multiplication properties of modulation spaces [10].

Weights. The treatment of the weights is always a bit technical. The weights in (8.15) are standard, but much more general weight functions can be used. For completeness we now add the precise definitions.

Definition 8.24. *We use two classes of weight functions. By v we denote a non-negative, even, and submultiplicative function on $\mathbb{R}^{2d}$, i.e., v satisfies the following properties: $v(0) = 1$, $v(\pm z_1, \pm z_2, \dots, \pm z_{2d}) = v(z_1, \dots, z_{2d})$, and $v(w+z) \le v(w)v(z), w, z \in \mathbb{R}^{2d}$. Without loss of generality we may assume that v is continuous.*

*Associated to a submultiplicative weight v, we define the class of v-*moderate *weights by*

$$\mathcal{M}_v = \left\{ m \ge 0 : \sup_{w \in \mathbb{R}^{2d}} \frac{m(w+z)}{m(w)} \le Cv(z), \forall\, z \in \mathbb{R}^{2d} \right\}. \tag{8.16}$$

Every weight of the form $v(z) = e^{a|z|^b}(1+|z|)^s \log^r(e+|z|)$ for parameters $a, r, s \ge 0$, $0 \le b \le 1$ is submultiplicative. A submultiplicative weight can grow at most exponentially. To see this, set $a = \sup_{|x| \le 1} v(x) \ge 1$. If $N-1 < |z| \le N$ for $N \in \mathbb{N}$, then $v(z) = v(N \cdot \frac{z}{N}) \le v(\frac{z}{N})^N \le a^N \le a^{|z|+1}$.

The polynomial weights

$$(1+|z|)^t \asymp (1+|z|^2)^{t/2} := \langle z \rangle$$

are submultiplicative for $t \ge 0$ and they are moderate with respect to the submultiplicative weight $v_s(z) = (1+|z|)^s$ if and only if $|t| \le s$. Note that if $m \in \mathcal{M}_v$, then $\frac{1}{Cv(z)} \le m(z) \le Cv(z)$ by (8.16). If v is submultiplicative, then $v^\alpha \in \mathcal{M}_v$ for $|\alpha| \le 1$.

Historical remarks. The modulation spaces were introduced by H. G. Feichtinger in two fundamental papers in 1980 and 1983. The original motivation was to investigate alternative concepts of smoothness that were not based on differentiation properties [15], [16], [17]. Thus in the beginning the theory of modulation spaces was developed in parallel to the theory of Besov spaces. Like Besov spaces, modulation spaces possess a thorough theory as function spaces, such as duality and interpolation theory, trace and restriction theorems, and atomic decompositions. Their subsequent evolution showed that modulation spaces arise naturally whenever a problem involving time-frequency shifts is analyzed rigorously.

In recent years modulation spaces have been used and applied in many problems of time-frequency analysis. We mention their appearance in the formulation of uncertainty principles [22], [4], in the problem of window design for Gabor frames and pulse shaping in OFDM [19], [31], [42], [55], in the characterization of time-frequency concentration by means of Gabor frames [19], [24], in the formulation of the Balian–Low Theorem [27], for nonlinear approximation of functions with Gabor frames and Wilson bases [32], and for sigma-delta-modulation with Gabor frames [60]. Last but not least, modulation spaces are used as symbol classes for pseudodifferential operators [28], [29], [47], [50], [52], [53], [57], [58]. The latter application is the focus of this chapter.

In this sense, we may say that

Modulation spaces are the appropriate function spaces for time-frequency analysis.

8.3.2 Basic Properties of Modulation Spaces

Next we state the main properties of modulation spaces and Banach spaces. This is usually a bit boring, and we will omit the rigorous proofs. For a detailed treatment from scratch we refer to [24, Ch.11]. Since the modulation spaces are defined by L^p-norms of an STFT, it is not surprising that they inherit several of their properties from the corresponding L^p-spaces, for instance, the duality and interpolation properties.

We first identify some modulation spaces with known spaces.

Lemma 8.25. (a) $M^2 = L^2(\mathbb{R}^d)$.

(b) *If* $m(x,\xi) = \langle x\rangle^s$ *for* $s \in \mathbb{R}$, *then*

$$M^2_m = L^2_s = \left\{ f \in \mathcal{S}' : \left(\int_{\mathbb{R}^d} |f(x)|^2 \langle x\rangle^{2s}\, dx \right)^{1/2} < \infty \right\}.$$

(c) *If* $m(x,\xi) = \langle \xi\rangle^s$, *then*

$$M^2_m = H^s = \left\{ f \in \mathcal{S}' : \left(\int_{\mathbb{R}^d} |\hat{f}(\xi)|^2 \langle \xi\rangle^{2s}\, d\xi \right)^{1/2} < \infty \right\}$$

(Bessel potential space).

(d) *The space* M^1 *is often denoted by* $S_0(\mathbb{R}^d)$ *and called the* Feichtinger algebra.

(e) *If* $v_s(x,\xi) = (1 + |x|^2 + |\xi|^2)^{s/2} = \langle z\rangle^s$, *then* $M^2_{v_s} = L^2_s \cap H^s = Q_s$ *where* Q_s *is the* Shubin class *[50], [3].*

(f) $\mathcal{S}(\mathbb{R}^d) = \bigcap_{s\geq 0} M^1_{v_s} = \bigcap_{s\geq 0} M^p_{v_s}$.

(g) $\mathcal{S}'(\mathbb{R}^d) = \bigcup_{s\geq 0} M^\infty_{1/v_s}$

Remark 8.26. (a) follows from the orthogonality relations (Theorem 8.12); (f) and (g) are reformulations of Theorem 8.16.

Next we present some basic properties of the modulation spaces.

Theorem 8.27. *Assume that* $1 \leq p, q \leq \infty$, *and* $m \in \mathcal{M}_v$.

(i) Equivalent norms: *If* $g \in M_v^1$, *then*

$$\|f\|_{M_m^{p,q}} \asymp \|V_g f\|_{L_m^{p,q}}.$$

Thus the definition of $M_m^{p,q}$ *is independent of the window* g *that is chosen to measure the time-frequency content of* f.

(ii) Duality: $(M_m^{p,q})' = M_{1/m}^{p',q'}$ *for* $1 \leq p,\, q < \infty$, *where* $p' = \frac{p}{p-1}$ *denotes the conjugate index.*

(iii) Inclusions: $M_m^{p,q} \subseteq M_m^{r,s} \Longleftrightarrow p \leq r$ *and* $q \leq s$.

(iv) *The* interpolation *of modulation spaces is similar to that for* L^p*-spaces, e.g.,*

$$[M^1, M^\infty]_\theta = M^p.$$

For proofs see [24] and the original literature. It suffices to say that the proofs rely heavily on the extended inversion formula (8.13) and (8.14).

8.4 Pseudodifferential Operators: Boundedness and Schatten Classes

In Section 8.2.4 we have investigated the formalism of linear operators and how they can be represented. We have put special emphasis on the use of time-frequency representations. In this part, we turn to a finer investigation of pseudodifferential operators and study their mapping and boundedness properties. The setup of time-frequency methods leads to surprisingly simple proofs of strong results; this is why the time-frequency approach may be called "pedestrian."

8.4.1 A Kernel Theorem

We first formulate a very general kernel theorem that extends and refines the classical Schwartz Kernel Theorem. The unweighted case was obtained in the important work of Feichtinger [14] and proved in [18]; weighted versions were investigated in [24, Ch. 14.3] and [59].

To distinguish the STFT of integral kernels and symbols on $\mathbb{R}^{2d}$ from the STFT of a distribution on $\mathbb{R}^d$, we will write $\mathcal{V}_\Phi \sigma$ and $\mathcal{V}_\Phi k$ when $\sigma, k \in \mathcal{S}'(\mathbb{R}^{2d})$.

Theorem 8.28. *Given a submultiplicative weight* v *on* $\mathbb{R}^{2d}$, *we define a submultiplicative weight* w *on* $\mathbb{R}^{4d}$ *by* $w(z_1, z_2, \zeta_1, \zeta_2) = v(z_1, \zeta_1)\, v(z_2, \zeta_2)$. *Then an operator* A *is continuous from* $M_v^1(\mathbb{R}^d)$ *into* $M_{1/v}^\infty = (M_v^1)'$ *if and only if there exists a kernel* $k \in M_{1/w}^\infty(\mathbb{R}^{2d})$ *such that* $\langle Af, g\rangle = \langle k, g \otimes \bar{f}\rangle$ *for* $f, g \in M_v^1$.

Proof (Sketch). We use a Gaussian window $\Phi(z) = e^{-\pi z^2} = \varphi(z_1)\varphi(z_2)$, which has the advantage that the window factors as $\Phi = \varphi \otimes \varphi$. As a consequence, the STFT of a product also factors and we have

$$\begin{aligned} \mathcal{V}_\Phi(f_1 \otimes f_2)(z_1, z_2, \zeta_1, \zeta_2) &= \big\langle f_1 \otimes f_2, M_{(\zeta_1,\zeta_2)} T_{(z_1,z_2)}(\varphi \otimes \varphi)\big\rangle \\ &= \langle f_1, M_{\zeta_1} T_{z_1}\varphi\rangle \, \langle f_2, M_{\zeta_2} T_{z_2}\varphi\rangle \\ &= V_\varphi f_1(z_1,\zeta_1)\, V_\varphi f_2(z_2,\zeta_2)\,. \end{aligned} \tag{8.17}$$

Consequently, the modulation space norm of tensor products factors and, using the definition of w, we have

$$\|f_1 \otimes f_2\|_{M^1_w(\mathbb{R}^{2d})} = \|f_1\|_{M^1_v(\mathbb{R}^d)} \, \|f_2\|_{M^1_v(\mathbb{R}^d)}\,. \tag{8.18}$$

Now assume that $k \in M^\infty_{1/w}$. Since this is the dual of M^1_w, the duality and (8.18) yield that

$$|\langle Af, g\rangle| = |\langle k, g \otimes \bar{f}\rangle| \le \|k\|_{M^\infty_{1/w}} \|g \otimes \bar{f}\|_{M^1_w} = \|k\|_{M^\infty_{1/w}} \, \|g\|_{M^1_v} \, \|f\|_{M^1_v}\,.$$

Since this holds for all $g \in M^1_v$, we conclude that $Af \in M^\infty_{1/v}$ with the norm estimate $\|Af\|_{M^\infty_{1/v}} \le \|k\|_{M^\infty_{1/w}} \, \|f\|_{M^1_v}$.

For the converse, we argue as follows. Suppose that we already know that there exists a kernel such that $\langle Af, g\rangle = \langle k, g \otimes \bar{f}\rangle$. For instance, if $v(z) = \mathcal{O}(|z|^N)$ for some $N \ge 0$, then $\mathcal{S} \subseteq M^1_v \subseteq M^\infty_{1/v} \subseteq \mathcal{S}'$, and thus the Schwartz Kernel Theorem guarantees the existence of such a kernel $k \in \mathcal{S}'(\mathbb{R}^{2d})$. For other growth conditions, one has to take recourse to the theory of ultra-distributions or use Wilson bases to establish the existence of a *distributional* kernel. Taking the existence of some k for granted, we will now show that $k \in M^\infty_{1/w}(\mathbb{R}^{2d})$. Using (8.17) we find that

$$\begin{aligned} \big\langle AM_{\zeta_2} T_{z_2}\varphi,\, M_{\zeta_1} T_{z_1}\varphi\big\rangle &= \big\langle k,\, M_{\zeta_1} T_{z_1}\varphi \otimes \overline{M_{\zeta_2} T_{z_2}\varphi}\big\rangle \\ &= \big\langle k,\, M_{\zeta_1} T_{z_1}\varphi \otimes M_{-\zeta_2} T_{z_2}\varphi\big\rangle \\ &= V_\Phi k(z_1, z_2, \zeta_1, -\zeta_2)\,. \end{aligned} \tag{8.19}$$

Since A maps M^1_v into $M^\infty_{1/v}$, we find that

$$\begin{aligned} \sup_{(z_1,\zeta_1)\in\mathbb{R}^{2d}} |\langle AM_{\zeta_2} T_{z_2}\varphi, M_{\zeta_1} T_{z_1}\varphi\rangle| \, \frac{1}{v(z_1,\zeta_1)} &= \|A\big(M_{\zeta_2} T_{z_2}\varphi\big)\|_{M^\infty_{1/v}} \\ &\le C\, \|M_{\zeta_2} T_{z_2}\varphi\|_{M^1_v} \\ &\le C\, v(z_2,\zeta_2)\, \|\varphi\|_{M^1_v}\,. \end{aligned}$$

Comparing with (8.19), we find that

$$\Big|\mathcal{V}_\Phi k(z,\zeta)\Big| \, \frac{1}{w(z,\zeta)} = \Big|\mathcal{V}_\Phi k(z,\zeta)\Big| \, \frac{1}{v(z_1,\zeta_1)\, v(z_2,\zeta_2)} \le C'\,,$$

and this means that $k \in M^\infty_{1/w}$. □

For the polynomial weights $v_s(z) = (1 + |z|^2)^{s/2}$ we obtain a slight improvement of the Schwartz Kernel Theorem.

Corollary 8.29. *If $A : \mathcal{S}(\mathbb{R}^d) \to \mathcal{S}'(\mathbb{R}^d)$, then there is $s \geq 0$ such that A maps $M^1_{v_s}(\mathbb{R}^d)$ into M^∞_{1/v_s}.*

Proof. By Theorem 8.16 $\mathcal{S}(\mathbb{R}^d)$ and $\mathcal{S}'(\mathbb{R}^d)$ can be represented as $\mathcal{S}(\mathbb{R}^d) = \bigcap_{s\geq 0} M^1_{v_s}$ and as $\mathcal{S}'(\mathbb{R}^d) = \bigcup_{s\geq 0} M^\infty_{1/v_s}$. Since

$$v_s(z,\zeta) \leq w_s(z,\zeta) := v_s(z_1,\zeta_1)v_s(z_2,\zeta_2) \leq v_{2s}(z,\zeta),$$

a symbol $\sigma \in \mathcal{S}'(\mathbb{R}^d)$ must be in M^∞_{1/w_s} for some $s \geq 0$. The easy part of Theorem 8.28 implies that A maps $M^1_{v_s}$ into M^∞_{1/v_s}. □

8.4.2 Boundedness

In classical analysis one uses the Hörmander classes $S^m_{\rho,\delta}$. A function σ on $\mathbb{R}^{2d}$ belongs to the symbol class $S^m_{\rho,\delta}$, if there exist constants such that

$$|D^\alpha_\xi D^\beta_x \sigma(x,\xi)| \leq C_{\alpha,\beta}(1+|\xi|)^{N+\delta|\beta|-\rho|\alpha|}$$

for all multi-indices $\alpha, \beta \geq 0$. In particular, these symbols are in $C^\infty(\mathbb{R}^{2d})$. This definition can be motivated as follows: A differential operator with variable coefficients $\sigma_\alpha \in C^\infty(\mathbb{R}^d)$ of the form

$$P(x,D)f(t) = \sum_{|\alpha|\leq m} \sigma_\alpha(x) D^\alpha f(t) = \int_{\mathbb{R}^d} \sum_{|\alpha|\leq m} \sigma_\alpha(x)\,(2\pi i\xi)^\alpha\, \hat{f}(\xi)\, e^{2\pi i \xi\cdot t}\, d\xi$$

has the symbol

$$\sigma(x,\xi) = \sum_{|\alpha|\leq m} \sigma_\alpha(x)\,(2\pi i \xi)^\alpha\,,$$

and so it is clear that

$$|D^\alpha_\xi \sigma(x,\xi)| \leq C_\alpha (1+|\xi|)^{m-|\alpha|}$$

for all α, in other words, $\sigma \in S^m_{1,0}$. A refinement of this observation leads to the Hörmander classes in a natural way.

It is well known that if $\sigma \in S^0_{\rho,\delta}$ and $0 \leq \delta \leq \rho \leq 1$ and $\delta < 1$, then σ^{KN} is bounded on $L^2(\mathbb{R}^d)$ [21], [40], [54].

The time-frequency approach leads to alternative boundedness conditions. These conditions could be useful in time-frequency analysis and for the analysis of time-varying systems. See [55], [56] for applications to mobile communication.

The following boundedness condition was discovered several times [52], [28] and several proofs are known [5], [24], [57], [26]. The evolution of this result suggests that modulation spaces are well suited as symbol classes for the analysis of time-varying systems (but perhaps not so much for PDEs).

Theorem 8.30. *If $\sigma \in M^{\infty,1}(\mathbb{R}^{2d})$, then σ^{KN} is bounded on $M^{p,q}$ for $1 \le p, q \le \infty$ and*

$$\|\sigma^{KN}\|_{op} \le C\|\sigma\|_{M^{\infty,1}} .$$

In particular, σ^{KN} is bounded on $L^2(\mathbb{R}^d)$.

Before proving this result, let us discuss its significance.

Remark 8.31. (a) Sjöstrand [52] uses the equivalent norm

$$\|\sigma\|_{M^{\infty,1}} = \int_{\mathbb{R}^d} \sup_{k\in\mathbb{Z}^d} |(T_k g \cdot \sigma)\hat{}(\xi)|\, d\xi$$

and derives the boundedness on $L^2(\mathbb{R}^d)$ as a consequence of a "symbolic calculus," using ideas from classical analysis. The symbol class $M^{\infty,1}$ is therefore often called the "Sjöstrand class."

(b) $M^{\infty,1}$ contains non-smooth symbols. For instance, doubly periodic symbols with an absolutely convergent Fourier series belong to $M^{\infty,1}$, but clearly such a series need not be differentiable. Consequently, $M^{\infty,1}$ is not comparable to any of the $S^m_{\rho,\delta}$.

(c) The definition of $M^{\infty,1}$ as $\|\sigma\|_{M^{\infty,1}} = \int_{\mathbb{R}^{2d}} \sup_{z\in\mathbb{R}^{2d}} |\mathcal{V}_\Phi\sigma(z,\zeta)|\, d\zeta < \infty$ looks definitely more complicated than the differentiability conditions in the Hörmander classes. The strength of Theorem 8.30 is revealed by the following corollaries.

The first consequence is an improvement of a famous boundedness theorem of A. Calderòn and R. Vaillancourt [6]; see also Folland's treatment of this result [21].

Corollary 8.32. (a) *If $\sigma \in C^{2d+1}(\mathbb{R}^{2d})$, i.e., if*

$$\sum_{|\alpha|\le 2d+1} \|D^\alpha\sigma\|_\infty < \infty ,$$

then σ^{KN} is bounded on $L^2(\mathbb{R}^d)$ and in fact on all $M^{p,q}$, $1 \le p, q \le \infty$.

(b) *If $\sigma \in C^s(\mathbb{R}^{2d})$ for $s > 2d$, then σ^{KN} is bounded on all $M^{p,q}$, $1 \le p, q \le \infty$* [35].

Proof. The statement follows from the embeddings $S^0_{0,0} \subseteq C^{2d+1}(\mathbb{R}^{2d}) \subseteq C^s(\mathbb{R}^{2d}) \subseteq M^{\infty,1}$ for $s > 2d$. See, for instance, [35] or [24, Ch. 14.5]. □

In a sense, $M^{\infty,1}$ seems to be a maximal extension of $S^0_{0,0}$.

In addition, Theorem 8.30 implies a result of Cordes [11].

Corollary 8.33. *If $D^\alpha\sigma \in L^\infty(\mathbb{R}^{2d})$ for all $\alpha \in \{0,1,2\}^{2d}$, then σ^{KN} is bounded on all $M^{p,q}$, $1 \le p, q \le \infty$.*

Proof. Again this follows from an embedding result and is similar to [24, Thm. 14.5.3]. □

Remark 8.34. The pedestrian approach also has its limitations. For the Kohn–Nirenberg correspondence the optimal results are known. Coifman and Meyer [7] have shown that σ^{KN} is bounded on $L^2(\mathbb{R}^d)$ if

$$|D_\xi^\alpha D_x^\beta \sigma(x,\xi)| \leq C\,\langle \xi \rangle^{\rho(|\beta|-|\alpha|)}$$

for some $\rho < 1$ and $|\alpha|, |\beta| \leq [n/2]$. But their proofs are rather sophisticated and no longer elementary.

8.4.3 Proof of the Boundedness

Before we give a detailed proof of Theorem 8.30, we sketch the idea and find the appropriate questions to ask.

In order to bring in modulation space norms, we use the identity (8.5) and the extended orthogonality relations (8.14). In this way we can write the action of σ^{KN} as follows:

$$\langle \sigma^{KN} f, g \rangle_{\mathbb{R}^d} = \langle \sigma, R(g,f) \rangle_{\mathbb{R}^{2d}} = \langle \mathcal{V}_\Phi \sigma, \mathcal{V}_\Phi (R(g,f)) \rangle_{\mathbb{R}^{4d}} \,. \tag{8.20}$$

This identity makes sense for $f, g \in \mathcal{S}(\mathbb{R}^d)$ and $\sigma \in \mathcal{S}'(\mathbb{R}^d)$. (Observe that we distinguish the STFT $V_g f$ for functions on $\mathbb{R}^d$ from STFT $\mathcal{V}_\Phi \sigma$ for functions on $\mathbb{R}^{2d}$.)

To proceed further, we need to understand the STFT of a Rihaczek distribution. The identity of the next lemma is a kind of "magic formula" and constitutes the technical backbone of any time-frequency proof. Other versions can be found in [24], [43], [41], [36].

Lemma 8.35 (Magic Formula). *Let* φ, ψ, f, $g \in L^2(\mathbb{R}^d)$ *and set* $\Phi = R(\varphi, \psi) \in L^2(\mathbb{R}^d)$. *Then, with* $z = (z_1, z_2)$, $\zeta = (\zeta_1, \zeta_2) \in \mathbb{R}^{2d}$, *we have*

$$\mathcal{V}_\Phi\big(R(g,f)\big)(z,\zeta) = e^{-2\pi i z_2\cdot\zeta_2}\, V_\varphi g(z_1, z_2+\zeta_1)\, \overline{V_\psi f(z_1+\zeta_2, z_2)}\,.$$

Proof. In contrast to similar formulas, this one is an exercise in bookkeeping. We first make the time-frequency shifts of the Rihaczek distribution explicit:

$$M_\zeta T_z R(\varphi,\psi)(x,\xi) = e^{2\pi i(\zeta_1\cdot x+\zeta_2\cdot\xi)}\, \varphi(x-z_1)\, \overline{\hat\psi(\xi - z_2)} e^{-2\pi i (x-z_1)\cdot(\xi-z_2)}\,.$$

Consequently, after a substitution,

$$\begin{aligned}\mathcal{V}_\Phi R(g,f)(z,\zeta) &= \langle R(g,f), M_\zeta T_z R(\varphi,\psi)\rangle \\ &= \int_{\mathbb{R}^d}\int_{\mathbb{R}^d} g(x)\overline{\hat f(\xi)}\, e^{-2\pi i x\cdot\xi}\, e^{-2\pi i(\zeta_1\cdot x+\zeta_2\cdot\xi)}\, \overline{\varphi(x-z_1)} \\ &\qquad \times \hat\psi(\xi-z_2)\, e^{2\pi i (x-z_1)\cdot(\xi-z_2)}\, dx\, d\xi\end{aligned}$$

$$= e^{2\pi i z_1 \cdot z_2} \int_{\mathbb{R}^d} g(x) \overline{\varphi(x - z_1)}\, e^{-2\pi i x \cdot (\zeta_1 + z_2)}\, dx$$
$$\times \int_{\mathbb{R}^d} \overline{\hat{f}(\xi)}\, \hat{\psi}(\xi - z_2)\, e^{-2\pi i \xi \cdot (z_1 + \zeta_2)}\, d\xi$$
$$= e^{2\pi i z_1 \cdot z_2}\, V_\varphi g(z_1, z_2 + \zeta_1)\, \overline{V_{\hat{\psi}} \hat{f}(z_2, -z_1 - \zeta_2)}$$
$$= e^{-2\pi i z_2 \cdot \zeta_2}\, V_\varphi g(z_1, z_2 + \zeta_1)\, \overline{V_\psi f(z_1 + \zeta_2, z_2)}\,.$$

In the last transformation we have used the fundamental formula (8.8). Since both $R(g, f)$ and $R(\varphi, \psi)$ are in $L^2(\mathbb{R}^{2d})$, the integral defining $\mathcal{V}_\Phi R(g, f)$ is absolutely convergent on $\mathbb{R}^{2d}$, and so the application of Fubini's Theorem is justified. □

We are now ready to prove Theorem 8.30.

Proof of Theorem 8.30. We apply Hölder's inequality in the form

$$|\langle F, G \rangle| \le \|F\|_{L^{\infty,1}}\, \|G\|_{L^{1,\infty}}$$

to (8.20) and obtain

$$|\langle \sigma^{KN} f, g \rangle| = \left| \left\langle \mathcal{V}_\Phi \sigma,\, \mathcal{V}_\Phi\big(R(g, f)\big) \right\rangle \right|$$
$$\le \|\mathcal{V}_\Phi \sigma\|_{L^{\infty,1}}\, \|\mathcal{V}_\Phi(R(g, f))\|_{L^{1,\infty}}$$
$$\le C\, \|\sigma\|_{M^{\infty,1}}\, \|R(g, f)\|_{M^{1,\infty}}\,.$$

To estimate the $M^{1,\infty}$-norm of $R(g, f)$, we use the magic formula of Lemma 8.35 with the window $\Phi = R(\varphi, \varphi)$ and apply Hölder's inequality twice:

$$\|\mathcal{V}_\Phi(R(g, f))\|_{L^{1,\infty}}$$
$$= \sup_{\zeta \in \mathbb{R}^{2d}} \int_{\mathbb{R}^{2d}} |\mathcal{V}_\Phi(R(g, f))(z, \zeta)|\, dz$$
$$= \sup_{\zeta \in \mathbb{R}^{2d}} \int_{\mathbb{R}^d} \int_{\mathbb{R}^d} \left| e^{-2\pi i z_2 \cdot \zeta_2}\, V_\varphi g(z_1, z_2 + \zeta_1)\, \overline{V_\varphi f(z_1 + \zeta_2, z_2)} \right| dz_1 dz_2$$
$$\le \sup_{\zeta \in \mathbb{R}^{2d}} \left(\int_{\mathbb{R}^d} \left(\int_{\mathbb{R}^d} |V_\varphi f(z_1 + \zeta_2, z_2)|^p\, dz_1 \right)^{q/p} dz_2 \right)^{1/q}$$
$$\times \left(\int_{\mathbb{R}^d} \left(\int_{\mathbb{R}^d} |V_\varphi g(z_1, z_2 + \zeta_1)|^{p'}\, dz_1 \right)^{q'/p'} dz_2 \right)^{1/q'}$$
$$= \|f\|_{M^{p,q}}\, \|g\|_{M^{p',q'}}\,.$$

In this way we have obtained for all $g \in M^{p',q'} = (M^{p,q})'$ that

$$\left| \langle \sigma^{KN} f, g \rangle \right| \le C\, \|\sigma\|_{M^{\infty,1}}\, \|f\|_{M^{p,q}}\, \|g\|_{M^{p',q'}}\,.$$

This inequality implies that $\sigma^{KN} f \in M^{p,q}$ and that

$$\|\sigma^{KN} f\|_{M^{p,q}} \le C \, \|\sigma\|_{M^{\infty,1}} \, \|f\|_{M^{p,q}} .$$

Strictly speaking, this argument covers only the cases when the duality of Theorem 8.27 is applicable and excludes the cases $(p,q) = (1,\infty)$ or $(p,q) = (\infty,1)$. This "gap" can be repaired with the observation that $M^{\infty,1}$ is the dual of the *norm-closure* of $\mathcal{H}_0$ in the $M^{1,\infty}$-norm and similarly for $M^{1,\infty}$. See [1] for the technical details of these exceptional cases. □

8.4.4 Schatten Classes

Recall the definition of the *Schatten classes* [51]. The singular values $s_j(A)$ of a compact operator on a Hilbert space $\mathcal{H}$ are the eigenvalues $\lambda_j(|A|)$ of the positive compact operator $|A| = (A^*A)^{1/2}$. A compact operator A belongs to the Schatten class $\mathcal{I}_p$ if and only if $(s_j(A)) \in \ell^p$. In this case, we write $A \in \mathcal{I}_p$ and impose the norm $\|A\|_{\mathcal{I}_p} = \left(\sum_{j=1}^{\infty} s_j(A)^p\right)^{1/p}$. The Schatten classes interpolate like ℓ^p, and so $[\mathcal{I}_1, \mathcal{B}(\mathcal{H})]_\theta = \mathcal{I}_p$ [51].

Hilbert–Schmidt Operators. We begin with the *Hilbert–Schmidt Operator Class* $\mathcal{I}_2$, because this is easy and a complete characterization is known.

Theorem 8.36 (Pool [46]). *A pseudodifferential operator σ^{KN} is Hilbert–Schmidt if and only if $\sigma \in L^2(\mathbb{R}^{2d})$. In this case $\|\sigma^{KN}\|_{\mathcal{I}_2} = \|\sigma\|_2$.*

Proof. It is well known [8] that an integral operator $Af(x) = \int_{\mathbb{R}^d} k(x,y) f(y)\, dy$ is Hilbert–Schmidt if and only if $k \in L^2(\mathbb{R}^{2d})$, and, in this case, $\|A\|_{\mathcal{I}_2} = \|k\|_2$. Since the kernel of σ^{KN} is obtained from its symbol by unitary transforms, namely, $k = \mathcal{T}'\mathcal{F}_2\sigma$ by Proposition 8.18, we obtain that $\|A\|_{\mathcal{I}_2} = \|k\|_2 = \|\mathcal{T}'\mathcal{F}_2\sigma\|_2 = \|\sigma\|_2$. □

Trace Class Operators. We first prove a simple result that goes back to H. Feichtinger. It was proved first in [23] (see also [28]) and independently in [13]. The following proof is new.

Theorem 8.37. *If $\sigma \in M^1(\mathbb{R}^{2d})$, then $\sigma^{KN} \in \mathcal{I}_1$.*

Proof. We apply the inversion formula (Corollary 8.14) to σ with the window $\Phi = R(\varphi, \varphi)$ and obtain

$$\sigma = \iint_{\mathbb{R}^{4d}} \mathcal{V}_\Phi \sigma(z,\zeta) \, M_\zeta T_z \Phi \, dz d\zeta .$$

Using the linearity of the Kohn–Nirenberg correspondence, σ^{KN} can be represented as a superposition of "elementary operators" as follows:

$$\sigma^{KN} = \iint_{\mathbb{R}^{4d}} \mathcal{V}_\Phi \sigma(z,\zeta) \, (M_\zeta T_z \Phi)^{KN} \, dz d\zeta .$$

If $\sigma \in M^1(\mathbb{R}^{2d})$, then $\mathcal{V}_\Phi \sigma \in L^1(\mathbb{R}^{4d})$, and so this integral is absolutely convergent, provided we can say something about the "elementary operators" $(M_\zeta T_z \Phi)^{KN}$. Using Lemma 8.35 and (8.5), we calculate that

$$\begin{aligned} \langle L_{M_\zeta T_z \Phi} f, g \rangle &= \langle M_\zeta T_z \Phi, R(g,f) \rangle \\ &= \overline{\mathcal{V}_\Phi (R(g,f))(z,\zeta)} \\ &= e^{2\pi i z_2 \cdot \zeta_2}\, V_\varphi f(z_1 + \zeta_2, z_2)\, \overline{V_\varphi g(z_1, z_2 + \zeta_1)} \\ &= e^{2\pi i z_2 \cdot \zeta_2}\, \langle f, M_{z_2} T_{z_1 + \zeta_2} \varphi \rangle \, \langle M_{z_2 + \zeta_1} T_{z_1} \varphi, g \rangle \,. \end{aligned}$$

So $(M_\zeta T_z \Phi)^{KN}$ is the rank one operator

$$f \to e^{2\pi i z_2 \cdot \zeta_2}\, \langle f, M_{z_2} T_{z_1+\zeta_2} \varphi \rangle \, M_{z_2+\zeta_1} T_{z_1} \varphi,$$

and its trace class norm is

$$\|(M_\zeta T_z \Phi)^{KN}\|_{\mathcal{I}_1} = \|\varphi\|_2^2 = 1,$$

independently of $z, \zeta \in \mathbb{R}^{2d}$. In conclusion, we obtain that

$$\|\sigma^{KN}\|_{\mathcal{I}_1} \le \iint_{\mathbb{R}^{4d}} |\mathcal{V}_\Phi \sigma(z,\zeta)|\, \|(M_\zeta T_z \Phi)^{KN}\|_{\mathcal{I}_1}\, dz d\zeta = \|\sigma\|_{M^1} \,. \qquad \square$$

Again, some known results follow from Theorem 8.37 by suitable embedding theorems.

Corollary 8.38 (Daubechies [12]). *If* $\sigma \in M^{2,2}_{v_s} = L^2_s \cap H^s$ *for* $s > 2d$, *then* $L_\sigma \in \mathcal{I}_1$.

Proof. Using $v_s(z,\zeta) = (1 + |z|^2 + |\zeta|^2)^{s/2}$, we find that

$$\|\sigma\|_{M^1} = \|\mathcal{V}_\Phi \sigma\, v_s\, v_s^{-1}\|_{L^1(\mathbb{R}^{4d})} \le \|\mathcal{V}_\Phi \sigma\, v_s\|_2\, \|v_s^{-1}\|_2 = C_s\, \|\sigma\|_{M^2_{v_s}}$$

with $C_s = \|v_s^{-1}\|_2 < \infty$ whenever $s > 2d$. Consequently, $M^2_{v_s} \hookrightarrow M^1$ for $s > 2d$ and Theorem 8.37 applies. $\square$

The following significant improvement was obtained by Heil, Ramanathan, and Topiwala in [35]; simplified proofs were given in [28] and [34].

Theorem 8.39. *If* $\sigma \in M^{2,2}_{v_s}$ *for* $s > d$, *then* $\sigma^{KN} \in \mathcal{I}_1$.

The proof is more subtle than the arguments above. It uses Gabor frames and is based on the observation that many of the elementary rank-one operators $(M_\zeta T_z \Phi)^{KN}$ have the same range and should be treated together.

Corollary 8.40 (Hörmander [38]). *If* $2k > d$, *then*

$$\|\sigma^{KN}\|_{\mathcal{I}_1} \le C \sum_{|\alpha| + \cdots + |\beta'| \le 2k} \|x^\alpha \xi^\beta D_x^{\alpha'} D_\xi^{\beta'} \sigma\|_2 \,.$$

Proof. The right-hand side dominates $\|\sigma\|_{L^2_{2k}} + \|\sigma\|_{H^{2k}} \asymp \|\sigma\|_{M^2_{v_{2k}}}$. So for $2k > d$, Theorem 8.39 applies. □

A more sophisticated trace-class result was proved by Hörmander in [37]. A special case says that if $\sigma \in S^m_{\rho,\delta} \cap L^1$, then $\sigma^{KN} \in \mathcal{I}_1$. This result is different; it does not imply Theorems 8.37 and 8.39, nor is it implied by them.

8.4.5 Modulation Spaces as Symbol Classes: The Complete Picture

In this section we use an arbitrary modulation space $M^{p,q}(\mathbb{R}^{2d})$ as a symbol for σ^{KN} and study the boundedness and Schatten class properties. It turns out that the two endpoint results in Theorems 8.30 and 8.37 suffice to obtain the complete picture. First, using the embedding and interpolation properties of modulation spaces stated in Theorem 8.27, we obtain the following refinement of Theorem 8.37.

Theorem 8.41. (a) *If $1 \le p, q \le 2$ and $\sigma \in M^{p,q}$, then $\sigma^{KN} \in \mathcal{I}_{\max\{p,q\}}$.*

(b) *If $p \ge 2$ and $q \le p'$, and if $\sigma \in M^{p,q}$, then $\sigma^{KN} \in \mathcal{I}_p$.*

Proof. (a) We first use interpolation. Fix $1 \le p \le 2$. Then by combining Theorem 8.27(iv) with the interpolation of the Schatten classes, we obtain that if $\sigma \in [M^{1,1}, M^{2,2}]_\theta = M^{p,p}$, then $\sigma^{KN} \in [\mathcal{I}_1, \mathcal{I}_2]_\theta = \mathcal{I}_p$.

Next, we use the inclusion relations. Fix $1 \le p, q \le 2$, and set $\mu = \max\{p, q\}$. Then by Theorem 8.27(iii), we have $M^{p,q} \subseteq M^{\mu,\mu}$ and therefore $\sigma^{KN} \in \mathcal{I}_\mu$, so part (a) is proved.

(b) We first use the interpolation $[M^{2,2}, M^{\infty,1}]_\theta = M^{p,p'}$, and then use inclusions. If $2 \le p \le \infty$ and $\sigma \in M^{p,p'} = [M^{2,2}, M^{\infty,1}]_\theta$, then $\sigma^{KN} \in [\mathcal{I}_2, \mathcal{B}(L^2)]_\theta = \mathcal{I}_p$. When $p \ge 2$ and $q \le p'$, we have $M^{p,q} \subset M^{p,p'}$, and consequently $\sigma^{KN} \in \mathcal{I}_p$. □

The complete picture for boundedness on $L^2(\mathbb{R}^d)$ can now be stated as follows.

Theorem 8.42. (a) *If $q \le 2$ and either $1 \le p \le 2$ or $p \le q'$ and if $\sigma \in M^{p,q}$, then σ^{KN} is a bounded operator on $L^2(\mathbb{R}^d)$.*

(b) *If $q > 2$ or if $p \ge 2$ and $p > q'$, then there exists $\sigma \in M^{p,q}$ such that σ^{KN} is unbounded on $L^2(\mathbb{R}^d)$.*

The sufficient conditions of part (a) are a consequence of Theorem 8.41, because operators in the Schatten class are in particular bounded on $L^2(\mathbb{R}^d)$. Part (b) can be shown by the construction of explicit counterexamples, see [30].

The following diagram taken from [29] (see Fig. 8.1) illustrates the region of boundedness. Each point in the unit square corresponds to a modulation space. Precisely, the point $(1/p, 1/q) \in [0,1]^2$ corresponds to $M^{p,q}$. According to Theorem 8.27(iii) the lexicographic ordering on $[0,1]^2$ corresponds to the inclusion of modulation spaces. This means that if a symbol class $M^{p,q}$ possesses certain properties, then all modulation spaces corresponding to points in the rectangle whose lower left vertex is $(1/p, 1/q)$ possess the same property. For the property "boundedness on $L^2(\mathbb{R}^d)$" this is shown in the diagram.

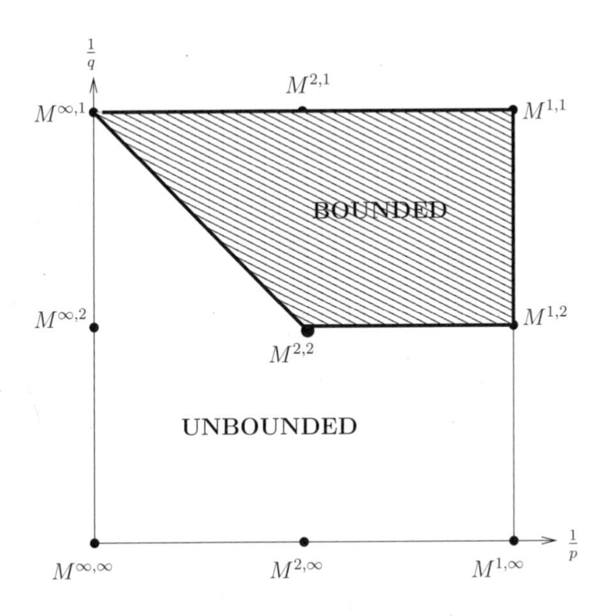

Fig. 8.1. Set of (p,q) for which $\sigma \in M^{p,q}(\mathbb{R}^{2d})$ implies that σ^{KN} is bounded or unbounded on $L^2(\mathbb{R}^d)$.

8.4.6 Further Topics

The mapping properties between different modulation spaces were studied in [29], [58], and mapping between weighted modulation spaces was studied in [59]. The study of time-frequency localization operators, also called Toeplitz operators, anti-Wick operators, or STFT multipliers, has greatly benefited from the time-frequency approach, and almost sharp results on Schatten class and boundedness are now known [9], [10], [58]. Algebra properties of $M^{\infty,1}$ and related spaces were studied in [25], [26], [44], [52].

Acknowledgments

I would like to express my gratitude to John J. Benedetto for his continued support and encouragement. Although I have never had the pleasure and privilege of collaborating with him, I have worked with four of his outstanding former students. Through this collaboration with his mathematical sons, I have learned enormously from and about John, and I consider him my mathematical uncle. This chapter, in particular, owes much to my joint work and many stimulating discussions with Chris Heil.

References

1. Á. Bényi, K. Gröchenig, C. Heil, and K. Okoudjou, Modulation spaces and a class of bounded multilinear pseudodifferential operators, *J. Operator Theory*, **54** (2005), pp. 389–401.
2. G. Björk, Linear partial differential operators and generalized distributions, *Ark. Mat.*, **6** (1966), pp. 351–407.
3. P. Boggiatto, E. Cordero, and K. Gröchenig, Generalized anti-Wick operators with symbols in distributional Sobolev spaces, *Integral Equations Operator Theory*, **48** (2004), pp. 427–442.
4. A. Bonami, B. Demange, and P. Jaming, Hermite functions and uncertainty principles for the Fourier and the windowed Fourier transforms, *Rev. Mat. Iberoamericana*, **19** (2003), pp. 23–55.
5. A. Boulkhemair, Remarks on a Wiener type pseudodifferential algebra and Fourier integral operators, *Math. Res. Lett.*, **4** (1997), pp. 53–67.
6. A.-P. Calderón and R. Vaillancourt, On the boundedness of pseudo-differential operators, *J. Math. Soc. Japan*, **23** (1971), pp. 374–378.
7. R. R. Coifman and Y. Meyer, *Au delà des opérateurs pseudo-différentiels*, Astérisque **57**, Société Mathématique de France, Paris, 1978.
8. J. B. Conway, *A Course in Functional Analysis*, Second Edition, Springer-Verlag, New York, 1990.
9. E. Cordero and K. Gröchenig, Time-frequency analysis of localization operators, *J. Funct. Anal.*, **205** (2003), pp. 107–131.
10. E. Cordero and K. Gröchenig, Necessary conditions for Schatten class localization operators, *Proc. Amer. Math. Soc.*, **133** (2005), pp. 3573–3579.
11. H. O. Cordes, On compactness of commutators of multiplications and convolutions, and boundedness of pseudodifferential operators, *J. Funct. Anal.*, **18** (1975), pp. 115–131.
12. I. Daubechies, On the distributions corresponding to bounded operators in the Weyl quantization, *Comm. Math. Phys.*, **75** (1980), pp. 229–238.
13. M. Dimassi and J. Sjöstrand, *Spectral Asymptotics in the Semi-Classical Limit*, London Mathematical Society Lecture Note Series, Vol. 268, Cambridge University Press, Cambridge, 1999.
14. H. G. Feichtinger, Un espace de Banach de distributions tempérées sur les groupes localement compacts abéliens, *C. R. Acad. Sci. Paris Sér. A-B*, **290** (1980), pp. A791–A794.

15. H. G. Feichtinger, On a new Segal algebra, *Monatsh. Math.*, **92** (1981), pp. 269–289.
16. H. G. Feichtinger, Modulation spaces on locally compact abelian groups, Technical report, University of Vienna, 1983.
17. H. G. Feichtinger, Modulation spaces on locally compact abelian groups, in: *Wavelets and Their Applications* (Chennai, January 2002), M. Krishna, R. Radha, and S. Thangavelu, eds., Allied Publ., New Delhi (2003), pp. 99–140.
18. H. G. Feichtinger and K. Gröchenig, Gabor wavelets and the Heisenberg group: Gabor expansions and short time Fourier transform from the group theoretical point of view, in: *Wavelets: A Tutorial in Theory and Applications*, C. K. Chui, ed., Academic Press, Boston, MA, 1992, pp. 359–397.
19. H. G. Feichtinger and K. Gröchenig, Gabor frames and time-frequency analysis of distributions, *J. Funct. Anal.*, **146** (1997), pp. 464–495.
20. H. G. Feichtinger and W. Kozek, Quantization of TF lattice-invariant operators on elementary LCA groups, in: *Gabor Analysis and Algorithms*, H. G. Feichtinger and T. Strohmer, eds., Birkhäuser, Boston, 1998, pp. 233–266.
21. G. B. Folland, *Harmonic Analysis in Phase Space*, Princeton Univ. Press, Princeton, NJ, 1989.
22. Y. V. Galperin and K. Gröchenig, Uncertainty principles as embeddings of modulation spaces, *J. Math. Anal. Appl.*, **274** (2002), pp. 181–202.
23. K. Gröchenig, An uncertainty principle related to the Poisson summation formula, *Studia Math.*, **121** (1996), pp. 87–104.
24. K. Gröchenig, *Foundations of Time-Frequency Analysis*, Birkhäuser, Boston, 2001.
25. K. Gröchenig, Composition and spectral invariance of pseudodifferential operators on modulation spaces, *J. Anal. Math.*, (2006), to appear.
26. K. Gröchenig, Time-frequency analysis of Sjöstrand's class, *Rev. Mat. Iberoamericana*, **22** (2006), to appear.
27. K. Gröchenig, D. Han, C. Heil, and G. Kutyniok, The Balian-Low theorem for symplectic lattices in higher dimensions, *Appl. Comput. Harmon. Anal.*, **13** (2002), pp. 169–176.
28. K. Gröchenig and C. Heil, Modulation spaces and pseudodifferential operators, *Integral Equations Operator Theory*, **34** (1999), pp. 439–457.
29. K. Gröchenig and C. Heil, Modulation spaces as symbol classes for pseudodifferential operators, In S. T. M. Krishna, R. Radha, editor, *Wavelets and Their Applications*, pages 151–170. Allied Publishers, Chennai, 2003.
30. K. Gröchenig and C. Heil, Counterexamples for boundedness of pseudodifferential operators, *Osaka J. Math*, **41** (2004), 681–691.
31. K. Gröchenig and M. Leinert, Symmetry and inverse-closedness of matrix algebras and symbolic calculus for infinite matrices, *Trans. Amer. Math. Soc.*, **358** (2006), pp. 2695–2711.
32. K. Gröchenig and S. Samarah, Non-linear approximation with local Fourier bases, *Constr. Approx.*, **16** (2000), pp. 317–331.
33. K. Gröchenig and G. Zimmermann, Hardy's theorem and the short-time Fourier transform of Schwartz functions, *J. London Math. Soc. (2)*, **63** (2001), pp. 205–214.
34. C. Heil, Integral operators, pseudodifferential operators, and Gabor frames, in: *Advances in Gabor Analysis*, H. G. Feichtinger and T. Strohmer, eds., Birkhäuser, Boston, 2003, pp. 153–169.

35. C. Heil, J. Ramanathan, and P. Topiwala, Singular values of compact pseudodifferential operators, *J. Funct. Anal.*, **150** (1997), pp. 426–452.
36. F. Hlawatsch, Regularity and unitarity of bilinear time-frequency signal representations, *IEEE Trans. Inform. Theory*, **38** (1992), pp. 82–94.
37. L. Hörmander, On the asymptotic distribution of the eigenvalues of pseudodifferential operators in $\mathbf{R}^n$, *Ark. Mat.*, **17** (1979), pp. 297–313.
38. L. Hörmander, The Weyl calculus of pseudodifferential operators, *Comm. Pure Appl. Math.*, **32** (1979), pp. 360–444.
39. L. Hörmander, *The Analysis of Linear Partial Differential Operators. I Distribution Theory and Fourier Analysis*, Second Edition, Springer-Verlag, Berlin, 1990.
40. L. Hörmander, *The analysis of linear partial differential operators. III Pseudodifferential operators*, Corrected reprint of the 1985 original, Grundlehren der Mathematischen Wissenschaften, Vol. 274, Springer-Verlag, Berlin, 1994.
41. P. Jaming, Principe d'incertitude qualitatif et reconstruction de phase pour la transformée de Wigner, *C. R. Acad. Sci. Paris Sér. I Math.*, **327** (1998), pp. 249–254.
42. A. J. E. M. Janssen, Duality and biorthogonality for Weyl-Heisenberg frames, *J. Fourier Anal. Appl.*, **1** (1995), pp. 403–436.
43. A. J. E. M. Janssen, Proof of a conjecture on the supports of Wigner distributions, *J. Fourier Anal. Appl.*, **4** (1998), pp. 723–726.
44. D. Labate, Pseudodifferential operators on modulation spaces, *J. Math. Anal. Appl.*, **262** (2001), pp. 242–255.
45. S. Pilipović and N. Teofanov, Pseudodifferential operators on ultra-modulation spaces, *J. Funct. Anal.*, **208** (2004), pp. 194–228.
46. J. C. T. Pool, Mathematical aspects of the Weyl correspondence, *J. Mathematical Phys.*, **7** (1966), pp. 66–76.
47. R. Rochberg and K. Tachizawa, Pseudodifferential operators, Gabor frames, and local trigonometric bases, in: *Gabor Analysis and Algorithms: Theory and Applications*, H. G. Feichtinger and T. Strohmer, eds., Birkhäuser, Boston, 1998, pp. 171–192.
48. H. H. Schaefer, *Topological Vector Spaces*, Third printing corrected, Springer-Verlag, New York, 1971.
49. L. Schwartz, Théorie des noyaux, in: *Proceedings of the International Congress of Mathematicians*, Cambridge, Mass., 1950, Vol. 1, Amer. Math. Soc., Providence, RI, 1952, pp. 220–230.
50. M. A. Shubin, *Pseudodifferential Operators and Spectral Theory*, Translated from the 1978 Russian original by Stig I. Andersson, Second Edition, Springer-Verlag, Berlin, 2001.
51. B. Simon, *Trace Ideals and Their Applications*, London Mathematical Society Lecture Note Series, Vol. 35, Cambridge University Press, Cambridge, 1979.
52. J. Sjöstrand, An algebra of pseudodifferential operators, *Math. Res. Lett.*, **1** (1994), pp. 185–192.
53. J. Sjöstrand, Wiener type algebras of pseudodifferential operators, in: *Séminaire sur les Équations aux Dérivées Partielles, 1994–1995*, Exp. No. IV, École Polytech., Palaiseau, 1995.
54. E. M. Stein, *Harmonic Analysis: Real-Variable Methods, Orthogonality, and Oscillatory Integrals*, Princeton Univ. Press, Princeton, NJ, 1993.
55. T. Strohmer, Approximation of dual Gabor frames, window decay, and wireless communications, *Appl. Comput. Harmon. Anal.*, **11** (2001), pp. 243–262.

56. T. Strohmer, Pseudodifferential operators and Banach algebras in mobile communications, *Appl. Comput. Harmon. Anal.*, **20** (2006), pp. 237–249.
57. J. Toft, Subalgebras to a Wiener type algebra of pseudo-differential operators, *Ann. Inst. Fourier (Grenoble)*, **51** (2001), pp. 1347–1383.
58. J. Toft, Continuity properties for modulation spaces, with applications to pseudo-differential calculus. I, *J. Funct. Anal.*, **207** (2004), pp. 399–42.
59. J. Toft, Continuity properties for modulation spaces, with applications to pseudo-differential calculus. II, *Ann. Global Anal. Geom.*, **26** (2004), pp. 73–106.
60. Ö. Yılmaz, Coarse quantization of highly redundant time-frequency representations of square-integrable functions, *Appl. Comput. Harmon. Anal.*, **14** (2003), pp. 107–132.

9

Linear Independence of Finite Gabor Systems

Christopher Heil

School of Mathematics, Georgia Institute of Technology, Atlanta, GA 30332, USA
heil@math.gatech.edu

Summary. This chapter is an introduction to an open conjecture in time-frequency analysis on the linear independence of a finite set of time-frequency shifts of a given L^2 function. Background and motivation for the conjecture are provided in the form of a survey of related ideas, results, and open problems in frames, Gabor systems, and other aspects of time-frequency analysis, especially those related to independence. The partial results that are known to hold for the conjecture are also presented and discussed.

9.1 Introduction

In 1987, John Benedetto introduced two of his young graduate students, David Walnut and myself, to a new mathematical research group that had been formed at the MITRE Corporation in McLean, Virginia. Later, as a postdoc at MIT, I met Jay Ramanathan[1] and Pankaj Topiwala,[2] then members of the main MITRE math research group in Bedford, Massachusetts. We began working together on problems in time-frequency analysis and wavelets. One direction followed a beautiful insight of Ramanathan's, applying Gabor frame expansions to derive boundedness and spectral results for pseudodifferential operators [70]. In another direction, we explored the basic structure of Gabor frames, which ultimately led us to make the following conjecture (sometimes called today the HRT Conjecture, the Linear Independence Conjecture for Time-Frequency Shifts, or the Zero Divisor Conjecture for the Heisenberg Group).

Conjecture 9.1. If $g \in L^2(\mathbf{R})$ is nonzero and $\{(\alpha_k, \beta_k)\}_{k=1}^N$ is any set of finitely many distinct points in $\mathbf{R}^2$, then $\{e^{2\pi i \beta_k x} g(x - \alpha_k)\}_{k=1}^N$ is a linearly independent set of functions in $L^2(\mathbf{R})$.

---

[1]Currently: Professor, Eastern Michigan University, Ypsilanti, Michigan 48197. Email: ramanath@emunix.emich.edu

[2]Currently: Founder, FastVDO LLC, 7150 Riverwood Drive, Columbia, Maryland 21046. Email: pnt@fastvdo.com

Despite the striking simplicity of the statement of this conjecture, it remains open today in the generality stated. This chapter provides some background and motivation for the conjecture in the form of a survey of related ideas, results, and open problems in frames, Gabor systems, and other aspects of time-frequency analysis, especially those related to independence. After a brief statement in Section 9.2 of some of the partial results that are known concerning the conjecture, Section 9.3 discusses some of the subtleties involved in the seemingly simple concepts of spanning and independence when dealing with infinite-dimensional spaces. In particular, this motivates the definition of frames. In Section 9.4 we specialize to the case of Gabor frames, whose elements consist of time-frequency shifts of a given function. We survey some topics in Gabor theory, including other open problems related to the Balian–Low Theorem and to Nyquist-type density phenomena for Gabor systems. The reader whose primary interest is the HRT Conjecture can skim or skip the surveys in Sections 9.3 and 9.4 and turn directly to Section 9.5, which discusses some of the partial results that have been obtained, and some of the ideas behind their proofs.

Throughout, the discussion is intended to be accessible to graduate students who have a little background in real and functional analysis and some familiarity with Hilbert spaces, especially $L^2(\mathbf{R})$ and $L^2[0,1]$. A summary of basic theorems from functional analysis can be found in the first chapter of [64]. Extensive references are given throughout, both to research papers and to textbooks or survey articles. The choice of references is usually made based more on their utility as background or additional reading than for historical completeness.

9.2 Statement of Partial Results

Despite attacks by a number of groups, the only published papers specifically about the HRT Conjecture appear to be [69], [96], and [88].[3] These will be quickly summarized now, with more details presented in Section 9.5.

The HRT Conjecture was first made in the paper [69], and some partial results were obtained there, including the following.

(a) If a nonzero $g \in L^2(\mathbf{R})$ is compactly supported, or just supported on a half-line, then the independence conclusion holds for any value of N.

(b) If $g(x) = p(x)\, e^{-x^2}$ where p is a nonzero polynomial, then the independence conclusion holds for any value of N.

(c) The independence conclusion holds for any nonzero $g \in L^2(\mathbf{R})$ if $N \leq 3$.

[3] A recent preprint by Radu Balan, "A noncommutative Wiener lemma and a faithful tracial state on Banach algebras of time-frequency shift operators," contains some new partial results.

(d) If the independence conclusion holds for a particular $g \in L^2(\mathbf{R})$ and a particular choice of points $\{(\alpha_k, \beta_k)\}_{k=1}^N$, then there exists an $\varepsilon > 0$ such that it also holds for any h satisfying $\|g - h\|_2 < \varepsilon$, using the same set of points.

(e) If the independence conclusion holds for one particular $g \in L^2(\mathbf{R})$ and particular choice of points $\{(\alpha_k, \beta_k)\}_{k=1}^N$, then there exists an $\varepsilon > 0$ such that it also holds for that g and any set of N points in $\mathbf{R}^2$ within ε of the original ones.

It is perhaps surprising that there are almost no partial results formulated in terms of smoothness or decay conditions on g. In particular, Conjecture 1 is open even if we impose the extra hypothesis that g lies in the Schwartz class $\mathcal{S}(\mathbf{R})$.

The next partial advance was made by Linnell in [96]. He used C^*-algebra techniques to prove that if the points $\{(\alpha_k, \beta_k)\}_{k=1}^N$ are a subset of some translate of a full-rank lattice in $\mathbf{R}^2$, then the independence conclusion holds for any nonzero g (such a lattice has the form $A(\mathbf{Z}^2)$ where A is an invertible matrix). Any three points in the plane always lie on a translate of some lattice, so this recovers and extends the partial result (b) above. However, four arbitrary points need not be contained in a translate of a lattice. Indeed, the case $N = 4$ of the conjecture is still open, and even the following special cases seem to be open.

Conjecture 9.2. If $g \in L^2(\mathbf{R})$ is nonzero, then each of the following is a linearly independent set of functions in $L^2(\mathbf{R})$:

(a) $\{g(x),\, g(x-1),\, e^{2\pi i x} g(x),\, e^{2\pi i \sqrt{2} x} g(x - \sqrt{2})\}$,

(b) $\{g(x),\, g(x-1),\, e^{2\pi i x} g(x),\, g(x - \pi)\}$.

Conjecture 9.2 remains open even if we impose the condition that g be continuous (or smoother). Recently Rzeszotnik has settled a different specific four-point case, showing that $\{g(x),\, g(x-1),\, e^{2\pi i x} g(x),\, g(x - \sqrt{2})\}$ is always linearly independent [106].

Finally, Kutyniok considered a generalized conjecture in [88], replacing the real line $\mathbf{R}$ by a locally compact Abelian group. Although the conjecture then becomes even more difficult to address, she was able to obtain some partial results. As we will mention later, even the seemingly trivial action of replacing $\mathbf{R}$ by $\mathbf{R}^n$ results in complications.

9.3 Spanning, Independence, Frames, and Redundancy

One motivation for the HRT Conjecture comes from looking at *frames*, which are possibly redundant or overcomplete collections of vectors in a Hilbert space which nonetheless provide basis-like representations of vectors in the space. Thus a frame "spans" the space in some sense, even though it may be "dependent." However, in infinite dimensions there are many shades of gray to the meanings of "spanning" and "independence." Some of the most important frames are "dependent" taken as a whole even though they have the property that every finite subset is linearly independent. One motivation for the HRT Conjecture is the question of whether the special class of *Gabor frames* (defined in Section 9.4) have this property.

In this section we provide some background on frames and the nuances of spanning and independence in infinite dimensions. For simplicity, the discussion will be kept to the setting of Banach spaces. Some of the definitions and discussion can be generalized easily to other settings, but other parts, especially the discussion of frames and their properties, are more explicitly Hilbert space theories, and there are subtleties in attempting to generalize those results beyond the Hilbert space setting.

We will state many facts but prove few of them; proofs and additional information can be found in the references provided in each section.

9.3.1 Spanning and Independence in Finite Dimensions

Spanning and independence are clear in finite dimensions. A set $\{f_1, \dots, f_M\}$ of M vectors in an N-dimensional vector space H *spans* H if for each vector $f \in H$ there exist scalars c_i (not necessarily unique) such that $f = c_1 f_1 + \cdots + c_M f_M$. This can only happen if $M \geq N$.

On the other hand, $\{f_1, \dots, f_M\}$ is *linearly independent* if whenever a vector $f \in H$ can be written as $f = c_1 f_1 + \cdots + c_M f_M$, it can only be so written in one way, i.e., the scalars c_i are unique if they exist at all. This can only happen if $M \leq N$.

When both of these happen simultaneously, we have a *basis*. In this case every $f \in H$ can be written as $f = c_1 f_1 + \cdots + c_M f_M$ for a unique choice of scalars c_i. This can only happen if $M = N$.

9.3.2 Spanning in Infinite Dimensions

For proofs, examples, and more information on bases, convergence of series, and related issues in normed spaces that are discussed in this section, we suggest the references [28], [64], [93], [98], [108], [114].

In a completely arbitrary vector space we can only define *finite* sums of vectors, because to define an infinite series we need a notion of convergence, and this requires a norm or metric or at least a topology. Thus, we define the *finite linear span* of a collection of vectors $\{f_\alpha\}_{\alpha \in J}$ to be

$$\operatorname{span}(\{f_\alpha\}_{\alpha \in J}) = \Big\{\sum_{i=1}^{N} c_i f_{\alpha_i} : N \in \mathbf{N},\, c_i \in \mathbf{C}, \alpha_i \in J\Big\}.$$

We say that $\{f_\alpha\}_{\alpha\in J}$ *spans* V if the finite span is all of V, i.e., every vector in V equals some finite linear combination of the f_α. We say that $\{f_\alpha\}_{\alpha\in J}$ is a *Hamel basis* if it both spans and is finitely linearly independent, or, equivalently, if every nonzero vector $f \in V$ can be written $f = \sum_{i=1}^N c_i f_{\alpha_i}$ for a unique choice of indices $\{\alpha_i\}_{i=1}^N$ and nonzero scalars $\{c_i\}_{i=1}^N$. For most vector spaces, Hamel bases are only known to exist because of the Axiom of Choice; in fact, the statement "Every vector space has a Hamel basis" is equivalent to the Axiom of Choice. Although Hamel bases are sometimes just called "bases," this is potentially confusing because if V is a normed space, then the word *basis* is usually reserved for something different (see Definition 9.3).

As soon as we impose a little more structure on our vector space, we can often construct systems which are much more convenient than Hamel bases. For example, in a Banach space we have a norm, and hence can form "infinite linear combinations" by using infinite series. In particular, given a collection $\{f_i\}_{i\in\mathbf{N}}$ indexed by the natural numbers and given scalars $\{c_i\}_{i\in\mathbf{N}}$, we say the series $f = \sum_{i=1}^\infty c_i f_i$ converges and equals f if $\|f - \sum_{i=1}^N c_i f_i\| \to 0$ as $N \to \infty$. Note that order in this series is important; if we change the order of indices we are not guaranteed that the series will still converge. If the convergence does not depend on the order it is called *unconditional convergence*, otherwise it is *conditional convergence.*

A related but distinct consequence of the fact that we have a norm is that we can form the closure of the finite linear span by constructing the set of all possible limits of finite linear combinations. This set is called the *closed span*, and is denoted $\overline{\operatorname{span}}(\{f_i\}_{i\in\mathbf{N}})$. Given $f \in \overline{\operatorname{span}}(\{f_i\}_{i\in\mathbf{N}})$, by definition there exist vectors $g_N \in \operatorname{span}(\{f_i\}_{i\in\mathbf{N}})$ which converge to f. However, this is not the same as forming infinite linear combinations. While each g_N is some finite linear combination of the f_i, it need not be true that we can write $g_N = \sum_{i=1}^N c_i f_i$ using a *single* sequence of scalars $\{c_i\}_{i\in\mathbf{N}}$.

Using these notions, we can form several variations on "spanning sets."

Definition 9.3. *Let $\{f_i\}_{i\in\mathbf{N}}$ be a countable sequence of vectors in a Banach space X.*

(a) $\{f_i\}_{i\in\mathbf{N}}$ *is* complete (*or* total *or* fundamental) *if* $\overline{\operatorname{span}}(\{f_i\}_{i\in\mathbf{N}}) = X$, *i.e., for each $f \in X$ and each $N \in \mathbf{N}$ there exist scalars $\{c_{N,i}(f)\}_{i\in\mathbf{N}}$ such that $\sum_{i=1}^N c_{N,i}(f)\, f_i \to f$ as $N \to \infty$.*

(b) $\{f_i\}_{i\in\mathbf{N}}$ *has* Property S *if for each $f \in X$ there exist scalars $\{c_i(f)\}_{i\in\mathbf{N}}$ such that*

$$f = \sum_{i=1}^\infty c_i(f) f_i. \tag{9.1}$$

(c) $\{f_i\}_{i\in\mathbf{N}}$ *is a* quasibasis *if it has Property S and for each $i \in \mathbf{N}$ the mapping $f \mapsto c_i(f)$ is linear and continuous (and hence defines an element of the dual space X^*).*

(d) $\{f_i\}_{i\in\mathbf{N}}$ *is a* basis *or* Schauder basis *if it has Property S and for each* $f \in X$ *the scalars* $\{c_i(f)\}_{i\in\mathbf{N}}$ *are unique.*

Completeness is a weak property. The definition says that there are finite linear combinations of the f_i that converge to f, but the scalars needed can change completely as the length N of the linear combination increases. On the other hand, unlike the other properties there exists a nice, simple characterization of complete sequences. For the case of a Hilbert space it is:

$$\{f_i\}_{i\in\mathbf{N}} \text{ is complete} \quad \Longleftrightarrow \quad \text{only } f = 0 \text{ is orthogonal to every } f_i$$

(for a general Banach space we just have to take f to lie in the dual space X^*). Consequently, if $\{f_i\}_{i\in\mathbf{N}}$ is complete, then every $f \in H$ is uniquely determined by the sequence of inner products $\{\langle f, f_i\rangle\}_{i\in\mathbf{N}}$, or in other words, the *analysis operator* $T(f) = \{\langle f, f_i\rangle\}_{i\in\mathbf{N}}$ is an injective mapping into the space of all sequences. However, this doesn't give us an algorithm for constructing f from those inner products, and in general there need not exist a stable way to do so, i.e., T^{-1} need not be continuous.

Property S seems to have no standard name in the literature (hence the uncreative name invented here), perhaps because it is not really a very useful concept by itself. In particular, the definition fails to provide us with a stable algorithm for finding a choice of coefficients $c_i(f)$ that can be used to represent f. The definition of quasibasis addresses this somewhat by requiring that each mapping $f \mapsto c_i(f)$ be continuous (for more details on quasibases, see [86] and the references therein). However, this is still not sufficient in most applications, as it is not so much the continuity of each individual map $f \mapsto c_i(f)$ that is important but rather the continuity of the mapping from f to the entire associated sequence $\{c_i(f)\}_{i\in\mathbf{N}}$. In other words, in concrete applications there is often some particular associated Banach space X_d of sequences (imposed by the context), and the mapping $f \mapsto \{c_i(f)\}_{i\in\mathbf{N}}$ must be a continuous linear map of X into X_d. Specializing to the Hilbert space case, this is one of the ideas behind the definition of *frames* (see Section 9.3.3).

Imposing uniqueness seems to be a natural requirement, and in fact, it can be shown that even though the definition of basis does not include the requirement that $f \mapsto c_i(f)$ be continuous, this follows automatically from the uniqueness assumption (and the fact that we are using *norm convergence*). Thus every basis is actually a quasibasis. Unfortunately, in many contexts uniqueness is simply too restrictive. For example, this is the case for Gabor systems (compare the Balian–Low Theorem, Theorem 9.16 below). The terms "basis" and "Schauder basis" are used interchangeably in the Banach space setting.

We can summarize the relations among the "spanning type" properties introduced so far by the following implications:

$$\text{basis} \;\; \begin{array}{c}\Longrightarrow \\ \not\Longleftarrow\end{array} \;\; \text{quasibasis} \;\; \Longrightarrow \;\; \text{Property S} \;\; \begin{array}{c}\Longrightarrow \\ \not\Longleftarrow\end{array} \;\; \text{complete.}$$

It seems unclear whether every system with Property S must actually be a quasibasis (compare [54]), but the other implications are known to not be reversible in general (even in a Hilbert space).

9.3.3 Frames

In the Hilbert space setting, *frames* are a class of sequences which not only are quasibases but also provide stable reconstruction formulas. Frames were first introduced by Duffin and Schaeffer [46], and that paper still provides instructive reading today. For proofs and more information on frames and the results presented in this section, we suggest [22], [28], [41], [55], [64], [71], [114]. For an interesting recent paper that deals with the issue of extending frames beyond the Hilbert space setting, see [58].

Definition 9.4. *A sequence $\mathcal{F} = \{f_i\}_{i\in\mathbf{N}}$ is a* frame *for a Hilbert space H if there exist constants A, $B > 0$, called frame bounds, such that*

$$\forall\, f \in H, \quad A\,\|f\|^2 \le \sum_{i=1}^{\infty} |\langle f, f_i\rangle|^2 \le B\,\|f\|^2. \tag{9.2}$$

The largest possible value for A and the smallest possible value for B are the optimal frame bounds. *If we can take $A = B$ in* (9.2)*, then we say the frame is* tight.

Every orthonormal basis is a tight frame, because an orthonormal basis satisfies the Parseval/Plancherel formula, which is exactly (9.2) for the case $A = B = 1$. However, not every frame is an orthonormal basis, even if $A = B = 1$. For example, if $\{e_i\}_{i\in\mathbf{N}}$ and $\{f_i\}_{i\in\mathbf{N}}$ are both orthonormal bases for H, then $\{\frac{1}{\sqrt{2}}e_i\}_{i\in\mathbf{N}} \cup \{\frac{1}{\sqrt{2}}f_i\}_{i\in\mathbf{N}}$ is a frame with $A = B = 1$ that is not an orthonormal basis. A frame for which we can take $A = B = 1$ is often called a *Parseval frame* or a *normalized tight frame* (but the latter term is confusing because some papers, such as [16], use it differently).

A frame need not be a basis (if not, it is said to be *redundant* or *overcomplete*). However, every frame is a quasibasis. A sketch of why this is true starts with the *analysis operator* $T\colon H \to \ell^2$ given by $T(f) = \{\langle f, f_i\rangle\}_{i\in\mathbf{N}}$ and the *frame operator* $S\colon H \to H$ given by $Sf = T^*Tf = \sum_i \langle f, f_i\rangle\, f_i$. The frame definition implies that T is a bounded injective mapping of H onto a closed subset of ℓ^2, and the inverse map.T^{-1} : range$(T) \to H$ is also bounded. Further, it can be shown that the series defining Sf converges for every f, and that S is actually a positive definite, invertible mapping of H onto itself. Writing out and rearranging the equalities $f = SS^{-1}f = S^{-1}Sf$ then gives the *frame expansions*

$$\forall\, f \in H, \quad f = \sum_{i=1}^{\infty} \langle f, \tilde{f}_i\rangle\, f_i = \sum_{i=1}^{\infty} \langle f, f_i\rangle\, \tilde{f}_i, \tag{9.3}$$

where $\tilde{f}_i = S^{-1}f_i$. Thus both the frame $\{f_i\}_{i\in\mathbf{N}}$ and its *canonical dual frame* $\{\tilde{f}_i\}_{i\in\mathbf{N}}$ are quasibases. We even have simple (and computable) formulas for the coefficients, namely, $c_i(f) = \langle f, \tilde{f}_i\rangle$ for the frame and $c_i(f) = \langle f, f_i\rangle$ for the dual frame. In general, however, these scalars need not be unique. Trivial examples of nonuniqueness are a frame which includes some zero vectors as elements, or the union of two orthonormal bases. For a nontrivial example, see Example 9.5.

Note that if we rearrange the elements of the frame then we still have a frame, because the series in (9.2) is a series of nonnegative numbers, and hence if it converges then any rearrangement also converges. It follows that the frame expansions in (9.3) converge *unconditionally*. This is one of many stability properties enjoyed by frames. As a consequence any countable index set can be used to specify the elements of a frame.

Although the scalars in the frame expansions in (9.3) need not be unique, out of all the possible sequences $\{c_i\}_{i\in\mathbf{N}}$ such that $f = \sum_i c_i f_i$, the frame coefficients have *minimal energy*, i.e., $\sum_i |\langle f, \tilde{f}_i\rangle|^2 \leq \sum_i |c_i|^2$, and equality holds only when $|c_i| = |\langle f, \tilde{f}_i\rangle|$ for all i. This does not imply that $\sum_i |c_i|^2$ is finite; compare [31], [63], [76]. In particular applications, we can make use of the fact that different choices of coefficients can be used to search for other noncanonical dual functions $\tilde{f}$ (possibly even from a larger space) that still provide frame expansions but may possess extra properties important for the application at hand. Some papers on noncanonical duals or on minimizing with respect to other criteria than energy are [27], [30], [92].

In finite dimensions, frames are easy to characterize:

- A collection $\{f_1, \dots, f_M\}$ is a frame for a finite-dimensional Hilbert space H if and only if $\{f_1, \dots, f_M\}$ spans H. Thus, in a finite-dimensional space, a collection is a basis if and only if it is a linearly independent frame.

In particular, every finite set of vectors $\{f_1, \dots, f_M\}$ in a Hilbert space is a frame for the subspace $S = \mathrm{span}(\{f_1, \dots, f_M\})$.

Finite-dimensional frames have many important applications, and there remain many deep and difficult mathematical questions concerning them, such as characterizing frames which have certain useful properties. We suggest [16], [24], [47], [109] as some interesting papers on "finite frames;" in particular, the last paper discusses links between finite frames and other areas such as discrete geometry and sphere packings.

The following illustrative example shows that the relationship between frames and linear independence is more complicated in infinite dimensions.

Example 9.5. Let $e_n(x) = e^{2\pi inx}$. The system of exponentials $\{e_n\}_{n\in\mathbf{Z}}$ is an orthonormal basis for the Hilbert space $L^2[0,1]$. What happens if we change the frequencies from integers n to integer multiples $n\beta$ of some $\beta > 0$? That is, setting $e_{n\beta}(x) = e^{2\pi in\beta x}$ but keeping the domain $[0,1]$, we ask what properties the new sequence $\{e_{n\beta}\}_{n\in\mathbf{Z}}$ has in $L^2[0,1]$.

Case $\beta > 1$. Note that each $e_{n\beta}$ is $\frac{1}{\beta}$-periodic and $\frac{1}{\beta} < 1$. Every finite linear combination will likewise be $\frac{1}{\beta}$-periodic, as will any element of the closed span. Hence the closed span cannot be all of $L^2[0,1]$, since there are many elements of $L^2[0,1]$ that are not $\frac{1}{\beta}$-periodic, such as $f(x) = x$. Also, the vector

$$f(x) = \begin{cases} 1, & 0 \le x < \varepsilon, \\ -1, & \frac{1}{\beta} \le x < \frac{1}{\beta} + \varepsilon, \\ 0, & \text{otherwise}, \end{cases}$$

is orthogonal to every $e_{n\beta}$. Thus $\{e_{n\beta}\}_{n\in\mathbf{Z}}$ is incomplete when $\beta > 1$.

Case $\beta < 1$. Suppose $\beta < 1$. Since $\{\sqrt{\beta}\, e^{2\pi i n\beta x}\}_{n\in\mathbf{Z}}$ is an orthonormal basis for the space $L^2[0,\frac{1}{\beta}]$, we have

$$\forall\, f \in L^2[0,\tfrac{1}{\beta}], \quad \sum_{n\in\mathbf{Z}} |\langle f, e_{n\beta}\rangle|^2 = \frac{1}{\beta}\, \|f\|_2^2. \tag{9.4}$$

Given $f \in L^2[0,1]$, extend it to $[0,\frac{1}{\beta}]$ by setting $f(x) = 0$ for $1 < x \le \frac{1}{\beta}$. Then we can apply (9.4), but because of the zero extension, the norm and inner product are from $L^2[0,1]$. In other words, (9.4) holds for $f \in L^2[0,1]$, so $\{e_{n\beta}\}_{n\in\mathbf{Z}}$ is a tight frame for $L^2[0,1]$ with frame bounds $A = B = \frac{1}{\beta}$. In fact, this set is the image of an orthonormal basis for $L^2[0,\frac{1}{\beta}]$ under the orthogonal projection $f \mapsto f\cdot\chi_{[0,1]}$. This can be generalized; from an operator theory viewpoint frames can be viewed in terms of projections and vice versa [61].

In any case, $\{e_{n\beta}\}_{n\in\mathbf{Z}}$ is a tight frame when $\beta < 1$, with frame bounds $A = B = \frac{1}{\beta}$ and frame operator $S = AI = \frac{1}{\beta}I$. Hence the dual frame elements are $\tilde{e}_{n\beta} = S^{-1}(e_{n\beta}) = \beta\, e_{n\beta}$, and frame expansions are trivial to compute. As a consequence, we can see directly that this frame is not a basis, because the coefficients in the frame expansion are not unique. For example, the constant function $e_0(x) = 1$ has two expansions:

$$\sum_{n\in\mathbf{Z}} \delta_n e_{n\beta} = e_0 = \sum_{n\in\mathbf{Z}} \langle e_0, \tilde{e}_{n\beta}\rangle\, e_{n\beta}, \tag{9.5}$$

where $\delta_n = 1$ when $n = 0$ and 0 otherwise. Since $\langle e_0, \tilde{e}_{n\beta}\rangle = \frac{1-e^{2\pi i n\beta}}{2\pi i n}$ for $n \ne 0$, the two expansions in (9.5) are in fact different.

By rearranging (9.5), we can write

$$e_0 = \sum_{n\ne 0} c_n e_{n\beta} \tag{9.6}$$

for appropriate scalars c_n, i.e., e_0 is expressible in terms of the other frame elements (this cannot happen in a basis). It can actually be shown that the proper subset $\{e_{n\beta}\}_{n\ne 0}$ of the original frame $\{e_{n\beta}\}_{n\in\mathbf{Z}}$ is still a frame for

$L^2[0,1]$, although it is no longer tight. Thus the original frame is redundant in the sense that a proper subset is still a frame. For this particular frame it can even be shown that there is an *infinite* set $J \subset \mathbf{Z}$ such that $\{e_{n\beta}\}_{n\in\mathbf{Z}\setminus J}$ is still a frame for $L^2[0,1]$, so in some sense the original frame is extremely redundant. Yet it is linearly independent using the standard abstract linear algebra definition of independence, namely, every finite subset of $\{e_{n\beta}\}_{n\in\mathbf{Z}}$ is linearly independent! For, if we write a finite linear combination as $\sum_{n=-N}^{N} c_n e^{2\pi i n\beta x} = \sum_{n=-N}^{N} c_n z^n$ where $z = e^{2\pi i\beta x}$, then we can apply the Fundamental Theorem of Algebra to conclude that such a polynomial cannot vanish for all x unless it is the trivial polynomial. Thus $\{e_{n\beta}\}_{n\in\mathbf{Z}}$ is very redundant taken as a whole even though every finite subset is independent.

Remark 9.6. a. The value $\frac{1}{\beta}$ is sometimes called the density of the system $\{e_{n\beta}\}_{n\in\mathbf{Z}}$. The value $\frac{1}{\beta} = 1$ is the *Nyquist density* for the exponentials; at the Nyquist density the system is an orthonormal basis, at lower densities ($\frac{1}{\beta} < 1$) it is incomplete and at higher densities ($\frac{1}{\beta} > 1$) it is an overcomplete frame. The Classical (or Shannon) Sampling Theorem for bandlimited signals is an immediate consequence of the frame properties of the exponentials; for more details on the Sampling Theorem we refer to Benedetto's text [13].

b. Many of the statements made about the system $\{e_{n\beta}\}_{n\in\mathbf{Z}}$ have analogues for "irregular" sequences of exponentials of the form $\{e_{\lambda_n}\}_{n\in\mathbf{Z}} = \{e^{2\pi i\lambda_n x}\}_{n\in\mathbf{Z}}$, where the λ_n are arbitrary points in $\mathbf{R}$. In these statements the *Beurling density* of the set $\{\lambda_n\}_{n\in\mathbf{Z}}$ replaces the value $\frac{1}{\beta}$. We suggest [114] and [60] as starting points for more details on this topic. In Section 9.4.1 below, we present some analogous results for irregular Gabor systems, and Beurling density is defined precisely there.

c. Frames of exponentials are very special types of frames, and only some of the statements made in Example 9.5 carry over to general frames. For an arbitrary frame, it can be shown that if $\{f_i\}_{i\in\mathbf{N}}$ is a redundant frame, then there exists at least a finite set F such that $\{f_i\}_{i\notin F}$ is still a frame. However, in general it need not be the case that infinitely many elements can be removed yet leave a frame [6], nor that there need be some subset of the frame that is a basis [29], [107]. The *Feichtinger Conjecture* is the statement that every frame $\mathcal{F} = \{f_i\}_{i\in I}$ satisfying $\inf \|f_i\| > 0$ can be written as a finite union of subsequences that are Riesz bases for their closed spans. This conjecture is open and has recently been shown to be equivalent to the deep and longstanding *Kadison–Singer Conjecture* in operator theory, which has been open since 1959 [85]; see [26], [23] for the proof of the equivalence.

9.3.4 Independence in Infinite Dimensions

We explore independence in more detail in this section. For proofs and more information, see [28], [64], [93], [98], [108], [114].

The following are several shades of gray in the possible definition of independence.

Definition 9.7. *Let $\{f_i\}_{i\in\mathbf{N}}$ be a countable sequence of elements in a Banach space X.*

(a) $\{f_i\}_{i\in\mathbf{N}}$ *is a* basis *or* Schauder basis *if for each $f \in X$ there exist unique scalars c_i such that $f = \sum_i c_i f_i$.*

(b) $\{f_i\}_{i\in\mathbf{N}}$ *is* minimal *if for each $j \in \mathbf{N}$, the vector f_j does not lie in $\overline{\mathrm{span}}(\{f_i\}_{i\neq j})$. Equivalently (via Hahn–Banach), there must exist a sequence $\{\tilde{f}_i\}_{i\in\mathbf{N}}$ in the dual space X^* that is* biorthogonal *to $\{f_i\}_{i\in\mathbf{N}}$, i.e., $\langle f_i, \tilde{f}_j\rangle = 1$ if $i = j$ and 0 if $i \neq j$.*

(c) $\{f_i\}_{i\in\mathbf{N}}$ *is* ω-independent *if the series $\sum_{i=1}^{\infty} c_i f_i$ can converge and equal the zero vector only when every $c_i = 0$.*

(d) $\{f_i\}_{i\in\mathbf{N}}$ *is* finitely independent *(or simply* independent*) if every finite subset is independent, i.e., for any N we have $\sum_{i=1}^{N} c_i f_i = 0$ if and only if $c_1 = \cdots = c_N = 0$.*

For example, consider the system of exponentials $\{e_{n\beta}\}_{n\in\mathbf{Z}}$ described in Example 9.5. We have already seen that the system is a basis only for $\beta = 1$. If $\beta > 1$ then it is not even complete, while if $\beta < 1$ then it is not a basis because we showed explicitly in (9.5) that the vector e_0 has two different series representations. Additionally, equation (9.6) implies that e_0 lies in the closure of $\mathrm{span}(\{e_{n\beta}\}_{n\neq 0})$, so the system is not minimal. Further, by subtracting e_0 from both sides of (9.6) we obtain a nontrivial infinite series that converges and equals the zero vector, so the system is not ω-independent. Even so, that system is finitely independent.

The following implications among these properties hold, none of which is reversible in general (even in a Hilbert space):

$$\text{basis} \quad \begin{array}{c}\Longrightarrow \\ \not\Longleftarrow\end{array} \quad \text{minimal} \quad \begin{array}{c}\Longrightarrow \\ \not\Longleftarrow\end{array} \quad \omega\text{-independent} \quad \begin{array}{c}\Longrightarrow \\ \not\Longleftarrow\end{array} \quad \begin{array}{c}\text{finitely} \\ \text{independent.}\end{array}$$

One technical point is that the definition of basis really combines aspects of both spanning and independence, i.e., a basis is necessarily complete and has Property S. Adding completeness doesn't change the implications above, e.g., every basis is both minimal and complete, but a minimal sequence that is complete need not be a basis (a sequence which is both minimal and complete is sometimes called an *exact sequence*). On the other hand, Property S is exactly what is missing for a minimal or ω-independent sequence to be a basis, for with either of those hypotheses, once we know that an infinite series $\sum_{i=1}^{\infty} c_i f_i$ converges, we can conclude that the coefficients are unique. However, as shown by the example of the exponentials, finite independence combined with Property S is *not* sufficient to ensure that we have a basis. Thus we have the following equivalences:

$$\text{basis} \iff \begin{array}{c}\text{minimal with}\\ \text{Property S}\end{array} \iff \begin{array}{c}\omega\text{-independent}\\ \text{with Property S,}\end{array}$$

and each of these implies finite independence, but not conversely: a finitely independent sequence which has Property S need not be a basis. Similarly, combining the various independence criteria with a frame hypothesis, we obtain the following result, which should be compared to Example 9.5, where we showed that a frame which is finitely independent need not be a basis.

Theorem 9.8. *Let $\{f_i\}_{i\in\mathbf{N}}$ be a countable sequence of elements in a Hilbert space H. Then the following statements are equivalent.*

(a) *$\{f_i\}_{i\in\mathbf{N}}$ is a frame and a basis for H.*

(b) *$\{f_i\}_{i\in\mathbf{N}}$ is a frame and a minimal sequence.*

(c) *$\{f_i\}_{i\in\mathbf{N}}$ is a frame and an ω-independent sequence.*

(d) *$\{f_i\}_{i\in\mathbf{N}}$ is a* Riesz basis *for H, i.e., it is the image of an orthonormal basis for H under a continuous linear bijection.*

(e) *$\{f_i\}_{i\in\mathbf{N}}$ is a* bounded unconditional basis *for H, i.e., for each $f \in H$ there are unique scalars c_i such that $f = \sum_i c_i f_i$, where the series converges unconditionally, and additionally $0 < \inf_i \|f_i\| \le \sup_i \|f_i\| < \infty$.*

The most common terms used to describe a frame which satisfies the equivalent conditions of Theorem 9.8 are *Riesz basis* and *exact frame*. Continuous linear bijections are known by a variety of names, including *topological isomorphisms*, *continuously invertible maps*, or even just *invertible maps*.

One set of extra hypotheses needed for a finitely independent set to be a Riesz basis is given in the following theorem quoted from [28, Prop. 6.1.2] and originally proved in [33], [87].

Theorem 9.9. *Let $\{f_i\}_{i\in\mathbf{N}}$ be a countable sequence of elements in a Hilbert space H, and let A_N be the optimal lower frame bound for $\{f_i\}_{i=1}^N$ as a frame for its span. Then the following statements are equivalent.*

(a) *$\{f_i\}_{i\in\mathbf{N}}$ is a Riesz basis for H.*

(b) *$\{f_i\}_{i\in\mathbf{N}}$ is finitely independent and* $\inf A_N > 0$.

(c) *$\{f_i\}_{i\in\mathbf{N}}$ is finitely independent and* $\lim_{N\to\infty} A_N$ *exists and is positive.*

If $\{e_n\}_{n\in\mathbf{N}}$ is an orthonormal basis for H, then the renormalized sequence $\{\frac{e_n}{n}\}_{n\in\mathbf{N}}$ is a trivial example of a Schauder basis that is not a Riesz basis. On the other hand, here is a nontrivial example where all the elements have the same norm.

Example 9.10. Fix $0 < \alpha < \frac{1}{2}$; then $|x|^\alpha$ and $|x|^{-\alpha}$ both belong to $L^2[-\frac{1}{2}, \frac{1}{2}]$. Hence $\{e^{2\pi inx}|x|^\alpha\}_{n\in\mathbf{Z}}$ and $\{e^{2\pi inx}|x|^{-\alpha}\}_{n\in\mathbf{Z}}$ are biorthogonal systems in $L^2[-\frac{1}{2}, \frac{1}{2}]$ and therefore are minimal. It is a much more difficult result, due to

Babenko [3], that these systems are actually Schauder bases for $L^2[-\frac{1}{2}, \frac{1}{2}]$ (see also the discussion in [108, pp. 351–354]). Since these systems are obtained by taking the orthonormal basis $\{e^{2\pi inx}\}_{n\in\mathbf{Z}}$ and performing an operation that is not a continuous bijection (i.e., multiplying by the function $|x|^\alpha$ which has a zero or by the unbounded function $|x|^{-\alpha}$), they are not Riesz bases. On the other hand, these systems do possess one but not both frame bounds. Specifically, $\{e^{2\pi inx}|x|^\alpha\}_{n\in\mathbf{Z}}$ is a Bessel sequence while $\{e^{2\pi inx}|x|^{-\alpha}\}_{n\in\mathbf{Z}}$ possesses a lower frame bound.

9.4 Gabor Frames

Now we will discuss the special class of frames known as *Gabor frames* (they are also often called *Weyl–Heisenberg frames*). This is one part of the broader field of *time-frequency analysis*. We will survey some results on Gabor frames, with the choice of topics inspired by issues relating to independence and the HRT Conjecture. As a result, many important topics and contributions by many founders of the field are not included. For a more complete introduction to time-frequency analysis we recommend Gröchenig's text [55], and for surveys and basic information on Gabor frames we suggest [22], [28], [41], [71], [81].

9.4.1 Density and Gabor Frames

In this section we give some background on Gabor systems and Gabor frames.

Definition 9.11. *Let $g \in L^2(\mathbf{R})$. The* translation *of g by $a \in \mathbf{R}$ is $T_a g(x) = g(x-a)$, and the* modulation *of g by $b \in \mathbf{R}$ is $M_b g(x) = e^{2\pi ibx} g(x)$. The compositions*

$$M_b T_a g(x) = e^{2\pi ibx} g(x-a)$$

and

$$T_a M_b(x) = e^{2\pi ib(x-a)} g(x-a) = e^{-2\pi iab} M_b T_a g(x)$$

are time-frequency shifts *of g. If $\Lambda \subset \mathbf{R}^2$ then the* Gabor system generated by g and Λ *is*

$$\mathcal{G}(g,\Lambda) = \{M_b T_a g\}_{(a,b)\in\Lambda}.$$

If a Gabor system is a frame for $L^2(\mathbf{R})$, then it is called a Gabor frame.

Note that Λ should technically be regarded as a sequence of points in $\mathbf{R}^2$ rather than a subset, because frames can allow duplicate elements, and so we should allow repetitions of points in Λ. However, for simplicity we usually just write $\Lambda \subset \mathbf{R}^2$ even though we mean that Λ is a sequence. Typically, we are interested in countable sets Λ, and the most common case is where Λ is the *rectangular lattice* $\Lambda = \alpha\mathbf{Z} \times \beta\mathbf{Z}$. The cases of general lattices $\Lambda = A(\mathbf{Z}^2)$ and nonlattice or "irregular" sets of time-frequency shifts Λ are also very interesting and important.

Example 9.12. (a) $\mathcal{G}(\chi_{[0,1)}, \mathbf{Z}^2)$ is a Gabor orthonormal basis for $L^2(\mathbf{R})$.

(b) Gabor studied the system $\mathcal{G}(e^{-x^2}, \mathbf{Z}^2)$ in the context of information theory [51]. It can be shown via Zak transform techniques that this system is not a frame, and is overcomplete by exactly one element [102], [77]. That is, one element may be removed and leave a complete set, but not two elements. Thus $\mathcal{G}(e^{-x^2}, \mathbf{Z}^2 \setminus (0,0))$ is minimal and complete. However, it is not a basis for $L^2(\mathbf{R})$ [49, p. 168]. The Zak transform is briefly discussed in Section 9.5.3.

(c) It is easy to create specific Gabor frames. For example, if $g \in L^2(\mathbf{R})$ is supported in $[0, \frac{1}{\beta}]$ and satisfies $\sum_{n\in\mathbf{Z}} |g(x-n\alpha)|^2 = 1$, then $\mathcal{G}(g, \alpha\mathbf{Z}\times\beta\mathbf{Z})$ is a tight frame. Note that this requires $\alpha\beta \le 1$; compare Theorem 9.13 below. If $\alpha\beta = 1$, then g will be discontinuous; compare Theorem 9.16 below. However, if $\alpha\beta < 1$, then we can create frames where g is as smooth as we like, even infinitely differentiable. These frames, and their wavelet analogues, are the "painless nonorthogonal expansions" of [42]. For recent higher-dimensional "irregular" frame constructions in a similar spirit, see [1].

Let us mention that Gabor systems have a natural connection to representation theory. The *time-frequency plane* $\mathbf{R}^2$ appearing in the definition of a Gabor system is really the *Heisenberg group* in disguise. One form of the Heisenberg group is $\mathbf{H} = \mathbf{T} \times \mathbf{R} \times \mathbf{R}$, where $\mathbf{T} = \{z \in \mathbf{C} : |z| = 1\}$, with a group operation that is induced by considering a point $(z, a, b) \in \mathbf{H}$ to correspond to the operator zM_bT_a defined by $(zM_bT_af)(x) = ze^{2\pi ibx}f(x-a)$. That is, $(z, a, b) \mapsto zM_bT_a$ is required to be an injective homomorphism of $\mathbf{H}$ into the set of unitary mappings of $L^2(\mathbf{R})$ onto itself. Since

$$(zM_bT_a) \circ (wM_dT_c) = e^{-2\pi iad} zwM_{b+d}T_{a+c},$$

the group operation on $\mathbf{H}$ is therefore

$$(z, a, b) \cdot (w, c, d) = (e^{-2\pi iad} zw, a + c, b + d).$$

This makes $\mathbf{H}$ a non-Abelian group, even though as a set it is simply the Cartesian product $\mathbf{T}\times\mathbf{R}\times\mathbf{R}$. To be a little more precise, the Heisenberg group is usually defined by considering compositions of "symmetric" time-frequency shifts $M_{\frac{b}{2}}T_aM_{\frac{b}{2}}$, and so the standard definitions differ in normalization from what is given here. For precise details we refer to [55, Ch. 9], [49]. The mapping $(z, a, b) \mapsto zM_bT_a$ is (except for normalization) the *Schrödinger representation* of $\mathbf{H}$. Although the unit modulus scalars z are needed to define the group operation in $\mathbf{H}$, they play no role in many parts of time-frequency analysis, and hence we often end up dealing with the time-frequency plane $\mathbf{R}^2$ rather than $\mathbf{H}$.

Gabor systems defined with respect to rectangular lattices are especially nice, and have connections to the theory of C^* and von Neumann algebras (for more on this connection see [44], [52], [61]). Such rectangular Gabor systems have properties that are very reminiscent of the system of exponentials discussed in Example 9.5. Specifically, the following result holds; note the similar role that the value $\alpha\beta$ plays as compared to the value of β in Example 9.5.

Theorem 9.13. *Let $g \in L^2(\mathbf{R})$ and let $\Lambda = \alpha\mathbf{Z}\times\beta\mathbf{Z}$ where $\alpha, \beta > 0$. Then the Gabor system $\mathcal{G}(g, \alpha\mathbf{Z}\times\beta\mathbf{Z}) = \{e^{2\pi i\beta nx}g(x-\alpha k)\}_{k,n\in\mathbf{Z}}$ satisfies the following.*

(a) *If $\mathcal{G}(g, \alpha\mathbf{Z}\times\beta\mathbf{Z})$ is a frame for $L^2(\mathbf{R})$, then $0 < \alpha\beta \leq 1$.*

(b) *If $\mathcal{G}(g, \alpha\mathbf{Z}\times\beta\mathbf{Z})$ is a frame for $L^2(\mathbf{R})$, then it is a Riesz basis if and only if $\alpha\beta = 1$.*

(c) *If $\alpha\beta > 1$, then $\mathcal{G}(g, \alpha\mathbf{Z}\times\beta\mathbf{Z})$ is incomplete in $L^2(\mathbf{R})$.*

Theorem 9.13 has a long history that we cannot do justice to here. We mention only the following facts, and for more detailed history and references refer to the expositions in [17], [41], [55], [81].

Part (c) of Theorem 9.13 was proved the case that $\alpha\beta$ is rational by Daubechies [40] and for arbitrary $\alpha\beta$ by Baggett [4]. Daubechies' proof relied on the Zak transform, while Baggett used the theory of von Neumann algebras. Daubechies also noted that a proof for general $\alpha\beta$ can be inferred from results of Rieffel [104] on C^* algebras. Another proof of part (c) based on von Neumann algebras is given in [44], and a new proof appears in [20].

Since every frame is complete, part (a) is of course a consequence of part (c), but we state it separately to emphasize the contrast with the case of irregular Gabor systems as stated in Theorem 9.14 below. A simple proof of part (a) was given by Janssen [79]. This proof relies on the algebraic structure of the rectangular lattice $\alpha\mathbf{Z}\times\beta\mathbf{Z}$ and the remarkable *Wexler–Raz Theorem* for Gabor frames $\mathcal{G}(g, \alpha\mathbf{Z}\times\beta\mathbf{Z})$. For more on Wexler–Raz see [79], the expositions in [55, Sec. 7.5], [81], [62], and the rigorous proofs in [80], [44].

In part (b), given a Gabor frame $\mathcal{G}(g, \alpha\mathbb{Z}\times\beta\mathbb{Z})$, it is easy to prove using the Zak transform that if $\alpha\beta = 1$, then this frame must must be a Riesz basis. However, the converse is not as easy, and was first proved by Ramanathan and Steger in [103] as a special case of a much more general result (Theorem 9.14) discussed below. Today there are "straightforward" proofs of part (b), again based on Wexler–Raz.

In Theorem 9.13, the value $\alpha\beta$ that distinguishes between the various cases is a measurement of the "size" of the lattice $\alpha\mathbf{Z}\times\beta\mathbf{Z}$, as it is the area of a fundamental domain for that lattice. In the irregular setting there is no analogue of fundamental domain, and instead it is the *Beurling density* of Λ that distinguishes between the various cases. Beurling density measures in some sense the average number of points inside unit squares. Because the points are not uniformly distributed, there is not a single definition, but rather we obtain lower and upper limits to the average density. More precisely, we count the average number of points inside squares of larger and larger radii and take the limit, yielding the definitions

$$D^-(\Lambda) = \liminf_{r\to\infty} \inf_{z\in\mathbf{R}^2} \frac{\#(\Lambda \cap Q_r(z))}{r^2},$$

$$D^+(\Lambda) = \limsup_{r\to\infty} \sup_{z\in\mathbf{R}^2} \frac{\#(\Lambda \cap Q_r(z))}{r^2},$$

for the *lower and upper Beurling densities of* Λ. Here $Q_r(z)$ is the square in $\mathbf{R}^2$ centered at z with side lengths r and $\#E$ denotes cardinality. Using this notation, we can give necessary conditions for a Gabor system to be a frame or Riesz basis for $L^2(\mathbf{R})$, as follows.

Theorem 9.14. *Let $g \in L^2(\mathbf{R})$ and let $\Lambda \subset \mathbf{R}^2$ be given. Then the Gabor system $\mathcal{G}(g, \Lambda)$ has the following properties.*

(a) *If $\mathcal{G}(g, \Lambda)$ is a frame for $L^2(\mathbf{R})$, then $1 \le D^-(\Lambda) \le D^+(\Lambda) < \infty$.*

(b) *If $\mathcal{G}(g, \Lambda)$ is a Riesz basis for $L^2(\mathbf{R})$, then $D^-(\Lambda) = D^+(\Lambda) = 1$.*

(c) *If $D^-(\Lambda) < 1$, then $\mathcal{G}(g, \Lambda)$ is not a frame for $L^2(\mathbf{R})$.*

The result above was first proved (under some extra hypotheses) by Landau [91] and Ramanathan and Steger [103]. Inspired by the paper [60], those extra hypotheses were removed in [32] (and the result was also extended to higher dimensions and multiple generators). The papers [8], [9] show among other results that Theorem 9.14 is not just a result about Gabor frames but can be extended to the much more general situation of *localized frames.* Moreover, new consequences follow even for Gabor frames, such as the fact that the index set of any tight Gabor frame must possess a certain amount of uniformity, in the sense that the upper and lower Beurling densities will coincide, i.e., $D^-(\Lambda) = D^+(\Lambda)$. Related results appear in [6], [7]. Localized frames were independently introduced by Gröchenig in [58]; among other results Gröchenig shows that localized frames are frames not merely for the underlying Hilbert space H but also for an entire family of associated Banach spaces. Additional papers on localized frames are [57], [36].

Since the Beurling density of a rectangular lattice is $D^-(\alpha\mathbf{Z} \times \beta\mathbf{Z}) = D^+(\alpha\mathbf{Z} \times \beta\mathbf{Z}) = \frac{1}{\alpha\beta}$, Theorem 9.14 almost, but not quite, recovers Theorem 9.13. One trivial difference is in part (b) of the two theorems: the implications proceed in both directions in Theorem 9.13(b) but only in one direction in Theorem 9.14(b). For a counterexample to the converse direction in Theorem 9.14(b), take a Gabor frame $\mathcal{G}(g, \Lambda)$ that happens to be a Riesz basis and add a single point, say λ, to Λ. Then $\mathcal{G}(g, \Lambda \cup \{\lambda\})$ is a redundant frame, but the Beurling density is the same, $D^-(\Lambda \cup \{\lambda\}) = D^+(\Lambda \cup \{\lambda\}) = 1$.

On the other hand, the difference between Theorem 9.13(c) and Theorem 9.14(c) is much more significant. Ramanathan and Steger conjectured in [103] that Theorem 9.14(c) should be improvable to say that if $D^-(\Lambda) < 1$ then $\mathcal{G}(g, \Lambda)$ is incomplete in $L^2(\mathbf{R})$, but this was shown in [17] to be false: for any $\varepsilon > 0$ there exists a function $g \in L^2(\mathbf{R})$ and a set $\Lambda \subset \mathbf{R}^2$ with

$D^+(\Lambda) < \varepsilon$ such that $\mathcal{G}(g, \Lambda)$ is complete. The counterexample built in a fundamental way on the work of Landau on the completeness of exponentials in $L^2(S)$ where S is a finite union of intervals. Another counterexample, in which Λ is a subset of a lattice, appears in [113]. In [99], [100], it is shown that there exist $g \in L^2(\mathbf{R})$ and Λ of the form $\Lambda = \{(\lambda_n, 0)\}_{n \in \mathbf{Z}}$ such that $\mathcal{G}(g, \Lambda) = \{T_{\lambda_n} g\}_{n \in \mathbf{Z}}$ is complete in $L^2(\mathbf{R})$ and $\{\lambda_n\}_{n \in \mathbf{Z}}$ is a perturbation of the integers $\mathbf{Z}$. Thus $D^+(\Lambda) = 0$ for this example.

We close this section with some remarks.

Remark 9.15. a. Theorems 9.13 and 9.14 provide necessary conditions for a Gabor system to be a frame. Some sufficient conditions are known; see for example [41, Sec. 3.4], [55, Sec. 6.5], [71, Sec. 4.1]. For the Gaussian function $g(x) = e^{-x^2}$, there is actually a complete characterization of when $\mathcal{G}(e^{-x^2}, \alpha\mathbf{Z} \times \beta\mathbf{Z})$ is a frame; specifically, this occurs if and only if $\alpha\beta < 1$; see the simple proof and additional references in [79].

The Gaussian is one of only three functions for which such a characterization is currently known (not counting trivial modifications such as translations, modulations, dilations). The other two are the hyperbolic secant $g(x) = \frac{1}{\cosh \pi x}$ and the two-sided exponential $g(x) = e^{-|x|}$ [83], [84]. The precise set of (α, β) for which $\mathcal{G}(\chi_{[0,1)}, \alpha\mathbf{Z} \times \beta\mathbf{Z})$ is a frame is not known, but surprisingly, it appears to be an extremely complicated set, called "Janssen's tie" [82]. The problem of characterizing those sets $E \subset \mathbf{R}$ such that $\mathcal{G}(\chi_E, \mathbf{Z}^2)$ is a frame has been shown to be equivalent to a longstanding open problem of Littlewood [25].

b. In concrete situations, a given Hilbert space often does not appear in isolation, but is associated with an extended family of Banach spaces. A classical example is $L^2(\mathbf{R})$ sitting inside the class of Lebesgue spaces $L^p(\mathbf{R})$, which are themselves contained in the extended family of Besov spaces $B_s^{p,q}(\mathbf{R})$ and Triebel–Lizorkin spaces $F_s^{p,q}(\mathbf{R})$. *Wavelet frames* (briefly discussed in Section 9.4.3) typically provide expansions of functions not only in the Hilbert space $L^2(\mathbf{R})$ but in all of the Besov and Triebel–Lizorkin spaces simultaneously. The norms of these spaces are related to smoothness properties of the function, and this smoothness information can likewise be identified by examining the coefficients in the wavelet expansion. Analogously, Gabor frames can provide expansions not only for $L^2(\mathbf{R})$ but for an appropriate extended family of Banach spaces $M_s^{p,q}(\mathbf{R})$ known as the *modulation spaces*. The modulation space norms quantify time-frequency concentration rather than smoothness. These spaces were introduced and extensively studied by Feichtinger and Gröchenig. An excellent textbook development of the modulation spaces appears in [55, Chs. 11–14]. We also mention the mostly expository paper [66], which surveys one application of the modulation spaces to the analysis of spectral properties of integral operators.

9.4.2 Gabor Riesz Bases and the Balian–Low Theorem

In this section we consider Gabor frames that are Riesz bases. At least for the case where Λ is a rectangular lattice, all Gabor Riesz bases are "bad." Paraphrased in a qualitative form, the *Balian–Low Theorem* or BLT[4] states that if $\mathcal{G}(g, \alpha\mathbf{Z} \times \beta\mathbf{Z})$ is a Gabor Riesz basis (which by Theorem 9.13 can only happen when $\alpha\beta = 1$), then g is either not smooth or decays poorly at infinity. Here are two precise variations on this theme. In these, $\hat{g}$ denotes the Fourier transform of g (we use Benedetto's preferred normalization $\hat{g}(\omega) = \int g(x)\, e^{-2\pi i \omega x}\, dx$).

Theorem 9.16 (Balian–Low Theorems).

(a) ***Classical BLT:*** *If $g \in L^2(\mathbf{R})$ is such that $\mathcal{G}(g, \alpha\mathbf{Z} \times \beta\mathbf{Z})$ is a Gabor Riesz basis for $L^2(\mathbf{R})$, then*

$$\left(\int_{-\infty}^{\infty} |tg(t)|^2\, dt\right) \left(\int_{-\infty}^{\infty} |\omega\hat{g}(\omega)|^2\, d\omega\right) = \infty. \tag{9.7}$$

(b) ***Amalgam BLT:*** *If $g \in L^2(\mathbf{R})$ is such that $\mathcal{G}(g, \alpha\mathbf{Z} \times \beta\mathbf{Z})$ is a Gabor Riesz basis for $L^2(\mathbf{R})$, then g, $\hat{g} \notin W(C_0, \ell^1)$, where*

$$W(C_0, \ell^1) = \left\{ \textit{continuous } f : \sum_{k=-\infty}^{\infty} \|f \cdot \chi_{[k,k+1]}\|_\infty < \infty \right\}. \tag{9.8}$$

Note that the quantity appearing on the left-hand side of (9.7) is the Heisenberg product that appears in the Classical Uncertainty Principle. In particular, the generator of any Gabor Riesz basis must maximize uncertainty. See [50] for a general survey on the uncertainty principle, [12] for some other connections between Gabor systems and uncertainty principles, and [56] for new uncertainty principles in time-frequency analysis.

The Classical BLT was introduced independently by Balian [10] and Low [97]. A gap in their proofs was later filled in Daubechies' influential article [40], which also contains many important results on frames, Gabor systems, and wavelets (much of which was incorporated into [41]). An exquisite proof of the Classical BLT for the case of orthonormal bases was given by Battle [11], based on the operator theory associated with the Classical Uncertainty Principle; this proof was extended to Riesz bases in [43]. Variations on these proofs that avoid the use of distributional differentiation are given in [17]. A survey of recent results on the BLT appears in this volume [38].

---

[4] Yes, that is a joke: in the United States a "BLT" is a Bacon, Lettuce, and Tomato sandwich. As far as I know, this acronym was first used in print in [17], due entirely to John Benedetto's wonderful sense of humor.

The Amalgam BLT was first published in [17], although it was proved earlier in [63]. It is shown in [17] that neither version of the BLT implies the other. The space $W(C_0, \ell^1)$ appearing in (9.8) is an example of a *Wiener amalgam space.* While specific amalgam spaces have been used often throughout the mathematical literature, Feichtinger introduced and extensively studied general amalgam spaces, whose norm combines a local criterion for membership with a global criterion. The article [65] is an expository introduction to these spaces, including many references as well as a simple proof of the Amalgam BLT.

The BLT emphasizes that for applications where Gabor systems are useful, it is *redundant* Gabor frames that will usually be most appropriate. On the other hand, there is a remarkable construction known as *Wilson bases* that are in the spirit of time-frequency constructions, are generated by "nice" functions, and are unconditional bases not only for $L^2(\mathbf{R})$ but also for the class of modulation spaces. Unfortunately, there is a cost in the form of increased technicality; Wilson bases do not have the simple form that Gabor systems have. For more on Wilson bases, we suggest [55, Sec. 8].

Let us close this section by pointing out some related open problems and questions.

- Does the BLT hold for lattices that are nonrectangular? This question was recently answered affirmatively for the case of the Classical BLT in one dimension in [59], but as soon as we move to higher dimensions, only partial results are known. In particular, it is shown in [59] that the Classical BLT generalizes to the case of *symplectic* lattices in higher dimensions, but for nonsymplectic lattices little is known (see also Section 9.5.1 below). Some weaker partial results are also known to hold for the analogue of the Classical BLT for irregular Gabor Riesz bases. For more on the BLT on symplectic lattices, see [59] and [15], and for other recent results on the BLT see [14], [53].

- The situation for the Amalgam BLT is even less clear: it is not known even in one dimension if the Amalgam BLT still holds if rectangular lattices are replaced by general lattices $A(\mathbf{Z}^2)$ or by irregular sets of time-frequency shifts.

- Little is known about Gabor systems that are Schauder bases but not Riesz bases for $L^2(\mathbf{R})$. One such example is $\mathcal{G}(g, \mathbf{Z}^2)$, where $g(x) = |x|^\alpha \chi_{[-\frac{1}{2},\frac{1}{2}]}(x)$ and $0 < \alpha < \frac{1}{2}$ (compare Example 9.10). It is not known if the BLT theorems hold if Gabor Riesz bases are replaced by Gabor Schauder bases, although it is interesting to note that the proof of the *Weak BLT* given in [17, Thm. 7.4] or [59, Thm. 8] generalizes from Riesz bases to Schauder bases.

- It was conjectured in [45] that Gabor Schauder bases follow the same Nyquist-type rules as Gabor Riesz bases, i.e., if $\mathcal{G}(g, \Lambda)$ is a Gabor Schauder basis then $D^-(\Lambda) = D^+(\Lambda) = 1$. Some partial results were obtained in

[45], but the conjecture remains open. More generally, is there a Nyquist density result for Gabor systems that are minimal and complete but not bases, such as $\mathcal{G}(e^{-x^2}, \mathbf{Z}^2 \setminus (0,0))$ from Example 9.12?

- In [115], [116], Zalik gave some necessary and some sufficient conditions on $g \in L^2(\mathbf{R})$ and countable subsets $\Gamma \subset \mathbf{R}$ such that $\{T_a g\}_{a\in\Gamma}$ is complete in $L^2(\mathbf{R})$; see also the recent constructions in [99], [100]. Olson and Zalik proved in [101] that no such system of pure translations can be a Riesz basis for $L^2(\mathbf{R})$, and conjectured that no such system can be a Schauder basis. This conjecture is still open. Since $\{T_a g\}_{a\in\Gamma} = \mathcal{G}(g, \Gamma \times \{0\})$ and $D^-(\Gamma \times \{0\}) = 0$, it follows from Theorem 9.14 that such a system can never be a frame for $L^2(\mathbf{R})$. Similarly, if the density theorem conjectured for Gabor Schauder bases in the preceding question could be proved, then the Olson–Zalik Conjecture would follow as a corollary.

9.4.3 The Zero Divisor Conjecture and a Contrast to Wavelets

One motivation for the HRT Conjecture is simply the question of how similar Gabor frames are to the system of exponentials presented in Example 9.5: Is every Gabor frame finitely linearly independent? Since this is a question about finite subsets, it leads directly to the statement of Conjecture 9.1.

As one motivation for why we might be interested in such a question, let us contrast the situation for a different class of objects, *wavelet systems*. For more background on wavelets, we suggest [41], [74], or, for more elementary introductions, [111], [19]. Many of the influential early wavelet papers and their precursors are reprinted in [72].

Just as Gabor systems are associated with the Schrödinger representation of the Heisenberg group, wavelets are associated with a representation of another group, the affine or $ax+b$ group. Instead of considering time-frequency shifts, consider *time-scale shifts*, which are compositions of the translation $T_a g(x) = g(x-a)$ with the *dilation*

$$D_r g(x) = r^{1/2} g(rx),$$

which has been normalized so that D_r is a unitary operator on $L^2(\mathbf{R})$. Specifically, given $g \in L^2(\mathbf{R})$ and a sequence $\Lambda \subset \mathbf{R} \times \mathbf{R}^+$, the *wavelet system generated by g and Λ* is the collection of time-scale shifts

$$\mathcal{W}(g, \Lambda) = \{T_a D_b g\}_{(a,b)\in\Lambda}.$$

The group underlying wavelet systems is the affine group, which is the set $\mathbf{A} = \mathbf{R}\times\mathbf{R}^+$ endowed with a group multiplication that makes the mapping $(a,b) \mapsto T_a D_b$ an injective homomorphism of $\mathbf{A}$ into the set of unitary mappings of $L^2(\mathbf{R})$ onto itself, and this mapping is the corresponding representation.

For wavelet systems, the analogue of Conjecture 9.1 fails. For example, a function φ is said to be *refinable*, or is called a *scaling function*, if there exist coefficients c_k such that

$$\varphi(x) = \sum_{k=-\infty}^{\infty} c_k \, \varphi(2x - k) \tag{9.9}$$

(often the term scaling function is reserved for functions that satisfy additional requirements beyond just (9.9), for example, they may be required to be associated with a multiresolution analysis, see Remark 9.17 below). If only finitely many coefficients c_k in (9.9) are nonzero, then we have an expression of finite linear dependence among the time-scale shifts of φ. That is, $\mathcal{W}(\varphi, \Lambda)$ is dependent with a finite Λ. It is not hard to construct such functions, for example, the box function $b = \chi_{[0,1)}$ satisfies the refinement equation

$$b(x) = b(2x) + b(2x - 1).$$

Thus the analogue of Conjecture 9.1 fails when the Heisenberg group is replaced by the affine group. This raises the fundamental question: what is the basic difference between the affine group and the Heisenberg group which makes their behavior with regard to this conjecture so different? While both the affine group and the Heisenberg group are non-Abelian, the Heisenberg group is "nearly Abelian" in contrast to the affine group. For example, both are locally compact topological groups, and the Heisenberg group is unimodular (left and right Haar measure coincide), as are all Abelian locally compact groups, whereas the affine group is nonunimodular (see [71]). Another difference is that the Heisenberg group has discrete subgroups (e.g., $\{1\} \times \mathbf{Z} \times \mathbf{Z}$) while the affine group does not. Which, if any, of these are the essential difference in regard to the HRT Conjecture?

More generally, we could replace the Heisenberg or affine groups by other groups, and consider representations of arbitrary topological groups on $L^2(\mathbf{R})$. This leads to a connection with the open *Zero Divisor Conjecture* in abstract algebra, introduced in [75]. For a discussion of the Zero Divisor Conjecture and the connection between zero divisors and independence of translates we refer to the papers by Linnell, including [94], [95], [96].

We conclude this section by pointing out a few other related questions and connections.

Remark 9.17. a. Refinable functions play a central role in several areas, including subdivision schemes in computer-aided geometric design and the construction of orthonormal wavelet bases [41]. The now-classical method for constructing a wavelet orthonormal basis for $L^2(\mathbf{R})$ begins with a scaling function φ which has the additional property that the collection of integer translates $\{\varphi(x-k)\}_{k\in\mathbf{Z}}$ is an orthonormal sequence in $L^2(\mathbf{R})$. Such a φ leads to a *multiresolution analysis* (MRA) for $L^2(\mathbf{R})$, from which is deduced the existence of a *wavelet* ψ which has the property that $\{2^{n/2}\psi(2^n x - k)\}_{n,k\in\mathbf{Z}}$ is an orthonormal basis for $L^2(\mathbf{R})$. Thus the scaling function φ, whose time-scale translates are finitely dependent, leads to another function ψ which generates an orthonormal basis for $L^2(\mathbf{R})$ via time-scale translates.

b. There are also non-MRA constructions of orthonormal wavelet bases; these are especially surprising in higher dimensions [5], [18], [39].

c. The refinement equation (9.9) implies that the graph of a refinable function φ possesses some self-similarity. This leads to connections to fractals and iterated function systems. The analysis of the properties of refinable functions is an interesting topic with a vast literature; we refer to the paper [21] for references.

d. While the analogue of the HRT Conjecture fails in general, Christensen and Lindner [34], [35] have interesting partial results on when independence holds, including estimates of the frame bounds of finite sets of time-frequency or time-scale shifts.

e. There are useful recent characterizations of frames that apply in both the wavelet and Gabor settings by Hernández, Labate, and Weiss [90], [73].

f. One fundamental difference between wavelet orthonormal bases or Riesz bases and their Gabor analogues is that the analogue of the Balian–Low Theorems fail for wavelets. For example, it is possible to find Schwartz-class functions ψ, or compactly supported $\psi \in C^{(n)}(\mathbf{R})$ with n arbitrarily large, which generate wavelet orthonormal bases for $L^2(\mathbf{R})$. Another fundamental difference occurs in regard to density phenomenon. While Theorem 9.14 shows that Gabor frames have a Nyquist density similar to the one obeyed by the system of exponentials discussed in Example 9.5, there is no exact analogue of the Nyquist density for wavelet frames. Analogues of the Beurling density appropriate for the affine group were introduced in the papers [67], [110] (see also [89] for a comparison of these definitions and [37] for density conditions on combined Gabor/wavelet systems). While there is no Nyquist phenomenon in the sense that wavelet frames can be constructed with any particular density, it was shown that wavelet frames cannot have zero or infinite density, and the details of the arguments suggest a surprising amount of similarity between the Heisenberg and affine cases, cf. [68].

9.5 Partial Results on the HRT Conjecture

Now we return to the HRT Conjecture itself, and present some of the partial results that are known and some of the techniques used to obtain them. Using the notation introduced so far, we can reword the conjecture as follows:

- If $g \in L^2(\mathbf{R})$ is nonzero and $\Lambda = \{(\alpha_k, \beta_k)\}_{k=1}^N$ is a finite set of distinct points in $\mathbf{R}^2$, then $\mathcal{G}(g, \Lambda)$ is linearly independent.

Before presenting the partial results themselves, let us make some general remarks on why the conjecture seems to be difficult. A profoundly deep explanation would be an interesting research problem in itself; we will only point out some particular difficulties.

One major problem is that the conjecture is resistant to transform techniques. For example, applying the Fourier transform simply interchanges the translations and modulations, converting the problem into one of exactly the same type. Another natural transform for time-frequency analysis is the *short-time Fourier transform* (STFT). Given a "window function" $\varphi \in L^2(\mathbf{R})$, the STFT of $g \in L^2(\mathbf{R})$ with respect to φ is

$$V_\varphi g(x,\omega) = \langle g, M_\omega T_x \varphi\rangle, \qquad (x,\omega) \in \mathbf{R}^2.$$

With φ fixed, the mapping $g \mapsto V_\varphi g$ is an isometry of $L^2(\mathbf{R})$ into $L^2(\mathbf{R}^2)$. However,

$$V_\varphi(M_b T_a g)(x,\omega) = e^{-2\pi i a b\omega}\, V_\varphi g(x-a, \omega-b).$$

Thus, the STFT converts translations and modulations of g into two-dimensional translations and modulations of $V_\varphi g$, again yielding a problem of the same type, except now in two dimensions. There are many closely related transforms, such as the Wigner distribution, and quadratic versions of these transforms such as $g \mapsto V_g g$, but all of these have related difficulties. On the other hand, as we will see below, for special cases transforms can yield useful simplifications.

9.5.1 Linear Transformations of the Time-Frequency Plane

There is a class of transformations that we can apply that will sometimes simplify the geometry of the set of points $\Lambda = \{(\alpha_k, \beta_k)\}_{k=1}^N$ appearing in Conjecture 9.1. For example, let A be the linear transformation $A(a,b) = (\frac{a}{r}, br)$ where $r > 0$ (note that $\det(A) = 1$). Then

$$D_r(M_{\beta_k} T_{\alpha_k} g)(x) = r^{1/2} e^{2\pi i \beta_k r x} g(rx - \alpha_k) = M_{\beta_k r} T_{\frac{\alpha_k}{r}} (D_r g)(x).$$

Since the dilation D_r is a unitary map, it preserves independence, and hence

$$\mathcal{G}(g,\Lambda) \text{ is independent} \quad \Longleftrightarrow \quad \mathcal{G}(D_r g, A(\Lambda)) \text{ is independent.}$$

Thus, we can change the configuration of the points from the set Λ to the set $A(\Lambda)$ at the cost of replacing the function g by a dilation of g.

In fact, if A is *any* linear transformation of $\mathbf{R}^2$ onto itself with $\det(A) = 1$, then there exists a unitary transformation $U_A \colon L^2(\mathbf{R}) \to L^2(\mathbf{R})$ such that

$$U_A(\mathcal{G}(g,\Lambda)) = \{U_A(M_b T_a g)\}_{(a,b)\in\Lambda} = \{c_A(a,b)\, M_v T_u (U_A g)\}_{(u,v)\in A(\Lambda)}, \tag{9.10}$$

where $c_A(a,b)$ are constants of unit modulus determined by A. The rightmost set in (9.10) is not a Gabor system, but is obtained from the Gabor system $\mathcal{G}(U_A g, A(\Lambda))$ by multiplying each element by a constant of unit modulus. Such multiplications do not change the two properties we are interested in here, namely, being a frame or being linearly independent. In particular,

$$\mathcal{G}(g, \Lambda) \text{ is independent} \quad \Longleftrightarrow \quad \mathcal{G}(U_A g, A(\Lambda)) \text{ is independent.}$$

As a specific example, consider the shear $A(a,b) = (a, b+ra)$. Then U_A is the modulation by a "chirp" given by $U_A f(x) = e^{\pi i r x^2} f(x)$. To see this, set $c_A(a,b) = e^{-\pi i r a^2}$ and verify that

$$c_A(a,b) M_{b+ra} T_a (U_A g)(x) = U_A (M_b T_a g)(x).$$

Another example is $A(a,b) = (b, -a)$, rotation by $\pi/2$, for which $U_A = \mathcal{F}$, the Fourier transform.

In addition to linear transformations of $\mathbf{R}^2$, we can also use rigid translations, for they correspond to replacing g by $M_d T_c g$ and Λ by $\Lambda - (c,d)$.

The operators U_A are called *metaplectic transforms*. Every linear transformation A of $\mathbf{R}^2$ with determinant 1, i.e., every element of the special linear group $SL(2,\mathbf{R})$, yields a metaplectic transform. Unfortunately, as soon as we move to higher dimensions, this is no longer true. Only matrices lying in the *symplectic group* $S_p(d)$, which is a subset of $SL(2d,\mathbf{R})$, yield metaplectic transforms (the symplectic group consists of those invertible matrices A which preserve the symplectic form $[z,z'] = x' \cdot \omega - x \cdot \omega'$, where $z = (x,\omega)$, $z' = (x',\omega') \in \mathbf{R}^{2d}$). This stems from properties of the Heisenberg group, and for more details we refer to [55, Ch. 9]. In any case, this means that some of the simplifications we apply below to prove some special cases of the HRT Conjecture may not apply in higher dimensions.

9.5.2 Special Case: Points on a Line

In this section we will prove a special case of the HRT Conjecture, assuming that $\Lambda = \{(\alpha_k, \beta_k)\}_{k=1}^N$ is a set of collinear points. By applying an appropriate metaplectic transform, we may assume that $\Lambda = \{(\alpha_k, 0)\}_{k=1}^N$. Then $\mathcal{G}(g,\Lambda) = \{g(x-\alpha_k)\}_{k=1}^N$ is a finite set of translations of g. Suppose that we have $\sum_{k=1}^N c_k\, g(x-\alpha_k) = 0$ a.e. for some scalars $c_1, \dots, c_N$. Taking the Fourier transform of both sides of this equation converts translations to modulations, resulting in the equation

$$\sum_{k=1}^N c_k\, e^{-2\pi i \alpha_k \xi}\, \hat{g}(\xi) = 0 \text{ a.e.}$$

Since $\hat{g} \neq 0$, it is nonzero on some set of positive measure. But then the *nonharmonic trigonometric polynomial* $m(\xi) = \sum_{k=1}^N c_k\, e^{-2\pi i \alpha_k \xi}$ must be zero on a set of positive measure. If the α_k are integer, it follows immediately from the Fundamental Theorem of Algebra that $c_1 = \cdots = c_N = 0$. This is still true for arbitrary α_k, since m can be extended from real values of ξ to complex values, and the extension is an analytic function. Thus, we conclude that $\mathcal{G}(g,\Lambda)$ is linearly independent when the points in Λ are collinear.

While the HRT Conjecture turns out to be trivial when we restrict to the case of pure translations of g, there are still many interesting remarks to be made about this case. Consider first the fact that if we replace $L^2(\mathbf{R})$ by $L^\infty(\mathbf{R})$ then finite sets of translations can be linearly dependent (for example, consider any periodic function). In one dimension, it can be shown that any finite set of translations of a nonzero function $g \in L^p(\mathbf{R})$ with $p < \infty$ are independent, but note that this is already a more difficult problem because the Fourier transform exists only distributionally when $p > 2$. Moreover, Rosenblatt and Edgar have shown that there is a surprise as soon as we move to higher dimensions: sets of translates can be dependent for finite p. The following result was proved in [48], [105].

Theorem 9.18.

(a) *If $g \in L^p(\mathbf{R}^d)$ is nonzero and $1 \le p \le \frac{2d}{d-1}$, $p \ne \infty$, then $\{g(x-\alpha_k)\}_{k=1}^N$ is linearly independent for any finite set of distinct points $\{\alpha_k\}_{k=1}^N$ in $\mathbf{R}^d$.*

(b) *If $\frac{2d}{d-1} < p \le \infty$, then there exists a nonzero $g \in L^p(\mathbf{R}^d)$ and distinct points $\{\alpha_k\}_{k=1}^N$ in $\mathbf{R}^d$ such that $\{g(x-\alpha_k)\}_{k=1}^N$ is linearly dependent.*

We close this section by noting that subspaces of $L^2(\mathbf{R})$ of the form $V = \overline{\operatorname{span}}(\{T_{\alpha_k} g\}_{k\in\mathbf{N}})$ generated from translations of a given function are important in a wide variety of applications. In particular, a subspace of the form $V = \overline{\operatorname{span}}(\{T_k g\}_{k\in\mathbf{N}})$ is invariant under integer translations, and hence is called a *shift-invariant space* (but it need not be *translation-invariant*, which means invariant under all translations). Shift-invariant spaces play key roles in sampling theory, the construction of wavelet bases and frames, and other areas. For a recent research-survey of shift-invariant spaces in sampling theory, we suggest [2].

9.5.3 Special Case: Lattices

Suppose that $\Lambda = \{(\alpha_k, \beta_k)\}_{k=1}^N$ is a finite subset of some lattice $A(\mathbf{Z}^2)$, where A is an invertible matrix. By applying a metaplectic transform, we may assume that Λ is a subset of a rectangular lattice $\alpha\mathbf{Z} \times \beta\mathbf{Z}$ with $\alpha\beta = |\det(A)|$. We say that it is a *unit lattice* if $\alpha\beta = 1$; in this case, by applying a dilation we can assume $\alpha = \beta = 1$. The HRT Conjecture is easily settled for this special case by applying the *Zak transform*, which is the unitary mapping of $L^2(\mathbf{R})$ onto $L^2([0,1)^2)$ given by

$$Zf(t,\omega) = \sum_{k\in\mathbf{Z}} f(t-k)\, e^{2\pi i k\omega}, \quad (t,\omega) \in [0,1)^2,$$

(the series converges in the norm of $L^2([0,1)^2)$). The Zak transform was first introduced by Gelfand (see [55, p. 148]), and goes by several names, including the Weil–Brezin map (representation theory and abstract harmonic analysis) and k-q transform (quantum mechanics). For more information, we refer to Janssen's influential paper [77] and survey [78], or Gröchenig's text [55]. For our purposes, the most important property of the Zak transform is that

$$Z(M_kT_ng)(t,\omega) = e^{2\pi int}e^{-2\pi ik\omega}Zg(t,\omega), \quad (k,n) \in \mathbf{Z}^2.$$

It follows easily from this that any finite linear combination of functions M_kT_ng with $(k,n) \in \mathbf{Z}^2$ is independent, using the fact that a nontrivial (two-dimensional) trigonometric polynomial cannot vanish on a set of positive measure. This settles the case $\alpha\beta = 1$.

This argument cannot be extended to more general rectangular lattices $\alpha\mathbf{Z}\times\beta\mathbf{Z}$ with $\alpha\beta \neq 1$ because $Z(M_{k\beta}T_{n\alpha}g)$ is no longer just a two-dimensional exponential times Zg. A fundamental obstacle is that the operators M_k, T_n commute when k, n are integers, but the operators $M_{k\beta}$, $T_{n\alpha}$ do not commute in general. On the other hand, the operators M_β, T_α generate a *von Neumann algebra*, and it is through this connection that Linnell was able to prove the HRT Conjecture for the special case that Λ is contained in an arbitrary lattice [96].

Although this is contained in Linnell's result, let us sketch a proof of the HRT Conjecture for the special case $N = 3$, since it reveals one of the difficulties in trying to prove the general case. If $N = 3$ then, by applying a metaplectic transform, we may assume that $\Lambda = \{(0,0), (a,0), (0,1)\}$, and hence $\mathcal{G}(g,\Lambda) = \{g(x), g(x-a), e^{2\pi ix}g(x)\}$. Suppose that

$$c_1g(x) + c_2g(x-a) + c_3e^{2\pi ix}g(x) = 0 \text{ a.e.}$$

If any one of c_1, c_2, c_3 is zero, then we are back to the collinear case, so we may assume they are all nonzero. Rearranging, we obtain

$$g(x-a) = m(x)\,g(x) \text{ a.e.}, \tag{9.11}$$

where $m(x) = -\frac{1}{c_2}(c_1 + c_3\,e^{2\pi ix})$. Note that m is 1-periodic. Iterating (9.11), we obtain for integer $n > 0$ that

$$|g(x-na)| = |g(x)| \prod_{j=0}^{n-1} |m(x-ja)| = |g(x)|\, e^{n\cdot\frac{1}{n}\sum_{j=0}^{n-1} p(x-ja)} \text{ a.e.} \tag{9.12}$$

where $p(x) = \ln|m(x)|$. Now, p is 1-periodic, so if a is irrational then the points $\{x - ja \bmod 1\}_{j=0}^\infty$ are a dense subset of $[0,1)$. In fact, they are "well distributed" in a technical sense due to the fact that $x \mapsto x + a \bmod 1$ is an *ergodic* mapping of $[0,1)$ onto itself (ergodic means that any subset of $[0,1)$ which is invariant under this map must either have measure 0 or 1). Hence the quantity $\frac{1}{n}\sum_{j=0}^{n-1} p(x-ja)$ is like a Riemann sum approximation to $\int_0^1 p(x)\,dx$, except that the rectangles with height $p(x-ja)$ and width $\frac{1}{n}$ are distributed "randomly" around $[0,1)$ instead of uniformly, possibly even with overlaps or gaps. Still, the ergodicity ensures that the Riemann sum analogy is a good one in the limit: the Birkhoff Ergodic Theorem implies that

$$\lim_{n\to\infty} \frac{1}{n}\sum_{j=0}^{n-1} p(x-ja) = \int_0^1 p(x)\,dx = C \text{ a.e.} \tag{9.13}$$

To be sure, we must verify that $C = \int_0^1 p(x)\,dx$ exists and is finite, but this can be shown based on the fact that any singularities of p correspond to zeros of the well-behaved function m. For more information on ergodic theory, we refer to [112].

Thus from (9.11) and (9.13) we see that $|g(x-na)| \approx e^{Cn}|g(x)|$. More precisely, if we fix $\varepsilon > 0$, then $\frac{1}{n}\sum_{j=0}^{n-1} p(x-ja) \geq (C-\varepsilon)$ for n large enough. Let us ignore the fact that "large enough" depends on x (or, by applying Egoroff's Theorem, restrict to a subset where the convergence in (9.13) is uniform). Substituting into (9.12) then yields $|g(x-na)| \geq e^{(C-\varepsilon)n}\,|g(x)|$ for n large. Considering x in a set of positive measure where g is nonzero and the fact that $g \in L^2(\mathbf{R})$, we conclude that $C - \varepsilon < 0$, and hence $C \leq 0$. A converse argument, based on the relation $g(x) = m(x+a)\,g(x+a)$ similarly yields the fact that $C \geq 0$. This still allows the possibility that $C = 0$, but a slightly more subtle argument presented in [69] also based on ergodicity yields the full result. The case for a rational is more straightforward, since in this case the points $x - ja \bmod 1$ repeat themselves.

The argument given above can be extended slightly: we could take $\Lambda = \{(0,j)\}_{j=0}^{N} \cup \{(a,0)\}$, for then we would still have (9.11) holding, with a different, but still 1-periodic, function m. However, the periodicity is critical in order to apply ergodic theory as we have done. An additional fundamental difficulty to extending further is that as soon as we have more than two distinct translates, we cannot rearrange a dependency relation into a form similar to (9.11) that can be easily iterated. For example, with three distinct sets of translations, instead of (9.11) we would have an equation like $g(x-a) = m(x)\,g(x) + k(x)\,g(x-b)$, which becomes extremely complicated to iterate.

9.5.4 Special Case: Compactly Supported Functions

Choose any finite set $\Lambda \subset \mathbf{R}^2$, and suppose that $g \in L^2(\mathbf{R})$ is compactly supported, or even just supported on a half-line. Given an arbitrary finite set Λ, write $\Lambda = \{(\alpha_k, \beta_{k,j})\}_{j=1,\dots,M_k,\,k=1,\dots,N}$, i.e., for each distinct translate group the corresponding modulates together. Given scalars $c_{k,j}$, suppose

$$0 = \sum_{k=1}^{N}\sum_{j=1}^{M_k} c_{k,j}\, M_{\beta_{k,j}} T_{\alpha_k} g(x) = \sum_{k=1}^{N} m_k(x)\, g(x-\alpha_k) \text{ a.e.}, \tag{9.14}$$

where $m_k(x) = \sum_{j=1}^{M_k} c_{k,j}\, e^{2\pi i \beta_{k,j} x}$. Since g is supported in a half-line, the supports of the functions $g(x-\alpha_k)$ overlap some places but not others. If we choose x in the appropriate interval then only one $g(x-\alpha_k)$ can be nonzero. For such x, the right-hand side of equation (9.14) will contain only one nonzero term, i.e., it reduces to $m_k(x)\,g(x-\alpha_k) = 0$ a.e. for some single k. We can find a subset of the support of $g(x-\alpha_k)$ of positive measure for which this is true, which implies the trigonometric polynomial $m_k(x)$ vanishes on a set of positive measure. But this cannot happen unless $c_{k,j} = 0$ for all j. We can then repeat this argument to obtain $c_{k,j} = 0$ for all j and k. For complete details, see [69].

9.5.5 Special Case: Hermite Functions

In this section we will prove the HRT Conjecture for the special case that $g(x) = p(x)\, e^{-x^2}$ where p is a nontrivial polynomial. Such functions are finite linear combinations of Hermite functions, and the collection of all such functions is dense in $L^2(\mathbf{R})$.

Given an arbitrary finite set Λ, write $\Lambda = \{(\alpha_k, \beta_{k,j})\}_{j=1,\dots,M_k,\, k=1,\dots,N}$, with $\alpha_1 < \cdots < \alpha_N$. If $N = 1$ then Λ is a set of collinear points, so we may assume $N > 1$. Given scalars $c_{k,j}$, suppose

$$\sum_{k=1}^{N} \sum_{j=1}^{M_k} c_{k,j}\, M_{\beta_{k,j}} T_{\alpha_k} g(x) = 0 \text{ a.e.}$$

Because of the special form of g, this simplifies to

$$e^{-x^2} \sum_{k=1}^{N} m_k(x)\, p(x - \alpha_k)\, e^{2\alpha_k x} = 0 \text{ a.e.}, \tag{9.15}$$

where $m_k(x) = \sum_{j=1}^{M_k} c_{k,j}\, e^{-\alpha_k^2}\, e^{2\pi i \beta_{k,j} x}$. Without loss of generality we may assume that m_1 and m_N are nontrivial, otherwise ignore those terms and reindex. Then dividing both sides of (9.15) by $e^{-x^2}\, e^{2\alpha_N x}$ and rearranging, we have

$$m_N(x)\, p(x - \alpha_N) = -\sum_{k=1}^{N-1} m_k(x)\, p(x - \alpha_k)\, e^{2(\alpha_k - \alpha_N)x} \text{ a.e.} \tag{9.16}$$

However, $\alpha_k - \alpha_N < 0$ for $k = 1, \dots, N-1$, so since each m_k is bounded, the right-hand side of (9.16) converges to zero as $x \to \infty$. On the other hand, as m_N is a nontrivial trigonometric polynomial and p is a nontrivial polynomial, the left-hand side does not converge to zero.

9.5.6 Special Case: Perturbations

Now that we have proved the HRT Conjecture for dense subsets of $L^2(\mathbf{R})$ such as the compactly supported functions, it is tempting to try to prove the general conjecture by applying some form of limiting argument. We will prove next a theorem in this spirit, and then see why this theorem fails to provide a proof of the full conjecture.

We will need to use the following lemma, which is actually just a special case of a general result that characterizes Riesz bases; compare [28, Thm. 3.6.6].

Lemma 9.19. *Let* $\{g_1, \dots, g_N\}$ *be a linearly independent set of vectors in a Hilbert space* H*. Let* A*,* B *be frame bounds for* $\{g_1, \dots, g_N\}$ *as a frame for its span* $S = \operatorname{span}(\{g_1, \dots, g_N\})$*. Then*

$$\forall\, c_1, \dots, c_N \in \mathbf{C}, \quad A \sum_{k=1}^{N} |c_k|^2 \le \Bigl\| \sum_{k=1}^{N} c_k g_k \Bigr\|^2 \le B \sum_{k=1}^{N} |c_k|^2. \tag{9.17}$$

Proof. Let $\{\tilde{g}_1, \dots, \tilde{g}_N\}$ be the canonical dual frame for $\{g_1, \dots, g_N\}$ in S. Given $c_1, \dots, c_N$, set $f = \sum_{k=1}^{N} c_k g_k$. The frame expansion of f is $f = \sum_{k=1}^{N} \langle f, \tilde{g}_k \rangle\, g_k$, and because of independence we must have $c_k = \langle f, \tilde{g}_k \rangle$. Since $\frac{1}{B}$, $\frac{1}{A}$ are frame bounds for the dual frame $\{\tilde{g}_1, \dots, \tilde{g}_N\}$, we have

$$\frac{1}{B}\, \|f\|^2 \le \sum_{k=1}^{N} |\langle f, \tilde{g}_k \rangle|^2 \le \frac{1}{A}\, \|f\|^2, \tag{9.18}$$

and rearranging (9.18) gives (9.17). □

We will also need to use the continuity of the operator groups $\{T_x\}_{x \in \mathbf{R}}$ and $\{M_\omega\}_{\omega \in \mathbf{R}}$. That is, we will need the fact that

$$\forall\, f \in L^2(\mathbf{R}), \quad \lim_{x \to 0} \|T_x f - f\|_2 = 0 = \lim_{\omega \to 0} \|M_\omega f - f\|_2. \tag{9.19}$$

The next theorem is stated in [69]. A proof of part (a) is given there, and will not be repeated here. The proof of part (b) is similar, but since it is not given in [69] and is slightly more complicated than part (a), we prove it here.

Theorem 9.20. *Assume that* $g \in L^2(\mathbf{R})$ *and* $\Lambda = \{(\alpha_k, \beta_k)\}_{k=1}^{N}$ *are such that* $\mathcal{G}(g, \Lambda)$ *is linearly independent. Then the following statements hold.*

(a) *There exists* $\varepsilon > 0$ *such that* $\mathcal{G}(h, \Lambda)$ *is independent for any* $h \in L^2(\mathbf{R})$ *with* $\|g - h\|_2 < \varepsilon$*.*

(b) *There exists* $\varepsilon > 0$ *such that* $\mathcal{G}(g, \Lambda')$ *is independent for any set* $\Lambda' = \{(\alpha_k', \beta_k')\}_{k=1}^{N}$ *such that* $|\alpha_k - \alpha_k'|,\ \ |\beta_k - \beta_k'| < \varepsilon$ *for* $k = 1, \dots, N$*.*

Proof. (b) Let A, B be frame bounds for $\mathcal{G}(g, \Lambda)$ as a frame for its span. Fix $0 < \delta < A^{1/2}/(2N^{1/2})$. Then by (9.19), we can choose ε small enough that

$$|r| \le \varepsilon \quad \Longrightarrow \quad \|T_r g - g\|_2 \le \delta,\ \|M_r g - g\|_2 \le \delta.$$

Now suppose that $|\alpha_k - \alpha_k'| < \varepsilon$ and $|\beta_k - \beta_k'| < \varepsilon$ for $k = 1, \dots, N$. Then for any scalars $c_1, \dots, c_N$ we have

$$\begin{aligned}
&\Bigl\| \sum_{k=1}^{N} c_k\, M_{\beta_k} T_{\alpha_k} g \Bigr\|_2 \\
&\qquad \le \Bigl\| \sum_{k=1}^{N} c_k\, M_{\beta_k} (T_{\alpha_k} h - T_{\alpha_k'}) g \Bigr\|_2 \\
&\qquad\qquad + \Bigl\| \sum_{k=1}^{N} c_k\, (M_{\beta_k} - M_{\beta_k'}) T_{\alpha_k'} g \Bigr\|_2 + \Bigl\| \sum_{k=1}^{N} c_k\, M_{\beta_k'} T_{\alpha_k'} g \Bigr\|_2
\end{aligned}$$

$$
\begin{aligned}
&\le \sum_{k=1}^{N} |c_k| \, \|T_{\alpha_k} g - T_{\alpha'_k} g\|_2 \, + \, \sum_{k=1}^{N} |c_k| \, \|(M_{\beta_k} - M_{\beta'_k}) T_{\alpha'_k} g\|_2 \\
&\qquad + \, \Bigl\| \sum_{k=1}^{N} c_k \, M_{\beta'_k} T_{\alpha'_k} g \Bigr\|_2 \\
&= \sum_{k=1}^{N} |c_k| \, \|T_{\alpha_k - \alpha'_k} g - g\|_2 \, + \, \sum_{k=1}^{N} |c_k| \, \|M_{\beta_k - \beta'_k} g - g\|_2 \\
&\qquad + \, \Bigl\| \sum_{k=1}^{N} c_k \, M_{\beta'_k} T_{\alpha'_k} g \Bigr\|_2 \\
&\le 2\delta \sum_{k=1}^{N} |c_k| \, + \, \Bigl\| \sum_{k=1}^{N} c_k \, M_{\beta'_k} T_{\alpha'_k} g \Bigr\|_2 \\
&\le 2\delta N^{1/2} \Bigl(\sum_{k=1}^{N} |c_k|^2 \Bigr)^{1/2} \, + \, \Bigl\| \sum_{k=1}^{N} c_k \, M_{\beta'_k} T_{\alpha'_k} g \Bigr\|_2 .
\end{aligned}
$$

However, we also have by Lemma 9.19 that

$$
A^{1/2} \Bigl(\sum_{k=1}^{N} |c_k|^2 \Bigr)^{1/2} \le \Bigl\| \sum_{k=1}^{N} c_k \, M_{\beta_k} T_{\alpha_k} g \Bigr\|_2 .
$$

Combining and rearranging these inequalities, we find that

$$
\Bigl(A^{1/2} - 2\delta N^{1/2} \Bigr) \Bigl(\sum_{k=1}^{N} |c_k|^2 \Bigr)^{1/2} \le \Bigl\| \sum_{k=1}^{N} c_k \, M_{\beta'_k} T_{\alpha'_k} g \Bigr\|_2 .
$$

Since $A^{1/2} - 2\delta N^{1/2} > 0$, it follows that if $\sum_{k=1}^{N} c_k \, M_{\beta'_k} T_{\alpha'_k} g = 0$ a.e. then $c_1 = \cdots = c_N = 0$. □

Unfortunately, Theorem 9.20 cannot be combined with the known special cases, such as for compactly supported g or for lattice Λ, to give a proof of the full HRT Conjecture. The problem is that ε in Theorem 9.20 depends on g and Λ. Analogously, a union of arbitrary open intervals $(r-\varepsilon_r, r+\varepsilon_r)$ centered at rationals $r \in \mathbf{Q}$ need not cover the entire line (consider $\varepsilon_r = |r-\sqrt{2}|$). What is needed is specific knowledge of how the value of ε depends on g and Λ. The proof above shows that the value of ε is connected to the lower frame bound for $\mathcal{G}(g, \Lambda)$ considered as a frame for its finite span in $L^2(\mathbf{R})$. This leads us to close this survey with the following fundamental problem.

- Given $g \in L^2(\mathbf{R})$ and a finite set $\Lambda \subset \mathbf{R}^2$, find explicit values for the frame bounds of $\mathcal{G}(g, \Lambda)$ as a frame for its span.

Explicit here means that the frame bounds should be expressed in some computable way as a function of g (or its properties, such as the size of its support) and the points in Λ. This is clearly an important problem with practical implications, since implementations will always involve finite sets. Surprisingly little is known regarding such frame bounds; the best results appear to be those of Christensen and Lindner [34]. In fact, even the seemingly "simpler" problem of computing explicit frame bounds for finite sets of exponentials $\{e^{2\pi i \alpha_k x}\}_{k=1}^N$ as frames for their spans in $L^2[0,1]$ is very difficult, but still of strong interest, see [33].

Acknowledgments

Deep and heartfelt thanks go to John Benedetto for introducing me to harmonic analysis, wavelets, and time-frequency analysis, and for his constant friendship and support. Thanks also to Jay Ramanathan and Pankaj Topiwala for their collaboration that led to our conjecture and partial results, to Palle Jorgensen for suggestions and advice on that initial paper, to Gestur Ólafsson and Larry Baggett for their interest in the conjecture and for encouraging me to write the notes that became this chapter, and to the many people with whom I have discussed this conjecture, including Radu Balan, Nick Brönn, Pete Casazza, Ole Christensen, Baiqiao Deng, Hans Feichtinger, Vera Furst, Karlheinz Gröchenig, Guido Janssen, Norbert Kaiblinger, Werner Kozek, Gitta Kutyniok, Demetrio Labate, David Larson, Chris Lennard, Peter Linnell, Yura Lyubarskii, Joseph Rosenblatt, Ziemowit Rzeszotnik, Andrew Stimpson, David Walnut, and Georg Zimmermann. The support of NSF grant DMS-0139261 is also gratefully acknowledged.

References

1. A. Aldroubi, C. A. Cabrelli, and U. M. Molter, Wavelets on irregular grids with arbitrary dilation matrices and frame atoms for $L^2(\mathbb{R}^d)$, *Appl. Comput. Harmon. Anal.*, **17** (2004), pp. 119–140.
2. A. Aldroubi and K. Gröchenig, Nonuniform sampling and reconstruction in shift-invariant spaces, *SIAM Review*, **43** (2001), pp. 585–620.
3. K. I. Babenko, On conjugate functions (Russian), *Doklady Akad. Nauk SSSR (N. S.)*, **62** (1948), pp. 157–160.
4. L. Baggett, Processing a radar signal and representations of the discrete Heisenberg group, *Colloq. Math.*, **60/61** (1990), pp. 195–203.
5. L. Baggett, H. Medina, and K. Merrill, Generalized multiresolution analyses, and a construction procedure for all wavelet sets in $\mathbf{R}^n$, *J. Fourier Anal. Appl.*, **6** (1999), pp. 563–573.
6. R. Balan, P. G. Casazza, C. Heil, and Z. Landau, Deficits and excesses of frames, *Adv. Comput. Math.*, **18** (2003), pp. 93–116.
7. R. Balan, P. G. Casazza, C. Heil, and Z. Landau, Excesses of Gabor frames, *Appl. Comput. Harmon. Anal.*, **14** (2003), pp. 87–106.

8. R. Balan, P. G. Casazza, C. Heil, and Z. Landau, Density, overcompleteness, and localization of frames, I. Theory, *J. Fourier Anal. Appl.*, **12** (2006), pp. 105–143.
9. R. Balan, P. G. Casazza, C. Heil, and Z. Landau, Density, overcompleteness, and localization of frames, II. Gabor frames, *J. Fourier Anal. Appl.*, to appear.
10. R. Balian, Un principe d'incertitude fort en théorie du signal ou en mécanique quantique, *C. R. Acad. Sci. Paris*, **292** (1981), pp. 1357–1362.
11. G. Battle, Heisenberg proof of the Balian–Low theorem, *Lett. Math. Phys.*, **15** (1988), pp. 175–177.
12. J. J. Benedetto, Gabor representations and wavelets, in: *Commutative Harmonic Analysis* (Canton, NY, 1987), Contemp. Math., Vol. 91, Amer. Math. Soc., Providence, RI, 1989, pp. 9–27.
13. J. J. Benedetto, *Harmonic Analysis and Applications*, CRC Press, Boca Raton, FL, 1997.
14. J. J. Benedetto, W. Czaja, P. Gadziński, and A. M. Powell, The Balian–Low Theorem and regularity of Gabor systems, *J. Geom. Anal.*, **13** (2003), pp. 217–232.
15. J. J. Benedetto, W. Czaja, and A. Ya. Maltsev, The Balian–Low theorem for the symplectic form on $\mathbf{R}^{2d}$, *J. Math. Phys.*, **44** (2003), pp. 1735–1750.
16. J. J. Benedetto and M. Fickus, Finite normalized tight frames, *Adv. Comput. Math.*, **18** (2003), pp. 357–385.
17. J. J. Benedetto, C. Heil, and D. F. Walnut, Differentiation and the Balian–Low Theorem, *J. Fourier Anal. Appl.*, **1** (1995), pp. 355–402.
18. J. J. Benedetto and M. Leon, The construction of single wavelets in D-dimensions, *J. Geom. Anal.*, **11** (2001), pp. 1–15.
19. A. Boggess and F. J. Narcowich, *A First Course in Wavelets with Fourier Analysis*, Prentice-Hall, Upper Saddle River, NJ, 2001.
20. M. Bownik and Z. Rzeszotnik, The spectral function of shift-invariant spaces, *Michigan Math. J.*, **51** (2003), pp. 387–414.
21. C. A. Cabrelli, C. Heil, and U. M. Molter, Self-similarity and multiwavelets in higher dimensions, *Memoirs Amer. Math. Soc.*, **170**, No. 807 (2004).
22. P. G. Casazza, The art of frame theory, *Taiwanese J. Math.*, **4** (2000), pp. 129–201.
23. P. G. Casazza, O. Christensen, A. M. Lindner, and R. Vershynin, Frames and the Feichtinger conjecture, *Proc. Amer. Math. Soc.*, **133** (2005), pp. 1025–1033.
24. P. G. Casazza, M. Fickus, J. Kovačević, M. T. Leon, and J. C. Tremain, A physical interpretation of tight frames, Chapter 4, this volume (2006).
25. P. G. Casazza and N. J. Kalton, Roots of complex polynomials and Weyl–Heisenberg frame sets, *Proc. Amer. Math. Soc.*, **130** (2002), pp. 2313–2318.
26. P. G. Casazza and J. C. Tremain, The Kadison–Singer problem in mathematics and engineering, *Proc. Natl. Acad. Sci. USA*, **103** (2006), pp. 2032–2039.
27. S. Chen, D. Donoho, and M. A. Saunders, Atomic decomposition by basis pursuit, *SIAM Review*, **43** (2001), pp. 129–157.
28. O. Christensen, *An Introduction to Frames and Riesz Bases*, Birkhäuser, Boston, 2003.
29. O. Christensen and P. G. Casazza, Frames containing a Riesz basis and preservation of this property under perturbations, *SIAM J. Math. Anal.*, **29** (1998), pp. 266–278.
30. O. Christensen and Y. C. Eldar, Oblique dual frames and shift-invariant spaces, *Appl. Comput. Harmon. Anal.*, **17** (2004), pp. 48–68.

31. O. Christensen and C. Heil, Perturbations of Banach frames and atomic decompositions, *Math. Nachr.*, **185** (1997), pp. 33–47.
32. O. Christensen, B. Deng, and C. Heil, Density of Gabor frames, *Appl. Comput. Harmon. Anal.*, **7** (1999), pp. 292–304.
33. O. Christensen and A. M. Lindner, Frames of exponentials: lower frame bounds for finite subfamilies, and approximation of the inverse frame operator, *Linear Algebra Appl.*, **323** (2001), pp. 117–130.
34. O. Christensen and A. M. Lindner, Lower bounds for finite wavelet and Gabor systems, *Approx. Theory Appl. (N.S.)*, **17** (2001), pp. 18–29.
35. O. Christensen and A. M. Lindner, Decompositions of wavelets and Riesz frames into a finite number of linearly independent sets, *Linear Algebra Appl.*, **355** (2002), pp. 147–159.
36. E. Cordero and K. Gröchenig, Localization of frames. II, *Appl. Comput. Harmon. Anal.*, **17** (2004), pp. 29–47.
37. W. Czaja, G. Kutyniok, and D. Speegle, Geometry of sets of parameters of wave packet framess, Appl. Comput. Harmon. Anal., **20** (2006), pp. 108–125.
38. W. Czaja and A. Powell, Recent developments in the Balian–Low Theorem, Chapter 5, this volume (2006).
39. X. Dai, D. R. Larson, and D. M. Speegle, Wavelet sets in $\mathbf{R}^n$, *J. Fourier Anal. Appl.*, **3** (1997), pp. 451–456.
40. I. Daubechies, The wavelet transform, time-frequency localization and signal analysis, *IEEE Trans. Inform. Theory*, **36** (1990), pp. 961–1005.
41. I. Daubechies, *Ten Lectures on Wavelets*, SIAM, Philadelphia, 1992.
42. I. Daubechies, A. Grossmann, and Y. Meyer, Painless nonorthogonal expansions, *J. Math. Phys.*, **27** (1986), pp. 1271–1283.
43. I. Daubechies and A. J. E. M. Janssen, Two theorems on lattice expansions, *IEEE Trans. Inform. Theory*, **39** (1993), pp. 3–6.
44. I. Daubechies, H. Landau, and Z. Landau, Gabor time-frequency lattices and the Wexler–Raz identity, *J. Fourier Anal. Appl.*, **1** (1995), pp. 437–478.
45. B. Deng and C. Heil, Density of Gabor Schauder bases, in: *Wavelet Applications in Signal and Image Processing VIII* (San Diego, CA, 2000), A. Aldroubi, A. Lane, and M. Unser, eds., Proc. SPIE Vol. 4119, SPIE, Bellingham, WA, 2000, pp. 153–164.
46. R. J. Duffin and A. C. Schaeffer, A class of nonharmonic Fourier series, *Trans. Amer. Math. Soc.*, **72** (1952), pp. 341–366.
47. K. Dykema, D. Freeman, K. Kornelson, D. Larson, M. Ordower, and E. Weber, Ellipsoidal tight frames and projection decomposition of operators, *Illinois J. Math.*, **48** (2004), pp. 477–489.
48. G. Edgar and J. Rosenblatt, Difference equations over locally compact abelian groups, *Trans. Amer. Math. Soc.*, **253** (1979), pp. 273–289.
49. G. B. Folland, *Harmonic Analysis on Phase Space*, Princeton Univ. Press, Princeton, NJ, 1989.
50. G. B. Folland and A. Sitaram, The uncertainty principle: A mathematical survey, *J. Fourier Anal. Appl.*, **3** (1997), pp. 207–238.
51. D. Gabor, Theory of communication, *J. Inst. Elec. Eng. (London)*, **93** (1946), pp. 429–457.
52. J.-P. Gabardo and D. Han, Aspects of Gabor analysis and operator algebras, in: *Advances in Gabor Analysis*, H. G. Feichtinger and T. Strohmer, eds., Birkhäuser, Boston, 2003, pp. 153–169.

53. J.-P. Gabardo and D. Han, Balian–Low phenomenon for subspace Gabor frames, *J. Math. Phys.*, **45** (2004), pp. 3362–3378.
54. B. R. Gelbaum, Notes on Banach spaces and bases, *An. Acad. Brasil. Ci.*, **30** (1958), pp. 29–36.
55. K. Gröchenig, *Foundations of Time-Frequency Analysis*, Birkhäuser, Boston, 2001.
56. K. Gröchenig, Uncertainty principles for time-frequency representations, in: *Advances in Gabor Analysis*, H. G. Feichtinger and T. Strohmer, eds., Birkhäuser, Boston, 2003, pp. 11–30.
57. K. Gröchenig, Localized frames are finite unions of Riesz sequences, *Adv. Comput. Math.*, **18** (2003), pp. 149–157.
58. K. Gröchenig, Localization of frames, Banach frames, and the invertibility of the frame operator, *J. Fourier Anal. Appl.*, **10** (2004), pp. 105–132.
59. K. Gröchenig, D. Han, C. Heil, and G. Kutyniok, The Balian–Low Theorem for symplectic lattices in higher dimensions, *Appl. Comput. Harmon. Anal.*, **13** (2002), pp. 169–176.
60. K. Gröchenig and H. Razafinjatovo, On Landau's necessary density conditions for sampling and interpolation of band-limited functions, *J. London Math. Soc. (2)*, **54** (1996), pp. 557–565.
61. D. Han and D. R. Larson, Frames, bases and group representations *Memoirs Amer. Math. Soc.*, **147**, No. 697 (2000).
62. E. Hayashi, S. Li, and T. Sorrells, Gabor duality characterizations, Chapter 7, this volume (2006).
63. C. Heil, Wiener Amalgam Spaces in Generalized Harmonic Analysis and Wavelet Theory, Ph.D. Thesis, University of Maryland, College Park, MD, 1990.
64. C. Heil, *A Basis Theory Primer*, manuscript, 1997. Electronic version available at `http://www.math.gatech.edu/~heil`.
65. C. Heil, An introduction to weighted Wiener amalgams, in: *Wavelets and their Applications* (Chennai, January 2002), M. Krishna, R. Radha and S. Thangavelu, eds., Allied Publishers, New Delhi, 2003, pp. 183–216.
66. C. Heil, Integral operators, pseudodifferential operators, and Gabor frames, in: *Advances in Gabor Analysis*, H. G. Feichtinger and T. Strohmer, eds., Birkhäuser, Boston, 2003, pp. 153–169.
67. C. Heil and G. Kutyniok, Density of weighted wavelet frames, *J. Geom. Anal.*, **13** (2003), pp. 479–493.
68. C. Heil and G. Kutyniok, The Homogeneous Approximation Property for wavelet frames, preprint (2005).
69. C. Heil, J. Ramanathan, and P. Topiwala, Linear independence of time-frequency translates, *Proc. Amer. Math. Soc.*, **124** (1996), pp. 2787–2795.
70. C. Heil, J. Ramanathan, and P. Topiwala, Singular values of compact pseudodifferential operators, *J. Funct. Anal.*, **150** (1996), pp. 426–452.
71. C. E. Heil and D. F. Walnut, Continuous and discrete wavelet transforms, *SIAM Review*, **31** (1989), pp. 628–666.
72. C. Heil and D. F. Walnut, eds., *Fundamental Papers in Wavelet Theory*, Princeton University Press, Princeton, NJ, 2006.
73. E. Hernández, D. Labate, and G. Weiss, A unified characterization of reproducing systems generated by a finite family, II, *J. Geom. Anal.*, **12** (2002), pp. 615–662.

74. E. Hernández and G. Weiss, *A First Course on Wavelets*, CRC Press, Boca Raton, FL, 1996.
75. G. Higman, The units of group-rings, *Proc. London Math. Soc. (2)*, **46** (1940), pp. 231–248.
76. J. R. Holub, Pre-frame operators, Besselian frames, and near-Riesz bases in Hilbert spaces, *Proc. Amer. Math. Soc.*, **122** (1994), pp. 779–785.
77. A. J. E. M. Janssen, Bargmann transform, Zak transform, and coherent states, *J. Math. Phys.*, **23** (1982), pp. 720–731.
78. A. J. E. M. Janssen, The Zak transform: a signal transform for sampled time-continuous signals, *Philips J. Res.*, **43** (1988), pp. 23–69.
79. A. J. E. M. Janssen, Signal analytic proofs of two basic results on lattice expansions, *Appl. Comput. Harmon. Anal.*, **1** (1994), pp. 350–354.
80. A. J. E. M. Janssen, Duality and biorthogonality for Weyl–Heisenberg frames, *J. Fourier Anal. Appl.*, **1** (1995), pp. 403–436.
81. A. J. E. M. Janssen, Representations of Gabor frame operators, in: *Twentieth Century Harmonic Analysis—A Celebration* (Il Ciocco, 2000), NATO Sci. Ser. II Math. Phys. Chem., **33**, Kluwer Acad. Publ., Dordrecht, 2001, pp. 73–101.
82. A. J. E. M. Janssen, Zak transforms with few zeros and the tie, in: *Advances in Gabor Analysis*, H. G. Feichtinger and T. Strohmer, eds., Birkhäuser, Boston, 2003, pp. 31–70.
83. A. J. E. M. Janssen, On generating tight Gabor frames at critical density, *J. Fourier Anal. Appl.*, **9** (2003), pp. 175–214.
84. A. J. E. M. Janssen and T. Strohmer, Hyperbolic secants yield Gabor frames, *Appl. Comput. Harmon. Anal.*, **12** (2002), pp. 259–267.
85. R. Kadison and I. Singer, Extensions of pure states, *Amer. J. Math.*, **81** (1959), pp. 383–400.
86. K. S. Kazarian, F. Soria, and R. E. Zink, On rearranges orthogonal systems as quasibases in weighted L^p spaces, in: *Interaction between Functional Analysis, Harmonic Analysis, and Probability* (Columbia, MO, 1994), N. Kalton, E. Saab, and S. Montgomery-Smith, eds., Lecture Notes in Pure and Appl. Math. Vol. 175, Dekker, New York, 1996, pp. 239–247,
87. H. O. Kim and J. K. Lim, New characterizations of Riesz bases, *Appl. Comput. Harmon. Anal.*, **4** (1997), pp. 222–229.
88. G. Kutyniok, Linear independence of time-frequency shifts under a generalized Schrödinger representation, *Arch. Math. (Basel)*, **78** (2002), pp. 135–144.
89. G. Kutyniok, Computation of the density of weighted wavelet systems, in: *Wavelets: Applications in Signal and Image Processing*, M. Unser, A. Aldroubi, and A. Laine eds., SPIE Proc. Vol. 5207, SPIE, San Diego, 2003, pp. 393–404.
90. D. Labate, A unified characterization of reproducing systems generated by a finite family, *J. Geom. Anal.*, **12** (2002), pp. 469–491.
91. H. Landau, On the density of phase-space expansions, *IEEE Trans. Inform. Theory*, **39** (1993), pp. 1152–1156.
92. S. Li and H. Ogawa, Pseudo-duals of frames with applications, *Appl. Comput. Harmon. Anal.*, **11** (2001), 289–304.
93. J. Lindenstrauss and L. Tzafriri, *Classical Banach Spaces, I*, Springer-Verlag, New York, 1977.
94. P. A. Linnell, Zero divisors and $L^2(G)$, *C. R. Acad. Sci. Paris Sér. I Math.*, **315** (1992), pp. 49–53.

95. P. A. Linnell, Analytic versions of the zero divisor conjecture, in: *Geometry and Cohomology in Group Theory*, P. H. Kropholler, G. A. Niblo, and R. Stöhr, eds., London Math. Soc. Lecture Note Series, Cambridge University Press, Cambridge, 1998, pp. 209–248.
96. P. A. Linnell, Von Neumann algebras and linear independence of translates, *Proc. Amer. Math. Soc.*, **127** (1999), pp. 3269–3277.
97. F. Low, Complete sets of wave packets, in: *A Passion for Physics—Essays in Honor of Geoffrey Chew*, C. DeTar, J. Finkelstein, and C. I. Tan, eds., World Scientific, Singapore, 1985, pp. 17–22.
98. J. Marti, *Introduction to the Theory of Bases*, Springer-Verlag, New York, 1969.
99. A. Olevskii, Completeness in $L^2(\mathbf{R})$ of almost integer translates, *C. R. Acad. Sci. Paris*, **324** (1997), pp. 987–991.
100. A. Olevskii and A. Ulanovskii, Almost integer translates. Do nice generators exist?, *J. Fourier Anal. Appl.*, **10** (2004), pp. 93–104.
101. T. E. Olson and R. A. Zalik, Nonexistence of a Riesz basis of translates, in: *Approximation Theory*, Lecture Notes in Pure and Applied Math., Vol. 138, Dekker, New York, 1992, pp. 401–408.
102. A. M. Perelomov, On the completeness of a system of coherent states (English translation), *Theoret. Math. Phys.*, **6** (1971), pp. 156–164.
103. J. Ramanathan and T. Steger, Incompleteness of sparse coherent states, *Appl. Comput. Harmon. Anal.*, **2** (1995), pp. 148–153.
104. M. Rieffel, Von Neumann algebras associated with pairs of lattices in Lie groups, *Math. Ann.*, **257** (1981), pp. 403–418.
105. J. Rosenblatt, Linear independence of translations, *J. Austral. Math. Soc. (Series A)*, **59** (1995), pp. 131–133.
106. Z. Rzeszotnik, private communication, 2004.
107. K. Seip, On the connection between exponential bases and certain related sequences in $L^2(-\pi,\pi)$, *J. Funct. Anal.*, **130** (1995), pp. 131–160.
108. I. Singer, *Bases in Banach Spaces* I, Springer-Verlag, New York, 1970.
109. T. Strohmer and R. W. Heath, Jr., Grassmannian frames with applications to coding and communication, *Appl. Comput. Harmon. Anal.*, **14** (2003), pp. 257–275.
110. W. Sun and X. Zhou, Density and stability of wavelet frames, *Appl. Comput. Harmon. Anal.*, **15** (2003), pp. 117–133.
111. D. F. Walnut, *An Introduction to Wavelet Analysis*, Birkhäuser, Boston, 2002.
112. P. Walters, *An Introduction to Ergodic Theory*, Springer-Verlag, New York, 1982.
113. Y. Wang, Sparse complete Gabor systems on a lattice, *Appl. Comput. Harmon. Anal.*, **16** (2004), pp. 60–67.
114. R. Young, *An Introduction to Nonharmonic Fourier Series*, Revised First Edition, Academic Press, San Diego, 2001.
115. R. A. Zalik, On approximation by shifts and a theorem of Wiener, *Trans. Amer. Math. Soc.*, **243** (1978), pp. 299–308.
116. R. A. Zalik, On fundamental sequences of translates, *Proc. Amer. Math. Soc.*, **79** (1980), pp. 255–259.

Part IV

Wavelet Theory

10

Explicit Cross-Sections of Singly Generated Group Actions

David Larson[1], Eckart Schulz[2], Darrin Speegle[3], and Keith F. Taylor[4]

[1] Department of Mathematics, Texas A&M University, College Station, TX 77843-3368, USA
larson@math.tamu.edu

[2] School of Mathematics, Suranaree University of Technology, 111 University Avenue, Nakhon Ratchasima, 30000, Thailand
eckart@math.sut.ac.th

[3] Department of Mathematics, Saint Louis University, 221 N Grand Blvd, St. Louis, MO 63103, USA
speegled@slu.edu

[4] Department of Mathematics and Statistics, Dalhousie University, Halifax, Nova Scotia, Canada B3H 3J5
keith.f.taylor@dal.ca

Summary. We consider two classes of actions on $\mathbb{R}^n$—one continuous and one discrete. For matrices of the form $A = e^B$ with $B \in M_n(\mathbb{R})$, we consider the action given by $\gamma \to \gamma A^t$. We characterize the matrices A for which there is a cross-section for this action. The discrete action we consider is given by $\gamma \to \gamma A^k$, where $A \in GL_n(\mathbb{R})$. We characterize the matrices A for which there exists a cross-section for this action as well. We also characterize those A for which there exist special types of cross-sections; namely, bounded cross-sections and finite-measure cross-sections. Explicit examples of cross-sections are provided for each of the cases in which cross-sections exist. Finally, these explicit cross-sections are used to characterize those matrices for which there exist minimally supported frequency (MSF) wavelets with infinitely many wavelet functions. Along the way, we generalize a well-known aspect of the theory of shift-invariant spaces to shift-invariant spaces with infinitely many generators.

Dedicated to John Benedetto.

10.1 Introduction

In discrete wavelet analysis on the line, the classical approach is to dilate and translate a single function, or *wavelet*, so that the resulting system is an orthonormal basis for $L^2(\mathbb{R})$. More precisely, a wavelet is a function $\psi \in L^2(\mathbb{R})$ such that

$$\{2^{j/2}\psi(2^j x + k) : k, j \in \mathbb{Z}\}$$

forms an orthonormal basis of $L^2(\mathbb{R})$.

In multidimensional discrete wavelet analysis, the approach is similar. Fix a matrix $A \in GL_n(\mathbb{R})$ and a full rank lattice Γ. A collection of functions $\{\psi^i : i = 1, \ldots, N\}$ is called an (A, Γ) *orthonormal wavelet of order* N if dilations by A and translations by Γ,

$$\{|\det A|^{j/2}\psi^i(A^j x + k) : i = 1, \ldots, N,\ j \in \mathbb{Z},\ k \in \Gamma\},$$

forms an orthonormal basis for $L^2(\mathbb{R}^n)$. In this generality, there is no characterization (in terms of A and Γ) of when wavelets exist. It was shown in [10] that, if A is expansive (that is, a matrix whose eigenvalues all have modulus greater than 1), then there does exist an orthonormal wavelet. A complete characterization of such wavelets in terms of the Fourier transform was given in [13]. The nonexpansive case remains problematic.

It is also possible to study the continuous version of wavelet analysis. Consider the full affine group of motions given by $GL_n(\mathbb{R}) \times \mathbb{R}^n$ with multiplication given by $(a, b)(c, d) = (ac, c^{-1}b + d)$. We are interested in subgroups of the full affine group of motions of the form

$$G = \{(a, b) : a \in D,\ b \in \mathbb{R}^n\},$$

where D is a subgroup of $GL_n(\mathbb{R})$. In this case, G is the semi-direct product $D \times_s \mathbb{R}^n$. Now, if we define the unitary operator T_g for $g \in G$ by

$$(T_g\psi)(x) = |\det a|^{-1/2}\psi(g^{-1}(x)),$$

then the continuous wavelet transform is given by

$$\langle f, \psi_g \rangle := \int_{\mathbb{R}^n} f(x)\,\overline{(T_g\psi)(x)}\,dx,$$

which is, of course, a function on G. The function ψ is a D-continuous wavelet if it is possible to reconstruct all functions f in $L^2(\mathbb{R}^n)$ via the following reconstruction formula:

$$f(x) = \int_G \langle f, \psi_g \rangle\, \psi_g(x)\, d\lambda(g),$$

where λ is Haar measure on G.

There is a simple characterization of continuous wavelets, given in [22].

Theorem 10.1 ([22]). *Let G be a subgroup of the full affine group of the form $D \times_s \mathbb{R}^n$. A function $\psi \in L^2(\mathbb{R}^n)$ is a D-continuous wavelet if and only if the Calderón condition*

$$\int_D |\hat{\psi}(\xi a)|^2\, d\mu(a) = 1 \text{ a.e. } \xi \text{ in } \widehat{\mathbb{R}^n} \tag{10.1}$$

holds, where μ is the left Haar measure for D.

In this paper, we will always assume one of the two following cases, which for our purposes will be the *singly generated subgroups* of $GL_n(\mathbb{R})$.

1. $D = \{A^k : k \in \mathbb{Z}\}$ for some $A \in GL_n(\mathbb{R})$, or
2. $D = \{A^t : t \in \mathbb{R}\}$ for some $A = e^B$, where $B \in M_n(\mathbb{R})$.

We will say that D is generated by the matrix A. Applying Theorem 10.1 to these cases gives the following characterizations.

Proposition 10.2. (a) *Let $A \in GL_n(\mathbb{R})$ and let D denote the dilation group $D = \{A^k : k \in \mathbb{Z}\}$. Then, $\psi \in L^2(\mathbb{R}^n)$ is a D-continuous wavelet if and only if*

$$\sum_{k\in\mathbb{Z}} |\hat{\psi}(\xi A^k)|^2 = 1$$

for almost all $\xi \in \widehat{\mathbb{R}^n}$.

(b) *Let $D = \{A^t : t \in \mathbb{R}\}$ for some $A = e^B$, where $B \in M_n(\mathbb{R})$. Then $\psi \in L^2(\mathbb{R}^n)$ is a D-continuous wavelet if and only if*

$$\int_{\mathbb{R}} |\hat{\psi}(\xi A^t)|^2 \, dt = 1$$

for almost all $\xi \in \widehat{\mathbb{R}^n}$.

In the case where D is generated by a single matrix as above, a complete characterization of matrices for which there exists a continuous wavelet is given in [17].

Theorem 10.3. *Consider the dilation group D as in Case 1 or 2 above. There exists a continuous wavelet if and only if $|\det(A)| \neq 1$.*

The wavelets constructed in [17] are of the form $\hat{\psi} = \chi_K$, for some set K. One drawback to the proof in [17] is that, while the proof is constructive, the sets K that are constructed are written as the countable union of set differences of sets consisting of those points whose orbits land in a prescribed closed ball a positive, finite number of times. Hence, it is not clear whether the set constructed in the end can be chosen to be "nice" or easily described.

The purpose of this chapter is two-fold. First, we will give explicitly defined, easily verified sets K such that χ_K is the Fourier transform of a continuous wavelet. Here, we will exploit the fact that we are in the singly generated group case to a very large extent. We will also obtain a characterization of matrices such that the set K can be chosen to be bounded as well as a characterization of matrices such that the only sets K that satisfy (10.1) have infinite measure.

Second, we will show how to use these explicit forms to characterize those matrices such that there exists a discrete wavelet of order infinity. Note that this seems to be a true application of the form of the sets K in Section 10.2, as it is not clear to the authors how to use the proof in [17] (or the related proof in [16]) to achieve the same result.

10.2 Cross-Sections

Throughout this section, we will use vector notation to denote elements of $\widehat{\mathbb{R}^n}$, and m will denote the Lebesgue measure on $\mathbb{R}^n$. Multiplication of a vector with a matrix will be given by γA, and we will reserve the notation A^t as "A raised to the t-power." In the few places for which we need the transpose of a matrix, we will give it a separate name.

Definition 10.4. *A Borel set $S \subset \widehat{\mathbb{R}^n}$ is called a* cross-section *for the continuous action $\gamma \to \gamma A^t$ $(t \in \mathbb{R})$ if*

(a) $\cup_{t\in\mathbb{R}} SA^t = \widehat{\mathbb{R}^n} \backslash N$ *for some set N of measure zero, and*
(b) $SA^{t_1} \cap SA^{t_2} = \emptyset$ *whenever $t_1 \neq t_2 \in \mathbb{R}$.*

Similarly, a Borel set $S \subset \widehat{\mathbb{R}^n}$ is called a cross-section *for the discrete action $\gamma \to \gamma A^k$ $(k \in \mathbb{Z})$ if*

(a) $\cup_{k\in\mathbb{Z}} SA^k = \widehat{\mathbb{R}^n} \backslash N$ *for some set N of measure zero, and*
(b) $SA^j \cap SA^k = \emptyset$ *whenever $j \neq k \in \mathbb{Z}$.*

Note that we have defined cross-sections using left products, which will eliminate the need for taking transposes in Section 10.3.

Note also that if S is a cross-section for the continuous action, then $\{\gamma A^t : \gamma \in S,\ 0 \le t < 1\}$ is a cross-section for the discrete action. Cross-sections are sometimes referred to as multiplicative tiling sets.

Remark 10.5. Let S be a cross-section for the action $\gamma \to \gamma A^k$. Then, SJ^{-1} is a cross-section for the action $\gamma \to \gamma JA^kJ^{-1}$, and similarly, for the continuous action $\gamma \to \gamma A^t$, where $A = e^B$, SJ^{-1} is a cross-section for the continuous action $\gamma \to \gamma \tilde{A}^t$, where $\tilde{A} = e^{JBJ^{-1}}$.

To begin with cross-sections for the continuous action, let $A = e^B \in GL_n(\mathbb{R})$ be given, where by the preceding remark we may assume that B is in real Jordan normal form. Then, B is a block diagonal matrix, and a block corresponding to a real eigenvalue α_i is of the form

$$B_i = \begin{pmatrix} \alpha_i & 1 & & (0) \\ & \ddots & \ddots & \\ & & \ddots & 1 \\ (0) & & & \alpha_i \end{pmatrix}$$

while a block corresponding to a complex pair of eigenvalues $\alpha_i \pm i\beta_i$ with $\beta_i \neq 0$ is of the form

$$B_i = \begin{pmatrix} D_i & I_2 & & (0) \\ & \ddots & \ddots & \\ & & \ddots & I_2 \\ (0) & & & D_i \end{pmatrix} \quad \text{with} \quad \begin{aligned} D_i &= \begin{pmatrix} \alpha_i & \beta_i \\ -\beta_i & \alpha_i \end{pmatrix}, \\ I_2 &= \begin{pmatrix} 1 & 0 \\ 0 & 1 \end{pmatrix}. \end{aligned}$$

In this basis, A^t is again a block diagonal matrix, and its blocks are of the form

$$A_i = e^{tB_i} = \begin{pmatrix} \lambda_i^t E_i(t) & t\lambda_i^t E_i(t) & \frac{t^2}{2!}\lambda_i^t E_i(t) & \cdot & \dots & \frac{t^{m-1}}{(m-1)!}\lambda_i^t E_i(t) \\ & \lambda_i^t E_i(t) & t\lambda_i^t E_i(t) & & & \vdots \\ & \ddots & \ddots & \ddots & & \vdots \\ & & \ddots & & t\lambda_i^t E_i(t) & \frac{t^2}{2!}\lambda_i^t E_i(t) \\ & & & & \lambda_i^t E_i(t) & t\lambda_i^t E_i(t) \\ (0) & & & & & \lambda_i^t E_i(t) \end{pmatrix}$$

with $\lambda_i = e^{\alpha_i}$ and $E_i(t) = 1$ or $E_i(t) = E_{\beta_i}(t) = \begin{pmatrix} \cos\beta_i t & \sin\beta_i t \\ -\sin\beta_i t & \cos\beta_i t \end{pmatrix}$ depending on whether this block corresponds to a real eigenvalue or a pair of complex eigenvalues of B. The eigenvalues of A are thus e^{α_i} and $e^{\alpha_i}e^{\pm i\beta_i}$, respectively.

For ease of notation, when referring to a specific block A_i of A we will drop the index i. Furthermore, $\mathbf{v}_1, \dots, \mathbf{v}_n$ will denote a Jordan basis of $\widehat{\mathbb{R}^n}$ chosen so that this block under discussion is the first block, and $(x_1, \dots, x_n)$ will denote the components of a vector γ in this basis.

Theorem 10.6. *Let $A = e^B$, where $B \in M_n(\mathbb{R})$ is in Jordan normal form. There exists a cross-section for the continuous action $\gamma \to \gamma A^t$ if and only if A is not orthogonal.*

Proof. Assume that A is not orthogonal. Then at least one of the following four situations, formulated in terms of the eigenvalues of B, will always apply.

Case 1: B has a real eigenvalue $\alpha \neq 0$. A corresponding block of A^t, which we may assume to be the first block, is of the form

$$\begin{pmatrix} \lambda^t & t\lambda^t & \dots & \frac{t^{m-1}}{(m-1)!}\lambda^t \\ & \ddots & \ddots & \\ & & \ddots & t\lambda^t \\ (0) & & & \lambda^t \end{pmatrix}$$

with $\lambda = e^\alpha \neq 1$. Set

$$S = \{\pm\mathbf{v}_1\} \times \operatorname{span}(\mathbf{v}_2, \dots, \mathbf{v}_n).$$

Then S is a cross-section and $\cup_{t\in\mathbb{R}} SA^t = \{(x_1, \dots, x_n) \in \widehat{\mathbb{R}^n} : x_1 \neq 0\}$.

Case 2: B has a complex pair of eigenvalues $\alpha \pm i\beta$ with $\alpha \neq 0$, $\beta > 0$. At least one block of A^t is then of the form

$$\begin{pmatrix} \lambda^t E_\beta(t) & t\lambda^t E_\beta(t) & \dots & \frac{t^{m-1}}{(m-1)!}\lambda^t E_\beta(t) \\ & \ddots & \ddots & \\ & & \ddots & t\lambda^t E_\beta(t) \\ (0) & & & \lambda^t E_\beta(t) \end{pmatrix}, \tag{10.2}$$

and replacing B with $-B$ if necessary, we may assume that $\lambda = e^\alpha > 1$. One easily checks that

$$S = \{s\mathbf{v}_1 : 1 \leq s < \lambda^{2\pi/\beta}\} \times \operatorname{span}(\mathbf{v}_3, \dots, \mathbf{v}_n)$$

is a cross-section and $\cup_{t\in\mathbb{R}} SA^t = \{(x_1, \dots, x_n) \in \widehat{\mathbb{R}^n} : x_1^2 + x_2^2 \neq 0\}$.

Case 3: B has an eigenvalue $\alpha = 0$ and at least one of the blocks of B belonging to this eigenvalue has a nontrivial nilpotent part. Then the corresponding block of A^t is of the form

$$\begin{pmatrix} 1 & t & & (*) \\ & \ddots & \ddots & \\ & & \ddots & t \\ (0) & & & 1 \end{pmatrix} \tag{10.3}$$

and is of at least size 2×2. We set

$$S = \{\ s\mathbf{v}_1 : s \in \mathbb{R}\backslash\{0\}\ \} \times \operatorname{span}(\mathbf{v}_3, \dots, \mathbf{v}_n)$$

so that S is a cross-section and $\cup_{t\in\mathbb{R}} SA^t = \{(x_1, \dots, x_n) \in \widehat{\mathbb{R}^n} : x_1 \neq 0\}$.

Case 4: B has a purely imaginary pair of eigenvalues $\pm i\beta$, $\beta > 0$, and at least one of the blocks of B belonging to this pair has a nontrivial nilpotent part. Then the corresponding block of A^t is of the form

$$\begin{pmatrix} E_\beta(t) & tE_\beta(t) & & (*) \\ & \ddots & \ddots & \\ & & \ddots & tE_\beta(t) \\ (0) & & & E_\beta(t) \end{pmatrix} \tag{10.4}$$

and is of at least size 4×4. Set

$$S = \left\{p\mathbf{v}_1 + q\mathbf{v}_3 + s\mathbf{v}_4 : p > 0,\ 0 \leq q < \frac{2\pi}{\beta}p,\ s \in \mathbb{R}\right\} \times \operatorname{span}(\mathbf{v}_5, \dots, \mathbf{v}_n).$$

Since this is the least intuitive case, let us verify in detail that S is a cross-section. For convenience, we group the first four coordinates of a vector $\gamma \in \widehat{\mathbb{R}^n}$ into two pairs, and write

$$\gamma = \big(\, (x_1, x_2), \; (x_3, x_4), \; x_5, \, x_6, \, \dots, \, x_n \big),$$

so that

$$\gamma A^t = \big(\, (x_1, x_2) E_\beta(t), \; t(x_1, x_2) E_\beta(t) + (x_3, x_4) E_\beta(t), \dots \big).$$

Now $E_\beta(t)$ acts by rotation through the angle βt, so whenever $x_1^2 + x_2^2 \neq 0$ there exists $t_1 \in \mathbb{R}$ such that

$$(x_1, x_2) E_\beta(t_1) = (p, 0)$$

for some $p > 0$. Then

$$\gamma A^{t_1} = (p, 0, t_1 p + y_3, y_4, \dots),$$

where $(y_3, y_4) = (x_3, x_4) E(t_1)$. So if we set $t_2 = t_1 + k \, \frac{2\pi}{\beta}$ for some integer k, then

$$\gamma A^{t_2} = \Big(p, 0, k \frac{2\pi p}{\beta} + t_1 p + y_3, y_4, \dots \Big).$$

Now there exists a k such that

$$0 \leq k \frac{2\pi p}{\beta} + t_1 p + y_3 < \frac{2\pi p}{\beta},$$

and for this choice of k, $\gamma A^{t_2} \in S$. We conclude that

$$\bigcup_{t \in \mathbb{R}} S A^t = \{(x_1, \dots, x_n) \in \widehat{\mathbb{R}^n} : x_1^2 + x_2^2 \neq 0\}.$$

Suppose now that

$$\gamma_1 A^{t_1} = \gamma_2 A^{t_2}$$

for some $\gamma_1, \, \gamma_2 \in S, \;\; t_1, \, t_2 \in \mathbb{R}$. Equivalently,

$$\gamma_1 = \gamma_2 A^t$$

for some t. If $\gamma_1 = (p_1, 0, q_1, s_1, \dots)$ and $\gamma_2 = (p_2, 0, q_2, s_2, \dots)$, then

$$\big(\, (p_1, 0), \; (q_1, s_1), \dots \big) = \big(\, (p_2, 0) E_\beta(t), \; t(p_2, 0) E_\beta(t) \; + \; (q_2, s_2) E_\beta(t), \dots \big)$$

so that

$$\begin{aligned} (p_1, 0) &= (p_2, 0) E_\beta(t), \\ (q_1, s_1) &= t(p_2, 0) E_\beta(t) \; + \; (q_2, s_2) E_\beta(t). \end{aligned}$$

The first equality gives $p_1 = p_2$ and $t = \frac{2\pi}{\beta} k$ for some integer k. Then the second equality reads

$$(q_1, s_1) = \Big(\frac{2\pi}{\beta} k p_1 \; + \; q_2, s_2 \Big),$$

which gives $s_1 = s_2$, and because $0 \leq q_2, q_1 < \frac{2\pi p}{\beta}$, also gives $k = 0$ and $q_1 = q_2$. Thus, S is indeed a cross-section.

Now suppose to the contrary that A is orthogonal, but there exists a cross-section S. Then

$$T = \{\gamma A^t : \gamma \in S,\ 0 \leq t < 1\}$$

is a cross-section for the discrete action of A on $\widehat{\mathbb{R}^n}$. Note that A maps the closed unit ball $B_1(0)$ onto itself, so if $T_o = T \cap B_1(0)$, then

$$B_1(0) = \bigcup_{k\in\mathbb{Z}} T_o A^k,$$

except for a set of measure zero, and this union is disjoint. Then

$$m\big(B_1(0)\big) = \sum_{k\in\mathbb{Z}} m(T_o A^k) = \sum_{k\in\mathbb{Z}} m(T_o) \in \{0, \infty\},$$

which is impossible. □

Remark 10.7. The cross-sections constructed in the proof above allow for a change of variables to integrate along the orbits.

For example, in Case 3, given $\gamma = (x_1, x_2, \ldots, x_n) \in \widehat{\mathbb{R}^n}$ with $x_1 \neq 0$, we set

$$\gamma = F(t, s, a_3, \ldots, a_n) = (s, 0, a_3, \ldots, a_n)A^t,$$

where $s \neq 0$. The Jacobian of this transformation is

$$\begin{vmatrix} (s,0,a_3,\ldots,a_n)B \\ \mathbf{v}_1 \\ \mathbf{v}_3 \\ \vdots \\ \mathbf{v}_n \end{vmatrix} \det(A)^t = \begin{vmatrix} 0 & s & * & \ldots & * \\ 1 & 0 & 0 & \ldots & 0 \\ 0 & 0 & 1 & \ldots & 0 \\ \vdots & & & & \\ 0 & 0 & 0 & \ldots & 1 \end{vmatrix} \det(A)^t = -s\delta^t \neq 0,$$

so that for $\hat{f} \in L^2(\widehat{\mathbb{R}}^n)$,

$$\int_{\widehat{\mathbb{R}^n}} \hat{f}(\gamma)\, d\gamma = \int_{\mathbb{R}^{n-2}} \int_{\mathbb{R}\setminus\{0\}} \int_{\mathbb{R}} \hat{f}(\,(s,0,a_3,\ldots,a_n)A^t\,)\, |s|\delta^t\, dt\, ds\, da_3 \cdots da_n.$$

In Case 4, given $\gamma = (x_1, x_2, \ldots, x_n) \in \widehat{\mathbb{R}^n}$ with $x_1^2 + x_2^2 \neq 0$, we set

$$\gamma = F(t, p, q, s, a_5, \ldots, a_n) = (p, 0, q, s, a_5, \ldots, a_n)A^t,$$

where $p > 0$, $0 \leq q < 2\pi p/\beta$. The Jacobian of this transformation is

$$\begin{vmatrix} (p,0,q,s,a_5,\ldots,a_n)B \\ \mathbf{v}_1 \\ \mathbf{v}_3 \\ \vdots \\ \mathbf{v}_n \end{vmatrix} \det(A)^t = \begin{vmatrix} 0 & \beta p & * & * & * & \ldots & * \\ 1 & 0 & 0 & 0 & 0 & \ldots & 0 \\ 0 & 0 & 1 & 0 & 0 & \ldots & 0 \\ \vdots & & & & & & \\ 0 & 0 & 0 & & 0 & \ldots & 1 \end{vmatrix} \det(A)^t = -\beta p\delta^t \neq 0$$

since $\beta \neq 0$. Thus,

$$\int_{\widehat{\mathbb{R}^n}} \hat{f}(\gamma)\, d\gamma = \int_{\mathbb{R}^{n-4}} \int_0^\infty \int_{\mathbb{R}} \int_0^{2\pi p/\beta} \int_{\mathbb{R}} \hat{f}(\,(p, 0, q, s, a_5, \ldots, a_n)A^t\,)\, |\beta| p \delta^t \\ dt\, dq\, ds\, dp\, da_5 \cdots da_n.$$

Any invertible matrix gives rise to a discrete action on $\widehat{\mathbb{R}^n}$, and nearly always there will exist a cross-section for this action.

Theorem 10.8. *Let $A \in GL_n(\mathbb{R})$ be in Jordan normal form, and consider the discrete action $\gamma \to \gamma A^k$.*

(a) *There exists a cross-section if and only if A is not orthogonal.*
(b) *There exists a cross-section of finite measure if and only if $|\det(A)| \neq 1$.*
(c) *There exists a bounded cross-section if and only if the (real or complex) eigenvalues of A have all modulus > 1 or all modulus < 1.*

Proof. To prove the first assertion, choose a Jordan basis $\mathbf{v}_1, \ldots, \mathbf{v}_n$ so that the Jordan block of A under discussion is the first block. Each Jordan block will be an upper diagonal matrix of the form

$$\begin{pmatrix} \lambda^k E_\beta(k) & \binom{k}{1}\lambda^{k-1} E_\beta(k-1) & \ldots & \ldots & \binom{k}{m-1}\lambda^{k-m+1} E_\beta(k-m+1) \\ & \ddots & & \ddots & \vdots \\ & & & \ddots & \binom{k}{1}\lambda^{k-1} E_\beta(k-1) \\ (0) & & & & \lambda^k E_\beta(k) \end{pmatrix},$$

where $E_\beta = 1$ if this block corresponds to a real eigenvalue λ, and E_β is a rotation if it belongs to a complex pair $\lambda e^{\pm i\beta}$ of eigenvalues. By a change of basis, we can always simplify this block to

$$\begin{pmatrix} \lambda^k E_\beta(k) & \binom{k}{1}\lambda^{k} E_\beta(k) & \ldots & \ldots & \binom{k}{m-1}\lambda^{k} E_\beta(k) \\ & \ddots & & \ddots & \vdots \\ & & & \ddots & \binom{k}{1}\lambda^{k} E_\beta(k) \\ (0) & & & & \lambda^k E_\beta(k) \end{pmatrix}. \tag{10.5}$$

Now if A is not orthogonal, then at least one of the following cases will be true.

Case 1: A has a real eigenvalue λ with $|\lambda| \neq 1$. Replacing A by A^{-1} if necessary we may assume that $|\lambda| > 1$. A corresponding block of A^k is an $m \times m$ upper diagonal matrix of the form (10.5) with $E_\beta = 1$, and one easily checks that

$$S = \{\, s\mathbf{v}_1 : 1 \leq |s| < |\lambda| \,\} \times \operatorname{span}(\mathbf{v}_2, \ldots, \mathbf{v}_n)$$

is a cross-section.

Case 2: A has a complex pair of eigenvalues $\lambda e^{\pm i\beta}$ with $\lambda \neq 1$, $0 < \beta < \pi$. We may again assume that $\lambda > 1$. A corresponding block of A^k is a $2m \times 2m$ upper diagonal matrix of form (10.5) with E_β a proper rotation. Then

$$S = \{\, s\mathbf{v}_1 \lambda^t E_\beta(t) : 1 \le s < \lambda^{2\pi/\beta},\ 0 \le t < 1 \,\} \times \operatorname{span}(\mathbf{v}_3, \dots, \mathbf{v}_n)$$

is a cross-section, which can be checked by using Case 2 in Theorem 10.6 and keeping in mind the note immediately following Definition 10.4.

Case 3: A has a real eigenvalue $\lambda = \pm 1$ and at least one of the blocks of A belonging to this eigenvalue has a nontrivial nilpotent part. Then the corresponding block of A^k is of the form (10.5) with $E_\beta = 1$ and is of at least size 2×2. One easily verifies that the set

$$S = \{\, s(\mathbf{v}_1 + t\mathbf{v}_2) : s \in \mathbb{R}\backslash\{0\},\ 0 \le t < 1 \,\} \times \operatorname{span}(\mathbf{v}_3, \dots, \mathbf{v}_n)$$

is a cross-section.

Case 4: A has a complex pair of eigenvalues $e^{\pm i\beta}$, $0 < \beta < \pi$, of modulus one and at least one of the blocks of A belonging to this pair has a nontrivial nilpotent part. Then the corresponding block of A^k is of the form (10.5), with $\lambda = 1$ and E_β a proper rotation, and

$$S = \Big\{ (p\mathbf{v}_1 + q\mathbf{v}_3 + s\mathbf{v}_4) \begin{pmatrix} E_\beta(t) & tE_\beta(t) \\ 0 & E_\beta(t) \end{pmatrix} \\ : p > 0,\ 0 \le q < \frac{2\pi}{\beta} p,\ s \in \mathbb{R},\ 0 \le t < 1 \Big\} \times \operatorname{span}(\mathbf{v}_5, \dots, \mathbf{v}_n)$$

is the desired cross-section, which can be checked by using Case 4 in Theorem 10.6 and keeping in mind the note immediately following Definition 10.4.

The argument at the end of the proof of Theorem 10.6 shows that if A is orthogonal, then a cross-section cannot exist. This proves the first assertion.

The remaining assertions are obvious if $n = 1$, or if $n = 2$ and A has complex eigenvalues. We thus can exclude this situation in what follows, so that the cross-section S constructed above has infinite measure.

Next let us prove the second assertion. In order to show that $|\det(A)| \neq 1$ is a sufficient condition, we only need to distinguish between the first two of the above cases.

We begin by considering the first case, and we may assume that $|\lambda| > 1$. Take the cross-section constructed above,

$$S = \{\, s\mathbf{v}_1 + \mathbf{v} : 1 \le |s| < |\lambda|,\ \mathbf{v} \in \operatorname{span}(\mathbf{v}_2, \dots, \mathbf{v}_n) \,\},$$

partition $\operatorname{span}(\mathbf{v}_2, \dots, \mathbf{v}_n)$ into a collection $\{T_k\}_{k=1}^\infty$ of measurable sets of positive, finite measure each, and set

$$S_k = \{\, s\mathbf{v}_1 + \mathbf{v} : 1 \le |s| < |\lambda|,\ \mathbf{v} \in T_k \,\}, \qquad k = 1, 2, \ldots.$$

Then $\{S_k\}_{k=1}^{\infty}$ is a partition of S into measurable subsets of positive, finite measure. Pick a collection of positive numbers $\{d_k\}_{k=1}^{\infty}$ so that $\sum_{k=1}^{\infty} d_k = 1$, and pick $n_k \in \mathbb{Z}$ such that $\delta^{n_k} \le \frac{d_k}{m(S_k)}$ where $\delta = |\det(A)|$. It follows that

$$\tilde{S} := \bigcup_{k=1}^{\infty} S_k A^{n_k}$$

is a cross-section for the discrete action such that

$$m(\tilde{S}) = \sum_{k=1}^{\infty} \delta^{n_k} m(S_k) \le \sum_{k=1}^{\infty} d_k = 1.$$

In the second case, we may assume that $\lambda > 1$. Start with the above constructed cross-section,

$$S = \{\, s\mathbf{v}_1 \lambda^t E_\beta(t) + \mathbf{v} : 1 \le s < \lambda^{2\pi/\beta},\ 0 \le t < 1,\ \mathbf{v} \in \operatorname{span}(\mathbf{v}_3, \ldots, \mathbf{v}_n) \,\},$$

partition $\operatorname{span}(\mathbf{v}_3, \ldots, \mathbf{v}_n)$ into a collection $\{T_k\}_{k=1}^{\infty}$ of measurable subsets of finite, positive measure each, and set

$$S_k = \{\, s\mathbf{v}_1 \lambda^t E_\beta(t) + \mathbf{v} : 1 \le s < \lambda^{2\pi/\beta},\ 0 \le t < 1,\ \mathbf{v} \in T_k \,\}, \qquad k = 1, 2 \ldots$$

so that $\{S_k\}_{k=1}^{\infty}$ is a partition of S into measurable subsets of positive, finite measure. Continuing as in the first case, we have shown sufficiency.

To prove the necessity implication, suppose that there exists a cross-section P of finite measure for the discrete action and $|\det(A)| = 1$. Let S denote the cross-section for the discrete action constructed in the proof of the first assertion above. Then,

$$\begin{aligned}
m(P) &= \int_{\widehat{\mathbb{R}^n}} \chi_P(\gamma)\, d\gamma \\
&= \sum_{i\in\mathbb{Z}} \int_S \chi_P(\gamma A^i)\, d\gamma \\
&= \sum_{i\in\mathbb{Z}} \int_{\widehat{\mathbb{R}^n}} \chi_P(\gamma A^i)\, \chi_S(\gamma)\, d\gamma \\
&= \sum_{i\in\mathbb{Z}} \int_{\widehat{\mathbb{R}^n}} \chi_P(\gamma)\, \chi_S(\gamma A^{-i})\, d\gamma \\
&= \sum_{i\in\mathbb{Z}} \int_P \chi_S(\gamma A^{-i})\, d\gamma \\
&= \int_{\widehat{\mathbb{R}^n}} \chi_S(\gamma)\, d\gamma = m(S) = \infty,
\end{aligned}$$

which is impossible. Thus, there cannot exist a cross-section of finite measure.

Finally, we will prove the last assertion. For sufficiency, it is enough to assume that all eigenvalues of A have modulus $|\lambda| < 1$ so that

$$\lim_{k\to\infty} \|A^k\| = 0.$$

Choosing each of the above sets T_k to be bounded we may assume that the sets S_k are bounded, so that there exist integers n_k such that $S_k A^{n_k}$ is contained in the unit ball. Then $\tilde{S} = \cup_{k=1}^{\infty} S_k A^{n_k}$ is the desired bounded cross-section.

For necessity, suppose to the contrary that there exists a bounded cross-section $\tilde{S}$, but A has an eigenvalue $|\lambda_1| < 1$ and an eigenvalue $|\lambda_2| \geq 1$. (The case where $|\lambda_1| \leq 1$ and $|\lambda_2| > 1$ is treated similarly.) Using the block decomposition of A it is easy to see that for almost all $\gamma \in \widehat{\mathbb{R}^n}$, either

$$\lim_{|k|\to\infty} \|\gamma A^k\| = \infty$$

or, in the special case where no eigenvalue of A lies outside of the unit circle,

$$\lim_{k\to-\infty} \|\gamma A^k\| = \infty$$

while $\{\gamma A^k : k \geq 0\}$ is bounded below away from zero. Thus, for almost all $\gamma \in \widehat{\mathbb{R}^n}$ there exists a constant $M = M(\gamma)$ so that

$$\|\gamma A^k\| > M \qquad\qquad \forall k \in \mathbb{Z}.$$

Fix any such γ. Then for sufficiently large scalars c, the orbit of $c\gamma$ does not pass through $\tilde{S}$, contradicting the choice of $\tilde{S}$. □

We note that in the proof of the second assertion, the sets T_k can be chosen so that the cross-section $\tilde{S}$ has unit measure.

Remark 10.9. In [17], it was obtained as a corollary of their general work that

(a) For $A \in GL_n(\mathbb{R})$ and $D = \{A^k : k \in \mathbb{Z}\}$, there is a continuous wavelet if and only if $|\det(A)| \neq 1$.
(b) For $A = e^B$ and $D = \{A^t : t \in \mathbb{R}\}$, there is a continuous wavelet if and only if $|\det(A)| \neq 1$.

It is possible to recover these results using the ideas in this section. We mention only how to do so in the case where continuous wavelets exist. Let $A \in GL_n(\mathbb{R})$, and let S be a cross-section of Lebesgue measure 1 for the discrete action $\gamma \to \gamma A^k$. Then, the function ψ whose Fourier transform equals χ_S is a continuous wavelet for the group $\{A^k : k \in \mathbb{Z}\}$. If in addition, $A = e^B$, then ψ is also a continuous wavelet for the group $\{A^t : t \in \mathbb{R}\}$ since

$$\int_{\mathbb{R}} |\chi_S(\gamma A^t)|^2 \, dt = \int_0^1 \sum_{k\in\mathbb{Z}} |\chi_S(\gamma A^t A^k)|^2 \, dt = \int_0^1 1 \, dt = 1.$$

We note here that the method of proof in [17], while ostensibly constructive, does not easily yield cross-sections of a desirable form such as the ones constructed above.

10.3 Shift-Invariant Spaces and Discrete Wavelets

Let $A \in GL_n(\mathbb{R})$ and $\Gamma \subset \mathbb{R}^n$ be a full-rank lattice. An (A, Γ) orthonormal [resp. Parseval, Bessel] wavelet of order N is a collection of functions $\{\psi^i\}_{i=1}^N$ (where here we allow the possibility of $N = \infty$) such that

$$\{|\det A|^{j/2}\psi^i(A^j \cdot +k) : j \in \mathbb{Z},\, k \in \Gamma,\, i = 1, \ldots, N\}$$

is an orthonormal basis [resp. Parseval frame, Bessel system] for $L^2(\mathbb{R}^n)$. There has been much work done on determining for which pairs (A, Γ) orthonormal wavelets of finite order exist, often with extra desired properties such as fast decay in time or frequency.

This is not necessary for the proofs that we present. Of particular importance in determining when orthonormal wavelets exist are the minimally supported frequency (MSF) wavelets, which are intimately related to wavelet sets. An (A, Γ) multi-wavelet set K of order L is a set that can be partitioned into subsets $\{K_i\}_{i=1}^L$ such that $\left\{\frac{1}{|\det(B)|^{1/2}}\chi_{K_i}\right\}_{i=1}^L$ is the Fourier transform of an (A, Γ) orthonormal wavelet, where $\Gamma = B\mathbb{Z}^n$, where B is an invertible matrix. When the order of a multi-wavelet set is 1, we call it a wavelet set. These have been studied in detail in [1], [2], [3], [10], [14], [15], [19], [21]. The following fundamental question in this area remains open, even in the case $L = 1$.

Question 10.10. For which pairs (A, Γ) and orders L do there exist (A, Γ) wavelet sets of order L?

It is known that if A is expansive and Γ is any full-rank lattice, then there exists an (A, Γ) wavelet set of order 1 [10]. One can also modify the construction to obtain (A, Γ) wavelet sets of any finite order along the lines of Theorem 10.23 below. Diagonal matrices A for which there exist $(A, \mathbb{Z}^n)$ multi-wavelet sets of finite order were characterized in [20]. Theorem 10.8, part (b) above implies that, in order for an (A, Γ) multi-wavelet set of finite order to exist, it is necessary that A not have determinant one. There is currently no good conjecture as to what the condition on (A, Γ) should be for wavelet sets to exist. It is known that $|\det(A)| \neq 1$ is not sufficient and that all eigenvalues greater than or equal to 1 in modulus is not necessary.

We begin with the following.

Theorem 10.11. *Let $A \in GL_n(\mathbb{R})$ and $\Gamma \subset \mathbb{R}^n$ be a full-rank lattice with dual Γ^*. The set K is a multi-wavelet set of order L if and only if*

$$\sum_{\gamma \in \Gamma^*} \chi_K(\xi + \gamma) = L \quad a.e.\ \xi \in \widehat{\mathbb{R}^n}, \tag{10.6}$$

$$\sum_{j \in \mathbb{Z}} \chi_K(\xi A^j) = 1 \quad a.e.\ \xi \in \widehat{\mathbb{R}^n}. \tag{10.7}$$

Proof. The forward direction is very similar to the arguments presented in [10], so we sketch the proof only. Let K be a multi-wavelet set of order L. Partition K into $\{K_i\}_{i=1}^L$ such that $\frac{1}{|\det(B)|^{1/2}}\chi_{K_i}$ is an (A, Γ) multi-wavelet of order L. Then, since $\chi_{K_i}(\xi A^j)$ is orthogonal to $\chi_{K_k}(\xi A^l)$ for each $(i,j) \neq (k,l)$, it follows that $K_i A^j \cap K_k A^l$ is a null set when $(i,j) \neq (k,l)$. Therefore, $\sum_{j\in\mathbb{Z}} \chi_K(\xi A^j) \leq 1$ a.e. $\xi \in \widehat{\mathbb{R}^n}$. Moreover, since every L^2 function can be written as the combination of functions supported on $\cup_{j=1}^{\infty} KA^j$, it follows that $\sum_{j\in\mathbb{Z}} \chi_K(\xi A^j) = 1$ a.e. $\xi \in \widehat{\mathbb{R}^n}$, proving (10.7). To see (10.6), since K_i is disjoint from $K_j A^k$ for all $(j,k) \neq (i,0)$, it follows that $\{\frac{1}{|\det(B)|^{1/2}} e^{2\pi i\langle\xi,\gamma\rangle} : \gamma \in \Gamma\}$ must be an orthonormal basis for $L^2(K_i)$. This implies (10.6).

For the reverse direction, it is clear that what is needed is to partition K into $\{K_i\}_{i=1}^L$ so that each K_i satisfies $\sum_{\gamma\in\Gamma^*} \chi_{K_i}(\xi+\gamma) = 1$ a.e. $\xi \in \widehat{\mathbb{R}^n}$. This will follow from repeated application of the following fact. Given a measurable set K such that $\sum_{\gamma\in\Gamma^*} \chi_K(\xi+\gamma) \geq 1$ a.e. $\xi \in \widehat{\mathbb{R}^n}$, there exists a set $U = U(K) \subset K$ such that

$$\sum_{\gamma\in\Gamma^*} \chi_U(\xi+\gamma) = 1, \quad a.e.\ \xi \in \widehat{\mathbb{R}^n}. \tag{10.8}$$

Now, let $\{V_i\}_{i=1}^{\infty}$ be a partition of $\widehat{\mathbb{R}^n}$ consisting of fundamental regions of Γ^*; that is, the sets V_i satisfy $\sum_{\gamma\in\Gamma^*} \chi_{V_i}(\xi+\gamma) = 1$ a.e. $\xi \in \widehat{\mathbb{R}^n}$. For a set $M \subset \widehat{\mathbb{R}^n}$ we define $M^t = \cup_{\gamma\in\Gamma^*}(M+\gamma)$. Let

$$L_0 = K.$$

Let

$$K_1 = (V_1 \cap L_0) \cup \big(U(L_0) \setminus (V_1 \cap L_0)^t\big),$$

where $U(L_0)$ is the subset of L_0 satisfying (10.8). Let $L_1 = L_0 \setminus K_1$, and notice that L_1 satisfies (10.6) with the right-hand side reduced by 1. In general, let

$$K_i = (V_i \cap L_{i-1}) \cup \big(U(L_{i-1}) \setminus (V_i \cap L_{i-1})^t\big),$$

and

$$L_i = L_{i-1} \setminus K_i.$$

In the case where L is finite, this procedure will continue for L steps, resulting in a partition of K with the desired properties. In this case, the initial partition $\{V_i\}$ was not necessary. In the case $L = \infty$, since the V_i's partition $\widehat{\mathbb{R}^n}$, the union of the K_i's will contain K. Since the K_i's were constructed to be disjoint and to satisfy (10.8), the proof is complete. □

There is also a soft proof of the reverse direction of Theorem 10.11, that yields slightly less information about wavelets, but provides some interesting facts about shift-invariant spaces. Before turning to the applications of Theorem 10.11, we provide this second proof.

When L is finite, we call an (A,Γ) orthonormal wavelet $\{\psi^i\}_{i=1}^L$ an (A,Γ) *combined MSF wavelet* if $\cup_{i=1}^L \mathrm{supp}(\hat{\psi}^i)$ has minimal Lebesgue measure. This terminology was introduced in [6], where it was shown that the minimal Lebesgue measure is L. It was also shown that if $\{\psi^i\}_{i=1}^L$ is a combined MSF wavelet, then there is a multi-wavelet set K of order L such that $K = \cup_{i=1}^L \mathrm{supp}(\hat{\psi}^i)$.

When $L = \infty$, it is not clear what the significance is for the union of the supports of $\hat{\psi}^i$ to have minimal Lebesgue measure. For this reason, we adopt the following definition. An (A,Γ) orthonormal wavelet $\{\psi^i\}_{i=1}^L$ is an (A,Γ) combined MSF wavelet if $K = \cup_{i=1}^L \mathrm{supp}(\hat{\psi}^i)$ is a multi-wavelet set of order L. This definition agrees with the previous definition in the case L is finite.

Let us begin by recalling some of the basic notions of shift-invariant spaces. A closed subspace $V \subset L^2(\mathbb{R}^n)$ is called *shift-invariant* if whenever $f \in V$ and $k \in \mathbb{Z}^n$, $f(x+k) \in V$. The shift-invariant space generated by the collection of functions $\Phi \subset L^2(\mathbb{R}^n)$ is denoted by $\mathcal{S}(\Phi)$ and given by

$$\overline{\mathrm{span}}\{\phi(x+k) : k \in \mathbb{Z}^n, \phi \in \Phi\}.$$

Given a shift-invariant space, if there exists a finite set $\Phi \subset L^2(\mathbb{R}^n)$ such that $V = \mathcal{S}(\Phi)$, then we say that V is finitely generated. In the case Φ can be chosen to be a single function, we say V is a principal shift-invariant (PSI) space. For further basics about shift-invariant spaces, we recommend [7], [11], [12]. We will follow closely the development in [7].

Proposition 10.12. *The map $\mathcal{T} : L^2(\mathbb{R}^n) \to L^2(\mathbb{T}^n, \ell^2(\mathbb{Z}^n))$ defined by*

$$\mathcal{T} f(x) = (\hat{f}(x+k))_{k\in\mathbb{Z}^n}$$

is an isometric isomorphism between $L^2(\mathbb{R}^n)$ and $L^2(\mathbb{T}^n, \ell^2(\mathbb{Z}^n))$, where $\mathbb{T}^n = \mathbb{R}^n/\mathbb{Z}^n$ is identified with its fundamental domain, e.g., $[0,1)^n$.

In what follows, as in Proposition 10.12, we will always assume that $\mathbb{T}^n = \mathbb{R}^n/\mathbb{Z}^n$ is identified with $[0,1)^n$.

A *range function* is a mapping

$$J : \mathbb{T}^n \to \{E \subset \ell^2(\mathbb{Z}^n) : E \text{ is a closed linear subspace}\}.$$

The function J is measurable if the associated orthogonal projections $P(x) : \ell^2(\mathbb{Z}^n) \to J(x)$ are weakly operator measurable. With these preliminaries, we can state an important theorem in the theory of shift-invariant spaces, which is due to Helson. The exact form of the theorem that we will use is taken from [7, Prop. 1.5].

Theorem 10.13. *A closed subspace $V \subset L^2(\mathbb{R}^n)$ is shift-invariant if and only if*

$$V = \{f \in L^2(\mathbb{R}^n) : \mathcal{T} f(x) \in J(x) \text{ for a.e. } x \in \mathbb{T}^n\},$$

where J is a measurable range function. The correspondence between V and J is one-to-one under the convention that the range functions are identified if they are equal a.e. Furthermore, if $V = \mathcal{S}(\Phi)$ for some countable $\Phi \subset L^2(\mathbb{R}^n)$, then

$$J(x) = \overline{\operatorname{span}}\{\mathcal{T}\phi(x) : \phi \in \Phi\}.$$

Definition 10.14. *The* dimension function *of a shift-invariant space V is the mapping* $\dim V : \mathbb{T}^n \to \mathbb{N} \cup \{0, \infty\}$ *given by*

$$\dim V(x) = \dim J(x), \tag{10.9}$$

where J is the range function associated with V. The spectrum *of V is defined by $\sigma(V) = \{x \in \mathbb{T}^n : J(x) \neq \{0\}\}$.*

We are now ready to state the main result from [7] that we will need in this paper.

Theorem 10.15. *Suppose V is a shift-invariant subspace of $L^2(\mathbb{R}^n)$. Then V can be decomposed as an orthogonal sum*

$$V = \bigoplus_{i \in \mathbb{N}} \mathcal{S}(\phi_i), \tag{10.10}$$

where $\{\phi_i(x + k) : k \in \mathbb{Z}^n\}$ is a Parseval frame for $\mathcal{S}(\phi_i)$ and $\sigma(\mathcal{S}(\phi_{i+1})) \subset \sigma(\mathcal{S}(\phi_i))$ for all $i \in \mathbb{N}$. Moreover, $\dim \mathcal{S}(\phi_i)(x) = \|\mathcal{T}\phi_i(x)\| \in \{0, 1\}$ *for $i \in \mathbb{N}$, and*

$$\dim V(x) = \sum_{i \in \mathbb{N}} \|\mathcal{T}\phi_i(x)\| \quad \text{for a.e. } x \in \mathbb{T}^n. \tag{10.11}$$

Finally, there is a folkloric fact about dimension functions that we recall here. See Theorem 3.1 in [8] for a discussion and references.

Proposition 10.16. *Suppose V is a shift-invariant space such that there exists a set Φ such that*

$$\{\phi(\cdot + k) : k \in \mathbb{Z}^n, \phi \in \Phi\}$$

is a Parseval frame for V. Then

$$\dim V(\xi) = \sum_{\phi \in \Phi} \sum_{k \in \mathbb{Z}^n} |\hat{\phi}(\xi + k)|^2. \tag{10.12}$$

The following theorem is a relatively easy application of Theorem 10.15, which was certainly known in the case $N < \infty$, and probably known to experts in the theory of shift-invariant spaces in this full generality. It seems to be missing from the literature, so we include a proof.

Theorem 10.17. *Let V be a shift-invariant subspace of $L^2(\mathbb{R}^n)$. There exists a collection $\Phi = \{\phi_i\}_{i=1}^N \subset L^2(\mathbb{R}^n)$ such that*

$$\{\phi_i(x+k) : i \in \{1, \ldots, N\}, k \in \mathbb{Z}^n\}$$

is an orthonormal basis for V if and only if $\dim V(x) = N$ *a.e.* $x \in \mathbb{T}^n$.

Proof. For the forward direction, it suffices to show that if $\{\phi_i(x+k) : k \in \mathbb{Z}^n, i = 1, \ldots, N\}$ is an orthonormal basis for the (necessarily shift-invariant) space V, then $\dim V(x) = N$ for a.e. $x \in \mathbb{T}^n$. It is easy to see that if $\{f(x+k) : k \in \mathbb{Z}^n\}$ is an orthonormal sequence, then $\sum_{k \in \mathbb{Z}^n} |\hat{f}(\xi + k)|^2 = 1$ a.e. Thus, by Proposition 10.16, $\dim V(x) = N$ a.e.

For the reverse direction, assume V is a shift-invariant space satisfying $\dim V(x) = N$ a.e. $x \in \mathbb{T}^n$. Let $\{\phi_i\}_{i=1}^\infty$ be the collection of functions such that (10.10) is satisfied. Using the facts that $\sigma(\mathcal{S}(\phi_{i+1})) \subset \sigma(\mathcal{S}(\phi_i))$ for all i, $\sigma(V) = \mathbb{T}^n$ and (10.11), it follows that

$$\sigma(\mathcal{S}(\phi_i)) = \begin{cases} \mathbb{T}^n & i \leq N, \\ 0 & i > N. \end{cases} \tag{10.13}$$

By Equation 10.10, we have for $1 \leq i \leq N$,

$$\dim_{\mathcal{S}(\phi_i)}(\xi) = 1 = \|\mathcal{T}\phi_i(\xi)\|^2 = \sum_{k \in \mathbb{Z}^n} |\hat{\phi}_i(\xi + k)|^2.$$

Thus, $\{\phi_i(x+k) : k \in \mathbb{Z}^n\}$ is an orthonormal basis for $\mathcal{S}(\phi_i)$. Since the spaces $\mathcal{S}(\phi_i)$ are orthogonal, $\{\phi_i(x+k) : i \in \{1, \ldots, N\}, k \in \mathbb{Z}^n\}$ is an orthonormal basis for V, as desired. □

We include a proof of the following proposition for completeness.

Proposition 10.18. *Let $V = \{f \in L^2(\mathbb{R}^n) : \operatorname{supp}(\hat{f}) \subset W\}$. Then, V is shift-invariant and* $\dim V(\xi) = \sum_{k \in \mathbb{Z}^n} \chi_W(\xi + k) = \#\{k \in \mathbb{Z}^n : \xi + k \in W\}$ *a.e.*

Proof. Clearly, V so defined is shift-invariant. Let $\{e_k : k \in \mathbb{Z}^n\}$ be the standard basis for $\ell^2(\mathbb{Z}^n)$, and let ψ_k be defined by $\hat{\psi}_k = \chi_{(\mathbb{T}^n + k) \cap W}$, again for $k \in \mathbb{Z}^n$. It is easy to see that $V = \mathcal{S}(\Psi)$, where $\Psi = \{\psi_k : k \in \mathbb{Z}^n\}$. Therefore, by Theorem 10.13, $J(\xi) = \overline{\operatorname{span}}\{\mathcal{T}\psi_k(\xi) : k \in \mathbb{Z}^n\} = \overline{\operatorname{span}}\{e_k : \xi + k \in W\}$. The result then follows from the definition of dimension function in (10.9). □

Corollary 10.19. *Let $A \in GL_n(\mathbb{R})$, and K be a measurable subset of $\widehat{\mathbb{R}^n}$. If*

$$\sum_{j \in \mathbb{Z}} \chi_K(\xi A^j) = 1 \quad \text{a.e. } \xi \text{ in } \widehat{\mathbb{R}^n}, \tag{10.14}$$

and

$$\sum_{k \in \mathbb{Z}^n} \chi_K(\xi + k) = N \quad \text{a.e. } \xi \text{ in } \widehat{\mathbb{R}^n},$$

then there is an $(A, \mathbb{Z}^n)$ orthonormal wavelet of order N with $\cup_{i=1}^N \operatorname{supp}(\hat{\psi}^i) = K$.

Proof. By Proposition 10.18 and Theorem 10.17, there exists $\Psi = \{\psi^i\}_{i=1}^N$ such that $\{M_k\hat{\psi}^i : k \in \mathbb{Z}^n,\ i = 1, \dots, N\}$ is an orthonormal basis for $L^2(K)$, where M_k denotes modulation by k. Thus, by (10.14), Ψ is an $(A, \mathbb{Z}^n)$ wavelet. □

The main theorem in this section is given in Theorem 10.23. Before stating this theorem, we give three results that will be useful in its proof.

Lemma 10.20. *Let $C \subset \mathbb{R}^n$ be a cone with nonempty interior, $\Gamma \subset \mathbb{R}^n$ be a full-rank lattice, and $T \in \mathbb{N}$. Then, the cardinality of $C \cap \Gamma \cap (\mathbb{R}^n \setminus B_T(0))$ is infinity.*

Proof. Let l be a line through the origin contained in the interior of C. The set $U = \{x \in \mathbb{R}^n : \mathrm{dist}(x, l) < \varepsilon\}$ is a centrally symmetric convex set, and

$$((C \cap B_T(0)) \setminus U) \text{ is bounded.} \tag{10.15}$$

By Minkowski's Theorem (see, for example Theorem 1, Chapter 2, Section 7 in [18] and discussion thereafter), the cardinality of $U \cap \Gamma$ is infinity. Hence, by (10.15), the result follows. □

The following proposition was proven in the setting of wave packets in $L^2(\mathbb{R})$ in [9]. We sketch the proof here in our setting of wavelets.

Proposition 10.21. *Suppose $A \in GL_n(\mathbb{R})$ has the following property: for all $Z \subset \widehat{\mathbb{R}^n}$ with positive measure and all $q \in \mathbb{N}$, there exist $x_1, \dots, x_q \in \mathbb{Z}$ such that*

$$m\left(\bigcap_{i=1}^q ZA^{x_i}\right) > 0.$$

Then, for every nonzero $\psi \in L^2(\mathbb{R}^n)$, ψ is not an $(A, \mathbb{Z}^n)$ Bessel wavelet.

Proof. Let $\psi \in L^2(\mathbb{R}^n)$, $\psi \neq 0$. Then there exists a set $Z \subset \widehat{\mathbb{R}^n}$ of positive measure such that $|\hat{\psi}(\xi)| \geq C > 0$ for all $\xi \in Z$. By reducing to a subset, we may assume that there exists a constant $K > 0$ such that, for every function $f \in L^2(\widehat{\mathbb{R}^n})$ with support in Z, we have

$$\sum_{k \in \mathbb{Z}^n} |\langle f, M_k\hat{\psi}\rangle|^2 \geq K\|f\|^2.$$

Since the operator $Df = |\det(A)|^{1/2} f(\cdot A)$ is unitary, for every $j \in \mathbb{Z}$ and for each function $f \in L^2(\widehat{\mathbb{R}^n})$ supported in $A^{-j}(Z)$, we obtain

$$\sum_{k \in \mathbb{Z}^n} |\langle f, D^j M_k\hat{\psi}\rangle|^2 \geq K\|f\|^2. \tag{10.16}$$

By hypothesis, there exist $x_1, \dots, x_q \in \mathbb{Z}$ such that for $U := \left(\cap_{i=1}^q ZA^{x_i}\right)$, we have $m(U) > 0$. This implies

$$\begin{aligned}\sum_{j\in\mathbb{Z},k\in\mathbb{Z}^n} |\langle \chi_U, D^j M_k \hat{\psi}\rangle|^2 &\geq \sum_{i=1}^{q} \sum_{k\in\mathbb{Z}^n} |\langle \chi_U, D^{x_i} M_k \hat{\psi}\rangle|^2 \\ &\geq \sum_{i=1}^{q} K \|\chi_U\|^2 \\ &= qK \|\chi_U\|^2.\end{aligned}$$

Thus, since q is arbitrary, ψ is not an $(A, \mathbb{Z}^n)$ Bessel wavelet. □

Theorem 10.22 (Bonferroni's Inequality). *If $\{A_i\}_{i=1}^N$ are measurable subsets of the measurable set B and k is a positive integer such that*

$$\sum_{i=1}^{N} |A_i| > k|B|,$$

then there exist $1 \leq i_1 < i_2 < \cdots < i_{k+1} \leq N$ such that

$$\left|\bigcap_{j=1}^{k} A_{i_j}\right| > 0.$$

Theorem 10.23. *Let $A \in GL_n(\mathbb{R})$ with real Jordan form J. The following statements are equivalent.*

(a) *For every full-rank lattice $\Gamma \subset \mathbb{R}^n$, there exists a (J, Γ) orthonormal wavelet of order ∞.*

(b) *There exists an $(A, \mathbb{Z}^n)$ orthonormal wavelet of order ∞.*

(c) *For every full-rank lattice $\Gamma \subset \mathbb{R}^n$, there exists an (A, Γ) orthonormal wavelet of order ∞.*

(d) *There exists a (nonzero) $(A, \mathbb{Z}^n)$ Bessel wavelet of order 1.*

(e) *J is not orthogonal.*

(f) *The matrix A is not similar (over $M_n(\mathbb{C})$) to a unitary matrix.*

Proof. Let us begin by summarizing the known results and obvious implications. The implication (b) $\Rightarrow$ (f) was proved in [16, Theorem 4.2]. The implications (c) $\Rightarrow$ (b) $\Rightarrow$ (d) are obvious, and (e) $\Leftrightarrow$ (f) is standard.

(e) $\Rightarrow$ (a). Let Γ be a full-rank lattice with convex fundamental region Y for Γ^*. By Theorem 10.11, it suffices to show that there is a measurable cross-section S for the discrete action $\xi \to \xi J^k$ satisfying (10.7). As in Theorems 10.6 and 10.8, we break the analysis into cases.

Case 1: There is an eigenvalue of J not equal to 1 in modulus. Without loss of generality, we assume that there is an eigenvalue of modulus greater than 1. In this case, J can be written as a block diagonal matrix

$$\begin{pmatrix} J_1 & 0 \\ 0 & J_2 \end{pmatrix}, \tag{10.17}$$

where J_1 is expansive, and we allow the possibility that $\operatorname{rank}(J_1) = \operatorname{rank}(J)$. Let S be an open cross-section for the discrete action $\xi \to \xi J_1^k$. Partition S into disjoint open subsets $\{S_i : i \in \mathbb{N}\}$. For each i, choose k_i such that there exists $\gamma_i \in \Gamma^*$ such that $S_i A^{k_i} \times \mathbb{R}^{\operatorname{rank}(J_2)} \supset (Y + \gamma_i)$. Then,

$$\bigcup_{i=1}^{\infty} (S_i A^{k_i} \times \mathbb{R}^{\operatorname{rank}(J_2)})$$

is a cross-section satisfying (10.7).

Case 2: All eigenvalues of J have modulus 1. This means that we are in Case 3 or Case 4 of Theorem 10.8. We show that in either of these cases, the cross-section exhibited in Theorem 10.8 satisfies (10.7). First, note that S in these cases is a cone of infinite measure with a dense, open subset S°. Let $B \subset S^\circ$ be an open ball bounded away from the origin satisfying $\overline{B} \subset S^\circ$. Let $\delta = \operatorname{diam}(Y)$. There exists a T such that

$$S_T := \{tb : t \geq T,\, b \in B\}$$

satisfies $\operatorname{dist}(S_T, \mathbb{R}^n \setminus S) > \delta$. By Lemma 10.20, $\Gamma^* \cap S_T$ has infinite cardinality, and by choice of δ, $Y + \gamma \subset S$ for each $\gamma \in \Gamma^* \cap S_T$. Therefore, S is a cross-section satisfying (10.7).

(a) $\Rightarrow$ (b) $\Rightarrow$ (c). This follows from the following two facts. First, Γ is a full-rank lattice if and only if there is an invertible matrix B such that $\Gamma = B\mathbb{Z}^n$. Second, if $B \in GL_n(\mathbb{R})$, Ψ is an (A, Γ) orthonormal wavelet if and only if $\Psi_B := \{\frac{1}{|\det(B)|^{1/2}} \psi(B^{-1}\cdot) : \psi \in \Psi\}$ is a $(BAB^{-1}, B\Gamma)$ orthonormal wavelet. Indeed, (a) $\Rightarrow$ (b) is then immediate.

To see (b) $\Rightarrow$ (c), recall that (b) $\Rightarrow$ (f). Thus, if (b) is satisfied, then $J = B^{-1}AB$ is not orthogonal. Let Γ be a full-rank lattice. There exists a $(J, B^{-1}\Gamma)$ orthonormal wavelet of order ∞, so there exists an (A, Γ) orthonormal wavelet of order ∞.

(d) $\Rightarrow$ (f). Suppose that the real Jordan form of A is orthogonal. Then, for any bounded set $Z \subset \widehat{\mathbb{R}^n}$, there exists M such that for every $k \in \mathbb{Z}, z \in Z$, we have $\|zA^k\| \leq M$. Furthermore, if Z has positive measure, then

$$\sum_{k \in \mathbb{Z}} m(ZA^k \cap B_M(0)) = \infty.$$

Therefore, by Bonferroni's inequality, for every $q \in \mathbb{N}$, there exist $k_1, \dots k_q$ such that

$$m\left(\bigcap_{j=1}^{q} ZA^{k_j}\right) > 0.$$

By Proposition 10.21, this says that for every nonzero ψ, ψ is not an $(A, \mathbb{Z}^n)$ Bessel wavelet. $\square$

Acknowledgments

The first author was partially supported by NSF grant DMS-0139386. The second author was supported by a research grant from Suranaree University of Technology.

References

1. L. Baggett, H. Medina, and K. Merrill, Generalized multiresolution analyses, and a construction procedure for all wavelet sets in $\mathbf{R}^n$, *J. Fourier Anal. Appl.*, **6** (1999), pp. 563–573.
2. J. J. Benedetto and M. Leon, The construction of single wavelets in D-dimensions, *J. Geom. Anal.*, **11** (2001), pp. 1–15.
3. J. J. Benedetto and M. T. Leon, The construction of multiple dyadic minimally supported frequency wavelets on $\mathbf{R}^d$, in: *The Functional and Harmonic Analysis of Wavelets and Frames* (San Antonio, TX, 1999), Contemp. Math., Vol. 247, Amer. Math. Soc., Providence, RI, 1999, pp. 43–74.
4. J. Benedetto and S. Li, The theory of multiresolution analysis frames and applications to filter banks, *Appl. Comput. Harmon. Anal.*, **5** (1998), pp. 389–427.
5. J. Benedetto and S. Sumetkijakan, A fractal set constructed from a class of wavelet sets, in: *Inverse Problems, Image Analysis, and Medical Imaging* (New Orleans, LA, 2001), Contemp. Math., Vol. 313, Amer. Math. Soc., Providence, RI, 2002, pp. 19–35.
6. M. Bownik, Combined MSF multiwavelets, *J. Fourier Anal. Appl.* **8** (2002), no. 2, pp. 201–210.
7. M. Bownik, The structure of shift-invariant subspaces of $L^2(\mathbb{R}^n)$, *J. Funct. Anal.*, **177** (2000), pp. 282–309.
8. M. Bownik and D. Speegle, The dimension function for real dilations and dilations admitting non-MSF wavelets, in: *Approximation Theory X* (St. Louis, MO 2001), Vanderbilt Univ. Press, Nashville, TN, 2002, pp. 63–85.
9. W. Czaja, G. Kutyniok, and D. Speegle, The geometry of sets of parameters of wave packet frames, *Appl. Comput. Harmon. Anal.*, **20** (2006), pp. 108–125.
10. X. Dai, D. R. Larson, and D. M. Speegle, Wavelet sets in $\mathbf{R}^n$, *J. Fourier Anal. Appl.*, **3** (1997), pp. 451–456.
11. C. de Boor, R. DeVore, and A. Ron, The structure of finitely generated shift-invariant spaces in $L_2(\mathbf{R}^d)$, *J. Funct. Anal.*, **119** (1994), pp. 37–78.
12. H. Helson, *Lectures on Invariant Subspaces*, Academic Press, New York–London, 1964.
13. E. Hernández, D. Labate, and G. Weiss, A unified characterization of reproducing systems generated by a finite family, II, *J. Geom. Anal*, **12** (2002), pp. 615–662.
14. E. Hernández, X. Wang, and G. Weiss, Smoothing minimally supported frequency wavelets, I, *J. Fourier Anal. Appl.*, **2** (1996), pp. 329–340.
15. E. Hernández, X. Wang, and G. Weiss, Smoothing minimally supported frequency wavelets, II, *J. Fourier Anal. Appl.*, **3** (1997), pp. 23–41.
16. E. Ionascu, D. Larson, and C. Pearcy, On the unitary systems affiliated with orthonormal wavelet theory in n-dimensions, *J. Funct. Anal.*, **157** (1998), pp. 413–431.

17. R. Laugesen, N. Weaver, G. Weiss, and N. Wilson, A characterization of the higher dimensional groups associated with continuous wavelets, *J. Geom. Anal.*, **12** (2002), pp. 89–102.
18. C. Lekkerkerker, *Geometry of Numbers*, Bibliotecha Mathematica, Vol. VIII, Wolters-Noordhoff Publishing, Groningen; North-Holland Publishing Co., Amsterdam–London, 1969.
19. G. Olafsson and D. Speegle, Wavelets, wavelet sets and linear actions on $\mathbb{R}^n$, in: *Wavelets, Frames and Operator Theory* (College Park, MD, 2003), Contemp. Math., Vol. 345, Amer. Math. Soc., Providence, RI, pp. 253–282.
20. D. Speegle, On the existence of wavelets for non-expansive dilation matrices, *Collect. Math.*, **54** (2003), pp. 163–179.
21. Y. Wang, Wavelets, tiling, and spectral sets, *Duke Math. J.* **114** (2002), pp. 43–57.
22. G. Weiss and E. Wilson, The mathematical theory of wavelets, in: *Twentieth Century Harmonic Analysis—A Celebration* (Il Ciocco, 2000), NATO Sci. Ser. II Math. Phys. Chem., Vol. 33, Kluwer Acad. Publ., Dordrecht, 2001, pp. 329–366.

11

The Theory of Wavelets with Composite Dilations

Kanghui Guo[1], Demetrio Labate[2], Wang–Q Lim[3], Guido Weiss[4], and Edward Wilson[5]

[1] Department of Mathematics, Southwest Missouri State University, Springfield, MO 65804, USA
kag026f@smsu.edu
[2] Department of Mathematics, North Carolina State University, Raleigh, NC 27695, USA
dlabate@math.ncsu.edu
[3] Department of Mathematics, Washington University, St. Louis, MO 63130, USA
wangQ@math.wustl.edu
[4] Department of Mathematics, Washington University, St. Louis, MO 63130, USA
guido@math.wustl.edu
[5] Department of Mathematics, Washington University, St. Louis, MO 63130, USA
enwilson@math.wustl.edu

Summary. A wavelet with composite dilations is a function generating an orthonormal basis or a Parseval frame for $L^2(\mathbb{R}^n)$ under the action of lattice translations and dilations by products of elements drawn from non-commuting sets of matrices A and B. Typically, the members of B are matrices whose eigenvalues have magnitude one, while the members of A are matrices expanding on a proper subspace of $\mathbb{R}^n$. The theory of these systems generalizes the classical theory of wavelets and provides a simple and flexible framework for the construction of orthonormal bases and related systems that exhibit a number of geometric features of great potential in applications. For example, composite wavelets have the ability to produce "long and narrow" window functions, with various orientations, well-suited to applications in image processing.

Dedicated to John J. Benedetto.

11.1 Introduction

We assume familiarity with the basic properties of separable Hilbert spaces. $L^2(\mathbb{R}^n)$, the space of all square integrable functions on $\mathbb{R}^n$, and $L^2(\mathbb{T}^n)$, the space of all square integrable $\mathbb{Z}^n$-periodic functions, are such spaces.

The construction and the study of orthonormal bases and similar collections of functions are of major importance in several areas of mathematics and applications, and have been a very active area of research in the last few decades.

Historically, there are two basic methods for constructing orthonormal bases of $L^2(\mathbb{R}^n)$. The most elementary approach is the following. Let $g = \chi_{[0,1)}$ and

$$\mathcal{G}(g) = \{e^{2\pi ikx}\, g(x-m) : k, m \in \mathbb{Z}\}.$$

It is easy to see that $\mathcal{G}(g)$ is an orthonormal basis for $L^2(\mathbb{R})$. More generally, let T_y, $y \in \mathbb{R}^n$, be the *translation operator*, defined by

$$T_y\, f(x) = f(x-y),$$

and M_ν, $\nu \in \mathbb{R}^n$, be the *modulation operator*, defined by

$$M_\nu\, f(x) = e^{2\pi i\nu x}\, f(x).$$

The *Gabor* or *Weyl–Heisenberg system* generated by $G = \{g_1, \dots, g_L\} \subset L^2(\mathbb{R}^n)$ is the family of the form

$$\mathcal{G}(G) = \{M_{bm}\, T_k\, g_\ell : k, m \in \mathbb{Z}^n,\ \ell = 1, \dots, L\}, \tag{11.1}$$

where $b \in GL_n(\mathbb{R})$. Then the basic question is: what are the sets of functions $G \subset L^2(\mathbb{R}^n)$ such that $\mathcal{G}(G)$ is an orthonormal basis for $L^2(\mathbb{R}^n)$? It turns out that one can construct several examples of such sets of functions G. However, a fundamental result in the theory of Gabor systems—the *Balian–Low Theorem*—shows that such functions are not very well behaved. In fact, if $G = \{g\}$, then g cannot have fast decay both in $\mathbb{R}^n$ and in $\widehat{\mathbb{R}}^n$; if $G = \{g_1, \dots, g_L\}$, then at least one function g_ℓ cannot have fast decay both in $\mathbb{R}^n$ and in $\widehat{\mathbb{R}}^n$ (see [1] for a nice overview of the Balian–Low Theorem, and [20] for the multiwindow case).

As we mentioned before, there is a second approach for constructing orthonormal bases of $L^2(\mathbb{R}^n)$. Let $A = \{a^i : i \in \mathbb{Z}\}$, where $a \in GL_n(\mathbb{R})$, and let us replace the modulation operator in (11.1) with the *dilation operator* D_a, defined by

$$D_a\, f(x) = |\det a|^{-1/2}\, f(a^{-1}x).$$

By doing this, we obtain the *affine* or *wavelet systems* generated by $\Psi = \{\psi_1, \dots, \psi_L\} \subset L^2(\mathbb{R}^n)$, which are the systems of the form

$$\mathcal{A}_A(\Psi) = \{D_{a^i}\, T_k\, \psi_\ell : a \in A,\ i \in \mathbb{Z},\ \ell = 1, \dots, L\}.$$

If $\mathcal{A}_A(\Psi)$ is an orthonormal basis for $L^2(\mathbb{R}^n)$, then Ψ is called a *multiwavelet* or, simply, a *wavelet* if $\Psi = \{\psi\}$. Even though the first example of a wavelet, the Haar wavelet, was discovered in 1909, the theory of wavelets was actually born in the beginning of the 1980s. Again, as in the case of Gabor systems, the basic mathematical question is: what are the wavelets $\Psi \subset L^2(\mathbb{R}^n)$? It turns out that there are "many" such functions and that, in a certain sense, they are "more plentiful" than the functions $G \subset L^2(\mathbb{R}^n)$ generating an orthonormal Gabor system. This fact, together with the ability of such bases to exhibit a number of properties which are useful in applications, in part explains the considerable success of wavelets in mathematics and applications in the last twenty years. We refer to [13] and [5] for more details, in particular regarding the comparison between Gabor and affine systems.

In the last few years, there has been a considerable interest, both in the mathematical and engineering literature, in the study of variants of the affine systems which contain basis elements with many more locations, scales and directions than the "classical" wavelets (see the papers in [19] for examples of such systems). The motivation for this study comes partly from signal processing, where such bases are useful in image compression and feature extraction, and partly from the investigation of certain classes of singular integral operators. The main subject of this paper is the study of a new class of systems, which we call *affine systems with composite dilations* and which have the form

$$\mathcal{A}_{AB}(\Psi) = \{D_a D_b T_k \Psi : k \in \mathbb{Z}^n, \, a \in A, \, b \in B\},$$

where $A, \, B \subset GL_n(\mathbb{R})$. If $\mathcal{A}_{AB}(\Psi)$ is an orthonormal basis, then Ψ will be called a *composite* or *AB-multiwavelet* (as before we use the term wavelet rather than multiwavelet if $\Psi = \{\psi\}$). As we will show, the theory of these systems generalizes the classical theory of wavelets and provides a simple and flexible framework for the construction of orthonormal bases and related systems that exhibit a number of geometric features of great potential in applications. For example, one can construct composite wavelets with good time-frequency decay properties whose elements contain "long and narrow" waveforms with many locations, scales, shapes and directions [8]. These constructions have properties similar to those of the *curvelets* [2] and *contourlets* [6], which have been recently introduced in order to obtain efficient representations of natural images. The theory of affine systems with composite dilations is more general. In fact, the contourlets can be described as a special case of these systems (see [9]). In addition, our approach extends naturally to higher dimensions and allows a multiresolution construction which is well suited to a fast numerical implementation.

It is of interest to point out that there exist affine systems with composite dilations $\mathcal{A}_{AB}(\Psi)$ that are orthonormal bases (as well as Parseval frames) for $L^2(\mathbb{R}^n)$ when the dilation set A is not known to be associated with an affine system $\mathcal{A}_A(\Psi)$ that is an orthonormal basis (or a Parseval frame) for $L^2(\mathbb{R}^n)$. For example, when $A = \{a^i : i \in \mathbb{Z}\}$, where $a = \begin{pmatrix} \lambda_1 & 0 \\ 0 & \lambda_2 \end{pmatrix}$, with $|\lambda_1| > 1 > |\lambda_2| > 0$, then, by a result in [16, Prop. 2.2], the Calderòn equation $\sum_{\ell=1}^{L} \sum_{i \in \mathbb{Z}} |\hat{\psi}_\ell(\xi a^i)|^2 = 1$ a.e. fails. This equation is one of the "traditional" equalities that characterize wavelets (see [15], [14]), and it is a necessary condition for MSF wavelets to exist (the MSF wavelets are those wavelets ψ such that $\hat{\psi} = \chi_T$, for some measurable set T). The situation for general wavelets is unknown, but it seems unlikely that wavelets could be obtained without satisfying the Calderòn condition. However, we will show in Section 11.2.2 that there are orthonormal bases of the form $\mathcal{A}_{AB}(\Psi)$ for $B = \{b^j : j \in \mathbb{Z}\}$ where $b = \begin{pmatrix} 1 & 1 \\ 0 & 1 \end{pmatrix}$.

11.1.1 Reproducing Function Systems

Before examining in more detail the properties of the affine systems with composite dilations and their variants, let us make a few observations about the general properties of the collections that form an orthonormal basis for a Hilbert space. We have the following simple proposition.

Proposition 11.1. *Let $\mathcal{H}$ be a separable Hilbert space, $T : \mathcal{H} \to \mathcal{H}$ be unitary and $\Phi = \{\phi_1, \dots, \phi_N\}$, $\Psi = \{\psi_1, \dots, \psi_M\} \subset \mathcal{H}$, where $N, M \in \mathbb{N} \cup \{\infty\}$. Suppose that $\{T^j \phi_k : j \in \mathbb{Z}, 1 \le k \le N\}$ and $\{T^j \psi_i : j \in \mathbb{Z}, 1 \le i \le M\}$ are orthonormal bases for $\mathcal{H}$. Then $N = M$.*

Proof. It follows from the assumptions that, for each $1 \le k \le N$,

$$\|\phi_k\|^2 = \sum_{j \in \mathbb{Z}} \sum_{i=1}^{M} |\langle \phi_k, T^j \psi_i \rangle|^2.$$

Thus,

$$\begin{aligned} N = \sum_{k=1}^{N} \|\phi_k\|^2 &= \sum_{k=1}^{N} \sum_{j \in \mathbb{Z}} \sum_{i=1}^{M} |\langle \phi_k, T^j \psi_i \rangle|^2 \\ &= \sum_{i=1}^{M} \sum_{j \in \mathbb{Z}} \sum_{k=1}^{N} |\langle T^{-j} \phi_k, \psi_i \rangle|^2 \\ &= \sum_{i=1}^{M} \|\psi_i\|^2 = M. \quad \square \end{aligned}$$

Using this proposition, we can now show that *there are no orthonormal bases for $L^2(\mathbb{R}^n)$ generated using only dilates of a finite family of functions.* Indeed, arguing by contradiction, suppose that there are finitely many functions $\{\phi^1, \dots, \phi^N\} \subset L^2(\mathbb{R}^n)$ such that $\{D_2^j \phi^\ell : j \in \mathbb{Z}, 1 \le \ell \le N\}$ is an orthonormal basis for $L^2(\mathbb{R}^n)$. On the other hand, it is known that there exist wavelets $\psi \in L^2(\mathbb{R}^n)$ for which $\{D_2^j T_k \psi : j \in \mathbb{Z}, k \in \mathbb{Z}^n\}$ is an orthonormal basis for $L^2(\mathbb{R}^n)$. This is a contradiction since, by Proposition 11.1, in the first case $N < \infty$, while, in the second case $M = \infty$, and, thus, $M \ne N$. The same argument applies to more general dilation matrices $A \in GL_n(\mathbb{R})$ (namely, all those for which wavelets exist). By applying a similar argument using the Gabor systems rather than the wavelets, one shows that *there are no orthonormal bases for $L^2(\mathbb{R}^n)$ generated using only modulations or only translations of a finite family of functions.* These observations, which we deduce from Proposition 11.1, are special cases of deeper and more general results obtained from the study of the *density* of Gabor and affine systems. Indeed, using this more general approach, one obtains similar results holding not only for orthonormal bases but even for frames (see [3], [12]).

In many situations, the notion of orthonormal basis turns out to be too restrictive and one can consider more general collections of functions, called *frames*, that preserve, as we will now show, many of the properties of bases.

A countable family $\{e_j : j \in \mathcal{J}\}$ of elements in a separable Hilbert space $\mathcal{H}$ is a *frame* if there exist constants $0 < \alpha \leq \beta < \infty$ satisfying

$$\alpha \left\|v\right\|^2 \leq \sum_{j \in \mathcal{J}} |\langle v, e_j \rangle|^2 \leq \beta \left\|v\right\|^2$$

for all $v \in \mathcal{H}$. The constants α and β are called *lower* and *upper frame bounds*, respectively. If the right-hand side inequality holds but not necessarily the left-hand side, we say that $\{e_j : j \in \mathcal{J}\}$ is a *Bessel system* with constant β. A frame is *tight* if α and β can be chosen so that $\alpha = \beta$, and is a *Parseval frame* if $\alpha = \beta = 1$. Thus, if $\{e_j : j \in \mathcal{J}\}$ is a Parseval frame in $\mathcal{H}$, then

$$\|v\|^2 = \sum_{j \in \mathcal{J}} |\langle v, e_j \rangle|^2$$

for each $v \in \mathcal{H}$. This is equivalent to the reproducing formula

$$v = \sum_{j \in \mathcal{J}} \langle v, e_j \rangle \, e_j \tag{11.2}$$

for all $v \in \mathcal{H}$, where the series in (11.2) converges in the norm of $\mathcal{H}$. Equation (11.2) shows that a Parseval frame provides a basis-like representation. In general, however, the elements of a frame need not be independent and a frame or Parseval frame need not be a basis. The elements of a frame $\{e_j\}_{j \in \mathcal{J}}$ must satisfy $\|e_j\| \leq \sqrt{\beta}$ for all $j \in \mathcal{J}$, as can easily be seen from

$$\|e_j\|^4 = |\langle e_j, e_j \rangle|^2 \leq \sum_{i \in \mathrm{i}} |\langle e_j, e_i \rangle|^2 \leq \beta \, \|e_j\|^2.$$

In particular, if $\{e_j\}_{j \in \mathcal{J}}$ is a Parseval frame, then $\|e_j\| \leq 1$ for all $j \in \mathcal{J}$, and the frame is an orthonormal basis for $\mathcal{H}$ if and only if $\|e_j\| = 1$ for all $j \in \mathcal{J}$. We refer the reader to [7] or [15, Ch. 8] for more details about frames.

11.1.2 Notation

It will be useful to establish the notation and basic definitions that will be used in this paper. We adopt the convention that $x \in \mathbb{R}^n$ is a column vector, i.e.,

$x = \begin{pmatrix} x_1 \\ \vdots \\ x_n \end{pmatrix}$, and that $\xi \in \widehat{\mathbb{R}}^n$ is a row vector, i.e., $\xi = (\xi_1, \ldots, \xi_n)$. Similarly for the integers, $k \in \mathbb{Z}^n$ is the column vector $k = \begin{pmatrix} k_1 \\ \vdots \\ k_n \end{pmatrix}$, and $\hat{k} \in \widehat{\mathbb{Z}}^n$ is the row vector $\hat{k} = (\hat{k}_1, \ldots, \hat{k}_n)$. A vector x multiplying a matrix $a \in GL_n(\mathbb{R})$ on the right is understood to be a column vector, while a vector ξ multiplying a on the left is a row vector. Thus, $ax \in \mathbb{R}^n$ and $\xi a \in \widehat{\mathbb{R}}^n$. The Fourier transform is defined as

$$\hat{f}(\xi) = (\mathcal{F} f)(x) = \int_{\mathbb{R}^n} f(x)\, e^{-2\pi i \xi x}\, dx,$$

where $\xi \in \widehat{\mathbb{R}}^n$, and the inverse Fourier transform is

$$\check{f}(x) = (\mathcal{F}^{-1} f)(x) = \int_{\widehat{\mathbb{R}}^n} f(\xi)\, e^{2\pi i x \xi}\, d\xi.$$

For any $E \subset \widehat{\mathbb{R}}^n$, we denote by $L^2(E)^\vee$ the space $\{f \in L^2(\mathbb{R}^n) : \operatorname{supp} \hat{f} \subset E\}$.

11.2 Affine Systems

Let $A \subset GL_n(\mathbb{R})$ be a countable set and $\Psi = \{\psi_1, \ldots, \psi_L\} \subset L^2(\mathbb{R}^n)$. We introduce the following extension of the previously defined notion of affine systems. A collection of the form

$$\mathcal{A}_A(\Psi) = \{D_a T_k \Psi : a \in A,\, k \in \mathbb{Z}^n\}$$

is a (generalized) *affine system.* A special role is played by those functions Ψ for which the system $\mathcal{A}_A(\Psi)$ is an orthonormal basis or, more generally, a Parseval frame for $L^2(\mathbb{R}^n)$. In particular, Ψ is an *orthonormal A-multiwavelet* if the set $\mathcal{A}_A(\Psi)$ is an orthonormal basis for $L^2(\mathbb{R}^n)$ and is a *Parseval frame A-multiwavelet* if $\mathcal{A}_A(\Psi)$ is a Parseval frame for $L^2(\mathbb{R}^n)$. The set A is *admissible* if a Parseval frame A-wavelet exists.

It will also be useful to consider A-multiwavelets which are defined on subspaces of $L^2(\mathbb{R}^n)$ of the form $L^2(S)^\vee$, where $S \subset \widehat{\mathbb{R}}^n$ is a measurable set with positive Lebesgue measure. As we will show, they will play a major role in the construction of the composite wavelets that we mentioned in Section 11.1. If $\mathcal{A}_A(\Psi)$ is a Parseval frame (resp., an orthonormal basis) for $L^2(S)^\vee$, then $\Psi \in L^2(\mathbb{R}^n)$ is called a Parseval frame A-multiwavelet (resp., an orthonormal A-multiwavelet) for $L^2(S)^\vee$. If such a multiwavelet exists, the set A is *S-admissible.*

The affine systems $\mathcal{A}_A(\{\psi\})$ where $|\hat{\psi}| = \chi_T$ for some measurable set $T \subseteq \widehat{\mathbb{R}}^n$ are called *minimally supported in frequency (MSF)* systems, and the corresponding function ψ is called an *MSF wavelet* for $L^2(S)^\vee$ if ψ is a Parseval frame A-wavelet for $L^2(S)^\vee$. One can show (see [10], [16]) that ψ is a Parseval frame A-wavelet for $L^2(S)^\vee$ if and only if $\Omega_T = \bigcup_{\hat{k}\in\widehat{\mathbb{Z}}^n}(T+\hat{k})$ is a disjoint union, modulo null sets, and $S = \bigcup_{a\in A}(T\,a^{-1})$ is also a disjoint union, modulo null sets. In this case, we say that the set T is both a $\widehat{\mathbb{Z}}^n$*-tiling set* for Ω_T and a A^{-1}*-tiling set* for S. In particular, $|T| \le 1$ since T is contained in a $\widehat{\mathbb{Z}}^n$-tiling set for $\widehat{\mathbb{R}}^n$.

In the following, we will examine the admissibility condition for different dilation sets A. In Section 11.2.1, we will recall the situation of the classical wavelets, where $A = \{a^i : i \in \mathbb{Z}\}$, for some $a \in GL_n(\mathbb{R})$. Next, in Section 11.2.2, we will introduce a new family of admissible dilation sets, where the dilations have the form $AB = \{ab : a \in A, b \in B\}$, and $A, B \subset GL_n(\mathbb{R})$.

11.2.1 Admissibility Condition: The Classical Wavelets

We consider first the case where $A = \{a^i : i \in \mathbb{Z}\}$, for some $a \in GL_n(\mathbb{R})$. This is the situation one encounters in the classical wavelet theory.

Recall that a matrix $a \in GL_n(\mathbb{R})$ is *expanding* if each eigenvalue λ of a satisfies $|\lambda| > 1$. Dai, Larson and Speegle [4] have shown that if $a \in GL_n(\mathbb{R})$ is expanding, then the set $A = \{a^i : i \in \mathbb{Z}\}$ is admissible (observe that, in this case, the set A is also a group). In fact, they have shown that, under this assumption on a, one can find a set $T \subseteq [-1/2, 1/2]^n$ which is both a $\widehat{\mathbb{Z}}^n$-tiling set for Ω_T and an A-tiling set for $\widehat{\mathbb{R}}^n$. Thus, $\psi = (\chi_T)^\vee$ is a tiling A-wavelet for $L^2(\mathbb{R}^n)$. In addition, they construct a set T' for which $(\chi_{T'})^\vee$ is an orthonormal A-wavelet. More generally, Wang [17] has shown that if a set $A \subset GL_n(\mathbb{R})$ admits an A^{-1}-tiling set and A contains an expanding matrix m for which $mA \subseteq A$, then A admits an orthonormal A-tiling wavelet.

Until very recently, all affine systems considered in the literature concerned dilation matrices which are expanding. In [14], however, one finds examples of admissible dilation sets $A = \{a^i : i \in \mathbb{Z}\}$, where the matrix $a \in GL_n(\mathbb{R})$ is not expanding. In particular, a theorem in [14] gives a set of equations characterizing Parseval frame A-wavelets, for a large class of matrices a, where a is not necessarily expanding.

Furthermore, observe that "general" dilation sets may fail to be admissible. Consider, for example, the set $A = \{2^j\, 3^i : i, j \in \mathbb{Z}\}$. Then one can show that there are no Parseval frame A-wavelets for this dilation set. In fact, since $\log 2^j 3^i = j \log 3 + i \log 2$ and $\log 3/\log 2$ is an irrational number, it follows that each nontrivial orbit of A is dense in $\mathbb{R}$, and this implies that there are no Parseval frame A-wavelets (see [9]) for more details). However, the set $\tilde{A} = \left\{ \begin{pmatrix} 2^j & 0 \\ 0 & 3^i \end{pmatrix} : i, j \in \mathbb{Z} \right\}$ is admissible. In fact, it is easy to see that, for any $\psi(x_1, x_2) = \psi_1(x_1)\, \psi_2(x_2)$, where ψ_1 is a (one-dimensional) dyadic wavelet and ψ_2 is a (one-dimensional) triadic wavelet, then ψ is an $\tilde{A}$-wavelet.

11.2.2 Admissibility Condition: The Composite Wavelets

Let $S_0 \subset \widehat{\mathbb{R}}^n$ be a region centered at the origin, and let $B \subset GL_n(\mathbb{R})$ be a set of matrices mapping S_0 into itself. In many situations, as we will show in more detail later, one can find a subregion $U_0 \subset S_0$ which is a B^{-1}-tiling region for S_0. This implies that

$$\{D_b\, T_k\, (\chi_{U_0})^\vee : b \in B,\, k \in \mathbb{Z}^n\}$$

is a Parseval frame for $L^2(S_0)^\vee$.

Next, consider a second set of matrices $A = \{a^i : i \in \mathbb{Z}\}$, where $a \in GL_n(\mathbb{R})$. If a is expanding in some direction, i.e., *some* of the eigenvalues λ of a satisfy $|\lambda| > 1$, and satisfies some other simple conditions to be specified later, then we can construct a sequence of B-invariant regions $S_i = S_0\, a^i$, $i \in \mathbb{Z}$, with $S_i \subset S_{i+1}$ and $\lim_{i\to\infty} S_i = \widehat{\mathbb{R}}^n$. This enables us to decompose the space $L^2(\mathbb{R}^n)$ into an exhaustive *disjoint* union of closed subspaces:

$$L^2(\mathbb{R}^n) = \bigoplus_{i\in\mathbb{Z}} L^2(S_{i+1} \setminus S_i)^\vee. \tag{11.3}$$

In general, one can find a region $R \subset S_1 \setminus S_0$ which is a B^{-1}-tiling region for $S_1 \setminus S_0$. If S_0 is sufficiently small, this implies that

$$\{D_b\, T_k\, (\chi_R)^\vee : b \in B,\, k \in \mathbb{Z}^n\}$$

is a Parseval frame for $L^2(S_1 \setminus S_0)^\vee$, and, as a consequence, the system

$$\{D_a^i\, D_b\, T_k\, (\chi_R)^\vee : b \in B,\, i \in \mathbb{Z},\, k \in \mathbb{Z}^n\}$$

is a Parseval frame for $L^2(\mathbb{R}^n)$.

This shows that the set

$$AB = \{a^i\, b : i \in \mathbb{Z},\, b \in B\}$$

is admissible, and that ψ is a Parseval frame AB-wavelet whenever ψ is a Parseval frame B-wavelet for $L^2(S_1 \setminus S_0)^\vee$.

In the following subsections, we will present several examples of such dilation sets which are admissible, and construct several examples of AB-wavelets.

Finite Group B

The first situation we consider is the case when B is a finite group. Since B is conjugate to a subgroup of the orthogonal group $O_n(\mathbb{R})$ (i.e., given any finite group B, there is a $P \in GL_n(\mathbb{R})$ and a $\tilde{B} \in O_n(\mathbb{R})$ such that $PBP^{-1} = \tilde{B}$), without loss of generality, we may assume that $B \subset O_n(\mathbb{R})$. Let $S_0 \subset \widehat{\mathbb{R}}^n$ be a compact region, starlike with respect to the origin, with the property that B maps S_0 into itself. In many situations, one can find a lattice $L \subset \widehat{\mathbb{R}}^n$

and a region $U_0 \subseteq S_0$ such that U_0 is both a B-tiling region for S_0 and a Λ-packing region for $\widehat{\mathbb{R}}^n$ (i.e., $\sum_{\lambda\in\Lambda}\chi_{U_0}(\xi+\lambda)\leq 1$ for a.e. $\xi\in\widehat{\mathbb{R}}^n$), where $\Lambda=\{\lambda\in\widehat{\mathbb{R}}^n : \lambda l\in\mathbb{Z}, \forall\, l\in L\}$ is the lattice dual to L. Then

$$\Phi_B = \{D_b\, T_l\, (\chi_{U_0})^\vee : b\in B,\, l\in L\}$$

is a Parseval frame for $L^2(S_0)^\vee$ (this fact is well known and can be found, for example, in [11]). Next suppose that $A=\{a^i : i\in\mathbb{Z}\}$, where $a\in GL_n(\mathbb{R})$ is expanding, $a\,B\,a^{-1}=B$ and $S_0\subseteq S_0\,a=S_1$. These assumptions imply that each region $S_i=S_0\,a^i$, $i\in\mathbb{Z}$, is B-invariant and the family of disjoint regions $S_i\subset S_{i+1}$, $i\in\mathbb{Z}$, tile the plane $\widehat{\mathbb{R}}^n$. Thus, one can decompose $L^2(\mathbb{R}^n)$ according to (11.3). As in the general situation described above, when S_0 is sufficiently small, one can find a region $R\subset S_1\setminus S_0$ which is a B-tiling region for $S_1\setminus S_0$, so that

$$\Psi_{AB} = \{D_a^i\, D_b\, T_l\, (\chi_R)^\vee : b\in B,\, i\in\mathbb{Z},\, l\in L\}$$

is a Parseval frame for $L^2(\mathbb{R}^n)$. In addition, when U_0 is a Λ-tiling region for $\widehat{\mathbb{R}}^n$ and $|\det a|\in\mathbb{N}$, one can decompose the region R into a disjoint union of regions $R_1,\dots,R_N$ in such a way that

$$\widetilde{\Psi}_{AB} = \{D_a^i\, D_b\, T_l\, (\chi_{R_\ell})^\vee : i\in\mathbb{Z},\, b\in B,\, l\in L,\, \ell=1,\dots,N\}$$

is not only a Parseval frame but also an orthonormal AB-multiwavelet for $L^2(\mathbb{R}^n)$. Moreover, in this case, the set Φ_B is an orthonormal basis for $L^2(S_0)^\vee$.

We will present two examples to illustrate the general construction that we have outlined above.

Example 11.2. The first example is illustrated in Fig. 11.1. Let B be the 8-element group consisting of the isometries of the square $[-1,1]^2$. Namely, $B=\{\pm b_0,\pm b_1,\pm b_2,\pm b_3,\}$ where $b_0=\begin{pmatrix}1&0\\0&1\end{pmatrix}$, $b_1=\begin{pmatrix}0&1\\1&0\end{pmatrix}$, $b_2=\begin{pmatrix}0&1\\-1&0\end{pmatrix}$, $b_3=\begin{pmatrix}-1&0\\0&1\end{pmatrix}$. Let U be the parallelogram with vertices $(0,0),(1,0),(2,1)$ and $(1,1)$ and $S_0=\bigcup_{b\in B}U\,b$ (see the snowflake region in Fig. 11.1). It is easy to verify that S_0 is B-invariant.

Now let $a=\begin{pmatrix}1&1\\-1&1\end{pmatrix}$, and $S_i=S_0\,a^i$, $i\in\mathbb{Z}$. Observe that a is expanding, $a\,B\,a^{-1}=B$ and $S_0\subsetneq S_0\,a=S_1$. In particular, the region $S_1\setminus S_0$ is the disjoint union $\bigcup_{b\in B}R\,b$, where the region R is the parallelogram illustrated in Fig. 11.1. Also observe that R is a fundamental domain. Thus, as in the general situation that we have described before, it turns out that the set

$$\{D_a^i\, D_b\, T_k\,\psi : i\in\mathbb{Z},\, b\in B,\, k\in\mathbb{Z}^2\},$$

where $\hat{\psi}=\chi_R$, is an orthonormal basis for $L^2(\mathbb{R}^2)$.

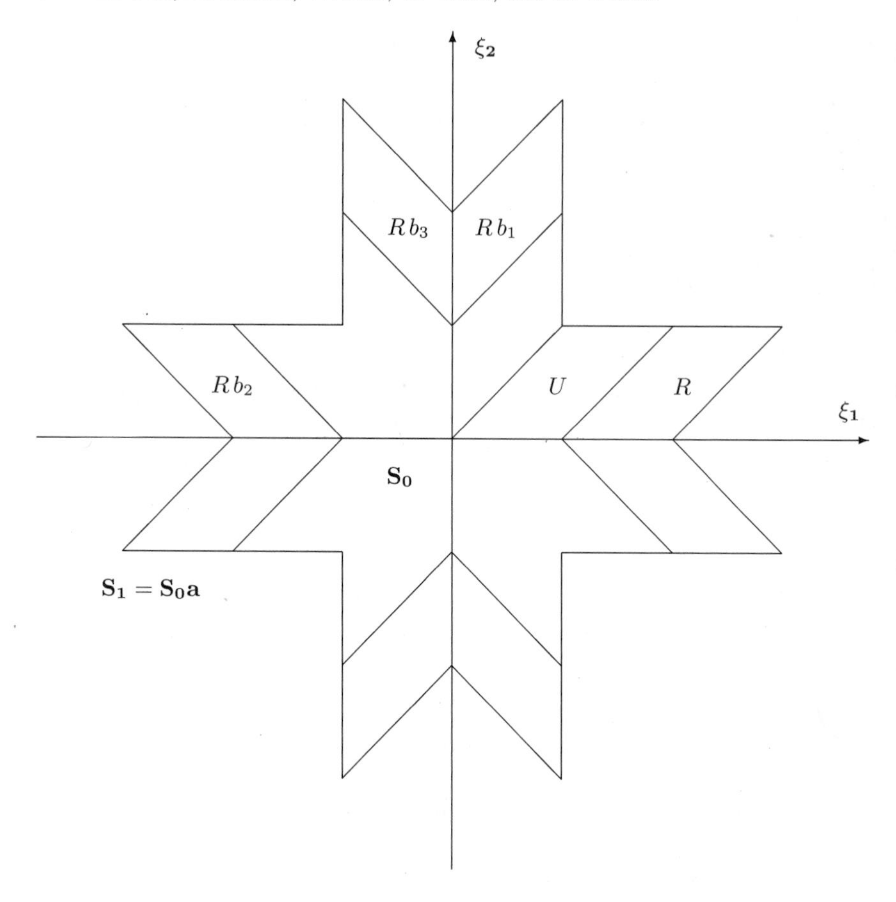

Fig. 11.1. Example 11.2. Admissible AB set; $A = \{a^i : i \in \mathbb{Z}\}$, where a is the quincunx matrix, and B is the group of isometries of the square $[-1,1]^2$.

Example 11.3. Next consider the situation where $a = \begin{pmatrix} 2 & 0 \\ 0 & 2 \end{pmatrix}$, and B is as in Example 11.2. Let U and S_i, $i \in \mathbb{Z}$, be defined as before. Also in this case, a is expanding, $a\,B\,a^{-1} = B$ and $S_1 = S_0 a \supset S_0$. A direct computation shows that the region $S_1 \setminus S_0$ is the disjoint union $\bigcup_{b \in B} Rb$, where $R = R_1 \bigcup R_2 \bigcup R_3$ and the regions R_1, R_2, R_3 are illustrated in Fig. 11.2. Observe that each of the regions R_1, R_2, R_3 is a fundamental domain. Thus, the system

$$\{D_a^i\, D_b\, T_k\, \psi^\ell : i \in \mathbb{Z},\, b \in B,\, k \in \mathbb{Z}^2,\, \ell = 1, \ldots, 3\},$$

where $\hat{\psi}^\ell = \chi_{R_\ell}$, $\ell = 1, 2, 3$, is an orthonormal basis for $L^2(\mathbb{R}^2)$.

The examples that we described are two-dimensional. It is clear, however, that not only are there many more examples in dimension $n = 2$, but also that

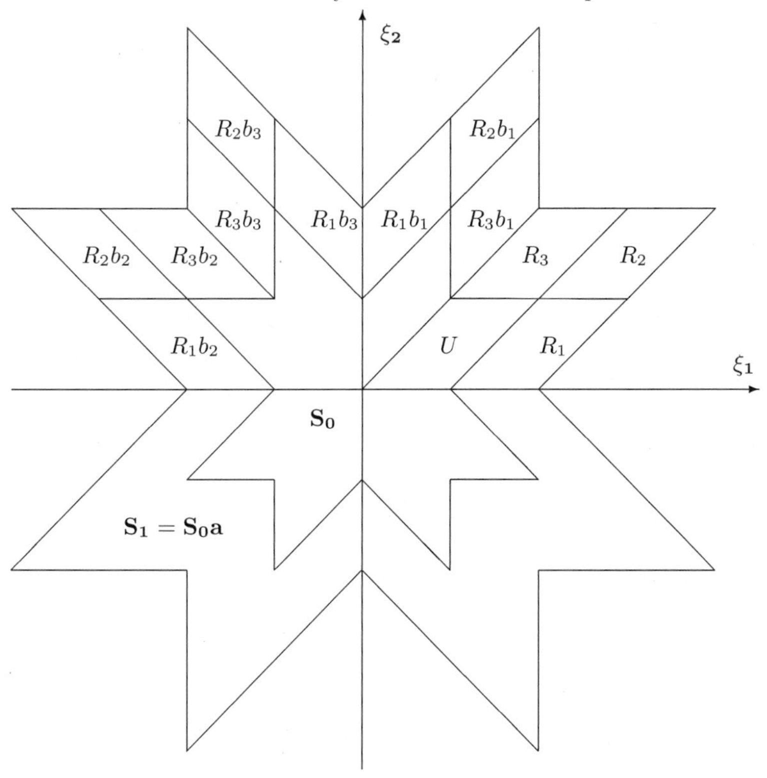

Fig. 11.2. Example 11.3. Admissible AB set; $A = \{a^i : i \in \mathbb{Z}\}$, where $a = 2I$, and B is the group of isometries of the square $[-1,1]^2$.

there are similar constructions for $n > 2$. For $n = 2$ or $n = 3$, in particular, every finite group of $O_n(\mathbb{R})$ acts by isometries on a regular polyhedron, and one can apply a construction very similar to the one described in these examples.

Shear Matrices B

We consider now the case of an infinite group. For simplicity, let us restrict ourselves, for the moment, to the two-dimensional situation and consider the group

$$B = \{b^j : j \in \mathbb{Z}\}, \qquad \text{where } b = \begin{pmatrix} 1 & 1 \\ 0 & 1 \end{pmatrix}.$$

Observe that $b^j = \begin{pmatrix} 1 & j \\ 0 & 1 \end{pmatrix}$. Since $(\xi_1, \xi_2)\, b^j = (\xi_1, \xi_2 + j\xi_1)$, the right action of the matrix $b^j \in B$ maps the line through the origin of slope m to the line through the origin of slope $m + j$. Observe that the matrices B are not expanding (all eigenvalues have magnitude one): they are called *shear matrices.* For $0 \le \alpha < \beta$, let $S(\alpha, \beta) = \{\xi = (\xi_1, \xi_2) \in \widehat{\mathbb{R}}^2 : \alpha \le |\xi_1| < \beta\}$ and let

$T(\alpha,\beta) = T^+(\alpha,\beta) \bigcup T^-(\alpha,\beta)$, where $T^+(\alpha,\beta)$ is the trapezoid with vertices $(\alpha,0)$, (α,α), $(\beta,0)$ and (β,β), and $T^-(\alpha,\beta) = \{\xi \in \widehat{\mathbb{R}}^2 : -\xi \in T^+(\alpha,\beta)\}$. A simple computation shows that $T(\alpha,\beta)$ is a B-tiling region for $S(\alpha,\beta)$, and, thus, for β sufficiently small, the function $\phi = (\chi_{T(\alpha,\beta)})^\vee$ is a Parseval frame B-wavelet for $L^2(S(\alpha,\beta))^\vee$. This shows that the set B is $S(\alpha,\beta)$-admissible.

Now fix $0 < c < 1$ and let $A = \{a^i : i \in \mathbb{Z}\}$, where $a = \begin{pmatrix} c^{-1} & a_{1,2} \\ 0 & a_{2,2} \end{pmatrix}$ and $a_{1,2}, a_{2,2} \in \mathbb{R}$, with $a_{2,2} \neq 0$. Let ψ be defined by $\hat{\psi} = \chi_{T(c,1)}$. Since $T(c,1)$ is a B-tiling region for $L^2(S(c,1))$ and a $\widehat{\mathbb{Z}}^n$-packing region for $\widehat{\mathbb{R}}^n$, it follows that ψ is a Parseval frame B-wavelet for $L^2(S(c,1))^\vee$. A direct computation shows that the sets $S(c,1)\,a^{-i}$, $i \in \mathbb{Z}$, tile $\widehat{\mathbb{R}}^2$. Hence

$$\Psi_{AB} = \{D_a^i\, D_b\, T_k\, \psi : i \in \mathbb{Z},\ b \in B,\ k \in \mathbb{Z}^2\}$$

is a Parseval frame for $L^2(\mathbb{R}^2)$. This shows that the set

$$AB = \{a^i\, b : i \in \mathbb{Z},\ b \in B\}$$

is admissible and ψ is a Parseval frame AB-wavelet for $L^2(\mathbb{R}^2)$. Observe that the Parseval frame AB-wavelet ψ depends only on $a_{11} = c^{-1}$ (on the other hand, the elements of the system Ψ_{AB}, when $i \neq 0$, do depend also on the other entries of the matrix a). Also observe that, if $|a_{22}| < 1$, no member of AB is an expanding matrix.

The following example shows how to obtain an orthonormal basis rather than a Parseval frame, by using a slightly different construction.

Example 11.4. Observe that, when $\alpha = 0$ and $\beta \in \mathbb{N}$, then $T(0,\beta)$ is the union of two triangles, which satisfies

$$\big(T^-(0,\beta) + (\beta, b)\big) \bigcup T^+(0,\beta) = [0,\beta)^2.$$

Let $\alpha = 0$ and $\beta = 1$. Then $U_0 = T(0,1)$ is a fundamental domain of $\mathbb{Z}^2$ and $\phi = (\chi_{U_0})^\vee$ is an orthonormal B-wavelet for $L^2(S(0,1))^\vee$ (see Fig. 11.3). Next let $A = \{a^i : i \in \mathbb{Z}\}$, where $a = \begin{pmatrix} 2 & a_{1,2} \\ 0 & a_{2,2} \end{pmatrix}$ and $a_{1,2}, a_{2,2} \in \mathbb{R}$, with $a_{2,2} \neq 0$. Similarly to the situation we described above, we have that the strip domains $S(1,2)\,a^{-i}$, $i \in \mathbb{Z}$, tile $\widehat{\mathbb{R}}^2$ and the set $T(1,2)$ is a B-tiling region for $L^2(S(1,2))$.

However, unlike the situation we described before, $T(1,2)$ is not a $\widehat{\mathbb{Z}}^n$-packing region for $\widehat{\mathbb{R}}^n$; in fact, its area is 3. On the other hand, we can split $T(1,2)$ into three regions of area one, which are fundamental domains of $\widehat{\mathbb{Z}}^2$. This can be done in several ways, for example, by introducing the trapezoids R_1, R_2, R_3 of Fig. 11.3, where $T(1,2) = R_1 \cup R_2 \cup R_3$. It then follows that the system

$$\Psi_{AB} = \{D_a^i\, D_b\, T_k\, \psi^\ell : i \in \mathbb{Z},\ b \in B,\ k \in \mathbb{Z}^2,\ \ell = 1, \ldots, 3\},$$

where $\hat{\psi}^\ell = \chi_{R_\ell}$, $\ell = 1, 2, 3$, is an orthonormal basis for $L^2(\mathbb{R}^2)$.

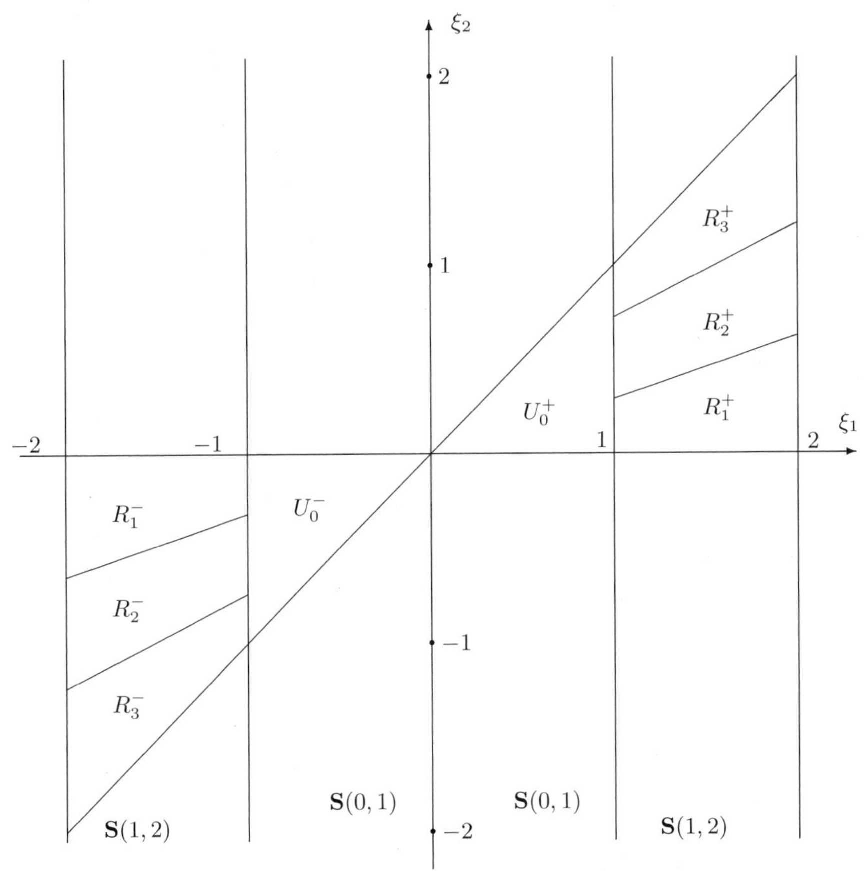

Fig. 11.3. Example 11.4. Admissible AB set; $A = \{a^i : i \in \mathbb{Z}\}$, where $a_{11} = 2$ and $a_{2,2} \neq 0$, and B is the group of shear matrices. In this figure, $U_0 = U_0^- \bigcup U_0^+$, $R_\ell = R_\ell^- \cup R_\ell^+$, $\ell = 1, 2, 3$, and $T(1, 2) = R_1 \cup R_2 \cup R_3$.

For $n > 2$, there are several generalizations of the constructions described in this section. For example, let $n = k + l$ and $B = \left\{ b_j = \begin{pmatrix} 1 & j \\ 0 & I_\ell \end{pmatrix} : j \in \mathbb{Z}^\ell \right\}$. Let $A = \{a^i : i \in \mathbb{Z}\}$, where $a = \begin{pmatrix} a_0 & a_1 \\ 0 & a_2 \end{pmatrix} \in GL_n(\mathbb{R})$, and $a_0 \in GL_k(\mathbb{R})$ is an expanding matrix. We then construct a 'trapezoidal' region T in $\widehat{\mathbb{R}}^n = \widehat{\mathbb{R}}^k \times \widehat{\mathbb{R}}^\ell$, which has for base the set $T_0 \subset \widehat{\mathbb{R}}^k$ and for which T is a B-tiling region for the strip $T_0 \times \widehat{\mathbb{R}}^\ell$. If T_0 is contained in sufficiently small neighborhood of the origin, we can ensure that the sets Sa^i, $i \in \mathbb{Z}$, tile $\widehat{\mathbb{R}}^n$ and thereby guarantee that $\psi = \chi_T^\vee$ is a Parseval frame AB-wavelet for $L^2(\mathbb{R}^n)$, where $AB = \{a^i b : i \in \mathbb{Z}, b \in B\}$ (thus, AB is admissible). In particular, the function $\psi = (\chi_R)^\vee$, where $R = \{(\xi_0, \ldots, \xi_\ell) \in \mathbb{R}^n : \frac{1}{2|a_0|} \leq |\xi_0| < \frac{1}{2}, 0 \leq$

$\xi_i/\xi_0 < 1, \text{ for } 1 \le i \le \ell\}$, is such a Parseval frame MRA AB wavelet. Observe that, as in the construction for $n = 2$, the set T and, thus, the wavelet ψ depends only on a_0 and B.

11.2.3 Further Remarks

Observe that, while all the examples of composite wavelets described in this section are of MSF type, that is, the magnitude of their Fourier transform is the characteristic function of a set, there are examples of composite wavelets that are not of this type. In [8], there are examples of composite wavelets $\psi \in L^2(\mathbb{R}^2)$ where $\hat{\psi}$ is in $C^\infty(\mathbb{R}^2)$. This implies that there is a $K_N > 0$ such that $|\psi(x)| \le K_N\,(1+|x|)^{-N}$, for any $N \in \mathbb{N}$ and, thus, ψ is well localized both in $\mathbb{R}^2$ and $\widehat{\mathbb{R}}^2$.

11.3 *AB*-Multiresolution Analysis

As in the classical theory of wavelets, it turns out that one can introduce a general framework based on a multiresolution analysis for constructing AB-wavelets. Let B be a countable subset of $\widetilde{SL}_n(\mathbb{Z}) = \{b \in GL_n(\mathbb{R}) : |\det b| = 1\}$ and $A = \{a^i : i \in \mathbb{Z}\}$, where $a \in GL_n(\mathbb{Z})$ (notice that a is an *integral* matrix). Also assume that *a normalizes B*, that is, $a\,b\,a^{-1} \in B$ for every $b \in B$, and that the quotient group $B/(aBa^{-1})$ has finite order. Then the sequence $\{V_i\}_{i\in\mathbb{Z}}$ of closed subspaces of $L^2(\mathbb{R}^n)$ is an *AB-multiresolution analysis (AB-MRA)* if the following holds:

(i) $D_b\,T_k\,V_0 = V_0$, *for any* $b \in B$, $k \in \mathbb{Z}^n$,
(ii) *for each* $i \in \mathbb{Z}$, $V_i \subset V_{i+1}$, *where* $V_i = D_a^{-i}\,V_0$,
(iii) $\bigcap V_i = \{0\}$ *and* $\overline{\bigcup V_i} = L^2(\mathbb{R}^n)$,
(iv) *there exists* $\phi \in L^2(\mathbb{R}^n)$ *such that* $\Phi_B = \{D_b\,T_k\,\phi : b \in B,\, k \in \mathbb{Z}^n\}$ *is a semi-orthogonal Parseval frame for* V_0*, that is,* Φ_B *is a Parseval frame for* V_0 *and, in addition,* $D_b\,T_k\,\phi \perp D_{b'}\,T_{k'}\,\phi$ *for any* $b \ne b'$, $b, b' \in B$, $k, k' \in \mathbb{Z}^n$.

The space V_0 is called an *AB scaling space* and the function ϕ is an *AB scaling function* for V_0. In addition, if Φ_B is an orthonormal basis for V_0, then ϕ is an *orthonormal AB scaling function.*

Observe that the main difference in the definition of AB-MRA with respect to the classical MRA is that, in the AB-MRA, the space V_0 is invariant with respect to the integer translations *and* with respect to the B-dilations. On the other hand, in the classical MRA, the space V_0 is only invariant with respect to the integer translation, and, as a consequence, the set of generators of V_0 is simply of the form $\Phi = \{T_k\,\phi : k \in \mathbb{Z}^n\}$.

As in the classical MRA, let W_0 be the orthogonal complement of V_0 in V_1, that is, $W_0 = V_1 \bigcap (V_o)^\perp$. Then, $V_1 = V_0 \bigoplus W_0$ and we have the following elementary result.

Theorem 11.5. (i) *Let $\Psi = \{\psi^1, \ldots, \psi^L\} \subset L^2(\mathbb{R}^n)$ be such that $\{D_b T_k \psi^\ell : b \in B, \ell = 1, \ldots, L, k \in \mathbb{Z}^n\}$ is a Parseval frame for W_0. Then Ψ is a Parseval frame AB-multiwavelet.*

(ii) *Let $\Psi = \{\psi^1, \ldots, \psi^L\} \subset L^2(\mathbb{R}^n)$ be such that $\{D_b T_k \psi^\ell : b \in B, k \in \mathbb{Z}^n, \ell = 1, \ldots, L\}$ is an orthonormal basis for W_0. Then Ψ is an orthonormal AB-multiwavelet.*

Proof. As in the classical MRA case, we define the spaces W_j as $W_j = V_{j+1} \bigcap (V_j)^\perp$, $j \in \mathbb{Z}$. It follows from the definition of AB-MRA that $L^2(\mathbb{R}^n) = \bigoplus_{j \in \mathbb{Z}} W_j$. Since $\{D_b T_k \psi^\ell : b \in B, \ell = 1, \ldots, L, k \in \mathbb{Z}^n\}$ is a Parseval frame for W_0, then $\{D_a^i D_b T_k \psi^\ell : b \in B, \ell = 1, \ldots, L, k \in \mathbb{Z}^n\}$ is a Parseval frame for W_i. Thus $\{D_a^i D_b T_k \psi^\ell : b \in B, i \in \mathbb{Z}, \ell = 1, \ldots, L, k \in \mathbb{Z}^n\}$ is a Parseval frame for $L^2(\mathbb{R}^n)$.

The proof for the orthonormal case is exactly the same. □

In the situation described by the hypotheses of Theorem 11.5 (where Ψ is not only a Parseval frame for $L^2(\mathbb{R}^n)$, but is also derived from an AB-MRA), we say that Ψ is a *Parseval frame MRA AB-wavelet* or an *orthonormal MRA AB-wavelet*, respectively.

It is easy to see that all the examples of wavelets with composite dilations described in Section 11.2 are indeed examples of MRA AB-multiwavelets. In Examples 11.2 and 11.3 from Section 11.2.2, the spaces $\{V_i = L^2(S_i)^\vee : i \in \mathbb{Z}\}$ form an AB-MRA, with orthonormal AB scaling function $\phi = (\chi_U)^\vee$, where U is a parallelogram (see Fig. 11.1 and Fig. 11.2). Observe that the AB scaling function ϕ as well as the AB scaling space $V_0 = L^2(S_0)^\vee$ are the same in both examples (the other spaces V_i, $i \neq 0$, are different because $S_i = S_0 a^i$, and a is different in the two examples). Similarly, in Example 11.4 from Section 11.2.2, the spaces $\{V_i = L^2(S_i)^\vee : i \in \mathbb{Z}\}$ form an AB-MRA, with orthonormal AB scaling function $\phi = (\chi_U)^\vee$, where U is the union of two triangles (see Fig. 11.3).

It turns out that, while it is possible to construct a Parseval frame AB-wavelet using a single generator $\Psi = \{\psi\}$ (examples of such singly generated AB-wavelets can be found in [8]), in the case of *orthonormal* MRA AB-multiwavelets, multiple generators are needed in general, that is, $\Psi = \{\psi^1, \ldots, \psi^L\}$, where $L > 1$. The following result establishes the number of generators needed to obtain an orthonormal MRA AB-wavelet.

Theorem 11.6. *Let $\Psi = \{\psi^1, \ldots, \psi^L\}$ be an orthonormal MRA AB-multiwavelet for $L^2(\mathbb{R}^n)$, and let $N = |B/aBa^{-1}|$ (= the order of the quotient group B/aBa^{-1}). Assume that $|\det a| \in \mathbb{N}$. Then $L = N\,|\det a| - 1$.*

Observe that this result generalizes the corresponding result for the classical MRA, where the number of generators needed to obtain an orthonormal MRA A-wavelet is given by $L = |\det a| - 1$ (cf., for example, [18]). As a simple application of Theorem 11.6, let us re-examine the examples described in Section 11.2. Also observe that, if B is a finite group, then $N = |B/aBa^{-1}| = 1$, and so, in this situation, $L = |\det a| - 1$. Applying Theorem 11.6 to Example 11.2 in Section 11.2.2, we obtain that the number of generators is $L = 1$ since a is the quincunx matrix and $\det a = 2$. In the case of Example 11.3, the number of generators is $L = 3$ since $a = 2I$ and $\det = 4$. In the case of Example 11.4, where B is the two-dimensional group of shear matrices and $A = \{a^i : i \in \mathbb{Z}\}$, with $a = \begin{pmatrix} 2 & a_{1,2} \\ 0 & a_{2,2} \end{pmatrix} \in GL_2(\mathbb{Z})$, a calculation shows that $|B/aBa^{-1}| = 2|a_{2,2}|^{-1}$ and, thus, the number of generators is $L = 2|a_{2,2}|^{-1}\, 2|a_{2,2}| - 1 = 3$.

In order to prove Theorem 11.6, we need some additional notation and construction.

We recall that a $\mathbb{Z}^n$*-invariant space* (or a *shift-invariant space*) of $L^2(\mathbb{R}^n)$ is a closed subspace $V \subset L^2(\mathbb{R}^n)$ for which $T_k V = V$ for each $k \in \mathbb{Z}^n$. For $\phi \in L^2(\mathbb{R}^n) \setminus \{0\}$, we denote by $\langle \phi \rangle$ the shift-invariant space generated by ϕ, that is,

$$\langle \phi \rangle = \overline{\operatorname{span}}\{T_k \phi : k \in \mathbb{Z}^n\}.$$

Given $\phi_1, \phi_2 \in L^2(\mathbb{R}^n)$, their *bracket product* is defined as

$$[\phi_1, \phi_2](x) = \sum_{k \in \mathbb{Z}^n} \phi_1(x-k)\, \overline{\phi_2(x-k)}. \tag{11.4}$$

The bracket product plays a basic role in the study of shift-invariant spaces. The following properties are easy to verify, and they can be found, for example, in [18, Sec. 3].

Proposition 11.7. *Let* ϕ, ϕ_1, $\phi_2 \in L^2(\mathbb{R}^n)$.

(i) *The series* (11.4) *converges absolutely a.e. to a function in* $L^1(\mathbb{T}^n)$.
(ii) *The spaces* $\langle \phi_1 \rangle$ *and* $\langle \phi_2 \rangle$ *are orthogonal if and only if* $[\hat{\phi}_1, \hat{\phi}_2](\xi) = 0$ *a.e.*
(iii) *Let* $V(\phi) = \{T_k \phi : k \in \mathbb{Z}^n\}$. *Then* $V(\phi)$ *is an orthonormal basis for* $\langle \phi \rangle$ *if and only if* $[\hat{\phi}, \hat{\phi}](\xi) = 1$ *a.e.*

The theory of shift-invariant spaces plays a basic role in the study of the MRA precisely because the scaling space V_0, as well as all other scaling spaces $V_i = D_a^{-i} V_0$, are shift-invariant spaces. In the AB-MRA, the AB-scaling space satisfies a different invariance property, and this motivates the following definition. If B is a subgroup of $\widetilde{SL}_n(\mathbb{Z})$, a $B \ltimes \mathbb{Z}^n$*-invariant space* of $L^2(\mathbb{R}^n)$ is a closed subspace $V \subset L^2(\mathbb{R}^n)$ for which $D_b T_k V = V$ for each $(b,k) \in B \ltimes \mathbb{Z}^n$, where $B \ltimes \mathbb{Z}^n$ is the semi-direct product of B and $\mathbb{Z}^n$ (that is, it is the group obtained from the set $B \times \mathbb{Z}^n$, with natural group action on $b \times \{0\}$ and $\{I_n\} \times \mathbb{Z}^n$, and with $(b,0)(I_n,k)(b,0)^{-1} = (I_n, bk)$).

For $b \in \widetilde{SL}_n(\mathbb{Z})$, we have

$$\{D_b T_k : k \in \mathbb{Z}^n\} = \{T_{k'} D_b : k' \in \mathbb{Z}^n\},$$

and, as a consequence, $D_b \langle \phi \rangle = \langle D_b \phi \rangle$ for each $\phi \in L^2(\mathbb{R}^n)$. We also have that $\widehat{\mathbb{Z}}^n b = \widehat{\mathbb{Z}}^n$ and, thus,

$$[\hat{D}_b \hat{\phi}_1, \hat{\phi}_2](\xi) = [\hat{\phi}_1, \hat{D}_{b^{-1}} \hat{\phi}_2](\xi b), \tag{11.5}$$

for each $\phi_1, \phi_2 \in L^2(\mathbb{R}^n)$ and $\xi \in \widehat{\mathbb{R}}^n$.

Let V be a $B \ltimes \mathbb{Z}^n$-invariant space of $L^2(\mathbb{R}^n)$. The set $\Phi = \{\phi^1, \ldots, \phi^N\}$, with $N \in \mathbb{N} \cup \{\infty\}$, is a $B \ltimes \mathbb{Z}^n$-*orthonormal set of generators* for V if the set $\{D_b T_k \phi^i : (b,k) \in B \ltimes \mathbb{Z}^n, 1 \leq i \leq N\}$ is an orthonormal basis for V. Equivalently, we have that $[\widehat{D}_b \hat{\phi}^i, \hat{\phi}^j] = \delta_{i,j} \, \delta_{b,I_n}$ a.e. In addition, if this is the case, we have that

$$V = \bigoplus_{j=1}^{N} \overline{\text{span}}\{D_b T_k \phi^j : b \in B, k \in \mathbb{Z}^n\}.$$

It follows from this observation and from the properties of shift-invariant spaces, that, if $f \in V$ and Φ is a $B \ltimes \mathbb{Z}^n$-orthonormal set of generators for V, then

$$\hat{f} = \sum_{j=1}^{N} \sum_{b \in B} [\hat{f}, \widehat{D}_b \hat{\phi}^j] \, \widehat{D}_b \hat{\phi}^j, \tag{11.6}$$

with convergence in $L^2(\mathbb{R}^n)$.

We can now show the following simple result, whose proof is similar to that of Proposition 11.1.

Proposition 11.8. *Let $\Phi = \{\phi^1, \ldots, \phi^N\}$ and $\Psi = \{\psi^1, \ldots, \psi^M\}$ be two $B \ltimes \mathbb{Z}^n$-orthonormal sets of generators for the same $B \ltimes \mathbb{Z}^n$-invariant spaces V. Then $M = N$.*

Proof. By (11.6), we have that

$$\hat{\psi}^i = \sum_{j=1}^{N} \sum_{b \in B} [\hat{\psi}^i, \widehat{D}_b \hat{\phi}^j] \, \widehat{D}_b \hat{\phi}^j,$$

with convergence in $L^2(\mathbb{R}^n)$, for each $i = 1, \ldots, M$. Thus,

$$1 = \|\psi^i\|^2 = \sum_{j=1}^{N} \|[\hat{\psi}^i, \hat{\phi}^j]\|^2_{L^2(\mathbb{T}^n)}$$

and, as a consequence,

$$M = \sum_{i=1}^{M} \|\psi^i\|^2 = \sum_{i=1}^{M} \sum_{j=1}^{N} \sum_{b \in B} \|[\hat{\psi}^i, \widehat{D}_b \hat{\phi}^j]\|^2_{L^2(\mathbb{T}^n)}. \tag{11.7}$$

On the other hand, using the same argument on ϕ^j, $1 \leq j \leq N$, we obtain

$$N = \sum_{j=1}^{N} \|\phi^j\|^2 = \sum_{i=1}^{M} \sum_{j=1}^{N} \sum_{b \in B} \|[\hat{\phi}^j, \widehat{D}_b \hat{\psi}^i]\|^2_{L^2(\mathbb{T}^n)}. \tag{11.8}$$

By (11.5), using the fact that B is a group, it follows that

$$\sum_{b\in B} \|[\hat{\phi}^j, \widehat{D}_b\,\hat{\psi}^i]\|^2_{L^2(\mathbb{T}^n)} = \sum_{b\in B} \|[\hat{\phi}^j, \widehat{D}_{b^{-1}}\,\hat{\psi}^i]\|^2_{L^2(\mathbb{T}^n)} = \sum_{b\in B} \|[\widehat{D}_b\,\hat{\phi}^j, \hat{\psi}^i]\|^2_{L^2(\mathbb{T}^n)},$$

and, thus, by comparing (11.7) and (11.8), it follows that $M = N$. □

We are now ready to prove Theorem 11.6.

Proof of Theorem 11.6. Let $\{V_i\}_{i\in\mathbb{Z}}$, where $V_i = D_a^{-i} V_0$, be an AB-MRA, and, for each $i \in \mathbb{Z}$, let $W_i = V_{i+1} \bigcap (V_i)^\perp$. It follows that $L^2(\mathbb{R}^n) = \bigoplus_{i\in\mathbb{Z}} W_i$. By the definition of AB-MRA, the space V_0 is $B \ltimes \mathbb{Z}^n$-invariant and there is an AB-scaling function ϕ which is an orthonormal $B \ltimes \mathbb{Z}^n$ generator of V_0. Since a normalizes B and a is an integral matrix, it follows that

$$\begin{aligned}\{D_b\, T_k\, D_a^{-1} : k \in \mathbb{Z}^n, b \in B\} &= \{D_a^{-1}\, D_{aba^{-1}}\, T_{ak} : k \in \mathbb{Z}^n, b \in B\} \\ &\subseteq \{D_a^{-1}\, D_b\, T_k : k \in \mathbb{Z}^n, b \in B\}.\end{aligned}$$

Thus, the spaces V_1 and W_0 are also $B \ltimes \mathbb{Z}^n$-invariant.

The functions $\psi^1, \ldots, \psi^L$ are $B \ltimes \mathbb{Z}^n$ orthonormal generators for W_0 and so $\{\phi, \psi^1, \ldots, \psi^L\}$ are $B \ltimes \mathbb{Z}^n$ orthonormal generators for $V_1 = V_0 \bigoplus W_0$.

Next, take a complete collection of distinct representatives $\beta_0, \ldots, \beta_{N-1}$ for $B/(a\,B\,a^{-1})$, where $N = |B/aBa^{-1}|$. Thus, each $b \in B$ uniquely determines $b' \in B$ and $j \in \{0, \ldots, N-1\}$ for which $b = (a\,b'\,a^{-1})\,\beta_j$. Then

$$D_a^{-1} D_b\,\langle\phi\rangle = D_{a^{-1}b}\,\langle\phi\rangle = D_{b'}\,D_{a^{-1}}D_{\beta_j}\,\langle\phi\rangle = D_{b'}\,D_{a^{-1}}\,\langle D_{\beta_j}\phi\rangle. \qquad (11.9)$$

Also, take a complete collection of distinct representatives $\alpha_0, \ldots, \alpha_{M-1}$ for $\mathbb{Z}^n/(a\,\mathbb{Z}^n)$, where $M = |\det a|$. Each $k \in \mathbb{Z}^n$ uniquely determines $k' \in \mathbb{Z}^n$ and $i \in \{0, \ldots, M-1\}$, for which $k = ak' + \alpha_i$. For any $\phi \in L^2(\mathbb{R}^n) \setminus \{0\}$, the space $D_a^{-1}\,\langle\phi\rangle$ is then the shift-invariant space generated by $\Phi = \{\phi^i = D_a^{-1}\,T_{\alpha_i}\,\phi : 0 \le i \le M-1\}$. Since D_a^{-1} is unitary, then Φ is a $\mathbb{Z}^n$-orthonormal generating set for $D_a^{-1}\,\langle\phi\rangle$ if and only if ϕ is a $\mathbb{Z}^n$-orthonormal generating set for $\langle\phi\rangle$, and this holds if and only if $[\hat{\phi}, \hat{\phi}] = 1$ a.e. Thus, if Φ is a $\mathbb{Z}^n$-orthonormal generating set for $D_a^{-1}\,\langle\phi\rangle$, we have

$$D_a^{-1}\,\langle\phi\rangle = \bigoplus_{i=0}^{M-1} \langle\phi^i\rangle. \qquad (11.10)$$

By equation (11.10), we have

$$D_{a^{-1}}\,\langle D_{\beta_j}\phi\rangle = \bigoplus_{i=0}^{M-1} \langle\phi_{i,j}\rangle,$$

where $\phi_{i,j} = D_a^{-1} D_{\alpha_i} D_{\beta_j}\,\phi$. Using the last equality and (11.9), we have

$$D_{a^{-1}} \overline{\text{span}}\{D_b T_k \phi : b \in B, k \in \mathbb{Z}^n\}$$
$$= \bigoplus_{i=0}^{M-1} \bigoplus_{j=0}^{N-1} \overline{\text{span}}\{D_{b'} T_{k'} \phi_{i,j} : b' \in B,\ k' \in \mathbb{Z}^n,\ 0 \le i \le M-1,\ 0 \le j \le N-1\}.$$

This shows that $\{\phi_{i,j} : 0 \le i \le M-1, 0 \le j \le N-1\}$ is a $B \ltimes \mathbb{Z}^n$ set of orthonormal generators for V_1, with NM elements. By Proposition 11.8, $NM = L+1$ and this completes the proof. □

References

1. J. J. Benedetto, C. Heil, and D. F. Walnut, Differentiation and the Balian–Low Theorem, *J. Fourier Anal. Appl.*, **1** (1995), pp. 355–402.
2. E. J. Candès and D. L. Donoho, New tight frames of curvelets and optimal representations of objects with C^2 singularities, *Comm. Pure Appl. Math.*, **57** (2004), pp. 219–266.
3. O. Christensen, B. Deng, and C. Heil, Density of Gabor frames, *Appl. Comput. Harmon. Anal.*, **7** (1999), pp. 292–304.
4. X. Dai, D. R. Larson, and D. M. Speegle, Wavelet sets in $\mathbb{R}^n$ II, in: *Wavelets, Multiwavelets and Their Applications* (San Diego, CA, 1997), Contemp. Math., Vol. 216, Amer. Math. Soc., Providence, RI, 1998, pp. 15–40.
5. I. Daubechies, *Ten Lectures on Wavelets*, SIAM, Philadelphia, 1992.
6. M. N. Do and M. Vetterli, Contourlets, in: *Beyond Wavelets*, G. V. Welland, ed., Academic Press, San Diego, 2003, pp. 83–106.
7. R. J. Duffin and A. C. Schaeffer, A class of nonharmonic Fourier series, *Trans. Amer. Math. Soc.*, **72** (1952), pp. 341–366.
8. K. Guo, W.-Q Lim, D. Labate, G. Weiss, and E. Wilson, Wavelets with composite dilations, *Electron. Res. Announc. Amer. Math. Soc.*, **10** (2004), pp. 78–87.
9. K. Guo, W.-Q Lim, D. Labate, G. Weiss, and E. Wilson, Wavelets with composite dilations and their MRA properties, *Appl. Comput. Harmon. Anal.*, **20** (2006), pp. 202-236.
10. Y.-H. Ha, H. Kang, J. Lee, and J. Seo, Unimodular wavelets for L^2 and the Hardy space H^2, *Michigan Math. J.*, **41** (1994), pp. 345–361.
11. D. Han and Y. Wang, Lattice tiling and the Weyl–Heisenberg frames, *Geom. Funct. Anal.*, **11** (2001), pp. 742–758.
12. C. Heil and G. Kutyniok, Density of weighted wavelet frames, *J. Geometric Analysis*, **13** (2003), pp. 479–493.
13. C. E. Heil and D. F. Walnut, Continuous and discrete wavelet transforms, *SIAM Review*, **31** (1989), pp. 628–666.
14. E. Hernández, D. Labate, and G. Weiss, A unified characterization of reproducing systems generated by a finite family, II, *J. Geom. Anal.*, **12** (2002), pp. 615–662.
15. E. Hernández and G. Weiss, *A First Course on Wavelets*, CRC Press, Boca Raton, FL, 1996.
16. D. Speegle, On the existence of wavelets for non-expansive dilation matrices, *Collect. Math*, **54** (2003), pp. 163–179.
17. Y. Wang, Wavelets, tilings and spectral sets, *Duke Math. J.*, **114** (2002), pp. 43–57.

18. G. Weiss, and E. Wilson, The mathematical theory of wavelets, in: *Twentieth Century Harmonic Analysis—A Celebration* (Il Ciocco, 2000), NATO Sci. Ser. II Math. Phys. Chem., Vol. 33, Kluwer Acad. Publ., Dordrecht, 2001, pp. 329–366.
19. G. V. Welland, ed., *Beyond Wavelets*, Academic Press, San Diego, 2003.
20. M. Zibulski and Y. Y. Zeevi, Analysis of multiwindow Gabor-type schemes by frame methods, *Appl. Comput. Harmon. Anal.*, **4** (1997), pp. 188–221.

Part V

Sampling Theory and Shift-Invariant Spaces

12

Periodic Nonuniform Sampling in Shift-Invariant Spaces

Jeffrey A. Hogan[1] and Joseph D. Lakey[2]

[1] Department of Mathematical Sciences, University of Arkansas, Fayetteville, AR 72701, USA
jeffh@uark.edu
[2] Department of Mathematical Sciences, New Mexico State University, Las Cruces, NM 88003-8001, USA
jlakey@nmsu.edu

Summary. This chapter reviews several ideas that grew out of observations of Djokovic and Vaidyanathan to the effect that a generalized sampling method for bandlimited functions, due to Papoulis, could be carried over in many cases to the spline spaces and other shift-invariant spaces. Papoulis' method is based on the sampling output of linear, time-invariant systems. Unser and Zerubia formalized Papoulis' approach in the context of shift-invariant spaces. However, it is not easy to provide useful conditions under which the Unser–Zerubia criterion provides convergent and stable sampling expansions. Here we review several methods for validating the Unser–Zerubia approach for periodic nonuniform sampling, which is a very special case of generalized sampling. The Zak transform plays an important role.

12.1 Introduction

To say that sampling has played an important role in John Benedetto's work would be a severe understatement. The references [6], [9], [11], [12], [14], [15], [18], [16] and [19] serve merely to give an idea of the breadth of applications to which Benedetto has applied sampling methods. In this chapter we review certain connections among sampling and shift-invariant spaces, including multiresolution spaces, that have also been a major component of Benedetto's recent work (e.g., [17]).

Starting in the early 1990s, several groups of researchers observed the possibility of abstracting useful ideas in the sampling of bandlimited functions to the context of shift-invariant spaces. Early results of Janssen [36], [37], Walter [45], and Aldroubi and Unser [4] illustrated how to build interpolating functions for principal shift-invariant (PSI) spaces from the generators of such spaces. A PSI space $V(\varphi) \subset L^2(\mathbb{R})$ is the closure in $L^2(\mathbb{R})$ of the span of the integer shifts $\varphi(\cdot - k)$ of the *generator* $\varphi \in L^2(\mathbb{R})$. We shall always assume that these shifts form a Riesz basis for $V(\varphi)$, that is, that there are constants A, B such that $A\|\{c_k\}\|_{\ell^2}^2 \leq \|\sum c_k\varphi(\cdot - k)\|_{L^2}^2 \leq B\|\{c_k\}\|_{\ell^2}^2$ whenever $\{c_k\} \in \ell^2(\mathbb{Z})$. When $\{\varphi(\cdot - k)\}$ are orthonormal we call φ an *orthogonal generator*. The Paley–Wiener space $\mathrm{PW}_{1/2}$ of L^2-functions bandlimited to $[-1/2, 1/2]$ is a PSI space *par excellence* with orthogonal generator $\varphi(x) = \mathrm{sinc}(x) = \frac{\sin \pi x}{\pi x}$. Here we are taking $\hat{f}(\xi) = \int f(t)e^{-2\pi i t\xi}\, dt$ to denote the Fourier transform on $L^1(\mathbb{R})$. In addition to being orthogonal to its shifts, the sinc function is cardinal, i.e., $\mathrm{sinc}(k) = \delta_{0k}$. As a consequence, the coefficients c_k of $f = \sum_k c_k \mathrm{sinc}(x-k)$ are the same as the integer samples of f, i.e., $f(k) = c_k$ when $f \in \mathrm{PW}_{1/2}$. Typically, an orthogonal generator φ of $V(\varphi)$ will not be cardinal. So, if $V(\varphi)$ contains a function S that interpolates the integer samples of f, meaning that any $f \in V(\varphi)$ can be written $f(t) = \sum_k f(k)S(t-k)$, then S will be different from φ. The aforementioned works [36], [37], [45], and [4] sought to identify S in terms of φ.

The scope of sampling in PSI spaces was broadened significantly by Aldroubi and Gröchenig ([2], [3]) who devised sufficient conditions and iterative reconstruction algorithms for *irregular sampling* in PSI spaces (see also [1], [25], [41] for a small subsample of related developments). We will focus on *periodic nonuniform sampling* (PNS). In the bandlimited case, such sampling is a special case of Papoulis' generalized sampling approach [38], whose extension to the PSI case (see [42], [44]) will also be reviewed. In contrast to the "fully nonuniform" case, which requires iterative reconstruction methods, the problem of reconstructing elements of a PSI space $V(\varphi)$ from its "periodic nonuniform" samples boils down to identifying a finite collection of functions whose shifts interpolate samples.

Djokovic and Vaidyanathan [28] first considered generalized sampling schemes, including local averaging and derivative sampling, in the PSI space setting. In each such case, one samples the output of an element of $V(\varphi)$ filtered through a linear time-invariant (LTI) system (see Benedetto [10] for basic definitions and properties of LTI systems). Just as Papoulis [38] had done in the bandlimited case, Unser and Zerubia [42] formalized the problem of characterizing PSI spaces $V(\varphi)$ and LTI systems under which any $f \in V(\varphi)$ can be interpolated from samples of its outputs. While general necessary and sufficient conditions for stable reproduction were provided in [42], just as in [28], it remained a challenge to provide practical conditions for the existence of interpolants and methods for computing them. Here we will report on recent progress in this direction, largely summarizing results in the papers [33] and [34]. The larger issues of (i) why one should consider general sampling methods in the first place and (ii) how to deal with the errors that inevitably arise will also be discussed.

Whether φ is a scaling function plays no role in formulating general statements about sampling in PSI spaces. However, it can play an important role in establishing the validity, i.e., convergence and/or stability, of specific sampling/interpolating schemes. As such, at several points in the ensuing discussion we will comment on extra benefits that pertain when φ is a scaling function.

Here is a brief outline of the remainder of the chapter. In Section 12.2 we discuss what is meant by generalized sampling and oversampling in the setting of PSI spaces. In Section 12.3 we specialize to the case of periodic, nonuniform sampling and oversampling. A few specific sampling schemes are discussed in detail, including what is required for the validity of these schemes. Finally, in Sections 12.4, 12.5 and 12.6 we discuss three types of errors that arise in the application of such schemes, including aliasing, noise—for example, quantization noise—and *delay* errors that can arise when a delayed version of $f \in V(\varphi)$—as opposed to f itself—is sampled. Delay errors do not arise in the bandlimited case because Paley–Wiener spaces are invariant under all translations—not just integer shifts. Proofs of most of the results that also appear in [33] and [34] will be omitted here. Two exceptions to this are Theorems 12.6 and 12.13. These proofs serve to illustrate the roles of scaling and orthogonality respectively and to give a flavor of how the other proofs proceed.

Many of the results discussed here have natural and important extensions to finitely generated shift-invariant spaces. We will forgo discussion of such extensions in order to avoid notational complications.

12.2 Generalized Sampling and Oversampling

12.2.1 Generalized Sampling of Bandlimited Signals

The inverse problem of recovering a signal f from its samples or samples of its image under a system of convolvers is, generically, ill-posed (e.g., [24], see also Walnut's contribution in this volume). However, the problem is *well-posed* under appropriate conditions on f and on the convolvers. The classical Shannon Sampling Theorem (e.g., [10]) provides a formula for interpolating f from its regularly spaced samples $f\left(\frac{k}{2\Omega}\right)$, $k \in \mathbb{Z}$, namely,

$$f(t) = \frac{1}{2\Omega} \sum_{k\in\mathbb{Z}} f\left(\frac{k}{2\Omega}\right) \frac{\sin 2\pi(t - \frac{k}{2\Omega})}{\pi(t - \frac{k}{2\Omega})}, \tag{12.1}$$

provided f belongs to the Paley–Wiener space PW_Ω of those $f \in L^2(\mathbb{R})$ such that $\hat{f}$ is supported in $[-\Omega, \Omega]$. The *Nyquist sampling rate* of 2Ω samples per unit time is critical: f cannot be recovered from samples taken at a lower rate.

Several years ago, Papoulis [38] proposed a general sampling method in which one samples the output of M convolvers μ_j, $1 \leq j \leq M$. One can recover f from the samples $(f * \mu_j)(Mk/2\Omega)$, $(k \in \mathbb{Z})$, provided the convolvers are independent in a suitable sense. Conditions on the convolvers reduce to finite-dimensional linear algebra.

There are several good reasons for phrasing the sampling problem in terms of samples of $f * \mu_j$ as opposed to the samples of f itself. Arguably, *measurements* that produce *samples* correspond to local averages, while the measurement process should be independent of time. Thus, one does not acquire actual sample values $f(k)$ so much as $(f * \mu)(k)$, where μ models the measurement process. In order to recover f one must be able to deconvolve. When f is bandlimited, this means that one should be able to divide by $\hat{\mu}$ on $[-\Omega, \Omega]$. This may not be convenient or possible with a single convolver μ, particularly when the bandwidth Ω is large. Nor may it be convenient or possible, or even desirable, to sample at the Nyquist rate $1/(2\Omega)$ with a single device. For these reasons and others, Papoulis proposed sampling at a rate of $(2\Omega)/M$ samples per unit time, using M sampling devices modelled by convolvers μ_j, $j = 1, \ldots, M$.

12.2.2 Sampling in PSI Spaces and Quasi-Inversion

PSI spaces form a useful alternative model to PW_Ω in several respects. Suppose that the temporal or spatial scale is normalized so that "representative" signals lie, or nearly lie, in $V(\varphi)$. Consider, for example, the case in which input signals are assumed to have some regularity; e.g., Hölder continuity of order α. If φ is as regular as the input space and generates a multiresolution analysis (MRA) then, as is well known, the orthogonal projection $P_j f$ of f onto the space $V_j(\varphi)$ spanned by $\{2^{j/2}\varphi(2^j \cdot -k)\}_{k\in\mathbb{Z}}$ satisfies $\|P_j f - f\|_2 \leq C2^{-j\alpha}$. Just as physical signals typically are not truly bandlimited, neither are they typically scale-limited, i.e., in $V_j(\varphi)$. Nevertheless, $V_j(\varphi)$ and, by rescaling, $V_0(\varphi)$, provides a reasonable model for *approximating* sufficiently slowly varying signals.

More recently, Unser and Zerubia [42] proposed a framework abstracting Papoulis' approach of sampling bandlimited output of linear systems and, at the same time, generalizing the specific sampling methods of Djokovic and Vaidyanathan [28]. We will extend their framework to encompass oversampling as well, after mentioning some important additional considerations in the PSI setting.

For one, unlike the bandlimited case, convolution does not typically preserve $V(\varphi)$ as it does PW_Ω. As such, given a system $\boldsymbol{\mu} = (\mu_1, \ldots, \mu_L)$ of convolvers, one should specify a priori an *input space* $\mathcal{H}_\mu$ of functions f such that, for each $j = 1, \ldots, M$, the coefficients $(f * \mu_j)(Mk)$ lie in $\ell^2(\mathbb{Z})$. On the other hand, while the convolvers serve to model some system of measurements, the space $V(\varphi)$ might serve to provide approximations of elements of $\mathcal{H}_\mu$.

With this in mind, there are two key compatibility properties that one might require of $f \in \mathcal{H}_\mu$ and any element of $V(\varphi)$ constructed from its generalized samples, namely: (i) the construction should reproduce f exactly if f already lies in $V(\varphi)$ and (ii) the construction should form a good approximation of any f nearly in $V(\varphi)$.

Sampling in $V(\varphi)$ and Quasi-Inversion

Here is the set-up for generalized sampling in a PSI space $V(\varphi)$.

Sample data. One is equipped with L channels generating measurement data

$$(f * \mu_j)(Pk) = \langle f, \tilde{\mu}_j(\cdot - Pk)\rangle, \quad (1 \le j \le L).$$

Here $\langle \cdot, \cdot \rangle$ always refers to the standard Hermitian form $\langle \psi, \sigma\rangle = \overline{\sigma(\overline{\psi})}$ whenever ψ belongs to a class of test functions on which the distribution σ is defined. Also, $\tilde{\mu}(x) = \overline{\mu(-x)}$ here. The convolvers $\mu_1, \ldots, \mu_L$ can be distributions, provided the input space $\mathcal{H}_\mu$ forms a space of test functions on which convolution with each μ_j is defined. One takes $L \ge P$ and refers to the case $L = P$ as the *critically sampled* case.

Reproduction of $V(\varphi)$ from samples. Reproducing elements of $V(\varphi)$ from their measurements $\{(f * \mu_j)(Pk)\}$ $(1 \le j \le L, k \in \mathbb{Z})$ amounts to identifying a sequence of *interpolating functions* $\{S_j\}_{j=1}^L \in V(\varphi)$ such that the mapping

$$f \mapsto \mathcal{S}(f) \; = \; \sum_{j=1}^{L} \sum_{k \in \mathbb{Z}} (f * \mu_j)(Pk)\, S_j(x - Pk) \tag{12.2}$$

defines a projection from $\mathcal{H}_\mu$ onto $V(\varphi)$, that is, $\mathcal{S}f = f$ whenever $f \in V(\varphi)$. To formulate a suitable reproduction criterion, one defines the matrix sequence

$$\{(A^{\mu\varphi})_{ji}\}(k) = (\varphi * \mu_j)(Pk - i + 1) \qquad (1 \le j \le L, 1 \le i \le P, k \in \mathbb{Z}). \tag{12.3}$$

The problem is to determine conditions under which $A^{\mu\varphi}$ possesses a convolution quasi-inverse.

If $f = \sum_k c_k \varphi(\cdot - k) \in V(\varphi)$, then a simple change of variables yields

$$\begin{aligned}(f * \mu_j)(Pk) &= \sum_{i=1}^{P} \sum_{\ell \in \mathbb{Z}} c_{P(k-\ell)+i-1}\, (\varphi * \mu_j)(P\ell - i + 1) \\ &= \sum_{i=1}^{P} \sum_{\ell \in \mathbb{Z}} A^{\mu\varphi}_{ji}(k - \ell)\, c_{P\ell+i-1}.\end{aligned}$$

Consider the space $V^{(P)}(\varphi)$ consisting of vector functions of the form $(f_0, \ldots, f_{P-1})^T$ in which each f_i has the form $f_i = \sum_{\ell \in \mathbb{Z}} d_{\ell,i} \varphi(x - P\ell - i)$. In turn, let $\ell^2(\mathbb{Z}, \mathbb{C}^P)$ denote the space of square summable vector sequences $\mathbf{d}(\ell) = (d_{\ell,0}, \ldots, d_{\ell,P-1})^T$. Since $\{\varphi(\cdot - k)\}$ forms a Riesz basis for $V(\varphi)$, $f = \sum_k c_k \varphi(\cdot - k) \mapsto \{c_k\}$ defines an isomorphism of $V(\varphi)$ with $\ell^2(\mathbb{Z})$. This mapping also induces an isomorphism between $V^{(P)}(\varphi)$ and $\ell^2(\mathbb{Z}, \mathbb{C}^P)$ that sends $(\sum c_{P\ell}\varphi(x - P\ell), \ldots, \sum c_{P\ell+P-1}\varphi(x - P\ell + 1 - P))$ to $\mathbf{c} = \{\mathbf{c}(\ell)\}_{\ell \in \mathbb{Z}}$ where $\mathbf{c}(\ell) = (c_{P\ell}, \ldots, c_{P\ell+P-1})^T$.

In turn, the mapping that sends $\mathbf{c}(\ell)$ corresponding to $f = \sum_k c_k\varphi(\cdot - k)$ to $\mathbf{b}(\ell) = ((f * \mu_1)(Pk), \ldots, (f * \mu_L)(Pk))^T$ is given by a matrix convolution,

$$\mathbf{b}(k) \;=\; \sum_{\ell\in\mathbb{Z}} A^{\mu\varphi}(k-\ell)\,\mathbf{c}(\ell).$$

Finding a quasi-inverse of $A^{\mu\varphi}$ then means finding an $L \times P$ matrix sequence $\{\widetilde{A^{\mu\varphi}}(\ell)\}$ such that

$$\sum_{\ell\in\mathbb{Z}} \left(\widetilde{A^{\mu\varphi}}\right)^T (k-\ell)\, A^{\mu\varphi}(\ell) \;=\; \delta_{0k} I_{P\times P}. \tag{12.4}$$

In the Fourier domain this amounts to pointwise quasi-inverting the matrix-valued trigonometric series $\sum_k A^{\mu\varphi}(k)e^{2\pi i k\xi}$.

12.2.3 Interpolating Functions

At this level of generality, it is difficult to formulate useful conditions for the existence of a quasi-inverse (see [42] for the critically sampled case), let alone one with desirable regularity properties. In the following sections we will formulate effective conditions for (quasi)-invertibility of $A^{\mu\varphi}$ when the convolvers μ_j are periodically, but nonuniformly, distributed point masses. But one important matter can be addressed in full generality, namely, how to produce the interpolating functions S_j for $V(\varphi)$ by means of which $f \in V(\varphi)$ is recovered from its (generalized) samples as in (12.2).

One seeks coefficients $\{s_{jk}\} \in \ell^2(\mathbb{Z})$ such that $S_j = \sum s_{jk}\varphi(\cdot - k)$. Interpolating $V(\varphi)$ then amounts to expressing the functions

$$\varphi(\cdot),\ \varphi(\cdot - 1),\ \ldots,\ \varphi(\cdot - P + 1)$$

as linear combinations of the functions S_j. In what follows, one denotes by $\widehat{A}(z)$ the z-transform of a matrix-valued sequence $A(k)$.

Proposition 12.1. *The $V(\varphi)$-coefficients $\{s_{jk}\}$ of the jth interpolating function S_j in (12.2) are given in terms of a convolution quasi-inverse $\left(\widetilde{A^{\mu\varphi}}\right)^T$ of $A^{\mu\varphi}$ in (12.3), via*

$$\sum_k s_{jk}\, z^k = (1\ z\ \cdots\ z^{1-P})\left(\widehat{\widetilde{A^{\mu\varphi}}}\right)^T(z^P). \tag{12.5}$$

Proof. By (12.2), for each $i = 1, \ldots, P$ one must have

$$\begin{aligned}\varphi(x-i+1) &= \sum_{j,\ell}(\varphi * \mu_j)(P\ell - i + 1)\, S_j(x - P\ell)\\ &= \sum_{j,\ell}(\varphi * \mu_j)(P\ell - i + 1)\sum_k s_{jk}\,\varphi(x - k - P\ell).\end{aligned}$$

Since the shifts of φ form a basis for $V(\varphi)$, for each $i = 1, \ldots, P$ one has

$$\sum_{j,\ell} (\varphi * \mu_j)(P\ell - i + 1)\, s_{jk} \;=\; \delta_{i-1,\, k+P\ell}. \tag{12.6}$$

Now choose $m \in \mathbb{Z}$ and $1 \leq i' \leq P$ so that $k = P(m - \ell) + i' - 1$ and define $B_{ij}(k) = s_{j,Pk+i-1}$. Then (12.6) can be rewritten as

$$\sum_{j,\ell} B_{i'j}(m - \ell)\, A_{ji}^{\mu\varphi}(\ell) = \delta_{ii'}\, \delta_{m0}.$$

That is, B is a convolution quasi-inverse for $A^{\mu\varphi}$. The representation (12.5) follows by passing to the z-domain. This proves the proposition. □

12.2.4 Sampling at Unit Rate

Output of a Single Channel

If a single function $S \in V(\varphi)$ is to interpolate integer sample outputs, then S must solve

$$\varphi(x) \;=\; \sum_k (\varphi * \mu)(k)\, S(x - k),$$

so $\widehat{S}(\xi) = \widehat{\varphi}(\xi)/m_\varphi(\xi)$ where $m_\varphi(\xi) = \sum_k (\varphi * \mu)(k) e^{-2\pi i k \xi}$. The $V(\varphi)$-coefficient s_{-k} of S is the kth Fourier coefficient of $1/m_\varphi$, provided the latter is defined.

Example: Derivative and Average Sampling

To illustrate the problem of determining a sense in which S is defined, consider the *simple* case $h = \delta'$. The space $V'(\varphi)$ of derivatives of $V(\varphi)$ coincides with the PSI space $V(\varphi')$ generated by the derivative of φ. Then $m_\varphi(\xi) = \sum_k \varphi'(k) e^{-2\pi i k \xi}$. Difficulty can arise in inverting m_φ, due to nonexistence or insufficient decay of φ', or to undesirable vanishing properties of m_φ. Actually, a scaling relation can be beneficial here. Suppose that φ satisfies

$$\widehat{\varphi}(2\xi) = H(\xi)\, \widehat{\varphi}(\xi) \tag{12.7}$$

in which $H(\xi)$ is divisible by $1 - e^{-2\pi i \xi}$. By Viete's formula (e.g., [27] p. 211), one can associate to φ another scaling function φ_- with scaling filter $H_-(z) = 2H(z)/(1 - z)$ such that $\varphi'(x) = \varphi_-(x) - \varphi_-(x - 1)$. Then $\sum_k \varphi'(k) e^{-2\pi i k \xi} = (1 - e^{2\pi i \xi}) \sum_k \varphi_-(k) e^{-2\pi i k \xi}$. For the problem of sampling local averages of $f \in V(\varphi)$ over $[k, k+1)$, one works in reverse, attaching to φ the scaling function $\varphi_+ = \varphi * \chi_{[0,1)}$, which satisfies $\varphi_+'(x) = \varphi(x) - \varphi(x - 1)$.

The Zak Transform

The Zak transform is useful for defining interpolating functions S in terms of φ. One defines the Zak transform Zf of f at the point (t,ξ) in $\mathbb{R}\times\mathbb{R}$ by

$$Zf(t,\xi)=\sum_k f(t+k)\,e^{2\pi ik\xi}$$

whenever the sum converges. It does so absolutely when f belongs to the Wiener amalgam space $W(L^\infty,\ell^1)$ of functions whose norms $\|f\|_{W(L^\infty,\ell^1)}=\sum_{k\in\mathbb{Z}}\sup_{t\in[k,k+1)}|f(t)|$ are finite. It is also simple to invert Zf, as

$$\int_0^1 Zf(t,\xi)\,d\xi=f(t) \tag{12.8}$$

whenever the integral converges. The quasi-periodicity relation

$$Zf(t+k,\xi+\ell)=e^{-2\pi i\ell\xi}\,Zf(t,\xi)\qquad(\ell,k\in\mathbb{Z})$$

allows one to extend Z to a unitary mapping from $L^2(\mathbb{R})$ to the space

$$\mathcal{Z}=\left\{F:\mathbb{R}^2\to\mathbb{C}:\begin{array}{c}F(t+k,\xi+\ell)=e^{-2\pi ik\xi}F(t,\xi)\,(k,\ell\in\mathbb{Z})\\ \text{and }\int_0^1\int_0^1|F(t,\xi)|^2\,dt\,d\xi<\infty\end{array}\right\}. \tag{12.9}$$

A familiar consequence of quasi-periodicity that is also important here is that if $Zf(t,\xi)$ is continuous, as it is when f is continuous and belongs to $W(L^\infty,\ell^1)$, then it has a zero in $Q=[0,1)\times[0,1)$ [29]. Further fundamental properties of the Zak transform are developed in Janssen's tutorial [36]; see [27] and [29], as well as Gabardo's chapter in this volume.

The Zak transform complexifies, at least formally, to a mapping $Z_{\mathbb{C}}f(t,z)$: $\mathbb{R}\times\mathbb{C}\to\mathbb{C}$ by taking the z-transform of the samples $\{f(t+k)\}_k$, i.e.,

$$Z_{\mathbb{C}}f(t,z)=\sum_k f(t+k)\,z^k.$$

The usual Zak transform corresponds to $z=e^{2\pi i\xi}$. If f is supported on $[0,M]$ ($M\in\mathbb{N}$), then, for each t, $Z_{\mathbb{C}}f(t,z)$ is a polynomial in z of degree at most $M-1$.

Using Z, the function $m_\varphi(\xi)$ can be written

$$m_\varphi(\xi)=\sum_k(\varphi*\mu)(k)\,e^{-2\pi ik\xi}=\int Z\overline{\varphi}(y,\xi)\,\mu(y)\,dy=\langle\mu,Z\tilde{\varphi}(\cdot,-\xi)\rangle,$$

where, again, $\tilde{\varphi}(t)=\overline{\varphi(-t)}$. Of basic importance is the sense, if any, in which m_φ can be inverted.

Offset Integer Sampling

When μ is a point mass δ_t, one has $m_\varphi(\xi) = Z\overline{\widetilde{\varphi}}(t,\xi)$. If $\varphi \in W(L^\infty, \ell^1)$, then $Z\overline{\widetilde{\varphi}}(t,\xi)$ is an absolutely convergent Fourier series for any t. By Wiener's lemma (e.g., [13]) the Fourier series of $1/m_\varphi$ is also absolutely convergent, *provided* $Z\overline{\widetilde{\varphi}}(t,\xi)$ does not vanish. Such inversion must fail for some t, possibly $t = 0$, so one may need to *offset* samples to invert m_φ. Assume for the moment that an offset t_0 can be chosen so that $\inf_\xi |Z\varphi(t_0,\xi)| \neq 0$.

In all that follows, we will use $C(\xi)$ to denote the Fourier series of the sequence $\{c_k\}$ of coefficients of a given function $f(t) = \sum_k c_k \varphi(t-k) \in V(\varphi)$. One has

$$Zf(t,\xi) \;=\; C(\xi)\, Z\varphi(t,\xi). \tag{12.10}$$

By (12.10) and (12.8),

$$\begin{aligned} f(t) &= \sum_k \left(\int_0^1 C(\xi)\, e^{-2\pi i k\xi}\, d\xi \right) \varphi(t-k) \\ &= \sum_k \left(\int_0^1 \frac{Zf(t_0,\xi)}{Z\varphi(t_0,\xi)}\, e^{-2\pi i k\xi}\, d\xi \right) \varphi(t-k) \\ &\equiv \sum_\ell f(t_0+\ell)\, S_{t_0}(t-\ell). \end{aligned} \tag{12.11}$$

The function $S_{t_0}(t) = \int_0^1 \frac{Z\varphi(t,\xi)}{Z\varphi(t_0,\xi)}\, d\xi$ is t_0-cardinal in the sense that $S_{t_0}(t_0+k) = \delta_k$. Several specific examples are computed by Janssen [37]. The role of offset sampling in minimizing *aliasing* effects will be considered in Section 12.4, see [37], [32], [31].

$V(\varphi)$ May Not Admit Offset Sampling

In what follows, we construct a φ that is continuous, supported in $[0,2]$, and has orthogonal shifts, but for which $Z\varphi(t,t) = 0$ for all t. For such φ, $Z\varphi(t,\cdot)$ cannot be inverted in $L^2[0,1)$ for any t, and so $V(\varphi)$ does not admit offset integer sampling.

Start with $f(t) = (\sum_k c_k e^{2\pi i k t})\chi_{[0,1]}(t)$ where $\{c_k\} \in \ell^1(\mathbb{Z})$ and $f(0^+) = \sum c_k = 0$. Then f is continuous. If $\varphi(t) = e^{2\pi i t} f(t) - f(t-1)$, then φ is supported in $[0,2]$ and $Z\varphi(t,\xi) = (e^{2\pi i t} - e^{2\pi i \xi}) Zf(t,\xi)$ vanishes on the diagonal $\xi = t$. In the particular case $f(t) = i \sin 2\pi t\, \chi_{[0,1)}(t)$, one has

$$\varphi(t) = e^{2\pi i t} f(t) - f(t-1) = i \sin 2\pi t \begin{cases} e^{2\pi i t}, & \text{if } 0 \le t < 1, \\ -1, & \text{if } 1 \le t \le 2, \\ 0, & \text{else.} \end{cases}$$

This φ is continuous, supported on $[0,2]$, has orthogonal (integer) shifts, and satisfies $\widehat{\varphi}(0) = -\frac{1}{2}$, and $Z\varphi(t,t) = 0$ for all t. This raises the following question.

Problem 12.2. Find conditions on a continuous orthogonal generator φ of $V(\varphi)$ under which the Zak transform $Z\varphi(t,\xi)$ has the property that there exists t_0 for which $\inf_\xi |Z\varphi(t_0,\xi)| > 0$.

It is by now a folklore conjecture that if φ is a well-behaved scaling function (e.g., orthogonal, compactly supported, and continuous), then, for some t_0, $\inf_\xi |Z\varphi(t_0,\xi)| > 0$. Rioul [39] makes the stronger conjecture that this is true whenever φ satisfies a dilation equation $\varphi(t/2) = \sum_k d_k \varphi(t-k)$ in which the Fourier series $D(\xi) = \sum_k d_k e^{2\pi i k\xi}$ has no pair of symmetric zeroes, i.e., there is no $0 \le \xi_0 < 1$ for which $D(\xi_0) = D(\xi_0 + 1/2) = 0$—a property satisfied by any quadrature mirror filter (QMF). This conclusion has further ramifications. For example, Rioul asserts that the conclusion implies "optimal regularity" for certain subdivision schemes.

Interpolating Functions Do Not Have Compact Support

Integer sampling cannot typically be accomplished with well-behaved sampling functions. Suppose, for example, that φ is continuous, has orthonormal shifts, is supported in $[0, M]$, and satisfies $\inf_\xi |\Phi(\xi)| > 0$, where $\Phi(\xi) = Z\varphi(0,\xi)$. Then the sampling function in (12.11) will not be compactly supported unless $|\Phi(\xi)|$ is constant. Since Φ is a trigonometric polynomial with $\Phi(0) = 1$, this implies that $\Phi(\xi) = e^{2\pi i P\xi}$ for some $P \in \{1, \ldots, M-1\}$. Then $\varphi(k) = \delta_{k-P}$, but Xia and Zhang [46] show that no such φ exists, although the Haar scaling function provides a discontinuous example. In the case of offset sampling, the sampling function S_{t_0} can only have compact support if $\varphi(t_0 + k) = \delta_{k-P}$. It is conjectured that no such continuous example exists.

Problem 12.3. Show that there is no continuous compactly supported orthogonal scaling function which is cardinal in the sense that $\varphi(t_0 + k) = \delta_{k-P}$ for some $P \in \mathbb{Z}$ and $t_0 \in \mathbb{R}$, or provide a counterexample.

12.3 Periodic Nonuniform Sampling

It is difficult to produce a simple joint criterion on convolvers $\{\mu_j\}$ and a generator φ under which the interpolating functions S_j in (12.2) have "nice" properties. For the sake of providing computable conditions for sampling, we will make a considerable specialization in what follows, taking the convolvers μ_j to be point masses, that is, $\mu_j = \delta_{-t_j}$. Then (12.2) specializes to

$$f \mapsto \mathcal{S}_{\mathbf{t}} f = \sum_{j=0}^{L-1} \sum_{k\in\mathbb{Z}} f(t_j + Pk)\, S_j(t - Pk), \qquad \mathbf{t} = (t_0, \ldots, t_{L-1}). \quad (12.12)$$

We will consider this case exclusively in what follows, but it is worth mentioning here that the study of $\mathcal{S}$ for convolvers $\{\mu_j\}$ that are anything but point masses remains an important open direction. Little is known about convergence and stability of the projection (12.2) in such cases.

Sampling along the lattice $\Lambda = \{t_j + P\ell\}, 0 \leq j < L, \ell \in \mathbb{Z}$ is referred to as *periodic nonuniform sampling*, since the pattern repeats after a period P, but the basic offsets $t_0, \ldots, t_{L-1}$ need not be regularly spaced. P is called the *sampling period*. We will always take $P \in \mathbb{N}$.

Much research has been carried out in this special case and for good reason: it is easier to quantify not only the nature of the sampling functions in this case, but also the nature of the errors—both from aliasing and from noise—that inevitably arise.

12.3.1 Bunched Sampling: History and Rationale

Periodic nonuniform sampling (PNS) of bandlimited functions seems first to have been proposed by Yen [47], who also determined a formula for interpolating such samples. Bracewell [22] also referred to such sampling as *bunched* sampling or *interlaced* sampling. As its name indicates, in bunched sampling several samples are taken over a short subset of the sampling period, and this is followed by a time gap in which no samples are taken. This description will apply to the specific PNS patterns that we consider in the following sections. There are several reasons why one might consider the use of such sampling patterns.

Noise coupling. One practical motivation for *bunched* sampling is to avoid the noise coupling that arises in passing sample data from an analog-to-digital convertor (ADC) on to a discrete signal processor (DSP) (e.g., [5]). This coupling can be avoided by having the ADC acquire and convert a burst of bunched samples during a time when the DSP is inactive. The DSP then processes during a second phase in which the ADC is inactive.

Effective bandwidth. A second practical use of PNS arises in the analysis of multiband signals. Such signals are bandlimited to $[-\Omega, \Omega]$ for some large Ω but, in fact, have spectra concentrated in a finite pairwise disjoint union $[\alpha_j, \beta_j]$ of intervals such that the *effective bandwidth* $B_{\rm eff} = \sum_j \beta_j - \alpha_j \ll 2\Omega$. Herley and Wong [30] have shown that one can recover such a signal from properly chosen periodic nonuniform samples taken at any average rate higher than *effective Nyquist* $1/B_{\rm eff}$ (see [23]). Naturally, issues of aliasing and stability are more subtle in this setting.

Staggered sampling. Regular sampling at a high rate with a single "device" is equivalent to sampling at a lower rate with several devices. However, the latter approach offers more flexibility in terms of the actual devices that are involved as well as their relative spacing. The sampling outputs of the individual devices are said to be *interlaced*.

These concerns apply to bandlimited signals, but PNS is readily adapted to signal recovery in PSI spaces $V(\varphi)$, provided one chooses offsets so as to avoid phase couplings among shifts of offsets of φ. What this means in specific situations will be considered in the next several sections.

12.3.2 Periodic Nonuniform Sampling in $V(\varphi)$

In order to focus ideas, as well as to juxtapose with the bandlimited case, in what follows *we shall always assume that φ has compact support in* $[0, M]$ and has well-defined values along the sampling lattice $\Lambda = \{t_j + P\ell\}$, $0 \le j < L, \ell \in \mathbb{Z}$ (see [33], [34], [26] for results and applications when φ is bandlimited). The PNS lattice Λ is then determined by an *offset vector* $\mathbf{t} = (t_0, t_1, \ldots, t_{L-1})$ in which $0 \le t_0 < t_1 < \cdots < t_{L-1} < P$, and one can write $\Lambda = \mathbf{t} + P\mathbb{Z}$. The average rate of L/P samples per unit time is called the *sampling rate. Critical sampling* refers to the case $L = P$; *oversampling* refers to $L > P$.

In the following sections we will review several particular PNS schemes that were proposed by Djokovic and Vaidyanathan [28]. Each such scheme gives rise to an interpolation formula as in (12.12).

Shift-Bunched Critical Sampling

By shift-bunched sampling we mean a PNS scheme in which there are M offsets $t_0 < t_1 < \cdots < t_{M-1}$ such that $t_{M-1} - t_0 < 1$ (e.g., [28], [43]). As before, $[0, M]$ supports φ. In the PSI space context, there is no loss in assuming that $t_0 \ge 0$ and $t_{M-1} < 1$. That is, all offsets are contained within one time shift. In order to maintain an average rate of one sample per unit time, the sampling period P must equal M. Thus one samples $f \in V(\varphi)$ at $(t_0, \ldots, t_{M-1}) + M\mathbb{Z}$. To recover f from its bunched samples one needs M interpolating functions in $V(\varphi)$ that we identify now.

The bunched samples of f satisfy

$$\begin{aligned} f(t_j + M\ell) &= \sum_k c_k \, \varphi(t_j + M\ell - k) \\ &= \sum_{k=0}^{M-1} c_{M\ell-k} \, \varphi(t_j + k) = \sum_{k=0}^{M-1} M^{\varphi}_{jk} \, c_{M\ell-k} \end{aligned}$$

in which M^{φ} is the $M \times M$ matrix with (j, k)-th entry

$$M^{\varphi}_{jk} = \varphi(t_j + k) \qquad (0 \le j, k \le M - 1). \tag{12.13}$$

If M^{φ} is invertible, then $c_{M\ell-k} = \sum_{j=0}^{M-1} (M^{\varphi})^{-1}_{kj} f(t_j + M\ell)$. Therefore,

$$
\begin{aligned}
f(t) &= \sum_k c_k\, \varphi(t-k) = \sum_\ell \sum_{q=0}^{M-1} c_{M\ell-q}\, \varphi(t+q-M\ell) \\
&= \sum_\ell \sum_{j=0}^{M-1} f(t_j+M\ell) \sum_{q=0}^{M-1} (M^\varphi)^{-1}_{qj}\, \varphi(t+q-M\ell) \\
&= \sum_\ell \sum_{j=0}^{M-1} f(t_j+M\ell)\, S_j(t-M\ell),
\end{aligned}
$$

where

$$
S_j(t) = \sum_{q=0}^{M-1} (M^\varphi)^{-1}_{qj}\, \varphi(t+q) \tag{12.14}
$$

is in $V(\varphi)$ and supported on $[1-M, M]$.

Examples of Shift-Bunched Sampling

***B*-splines.** When φ is the quadratic B-spline supported on $[0,3]$, $\det M^\varphi$ is a Vandermonde determinant, and hence does not vanish as long as the offsets t_0, t_1, t_2 are distinct, as Djokovic and Vaidyanathan [28] verified. Ordinary integer sampling is not valid in this case.

Daubechies' 4-coefficient scaling functions. Critical shift-bunched sampling is also possible for the Daubechies scaling functions φ_ν $(-1<\nu<0)$ supported on $[0,3]$. By $\varphi = \varphi_\nu$ we mean the scaling function defined by the QMF (see (12.22)) $H(z) = \sum h_k z^k$ with coefficients $h_0 = \frac{\nu(\nu-1)}{2(1+\nu^2)}$, $h_1 = \frac{1-\nu}{2(1+\nu^2)}$, $h_2 = \frac{\nu+1}{2(1+\nu^2)}$, $h_3 = \frac{\nu(\nu+1)}{2(1+\nu^2)}$. When samples are taken at $\{\frac{1}{4}, \frac{1}{2}, \frac{3}{4}\} + 3\mathbb{Z}$, M^φ becomes

$$
M^\varphi = \begin{pmatrix} \varphi(1/4) & \varphi(5/4) & \varphi(9/4) \\ \varphi(1/2) & \varphi(3/2) & \varphi(5/2) \\ \varphi(3/4) & \varphi(7/4) & \varphi(11/4) \end{pmatrix}.
$$

The scaling relation (12.7) determines the values $\varphi(1) = \frac{\nu-1}{2\nu}$, $\varphi(2) = \frac{1+\nu}{2\nu}$ as well as at the dyadic rationals. In this case, one obtains $\det(M^\varphi) = \frac{(1-\nu)^3}{4(1+\nu^2)^4}$, which is non-zero whenever $-1<\nu<0$. Thus, a dyadic scaling relation can enable one to choose suitable dyadic offsets for bunched sampling.

Support-Bunched Sampling

By support-bunched sampling we mean PNS in which the L offsets all lie within $[0,M]$, but the sampling period P exceeds M. We will only discuss a special case of critical sampling in which $L = P = NM-1$ for some $N \in \mathbb{N}$. To fix ideas, in the ensuing discussion we will always take $N=2$.

Thus, $f \in V(\varphi)$ is sampled at $\{t_1, t_2, \ldots, t_{2M-1}\} + (2M-1)\mathbb{Z}$, as in [28]. If $f(t) = \sum_k c_k \varphi(t-k)$, then, for each integer ℓ,

$$f(t_j + (2M-1)\ell) = \sum_k c_k\, \varphi(t_j + (2M-1)\ell - k)$$
$$= \sum_{k=1-M}^{M-1} c_{(2M-1)\ell-k}\, \varphi(t_j + k). \qquad (12.15)$$

Define vectors $\mathbf{f}^{(\ell)}$, $\mathbf{c}^{(\ell)} \in \mathbb{C}^{2M-1}$ and a $(2M-1)\times(2M-1)$ matrix N^{φ} by $\mathbf{f}_j^{(\ell)} = f(t_j + (2M-1)\ell)$, $\mathbf{c}_k^{(\ell)} = c_{(2M-1)\ell-k}$, and

$$N_{jk}^{\varphi} = \varphi(t_j + k) \quad (1-M \le k \le M-1, 1 \le j \le 2M-1). \qquad (12.16)$$

Then (12.15) may be expressed as the matrix equation $\mathbf{f}^{(\ell)} = N^{\varphi}\mathbf{c}^{(\ell)}$. If N^{φ} is invertible, the coefficients c_k are determined *bunchwise* via $\mathbf{c}^{(\ell)} = (N^{\varphi})^{-1}\mathbf{f}^{(\ell)}$, that is,

$$c_{(2M-1)\ell-k} = \sum_{j=1}^{2M-1} (N^{\varphi})_{kj}^{-1}\, f(t_j + (2M-1)\ell).$$

Consequently,

$$f(t) = \sum_{\ell} \sum_{k=1-M}^{M-1} c_{(2M-1)\ell-k}\, \varphi(t + k - (2M-1)\ell)$$
$$= \sum_{\ell} \sum_{k=1-M}^{M-1} \sum_{j=1}^{2M-1} (N^{\varphi})_{kj}^{-1}\, f(t_j + (2M-1)\ell)\, \varphi(t + k - (2M-1)\ell)$$
$$= \sum_{\ell} \sum_{j=1}^{2M-1} f(t_j + (2M-1)\ell)\, S_j(t - (2M-1)\ell),$$

where the sampling functions $S_j(t)$ in (12.12) now take the form

$$S_j(t) = \sum_{k=1-M}^{M-1} (N^{\varphi})_{kj}^{-1}\, \varphi(t+k) \in V(\varphi). \qquad (12.17)$$

When N^{φ} is invertible, S_j is defined and supported in $[1-M, 2M-1]$.

Examples of Support-Bunched Sampling

Daubechies' 4-coefficient scaling functions. Support-bunched sampling is valid with offset vector $\mathbf{t} = (\frac{1}{2}, 1, \frac{3}{2}, 2, \frac{5}{2})$ for the family φ_ν $(-1 < \nu < 0)$ of Daubechies scaling functions supported on $[0, 3]$. In this case, N^{φ} takes the form

$$N^{\varphi} = \begin{pmatrix} 0 & 0 & \varphi(\frac{1}{2}) & \varphi(\frac{3}{2}) & \varphi(\frac{5}{2}) \\ 0 & 0 & \varphi(1) & \varphi(2) & 0 \\ 0 & \varphi(\frac{1}{2}) & \varphi(\frac{3}{2}) & \varphi(\frac{5}{2}) & 0 \\ 0 & \varphi(1) & \varphi(2) & 0 & 0 \\ \varphi(\frac{1}{2}) & \varphi(\frac{3}{2}) & \varphi(\frac{5}{2}) & 0 & 0 \end{pmatrix}.$$

Again, the scaling relation determines the relevant values of φ and one obtains $\det(N^{\varphi}) = -\frac{(\nu^4+1)^4}{16\nu^3(\nu^2+1)^3} \neq 0$ for ν in the range $-1 < \nu < 0$.

B-splines. When φ is the quadratic B-spline and, again, $\mathbf{t} = (\frac{1}{2}, 1, \frac{3}{2}, 2, \frac{5}{2})$, one has $\det(N^{\varphi}) = 8^{-3}$.

Certain other scaling functions. In the examples considered thus far, φ is always a scaling function. The following theorem provides a more general and natural explanation of how and why a scaling relation can be used to validate bunched sampling.

Theorem 12.4. *Let φ be a continuous, orthogonal scaling function supported on $[0, M]$ such that $\varphi(1) \neq 0 \neq \varphi(M-1)$. Suppose that the polynomial $\Phi_{\mathbb{C}}(z) = \sum_k \varphi(k) z^k$ has no pair of roots of the form $\{z_0,\ z_0^2\}$. Let $t_j = \frac{j}{2}$ $(1 \leq j \leq 2M-1)$ be offsets, and let N^{φ} be the $(2M-1) \times (2M-1)$ matrix defined by $(N^{\varphi})_{jk} = \varphi(t_j + k)$ $(1 - M \leq k \leq M-1,\ 1 \leq j \leq 2M-1)$. Then N^{φ} is invertible.*

The proof in [33] uses ideas in Section 12.3.4 to show that the hypothesis that N^{φ} is singular leads to a contradiction on the zeroes of Φ. This theorem applies to the Daubechies family φ_{ν}. However, the condition on the zeroes of $\Phi_{\mathbb{C}}$ need not apply to an arbitrary scaling function.

12.3.3 Periodic Oversampling and Coprimality

In PSI spaces, critical bunched sampling allows more flexibility than integer or offset integer sampling. However, there can still be obstacles to effective reconstructions and stability. As in the case of Paley–Wiener spaces, one can often achieve better results if one is able to oversample. In this section we consider conditions under which periodic oversampling can enable effective reconstructions. In what follows we refer to a family of polynomials as being coprime on a set S provided that they have no common zeroes in S.

Coprimality on $\mathbb{T}$

In what follows we will write $\mathbf{Z}\varphi(\mathbf{t}, \xi) = (Z\varphi(t_0, \xi), \ldots, Z\varphi(t_{L-1}, \xi))$ where $\mathbf{t}$ is a fixed offset vector with $0 \leq t_0 < \cdots < t_{L-1} < 1$. When φ is bounded with compact support, coprimality of the trigonometric polynomials $Z\varphi(t_j, \cdot)$ on the unit circle $\mathbb{T}$ is equivalent to existence of some c such that

$$\|\mathbf{Z}\varphi(\mathbf{t}, \xi)\| \geq c > 0, \qquad (\xi \in \mathbb{T}), \tag{12.18}$$

where $\|\mathbf{a}\|$ denotes the Euclidean norm of the vector $\mathbf{a} = (a_0, \ldots, a_{L-1})$. Let $f(t) = \sum_k c_k \varphi(t-k) \in V(\varphi)$ with $Zf(t,\xi) = C(\xi) Z\varphi(t,\xi)$. In our vector notation,

$$\mathbf{Z}f(\mathbf{t},\xi) \cdot \mathbf{Z}\varphi(\mathbf{t},\xi) = C(\xi)\, \|\mathbf{Z}\varphi(\mathbf{t},\xi)\|^2,$$

where the dot product is the Hermitian inner product on $\mathbb{C}^L$. Assuming (12.18), one obtains

$$C(\xi) = \frac{\mathbf{Z}f(\mathbf{t},\xi) \cdot \mathbf{Z}\varphi(\mathbf{t},\xi)}{\|\mathbf{Z}\varphi(\mathbf{t},\xi)\|^2}.$$

Since $c_k = \int_0^1 C(\xi) e^{-2\pi i k\xi}\, d\xi$, one also obtains

$$f(t) = \sum_{j=0}^{L-1} \sum_m f(t_j + m)\, S_j(t-m)$$

in which

$$S_j(t) = \sum_k \left\{ \int_0^1 \frac{\overline{Z\varphi(t_j,\xi)}}{\|Z\varphi(\mathbf{t},\xi)\|^2} e^{-2\pi i k\xi}\, d\xi \right\} \varphi(t-k) \in V(\varphi). \tag{12.19}$$

In Section 12.3.4 we give conditions under which the assumption (12.18) is valid, along with estimates of the required sampling rate L.

Coprimality on $\mathbb{C}$

As before, we let $\mathbf{t} = (t_0, \ldots, t_{L-1})$ where $0 \le t_0 < t_1 < \cdots < t_{L-1} < 1$, but now we impose the stronger condition that the polynomials $\{Z_{\mathbb{C}}\varphi(t_j, z)\}_{j=0}^{L-1}$ have no common zero in the whole plane—not just on $\mathbb{T}$. We will provide conditions that guarantee coprimality in Section 12.3.4, but here we consider its consequences. Since the polynomial entries of $\mathbf{Z}\varphi(\mathbf{t}, z)$ all have degree at most $M-1$ and are coprime, Euclid's algorithm (e.g., [35]) furnishes a vector polynomial $\mathbf{P}(z) = (P_0(z), P_1(z), \ldots, P_{L-1}(z))$ with entries of degree at most $M-2$ such that

$$\mathbf{P}(z) \cdot \mathbf{Z}\varphi(\mathbf{t}, z) = 1 \tag{12.20}$$

for all $z \in \mathbb{C}$. Thus, if $f(t) = \sum_k c_k \varphi(t-k) \in V(\varphi)$, so that $Zf(\cdot,\xi) = C(\xi) Z\varphi(\cdot,\xi)$, (12.20) yields

$$C(z) = C(z)\mathbf{P}(z) \cdot \mathbf{Z}\varphi(\mathbf{t}, z) = \mathbf{P}(z) \cdot \mathbf{Z}f(\mathbf{t}, z).$$

Writing $P_j(z) = \sum_{m=0}^{M-2} p_{jm}\, z^m$, the scaling coefficients of f are recovered by

$$c_k = \int_0^1 \mathbf{P}(\xi) \cdot \mathbf{Z}f(\mathbf{t},\xi)\, e^{-2\pi i k\xi}\, d\xi = \sum_{j=0}^{L-1} \sum_{m=0}^{M-2} p_{jm}\, f(t_j + k - m).$$

Consequently, f can be reconstructed from its samples along $\mathbf{t} + \mathbb{Z}$ by

$$
\begin{aligned}
f(t) &= \sum_k \sum_{j=0}^{L-1} \sum_{m=0}^{M-2} p_{jm}\, f(t_j+k)\, \varphi(t-j-k) \\
&= \sum_k \sum_{j=0}^{L-1} f(t_j+k) \sum_{m=0}^{M-2} p_{jm}\, \varphi(t-m-k) \\
&= \sum_k \sum_{j=0}^{L-1} f(t_j+k)\, S_j(t-k).
\end{aligned}
$$

Here, $S_j(t) = \sum_{m=0}^{M-2} p_{jm}\varphi(t-m) \in V(\varphi)$ is supported on $[0, 2M-2]$.

Shorter Supports for Higher Rates: Weighted Average Sampling

Polynomials of lower degree, and hence sampling functions of smaller support, are possible at the expense of higher sampling rates. Suppose that $0 \le t_0 < t_1 < \cdots < t_{L-1} < 1$, φ is supported on $[0, M]$, and $\{Z_{\mathbb{C}}\varphi(t_\ell, \cdot)\}_{\ell=0}^{L-1}$ are coprime. We seek polynomials $P_0, P_1, \ldots, P_{L-1}$ of degree less than or equal to D such that (12.20) holds for all $z \in \mathbb{C}$. Equating coefficients on both sides of (12.20) yields $M+D$ equations in the $L(D+1)$ coefficients of the P_ℓ.

In the extreme case $D = 0$, one seeks $a_0, a_1, \ldots, a_{L-1} \in \mathbb{C}$ such that $\sum_{\ell=0}^{L-1} a_\ell Z_{\mathbb{C}}\varphi(t_\ell, z) = 1$ for all $z \in \mathbb{C}$. Equivalently, $\sum_{\ell=0}^{L-1} a_\ell \varphi(t_\ell + k) = \delta_k$. In the case $L = M$, a solution exists precisely when one can find $\mathbf{a} = (a_0, a_1, \ldots, a_{L-1})$ satisfying $\mathbf{a}M^\varphi = (1, 0, \ldots, 0)$ where M^φ is defined in (12.13). In particular, this is the case when M^φ is invertible. Then, for $f(t) = \sum_k c_k \varphi(t-k) \in V(\varphi)$, $C(z) = \sum_{\ell=0}^{M-1} a_\ell Z_{\mathbb{C}} f(t_\ell, z)$ and hence $c_k = \sum_{\ell=0}^{M-1} a_\ell f(t_\ell + k)$. That is, the coefficients c_k are simply weighted averages of samples of f taken on $[k, k+1)$ and

$$
f(t) = \sum_{\ell=0}^{M-1} a_\ell \left(\sum_k f(t_\ell + k)\varphi(t-k) \right). \tag{12.21}
$$

However, the local sampling rate is proportional to the support length of φ.

12.3.4 Validation of Oversampling Techniques

In this section we will provide some sufficient conditions guaranteeing the validity of periodic oversampling in $V(\varphi)$. We will only provide sketchy arguments to give a flavor of how validity is established. Details and related results can be found in [33] and [34].

Continuity Validates Shift-Bunched Oversampling at a High Rate

For a high enough rate L/M, coprimality on $\mathbb{C}$ follows from a mild continuity condition on φ, provided φ is orthogonal to its integer shifts.

Theorem 12.5. *Let φ be compactly supported in $[0, M]$, Lipschitz continuous of order $\alpha > 0$, and orthogonal to its integer shifts. Let $L \geq 2$ be an integer, $0 \leq t_0 < t_1 < \cdots < t_{L-1} < 1$, $\mathbf{t} = (t_0, t_1, \ldots, t_{L-1})$, and $\Delta(\mathbf{t}) = \max_{1 \leq l \leq L-1}\{t_0, t_\ell - t_{\ell-1}, 1 - t_{L-1}\}$. Then $\|Z_{\mathbb{C}}\varphi(\mathbf{t}, z)\|^2 \geq c > 0$ for all $z \in \mathbb{C}$, provided $\Delta(\mathbf{t})$ is sufficiently small.*

The required oversampling rate in Theorem 12.5 depends on quantities like Lipschitz constants that can be difficult to compute. When, in addition to being orthogonal to its shifts, φ satisfies a scaling relation (12.7), much more effective estimates on the sampling rate L can be deduced from information on the zeroes of the QMF H. Theorems to this effect are based on a formulation of the fundamental scaling relationship in terms of the Zak transform, first noted by Janssen [37].

Scaling in the Zak Domain

If φ is a scaling function, then the scaling relation can be used to verify effective sampling rates for coprimality. Rewriting the scaling relation in the Zak domain aids in doing so. We will assume in what follows that the reader is familiar with the concept of a multiresolution analysis (e.g., [17]). For us, the scaling relation (12.7) with filter $H(\xi) = \sum_k h_k e^{-2\pi i k\xi}$ remains the central point of emphasis. Assuming that φ is orthogonal to its shifts, one obtains $h_k = \frac{1}{2}\int_{-\infty}^{\infty} \varphi(\frac{t}{2})\overline{\varphi(t-k)}\, dt$ and H is a *quadrature mirror filter* (QMF), i.e.,

$$|H(\xi)|^2 + |H(\xi + \frac{1}{2})|^2 \equiv 1. \tag{12.22}$$

For the z-transform $H_{\mathbb{C}}(z) = \sum_k h_k z^k$ ($z \in \mathbb{C}$) of $\{h_k\}_k$, the QMF condition (12.22) extends to

$$H_{\mathbb{C}}(z)\,\overline{H_{\mathbb{C}}(z^*)} + H_{\mathbb{C}}(-z)\,\overline{H_{\mathbb{C}}(-z^*)} = 1 \quad (z \neq 0), \tag{12.23}$$

where, as before, $z^* = 1/\overline{z}$. In order that a trigonometric polynomial H generates an orthogonal scaling function via (12.7), H must satisfy the *τ-cycle condition*: there is no non-trivial τ-cycle $\{\xi_1, \xi_2, \ldots, \xi_n\} \subset \mathbb{T}$ ($\tau\xi_i \equiv \xi_i^2 = \xi_{i+1}$ with $\xi_1 = \xi_n$) such that $|H_{\mathbb{C}}(\xi_i)| = 1$ for $1 \leq i \leq n$ (e.g., [27], p. 188).

Janssen [37] observed that, upon setting

$$T_H f(z^2) = H(z)f(z) + H(-z)f(-z), \tag{12.24}$$

the scaling relation (12.7) can be written

$$Z_{\mathbb{C}}\,\varphi(x, z) = T_H\,(Z_{\mathbb{C}}\,\varphi(2t, \cdot))(z), \qquad \text{or} \quad Z\varphi(t, \xi) = T_H\,(Z\varphi(2t, \cdot))(\xi). \tag{12.25}$$

Thus, one can use T_H to compute the values of $Z\varphi$ along the line $t = t_0$ from those along $t = 2t_0$. Letting M denote the operator of multiplication by z^{-1}, i.e., $Mf(z) = z^{-1}f(z)$ for Laurent series f, the quasi-periodicity of Zak transforms extends to $Z_{\mathbb{C}}f(t+1,z) = M(Z_{\mathbb{C}}f(t,\cdot))(z)$. Hence, for a dyadic rational $p/2^j$ $(p, j \in \mathbb{Z},\ j \geq 0)$, multiple applications of (12.25) yield

$$Z_{\mathbb{C}}\varphi\left(\frac{p}{2^j}, z\right) = T_H^j\left(Z_{\mathbb{C}}\varphi(p,\cdot)\right)(z) = T_H^j\ M^p\,\Phi_{\mathbb{C}}(z), \tag{12.26}$$

where $\Phi_{\mathbb{C}}(z) = Z_{\mathbb{C}}\varphi(0,z) = \sum_k \varphi(k)z^k$. Consequently, given Φ and H, (12.26) allows one to calculate (iteratively) the values of $Z_{\mathbb{C}}\varphi(t,z)$ for all z and all dyadic rationals t.

Coprimality for Scaling Functions

We include a proof of the following theorem to give a flavor of the techniques used to deduce effective sampling rates from QMF conditions.

Theorem 12.6. *Suppose that φ is a continuous scaling function supported on $[0, 2^J+1]$ whose QMF vanishes on $\mathbb{T}$ only at $\xi = \frac{1}{2}$. Then the polynomials $Z\varphi(\ell/2^J, \xi)$, $\ell = 0, \dots, 2^J-1$, have no common zeroes on $\mathbb{T}$.*

Proof. As before, let $\Phi(\xi) = Z\varphi(0,\xi)$. By (12.26),

$$\begin{aligned} Z\varphi\left(\frac{\ell}{2^J}, \xi\right) = T^J M^\ell \Phi(\xi) &= \sum_{j=0}^{2^J-1} e^{-2\pi i j\ell/2^J}\,\Phi\left(\frac{\xi+j}{2^J}\right)\prod_{p=1}^{J} H\left(\frac{\xi+j}{2^p}\right) \\ &= 2^{J/2}\mathcal{F}_{2^J}\left(\Phi\left(\frac{\xi+\cdot}{2^J}\right)\prod_{p=1}^{J} H\left(\frac{\xi+\cdot}{2^p}\right)\right)(\ell), \end{aligned}$$

where $\mathcal{F}_{2^J}$ is the 2^J-point discrete Fourier transform. Therefore, if $Z\varphi(\frac{\ell}{2^J}, \eta) = 0$ for all ℓ then, by Fourier uniqueness, for $1 - 2^{J-1} \leq k \leq 2^{J-1}$,

$$\Phi\left(\frac{\eta+k}{2^J}\right)\prod_{p=1}^{J} H\left(\frac{\eta+k}{2^p}\right) = 0. \tag{12.27}$$

Since φ is supported on $[0, 2^J+1]$, $\Phi_{\mathbb{C}}$ has degree 2^J and has a zero at $z=0$. Hence $\Phi_{\mathbb{C}}$ can have at most $2^J - 1$ zeroes on the unit circle. Consequently, (12.27) implies that $\prod_{p=1}^{J} H(\frac{\eta+k}{2^p}) = 0$ for at least one value of k (say $k = \kappa$). Since H has zeroes on the unit circle at $\xi = \frac{1}{2}$ only, we have $\frac{\eta+\kappa}{2^p} = Q + \frac{1}{2}$ for some integer Q and $1 \leq p \leq J$. Hence $\eta = 2^p(Q+\frac{1}{2}) - \kappa = 2^pQ + 2^{p-1} - \kappa \in \mathbb{Z}$. But this means that $\Phi(2^pQ + 2^{p-1} - \kappa) = 0$, a contradiction since $\Phi(m) = 1$ for all integers m. $\square$

The Daubechies scaling functions φ_ν are supported on $[0,3]$ and satisfy the condition of the theorem with $J=1$, so sampling at half-integers provides an effective oversampling scheme. For QMFs having zeroes away from $\xi = 1/2$, it may be necessary to sample at a higher rate (see [33]).

Except in special cases in which the sum in (12.18) is constant, coprimality on $\mathbb{T}$ does not suffice to yield compactly supported interpolating functions. To obtain local interpolators one needs coprimality on all of $\mathbb{C}$. In the case of orthogonal MRA scaling functions this can be done, but at the cost of a sampling rate that grows geometrically with the support length M of φ.

Theorem 12.7. *Suppose that φ is a continuous scaling function for an MRA supported on $[0, M]$. Then the polynomials $\{Z_{\mathbb{C}}\varphi(\frac{\ell}{2^{M-2}}, z)\}_{\ell=0}^{2^{M-2}-1}$ have no common zeroes.*

The proof in [33] uses the fact that the zeroes of H must have some geometric structure. It is worth reviewing one ingredient of the proof here, that illustrates the utility of (12.25) and will also be useful in the next section.

Lemma 12.8. *Suppose that φ is an orthogonal scaling function such that $\Phi(z) = Z_{\mathbb{C}}\varphi(0, z)$ has no pair of zeroes of the form $(w, w^2) \in \mathbb{C}^2 \setminus \{(0, 0)\}$. Then $\Phi(z)$ and $Z_{\mathbb{C}}\varphi(\frac{1}{2}, z)$ are coprime.*

Proof. If the conclusion fails, i.e., if $\Phi(z_0) = Z_{\mathbb{C}}\varphi(\frac{1}{2}, z_0) = 0$ for some $0 \neq z_0 \in \mathbb{C}$, then by (12.25) with $z_1^2 = z_0$,

$$0 = \Phi(z_0) = T_H\, \Phi(z_0) = H(z_1)\, \Phi(z_1) + H(-z_1)\, \Phi(-z_1)$$

$$0 = Z_{\mathbb{C}}\left(\frac{1}{2}, z_0\right) = T_H\ M\Phi(z_0) = \frac{1}{z_1}\left(H(z_1)\, \Phi(z_1) - H(-z_1)\, \Phi(-z_1)\right),$$

where H is the QMF of φ. Together, these equations imply that $H(z_1)\Phi(z_1) = H(-z_1)\Phi(-z_1) = 0$. By (12.23), $H(z_1)$ and $H(-z_1)$ cannot both vanish. Relabelling if necessary, we conclude that $\Phi(z_1) \neq 0$, thus contradicting the hypothesis on the roots of Φ. The lemma follows. □

12.3.5 Design of Scaling Functions for Oversampling

In [34] we proposed a method for parameterizing scaling functions in terms of their sample values. Here we will make use of the ideas underlying that construction to see how to manufacture orthogonal scaling functions φ for which $V(\varphi)$ admits local interpolation via periodic oversampling at a *low* rate. In terms of the coprimality criterion, given an offset vector $\mathbf{t} = (t_0, \ldots t_{L-1}) \in [0, 1)^L$, with L small, one seeks to construct a (scaling) function supported in $[0, M]$ such that the polynomials $\{Z\varphi(t_j, z)\}_{j=0}^{L-1}$ are coprime.

For purposes of illustration, we will take $L = 2$ with $t_0 = 0$ and $t_1 = \frac{1}{2}$. One wishes to construct an orthogonal scaling function φ supported on $[0, M]$ whose integer values $\varphi(1), \ldots, \varphi(M-1)$ are defined by a given vector $\mathbf{a} = (a_1, a_2, \ldots, a_{M-1}) \in \mathbb{C}^{M-1}$, i.e., $\varphi(k) = a_k$. Then, necessarily,

(i) $\sum_{k=1}^{M-1} a_k = 1$,

while it is convenient to assume that

(ii) $\sum_{k=1}^{M-1} a_k \overline{a_{k-2m}} = \frac{c}{2}\delta_m$ for some $c \geq 1$.

If, in addition, $\Phi(z) = \sum_k a_k z^k$ has no pair of zeroes of the form $(w, w^2) \in \mathbb{C}^2 \setminus \{(0,0)\}$ then, by Lemma 12.8, one will be able to interpolate elements of $V(\varphi)$ from their half-integer samples as in Section 12.3.3.

Now specialize further to $M = 5$. Then $\Phi(z) = \sum \varphi(k) z^k = \sum_{k=1}^{4} \phi(k) z^k$ can be written $\Phi(z) = czP(z)$ where P has degree at most 3. Suppose, additionally, that Φ is to have real coefficients and P has a complex conjugate pair of roots. Then $\Phi(z) = cz(z - z_0)(z - z_1)(z - \overline{z_1})$ with $z_0 \in \mathbb{R} \setminus \{0\}$ and z_1 complex. The condition $\Phi(1) = \sum_{k=1}^{4} \varphi(k) = 1$ implies that $c = [(1 - z_0)(1 - 2\Re(z_1) + |z_1|^2)]^{-1}$. As observed in [34], if

$$\Phi(z)\,\overline{\Phi(z^*)} + \Phi(-z)\,\overline{\Phi(-z^*)} = \text{const}\,, \tag{12.28}$$

then a QMF can be defined by taking

$$H(z) = \text{const}\,\big(\Phi(z^2)\,\overline{\Phi(z^*)} + z^Q\,\Phi(-z)\,\overline{\Phi(-(z^*)^2)}\tag{12.29}$$

for an appropriate $Q \in \mathbb{Z}$.

In our case, (12.28) is equivalent to $\varphi(1)\varphi(3) + \varphi(2)\varphi(4) = 0$. Together with $\Phi(1) = 1$, this implies that

$$2\Re(z_1)(|z_1|^2 + 1) = \frac{-(z_0^2 + 1)}{z_0}. \tag{12.30}$$

Therefore, z_0 and $\Re(z_1)$ have opposite signs and z_1 cannot be purely imaginary. For the sake of concreteness, let $z_0 = -2$ and $z_1 = re^{i\theta}$. Then (12.30) yields $r^2 \cos\theta - \frac{5r}{4} + \cos\theta = 0$. Choose $\theta = \frac{\pi}{3}$. Then $r = 2$ or $r = 1/2$. Choose $r = 2$. Then $z_1 = 1 + i\sqrt{3}$ and $z_2 = 1 - i\sqrt{3}$, so $\Phi(z) = \frac{1}{9}(2z - 3z^2 + 6z^3 + 4z^4)$.

To construct H, set $Q = -3$ in (12.29). Then, normalizing so that $H(1) = 1$, from (12.29) one obtains

$$H(z) = \frac{1}{130}\big(-30 + 20z + 75z^2 + 15z^3 + 20z^4 + 30z^5\big).$$

The QMF condition (12.22) of H is readily checked by hand or computer algebra. Applying (12.26), we also obtain $Z_{\mathbb{C}}\varphi(\frac{1}{2}, z) = T_H M\Phi(z) = \frac{1}{117}(-12 - 18z + 105z^2 + 18z^3 + 24z^4)$.

To build interpolating functions, one seeks polynomials P_0, P_1 of degree no greater than 3 such that $P_0(z)\Phi(z) + P_1(z)Z_{\mathbb{C}}\varphi(\frac{1}{2}, z) = 1$. P_0 and P_1 can be computed by the Euclidean algorithm or by computer algebra. The latter approach yields

$$\begin{aligned} P_0(z) &= -2.9261 + 30.353\,z + 5.334\,z^2 + 6.7126\,z^3 \\ P_1(z) &= -9.75 + 8.2851\,z - 22.465\,z^2 - 14.544\,z^3. \end{aligned}$$

The corresponding interpolating functions $S_0(t)$ and $S_1(t)$ are determined by $S_j(t) = \sum_k s_{jk}\varphi(t - k)$ where s_{jk} is the coefficient of z^k in P_j, $j = 0, 1$. Any $f \in V(\varphi)$ can be recovered from its half-integer samples by

$$f(t) = \sum_k f(k)S_0(t-k) + f\big(k+\frac{1}{2}\big)S_1(t-k).$$

The two interpolating functions S_0 and S_1 are plotted in Fig. 12.1.

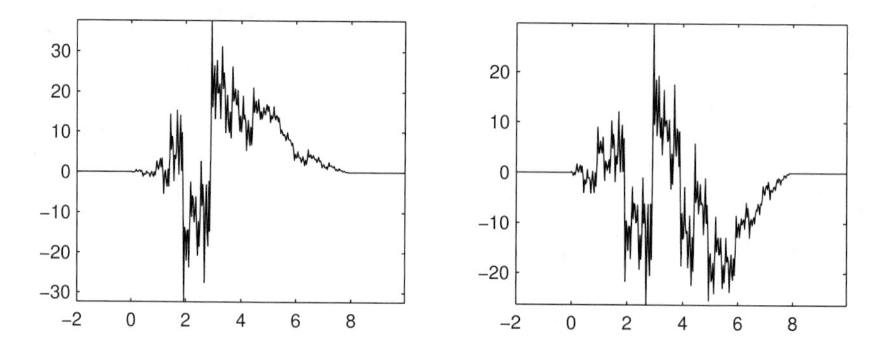

Fig. 12.1. Interpolating functions S_0 and S_1 for half-integer oversampling.

The method that we have just outlined for building a scaling function amenable to half-integer sampling has no provision for regularity of φ. In fact, although its integer values are well-defined, the scaling function φ just constructed is discontinuous, particularly at $t = 1$ and $t = 2$. This poses serious instability problems for sampling. It turns out that the scaling function φ_{perm} with filter $H_{\text{perm}} = \frac{1}{130}(20 + 30z + 75z^2 + 15z^3 - 30z^4 + 20z^5)$ obtained by transposing the coefficients of z^0 and z^4 and of z^1 and z^5 of H defines an orthogonal scaling function for which Φ_{perm} and $Z\varphi_{\text{perm}}(1/2, z)$ are also coprime (see Fig. 12.2). This works for reasons that go beyond the present discussion. This brings us to an important open problem concerning the construction of continuous scaling functions amenable to sampling.

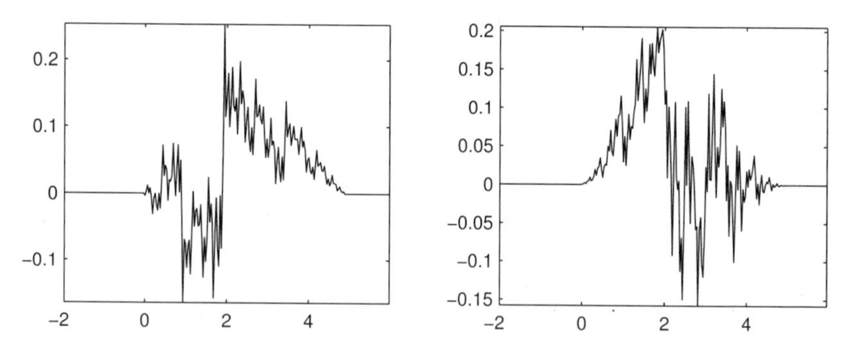

Fig. 12.2. Scaling function for half-integer oversampling and its permuted version.

Problem 12.9. Construct, for a given offset vector $\mathbf{t} = (t_0, \ldots, t_{L-1})$ and support M, an orthogonal scaling function φ having optimal smoothness such that $\|\mathbf{Z}\varphi(\mathbf{t}, z)\|^2 = \sum_{j=0}^{L-1} |Z\varphi(t_j, z)|^2$ has no zero on the complex plane.

12.4 Aliasing

12.4.1 Sampling Operators and MRA

The operator $\mathcal{S} = \mathcal{S}_{\mathbf{t}}$ in (12.12) defines an oblique projection onto $V(\varphi)$ in the sense that $\mathcal{S}f = f$ whenever $f \in V(\varphi)$. When φ is a scaling function for an MRA there is a natural, if not canonical, ambient space $\mathcal{H}$—namely the space V_1 of 2-scale dilates of elements of $V(\varphi) = V_0$—to consider as a space of inputs for the sampling projection $\mathcal{S}$. Janssen [37] proposed this choice of $\mathcal{H}$ by analogy with the Shannon-MRA case in which V_0 is the Paley–Wiener space $\mathrm{PW}_{1/2}$, i.e., bandlimited to $[-1/2, 1/2]$, and V_1 is then the space PW_1. In that case, aliasing refers to the error resulting from sampling at only half the Nyquist rate.

Actually, aliasing can be expressed in terms of the norm of $\mathcal{S}$ when thought of as a mapping from any PSI space $V(\psi)$ to $V(\varphi)$. In the MRA case, one has $V_1 = V_0 \oplus W_0$. Although V_1 is not itself a PSI space, it is the fact that W_0 may not be in the kernel of $\mathcal{S}$ that gives aliasing its meaning in this case. Thus, it makes sense to think of aliasing in terms of $\mathcal{S}$ as a mapping from W_0 to V_0 in this case. Despite its nebulous interpretation in terms of signals, the mathematical formulation of aliasing in the PSI context still parallels the Paley–Wiener case closely. Janssen's aliasing is the special case in which φ is an MRA scaling function and ψ is an associated wavelet.

Definition 12.10. *The aliasing norm $N(\mathcal{S}, \psi, \varphi)$ of a sampling operator $\mathcal{S}$ is its norm as a linear operator from $V(\psi)$ to $V(\varphi)$, that is,*

$$N(\mathcal{S}, \psi, \varphi) = \sup\left\{\frac{\|\mathcal{S}f\|_2}{\|f\|_2};\ f \in V(\psi)\right\}. \tag{12.31}$$

When φ, ψ are understood, we write $N(\mathcal{S})$ for short. The aliasing norm is a measure of the obliqueness of $\mathcal{S}$. If $\mathcal{S}$ is orthogonal and if $V(\psi) \subset V(\varphi)^{\perp}$, then there is no aliasing. A large aliasing norm indicates that $\mathcal{S}$ does not distinguish $V(\psi)$ from $V(\varphi)$ effectively. In this section we will present explicit formulas for aliasing norms associated with the sampling schemes outlined above. The formulas involve the offset parameters of those schemes explicitly and can be optimized accordingly.

12.4.2 Critical Offset Sampling

The norm of Janssen's [37] offset sampling operator $\mathcal{S}_{t_0}$ defined in (12.11) is

$$N(\mathcal{S}_{t_0}, \psi, \varphi) = \sup_{\|D\|_2=1} \int_0^1 |D(\xi)|^2 \left| \frac{Z\psi(t_0,\xi)}{Z\varphi(t_0,\xi)} \right|^2 d\xi = \sup_\xi \left| \frac{Z\psi(t_0,\xi)}{Z\varphi(t_0,\xi)} \right|^2 . \tag{12.32}$$

This was first observed by Walter [45] in the case $t_0 = 0$ and by Janssen [37] in the general case. One can regard t_0 as a parameter over which (12.32) can be minimized. In Fig. 12.3, $N(\mathcal{S}_{t_0})$ is computed over a range of t_0's for the Daubechies D_4 scaling function φ and corresponding wavelet ψ. There, the offset aliasing norm is minimized at around $t_0 \approx 0.62$. The value of $N(\mathcal{S}_{t_0})$ blows up for $t_0 \approx 0.2$, reflecting the fact that the Zak transform of the D_4 scaling function has a zero near $(0.2, 0.5)$.

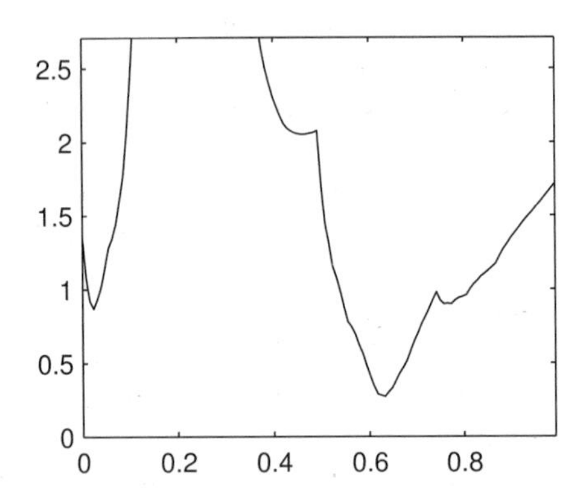

Fig. 12.3. Maximum magnitude of $Z\psi(t_0,\cdot)/Z\varphi(t_0,\cdot)$ as it depends on t_0 for D_4.

12.4.3 Aliasing Norms for Bunched Sampling

A judicious choice of offsets can lead to better control of aliasing. In the case of shift-bunched sampling, the interpolating functions are given by (12.14).

Theorem 12.11. *Let φ and ψ be orthogonal generators, both supported in $[0, M]$, for the respective PSI spaces $V(\varphi)$ and $V(\psi)$. Also, let $\mathcal{S}_\mathbf{t}$ be the sampling operator for critical shift-bunched sampling associated with $V(\varphi)$ and offset vector $\mathbf{t} = (t_0, t_1, \ldots, t_{M-1})$. Let M^ψ be the $M \times M$ matrix defined by $M^\psi_{jk} = \psi(t_j + k)$ $(0 \le j, k \le M-1)$, with M^φ defined accordingly. Then the aliasing norm of $\mathcal{S}_\mathbf{t}$ defined by (12.31) is*

$$N(\mathcal{S}_\mathbf{t}) = \|(M^\varphi)^{-1} M^\psi\|. \tag{12.33}$$

Figure 12.4 shows the aliasing norms of critical shift-bunched sampling operators for the Daubechies D_4 scaling function and for φ_ν with $\nu = -0.001$ respectively. Here $t_0 = 0$, $t_1 = 0.5$, and $t_2 = t$ is the variable $(0 \le t \le 0.45)$.

It illustrates that the aliasing norm can be quite large—the local minimum being about 25 for D_4. It can be made smaller by taking t_2 closer to 1, but the smallest aliasing norm that can be achieved with shift-bunched sampling is still much larger than that achieved with optimized offset integer sampling, at least in the D_4 case. In this case, large offset samples of $g \in W_0$ can coincide with peaks of interpolating functions. When $\nu = -0.001$, φ_ν is close to a Haar function. Aliasing performance is better than for D_4, although φ_ν has little regularity in this case.

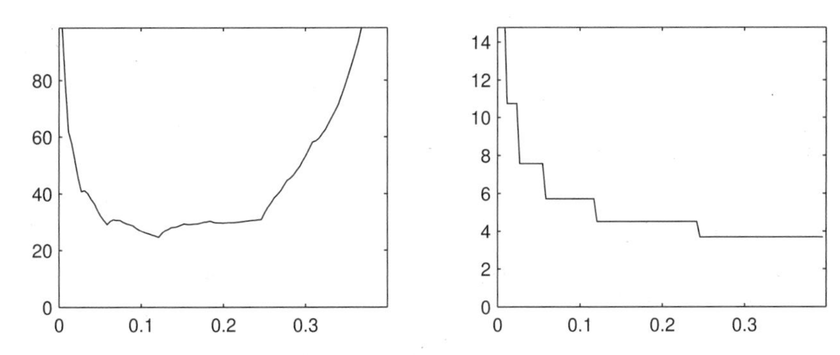

Fig. 12.4. Aliasing norm for staggered sampling with D_4 and φ_ν, $\nu = -0.001$.

A corresponding representation of the aliasing norm applies to support-bunched critical sampling. As before, let φ be an orthogonal generator supported on $[0, M]$ and ψ a generator of $V(\psi)$ also supported in $[0, M]$. Let $\mathbf{t} = (t_0, t_1, \ldots, t_{2M-1})$ be an offset vector with $0 \leq t_1 < t_2 < \cdots < t_{2M-1} < M$ and let N^φ be the $(2M-1) \times (2M-1)$ defined in (12.16) with corresponding interpolating functions as in (12.17) when N^φ is invertible. We have the following representation of the aliasing norm; see [34] for details.

Theorem 12.12. *Let φ, ψ be orthogonal generators of $V(\varphi)$ and $V(\psi)$ respectively, both supported in $[0, M]$. Let $\mathcal{S}_\mathbf{t}$ be the sampling operator associated to φ and let N^φ (and N^ψ) be defined as in* (12.16). *Then the aliasing norm of the sampling operator $\mathcal{S}_\mathbf{t}$ defined in* (12.12) *is*

$$N(\mathcal{S}_\mathbf{t}) = \|(N^\varphi)^{-1} N^\psi\|. \tag{12.34}$$

12.4.4 Aliasing Reduction by Oversampling

Oversampling theorems for the Paley–Wiener space provide some hope that oversampling can lead to reduction or elimination of aliasing in $V(\varphi)$. Here we will analyze the performance of oversampling in terms of aliasing when one of the coprimality conditions (12.18) or (12.20) is satisfied.

Coprimality on $\mathbb{T}$

If the polynomials $Z\varphi(t_j,\cdot)$, $j=0,\dots,L-1$ are coprime on $\mathbb{T}$, then the coefficient sequences $\{c_{jk}\}$ defined by

$$c_{jk}=\int_0^1 \frac{\overline{Z\varphi(t_j,\xi)}}{\|\mathbf{Z}\varphi(\mathbf{t},\xi)\|^2}\, e^{-2\pi ik\xi}\, d\xi$$

converge in $\ell^2(\mathbb{Z})$ for each $j=0,\dots,L-1$, thus giving rise to interpolating functions $S_j(t)=\sum_k c_{jk}\varphi(t-k)\in V(\varphi)$. Here we will write

$$f\mapsto \mathcal{O}_{\mathbf{t}}f(t)=\sum_{j=0}^{L-1}\sum_{\ell} f(t_j+\ell)\, S_j(t-\ell) \tag{12.35}$$

for the sampling projection (see (12.12)), to emphasize that we are oversampling.

Theorem 12.13. *Let φ, ψ be orthogonal generators of PSI spaces $V(\varphi)$ and $V(\psi)$ respectively. Let $\mathcal{O}_{\mathbf{t}}$ be the oversampling operator associated to φ and offset vector* $\mathbf{t}$ *as in* (12.35). *Set*

$$A_{\varphi,\psi}(\mathbf{t},\xi)=\frac{\mathbf{Z}\psi(\mathbf{t},\xi)\cdot\mathbf{Z}\varphi(\mathbf{t},\xi)}{\|\mathbf{Z}\varphi(\mathbf{t},\xi)\|^2}. \tag{12.36}$$

Then $\|\mathcal{O}_{\mathbf{t}}\|_{V(\psi)\to V(\varphi)}=\sup_\xi |A_{\varphi,\psi}(\mathbf{t},\xi)|$ while, if $A_{\varphi,\psi}(\mathbf{t},\cdot)$ is nonvanishing on $\mathbb{T}$, then $\|(\mathcal{O}_{\mathbf{t}})^{-1}\|_{V(\psi)\to V(\varphi)}=(\inf_\xi |A_{\varphi,\psi}(\mathbf{t},\xi)|)^{-1}$.

We include the proof to illustrate how orthonormality is used.

Proof. If $f(t)=\sum_k b_k\psi(t-k)\in V(\psi)$, then

$$\mathcal{O}_{\mathbf{t}}f(t)=\sum_{j=0}^{L-1}\sum_{\ell}\sum_{k} b_k\,\psi(t_j-k+\ell)\,S_j(t-\ell).$$

Taking the Fourier transform of both sides and using (12.19) yields

$$\begin{aligned}
\widehat{\mathcal{O}_{\mathbf{t}}f}(\xi)&=\sum_{j=0}^{L-1}\sum_{\ell}\sum_{k} b_k\,\psi(t_j+\ell-k)\,e^{-2\pi i\ell\xi}\,\widehat{S_j}(\xi)\\
&=B(-\xi)\sum_{j=0}^{L-1} Z\psi(t_j,-\xi)\,\widehat{S_j}(\xi)\\
&=B(-\xi)\left(\frac{\mathbf{Z}\psi(\mathbf{t},-\xi)\cdot\mathbf{Z}\varphi(\mathbf{t},-\xi)}{\|\mathbf{Z}\varphi(\mathbf{t},-\xi)\|^2}\right)\hat{\varphi}(\xi)=B(-\xi)\,A_{\varphi,\psi}(\mathbf{t},-\xi)\hat{\varphi}(\xi)
\end{aligned}$$

with $B(\xi)=\sum_k b_k e^{2\pi ik\xi}$ and $A_{\varphi,\psi}(\mathbf{t},\xi)$ as in (12.36). Since φ has orthonormal shifts,

$$\|\mathcal{O}_{\mathbf{t}} f\|_2^2 = \|(\mathcal{O}_{\mathbf{t}} f)^\wedge\|_2^2 = \int_0^1 |B(\xi)|^2 \, |A_{\varphi,\psi}(\mathbf{t},\xi)|^2 \, d\xi.$$

Since ψ has orthonormal shifts, $\|f\|_2^2 = \int_0^1 |B(\xi)|^2 \, d\xi$ so

$$\inf_\xi |A_{\varphi,\psi}(\mathbf{t},\xi)| \, \|f\|_2 \le \|\mathcal{O}_{\mathbf{t}} f\|_2 \le \sup_\xi |A_{\varphi,\psi}(\mathbf{t},\xi)| \, \|f\|_2$$

with sharp bounds, as in (12.32). This proves the theorem. $\square$

The vector $\mathbf{t}$ can be adjusted to minimize aliasing. In fact, in the case of the Shannon scaling function ($V(\varphi) = \mathrm{PW}_{1/2}$) and Shannon wavelet ψ, the aliasing norm of any offset (critical) sampling operator is unity. In the case of oversampling with any offset vector of the form $(t_0, t_0 + 1/2)$, i.e., Nyquist for PW_1, $A_{\varphi,\psi}(\mathbf{t},\xi) = 0$, that is, $\mathcal{O}_{\mathbf{t}}$ annihilates W_0 (see [34] for details).

Coprimality on $\mathbb{C}$

The previous theorem takes no advantage of support properties of φ. Suppose now that φ is an orthogonal generator supported on $[0, M]$ and

$$0 \le t_0 < t_1 < \cdots < t_{L-1} < 1$$

are sampling nodes for which the polynomials $\{Z_{\mathbb{C}}\varphi(t_n, z)\}_{n=0}^{L-1}$, whose degrees are at most $M-1$, are coprime on $\mathbb{C}$. Euclid's algorithm then ensures the existence of polynomials $P_0, P_1, \ldots, P_{N-1}$ of degree at most $M-2$ such that (12.20) holds. Then the sampling functions defined in Section 12.3.3 have compact support in $[0, 2M-2]$. The aliasing performance of $\mathcal{O}_{\mathbf{t}}$ in (12.35) is determined in the following result whose proof can be found in [34].

Theorem 12.14. *Let φ, ψ be orthogonal generators of $V(\varphi)$ and $V(\psi)$ respectively. For an offset vector $\mathbf{t} = (t_0, \ldots, t_{L-1})$ let the vector polynomial $\mathbf{P} = (P_0, \ldots, P_{L-1})$ be as in* (12.20) *and $\mathcal{O}_{\mathbf{t}}$ be as in* (12.35). *Set*

$$A_{\mathbf{t}}(\xi) = \mathbf{Z}\psi(\mathbf{t},\xi) \cdot \mathbf{P}(\xi). \tag{12.37}$$

Then $\|\mathcal{O}_{\mathbf{t}}\|_{V(\psi)\to V(\varphi)} = \sup_\xi |A_{\mathbf{t}}(\xi)|$. Additionally, if $A_{\mathbf{t}}$ does not vanish on $\mathbb{T}$, then $\|(\mathcal{O}_{\mathbf{t}})^{-1}\|_{V(\varphi)\to V(\psi)} = 1/(\inf_\xi |A_{\mathbf{t}}(\xi)|)$.

In Fig. 12.5, the aliasing norms of the oversampling operators $\mathcal{O}_{(t,t+1/2)}$ are plotted for the Daubechies D_4 scaling function and wavelet. The horizontal axis is the offset parameter t. The aliasing norm is very small when $t \approx 0.09$.

12.4.5 Eliminating Aliasing

In Section 12.3.3 we saw that coefficients c_k of $f = \sum c_k \varphi(\cdot - k)$ could be expressed as weighted averages of samples of f taken at a sufficiently high

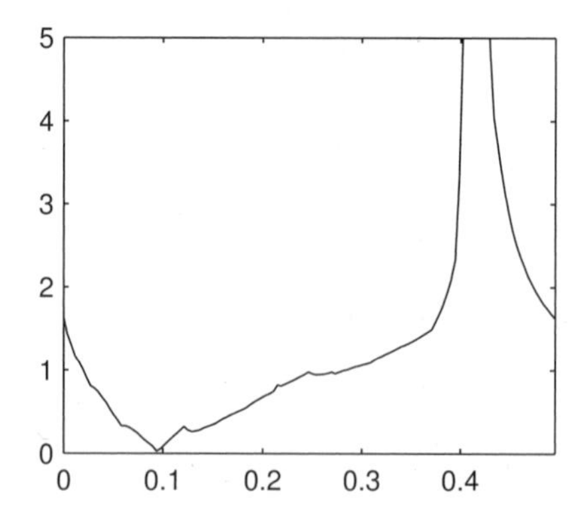

Fig. 12.5. Aliasing norm $N(\mathcal{O}_{(t,t+1/2)})$, $0 \le t \le 1/2$, for the D_4 scaling function.

rate. Similar ideas apply to the problem of eliminating aliasing. The latter requires

$$A_{\mathbf{t}}(z) = \mathbf{P}(z) \cdot \mathbf{Z}_{\mathbb{C}}\psi(\mathbf{t}, z) \equiv 0. \tag{12.38}$$

Oversampling at a high enough rate can eliminate aliasing altogether. Suppose that (12.20) can be obtained in such a way that the components of $\mathbf{P} = (P_0, P_1, \ldots, P_{L-1})$ have degree at most D. The polynomial $\mathbf{P}(z) \cdot \mathbf{Z}_{\mathbb{C}}\varphi(\mathbf{t}, z)$ has degree at most $D + M - 1$ and therefore has $D + M$ coefficients. Writing $P_j(z) = \sum_{k=0}^{D} p_{jk} z^k$ and assuming, as before, that both φ, ψ are supported in $[0, M]$, a joint solution of (12.20) and (12.38) imposes $2(D + M)$ constraints on the $L(D+1)$ coefficients p_{jk}. Thus, to ensure the existence of simultaneous solutions of (12.20) and (12.38) one should have $2(D + M) \le L(D + 1)$. If $M = 3$, as in the case of the D_4 scaling function, with ψ an associated D_4 wavelet, one needs $L \ge 6$ when $D = 0$, while if $M = D = 3$ one just needs $L \ge 3$.

When $D = 0$, a common solution $\mathbf{a} = (a_0, a_1, \ldots, a_{L-1})$ of (12.20) and (12.38) i.e., of

$$\mathbf{a} \cdot \mathbf{Z}_{\mathbb{C}}\varphi(\mathbf{t}, z) \equiv 1, \qquad \mathbf{a} \cdot \mathbf{Z}_{\mathbb{C}}\psi(\mathbf{t}, z) \equiv 0, \tag{12.39}$$

requires $L \ge 2M$. Then

$$\mathcal{O}_{\mathbf{t}} f(t) = \sum_k \left(\sum_{j=0}^{L-1} f(t_j + k) a_j \right) \varphi(t - k).$$

As in Section 12.3.3, the interpolating functions then coincide with the scaling functions while the coefficients c_k of $f = \sum_k c_k \varphi(\cdot - k)$ are obtained via quadrature: $c_\ell = \sum_{j=0}^{L-1} a_j f(t_j + \ell)$. When $\mathbf{a}$ satisfies (12.39), $\mathcal{O}_{\mathbf{t}}$ will annihilate $V(\psi)$.

12.5 Noise

Quantization error in analog-to-digital conversion is usually thought of in terms of noise—particularly as high frequency noise. As such, quantization error is often treated similarly to aliasing error. Consider the problem of adding noise to a signal in PW_Ω, then sampling along $\mathbf{t} + \frac{1}{2\Omega}\mathbb{Z}$, where $\mathbf{t} = (t_0, \ldots, t_{L-1}) \subset [0, 1/(2\Omega)]$. One can model *sampling noise* (due to quantization and other effects) as white noise: the error added to each sample is mean zero with the same variance σ^2 for each offset (e.g., [20]). This variance should not depend on $\mathbf{t}$. However, just as different offsets have different effects on aliasing performance, the *reconstruction error* rec err(t) *will* depend on $\mathbf{t}$.

Seidner and Feder [40] quantified reconstruction error in terms of a *noise amplification factor*

$$A_\varepsilon = \frac{\mathrm{Ave}(E\{|\mathrm{rec\ err(t)}|^2\})}{\sigma^2}$$

in which Ave denotes time averaging, while $E\{|\mathrm{rec\ err(t)}|^2\}$ is the variance of the reconstruction error at a particular time. Defining

$$S_j(t) = \prod_{i \neq j} \sin^2 \frac{\Omega}{L}(t - t_j),$$

Seidner and Feder found that

$$A_\varepsilon = \frac{1}{L} \sum_{j=0}^{L-1} \sum_{k=0}^{L-1} \frac{S_j(k/\Omega)}{S_j(t_j)}$$

and noted that A_ε is minimized over all generating vectors $\mathbf{t}$ precisely when the t_j are evenly spaced.

Bölcskei [21] derived a formula for reconstruction error in the context of PSI spaces that is reminiscent of the aliasing quantities considered in Section 12.4.4. Here, one supposes that the input to the synthesis filter bank is

$$\tilde{f}(M\ell + t_j) = f(M\ell + t_j) + q_j(\ell)$$

in which $q_j(\ell)$ is additive, zero mean Gaussian noise with variance $\sigma_q^2 = E\{|q_j(\ell)|^2\}$, independent of j. One defines the polyphase matrix $A^\varphi_{jk}(z)$ with entries $A^\varphi_{jk}(z) = \sum_\ell \varphi(t_j + M\ell - k)z^{-\ell}$. As before, one assumes that $\mathrm{supp}\,\varphi \subset [0, \mathrm{M}]$. Bölcskei [21] asserted that the variance σ_{re} of the reconstruction error engendered by interpolating samples when $f \in V(\varphi)$ is minimized when the interpolating functions are defined, analogously to (12.5), in terms of the pseudoinverse $R^\dagger(z) = (A^{\varphi *}(z)A^\varphi(z))^{-1}A^{\varphi *}(z)$. Here $A^*(z)$ denotes the conjugate transpose of $A(z^*)$.

Importantly, Bölcskei went on to quantify this noise amplification in terms analogous to those encountered in Section 12.4.4. Setting $z = e^{2\pi i \xi}$ and abusing notation as before, denote by $\Lambda(\xi)$ the spectrum of the self-adjoint matrix

$A^{\varphi *}(\xi)A^{\varphi}(\xi)$. Set $\alpha = \inf_\xi \min \Lambda(\xi)$ and $\beta = \sup_\xi \max \Lambda(\xi)$. Error amplification bounds can be expressed in terms of α, β as

$$\frac{1}{\beta} \leq \frac{\sigma_{\text{re}}^2}{\sigma_q^2} \leq \frac{1}{\alpha}.$$

In practice, the ratio $\rho = \beta/\alpha$ can be thought of as a type of *condition number*, representing the stability of the reconstruction via interpolation of samples. The quantity ρ can be expressed in now familiar terms (see [21]) as

$$\rho = \frac{\beta}{\alpha} = \frac{\sup_\xi \left(\|\mathbf{Z}\varphi(\mathbf{t}, \xi)\|^2 \right)}{\inf_\xi \left(\|\mathbf{Z}\varphi(\mathbf{t}, \xi)\|^2 \right)}.$$

In contrast to the Paley–Wiener case, ρ is not necessarily minimized when the entries of $\mathbf{t}$ are evenly spaced.

12.6 Translation-Invariance and Lack Thereof

A subspace $V \subset L^2(\mathbb{R})$ is *translation-invariant* if every translation operator $\tau_a : f(t) \mapsto f(t-a)$ preserves V. The only closed, translation-invariant subspaces of $L^2(\mathbb{R})$ are the subspaces $V_E = \{(\hat{f}\chi_E)^\vee : f \in L^2(\mathbb{R})\}$ where $E \subset \mathbb{R}$ is measurable. If V_E is a closed, translation-invariant subspace of $L^2(\mathbb{R})$, then it makes no difference whether one samples $f \in V_E$ along $\{t_k\}$ or at an arbitrary translate $\{t_k - a\}$ thereof. In the case of sampling in general PSI spaces this is no longer true, since PSI spaces are only invariant under integer shifts—not arbitrary ones. This brings up an interesting question concerning sampling in PSI spaces. Suppose that one has samples of some delayed version $\tau_a f$ of a signal f that belongs to or nearly belongs to $V(\varphi)$. How can one determine the delay, a?

12.6.1 Delay Discrepancy

Given $f \in V(\varphi)$, $a \in \mathbb{R}$, consider the quantity

$$d_\varphi(f, a) = \|\tau_a f - \mathcal{P}_\varphi(\tau_a f)\|_2^2,$$

where $\mathcal{P}_\varphi$ denotes the *orthogonal* projection onto $V(\varphi)$. If $\tau_a f \in V(\varphi)$, then $d(f, a) = 0$, while if $\tau_a f \in V(\varphi)^\perp$, then $d(f, a) = \|f\|_2^2$. Hence $d(f, a)$ measures the energy of f that "leaks out" of $V(\varphi)$ when f is translated (delayed) by a. The non-invariance of the space $V(\varphi)$ under the translation τ_a may be measured by

$$d_\varphi(a) = \sup_{f \in V(\varphi),\ \|f\|_2 = 1} d(f, a)$$

and as a general measure of non-invariance of $V(\varphi)$ under translations we consider norms of d_φ such as the *discrepancies*

$$d_\varphi = d_\varphi^\infty = \sup_{0 \le a < 1} d_\varphi(a), \quad \text{or} \quad d_\varphi^1 = \int_0^1 d_\varphi(a)\, da.$$

We will consider only the maximum discrepancy d_φ; see Bastys [7], [8] for related measures.

A large value of $d_\varphi(a)$ (i.e., $d_\varphi(a) \sim 1$) means that there is a signal $f \in V(\varphi)$ such that $\tau_a f$ has most of its energy in $V(\varphi)^\perp$. In fact, we have the following somewhat surprising result, whose proof can be found in [34].

Theorem 12.15. *Let $\varphi \in W(L^\infty, \ell^1)$ be an orthogonal generator for the PSI space $V(\varphi)$. Then there exists $a \in (0,1)$ with $d_\varphi(a) = 1$, i.e., $d_\varphi = 1$. If, in addition, φ is the scaling function of an MRA, then $d_\varphi(1/2) = 1$.*

The theorem serves as a warning about thinking of $V(\varphi) = V_0$ as a space of averages, with wavelet spaces containing details, when $V_0 = V(\varphi)$ is the base space of an MRA: in that case the delay of some *average* by $1/2$ consists (almost) entirely of *details*.

The hypothesis that $\varphi \in W(L^\infty, \ell^1)$ is needed: there is no delay discrepancy in the Paley–Wiener space $\mathrm{PW}_{1/2}$. However, $\mathrm{sinc}(t) \notin W(L^\infty, \ell^1)$.

The proof of Theorem 12.15 hinges on the *temporal autocorrelation* of the Zak transform, namely

$$B_\varphi(a, \xi) = \int_0^1 Z\varphi(s - a, \xi)\, \overline{Z\varphi}(s, \xi)\, ds.$$

For $f \in V(\varphi)$ and $C(\xi) = Zf(a,\xi)/Z\varphi(a,\xi)$,

$$\|\mathcal{P}_\varphi(\tau_a f)\|_2^2 = \int_0^1 C(\xi)\, |B_\varphi(a, \xi)|^2.$$

As the Zak transform of $\varphi^* * \varphi(-s)$, B_φ has a zero in the unit square. The first statement follows by choosing C to be localized near such a zero. When φ is a scaling function, algebraic properties inherited by $Z\varphi$ imply that $B(1/2, \xi) = 0$ at those ξ for which the QMF H satisfies $|H(\xi/2)|^2 = 1/2$.

12.6.2 Delay Errors and Their Reduction by Oversampling

Suppose that integer sampling is valid in $V(\varphi)$. That is, $V(\varphi)$ possesses an interpolating function S such that any $f \in V(\varphi)$ has an expansion $f(t) = \sum_k f(k) S(t-k)$. Suppose that a receiver acquires samples of a signal g of the form $g = \tau_a f$ for some $f \in V(\varphi)$. Theorem 12.15 states that, if $\varphi \in W(L^\infty, \ell^1)$ is an orthogonal generator of $V(\varphi)$, then an error in the assumed value of a can lead to an arbitrarily large error in the sampling reconstruction. As such, it is important to be able to determine the delay a when the hypothesis $f \in V(\varphi)$ applies. Oversampling can be helpful in determining the amount a of the delay.

Suppose that we wish to use periodic oversampling (i.e., (12.35)) to determine $f \in V(\varphi)$ from its samples $f(t_\ell + k)$. As before, $k \in \mathbb{Z}$ and

$t_0 < t_1 < \cdots < t_{L-1}$ defines an offset vector $\mathbf{t}$. Previously, all entries of $\mathbf{t}$ were taken in $[0,1)$. Suppose now that, instead of inputting the samples of f, we input those of $\tau_a f$. That is, our input data is $\{f(t_\ell + k - a)\}$ for some $a \in (0,1)$. To estimate the discrepancy between f and our reconstruction, we compute the difference in L^2-norm between f and $\mathcal{O}_{\mathbf{t}}\tau_a f$.

Theorem 12.16. *Let* φ, $\mathbf{t} = (t_0, t_1, \ldots, t_{L-1})$ *be as above and set* $\tau_a \mathbf{t} = (t_0 - a, \ldots, t_{L-1} - a)$ $(0 \le a < 1)$. *Set*

$$D_\varphi(a, \mathbf{t}, \xi) = \frac{\mathbf{Z}\varphi(\tau_a \mathbf{t}, \xi) \cdot \mathbf{Z}\varphi(\mathbf{t}, \xi)}{\|\mathbf{Z}\varphi(\mathbf{t}, \xi)\|^2} = A_{\varphi, \tau_a \varphi}(\mathbf{t}, \xi). \tag{12.40}$$

Then

$$\|I - \mathcal{O}_{\mathbf{t}}\tau_a\|_{V(\varphi)\to V(\varphi)} = \sup_\xi |D_\varphi(y, a, \mathbf{t}, \xi) - 1|.$$

One would like $D_\varphi(y, a, \mathbf{t}, \xi)$ to be close to one for all ξ. In this situation, shifting the data by a causes little error in the reconstruction. Just as regularity could be used to validate oversampling in $V(\varphi)$ at a high enough rate, one can also guarantee low discrepancy at a high enough sampling rate. As before, though, algebraic properties of φ might provide more leverage in guaranteeing small errors. In fact, algebraic properties can be used to determine a when $f \in V(\varphi)$ and $\tau_a f$ is periodically sampled at a high enough rate.

To keep notational complexity to a minimum, assume that φ has compact support and that one can find $0 \le t_0 < t_1 < 1$ such that the polynomials $Z\varphi(b + t_\varepsilon, \xi)$ $(\varepsilon = 0, 1)$ are coprime on $\mathbb{T}$ for all b. If, in addition, φ is a scaling function with some regularity, then one is able to determine that $a = b$ as follows.

Theorem 12.17. *Let* φ *be a compactly supported orthogonal scaling function in Lip*(α) $(\alpha > 0)$ *and let* $0 \le t_0 < t_1 < 1$. *Suppose that* $Z_{\mathbb{C}}\varphi(b + t_0, \cdot)$ *and* $Z_{\mathbb{C}}\varphi(b + t_1, \cdot)$ *are coprime for all* b. *For* $a \in [0,1)$ *and* $f \in V(\varphi)$, $f \neq 0$, *set*

$$F(b,p) = \sum_\ell \big[f(a + t_0 + \ell)\varphi(b + t_1 + p - \ell) - f(a + t_1 + \ell)\varphi(b + t_0 + p - \ell)\big].$$

Then $a = b$ *if and only if* $F(b,p) = 0$ *for all* $p \in \mathbb{Z}$.

The scaling property and continuity play decisive roles in the proof. Rather than reviewing it (see [34]), we illustrate its use in the case of the Daubechies D_4 scaling function. The coprimality condition is satisfied when $t_0 = 0$ and $t_1 = 1/2$. On the left-hand side of Fig. 12.6, we consider a signal f in the multiresolution space $V(\varphi)$ supported on $[0, 100]$. The signal was translated by $1/4$ then sampled at the integers. On the right, $\|F(b,\cdot)\|_2$ is computed for this data using $t_0 = 0$ and $t_1 = 1/2$. As indicated, F has a zero at $\xi = 1/4$, thus demonstrating the effectiveness of the algorithm in determining the delay of sampled data.

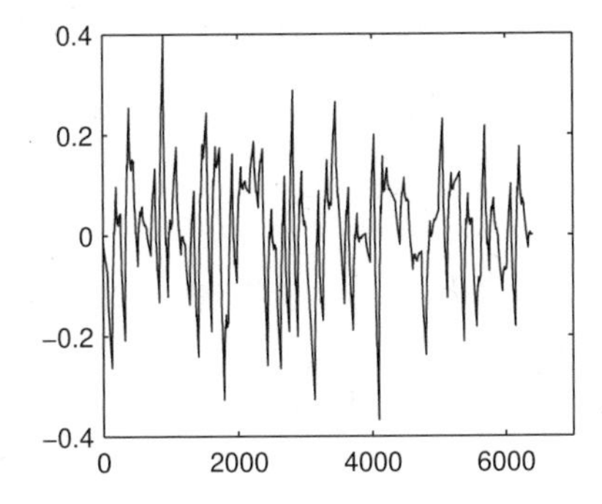

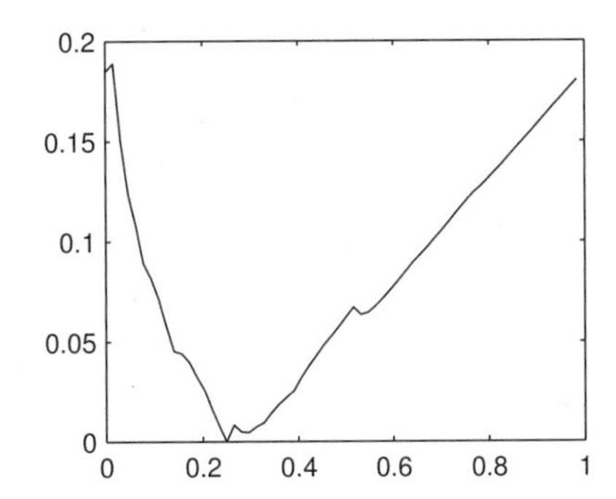

Fig. 12.6. Synthetic signal in $V(\varphi)$ with φ the Daubechies D_4 scaling function. On the right is the shift determination function $\|F(b, \cdot)\|_2$ for this data.

Acknowledgments

John Benedetto is in our "Ted Williams" category of mathematicians. His panoramic vision of harmonic analysis, deeply relevant insights, and prolific output have been and continue to be an inspiration. But more importantly to us, John's generosity has been instrumental in shaping our careers in positive ways.

The work presented here has also benefited from conversations and correspondence with John Gilbert, Karlheinz Gröchenig, Cristina Pereyra, Xiaoping Shen, and Gilbert Walter. We would like to thank Chris Heil for encouraging us to write this chapter in John Benedetto's honor.

Lakey gratefully acknowledges support through ARO contract DAAD19-02-1-0211.

References

1. A. Aldroubi and H. Feichtinger, Exact iterative reconstruction algorithm for multivariate irregularly sampled functions in spline-like spaces: the L^p-theory, *Proc. Amer. Math. Soc.*, **126** (1998), pp. 2677–2686.
2. A. Aldroubi and K. Gröchenig, Nonuniform sampling and reconstruction in shift-invariant spaces, *SIAM Review*, **43** (2001), pp. 585–620.
3. A. Aldroubi and K. Gröchenig, Beurling-Landau-type theorems for non-uniform sampling in shift invariant spline spaces, *J. Fourier Anal. Appl.*, **6** (2000), pp. 93–103.
4. A. Aldroubi and M. Unser, Projection based prefiltering for multiwavelet transforms, *IEEE Trans. Signal Proc.*, **46** (1998), pp. 3088–3092.
5. X. Aragones, J. L. Gonzalez and A. Rubio, *Analysis and Solutions for Switching Noise in Coupling Mixed Signal ICs*, Kluwer, Dordrecht, 1999.
6. N. Atreas, J. J. Benedetto and C. Karanikas, Local sampling for regular wavelet and Gabor expansions, *Sampl. Theory Signal Image Process.*, **2** (2003), pp. 1–24.

7. A. J. Bastys, Translation invariance of orthogonal multiresolution analyses of $L^2(\mathbb{R})$, *Appl. Comput. Harmon. Anal.*, **9** (2000), pp. 128–145.
8. A. J. Bastys, Orthogonal and biorthogonal scaling functions with good translation invariance characteristic, in: Proc. SampTA, Aveiro (1997), pp. 239–244.
9. J. J. Benedetto, Frames, sampling, and seizure prediction, in: *Advances in Wavelets* (Hong Kong, 1997), Springer, Singapore, 1999, pp. 1–25.
10. J. J. Benedetto, *Harmonic Analysis and its Applications*, CRC Press, Boca Raton, 1997.
11. J. J. Benedetto, Frame decompositions, sampling, and uncertainty principle inequalities, in: *Wavelets: Mathematics and Applications*, CRC Press, Boca Raton, 1994, pp. 247–304.
12. J. J. Benedetto, Irregular sampling and frames, in: *Wavelets: A Tutorial in Theory and Applications*, C. K. Chui ed., CRC Press, Boca Raton, 1992, pp. 445–507.
13. J. J. Benedetto, *Spectral Synthesis*, Academic Press, New York, 1975.
14. J. J. Benedetto and P. J. S. G. Ferreira, Introduction, in: *Modern Sampling Theory*, J. J. Benedetto and P. J. S. G. Ferreira, eds., Birkhäuser, Boston, 2001, pp. 1–26.
15. J. J. Benedetto and W. Heller, Irregular sampling and the theory of frames. I, *Note Mat.*, **10**, pp. 103–125.
16. J. J. Benedetto and S. Scott, Frames, irregular sampling, and a wavelet auditory model, in: *Nonuniform Sampling*, Kluwer/Plenum, New York, 2001, pp. 585–617.
17. J. J. Benedetto and O. Treiber, Wavelet frames: multiresolution analysis and extension principles, in: *Wavelet Transforms and Time-Frequency Signal Analysis*, L. Debnath, ed., Birkhäuser, Boston, 2001, pp. 3–36.
18. J. J. Benedetto, Ö. Yilmaz, and A. M. Powell, Sigma-Delta quantization and finite frames, in: Proc. IEEE Int. Conf. on Acoustics, Speech and Signal Processing (ICASSP), Vol. 3, Montreal, Canada, 2004, pp. 937–940.
19. J. J. Benedetto and G. Zimmermann, Sampling operators and the Poisson summation formula, *J. Fourier Anal. Appl.*, **3** (1997), pp. 505–523.
20. W. R. Bennett, Spectra of quantized signals, *Bell System Tech. J.*, **27** (1948), pp. 446–472.
21. H. Bölcskei, Oversampling in wavelet subspaces, in: Proc. IEEE-SP 1998 Int. Sympos. Time-Frequency Time-Scale Analysis, Pittsburgh, 1998, pp. 489–492.
22. R. Bracewell, *The Fourier Transform and its Applications*, McGraw-Hill, New York, 1965.
23. Y. Bresler and R. Venkataramani, Sampling theorems for uniform and periodic nonuniform MIMO sampling of multiband signals, *IEEE Trans. Signal Process.*, **51** (2003), pp. 3152–3163.
24. S. D. Casey and D. F. Walnut, Systems of convolution equations, deconvolution, Shannon sampling, and the wavelet and Gabor transforms, *SIAM Review*, **36** (1994), pp. 537–577.
25. W. Chen, S. Itoh and J. Shiki, Irregular sampling theorems for wavelet subspaces, *IEEE Trans. Inform. Theory*, **44** (1998), pp. 1131–1142.
26. M. Croft and J. A. Hogan, Wavelet-based signal extrapolation, in: Proc. Fourth Int. Symp. Sig. Proc. and Appl., Gold Coast, Australia, **2** (1996), pp. 752–755.
27. I. Daubechies, *Ten Lectures on Wavelets*, SIAM, Philadelphia, 1992.
28. I. Djokovic and P. P. Vaidyanathan, Generalized sampling theorems in multiresolution subspaces, *IEEE Trans. Signal Proc.*, **45** (1997), pp. 583–599.

29. C. Heil and D. Walnut, Continuous and discrete wavelet transforms, *SIAM Review*, **31** (1989), pp. 628–666.
30. C. Herley and P.-W. Wong, Minimum rate sampling and reconstruction of signals with arbitrary frequency support, *IEEE Trans. Inform. Theory*, **45** (1999), pp. 1555–1564.
31. J. A. Hogan and J. D. Lakey, Sampling and aliasing without translation-invariance, in: Proc. SampTA, Orlando, 2001, pp. 61–66.
32. J. A. Hogan and J. D. Lakey, Sampling for shift-invariant and wavelet subspaces, in: "Wavelet Applications in Signal and Image Processing VIII" (San Diego, CA, 2000), Proc. SPIE **4119**, A. Aldroubi, A. F. Laine, and M. A. Unser, eds., SPIE, Bellingham, WA, 2000, pp. 36–47.
33. J. A. Hogan and J. D. Lakey, Sampling and oversampling in shift-invariant and multiresolution spaces I: Validation of sampling schemes, *Int. J. Wavelets Multiresolut. Inf. Process.*, **3** (2005), pp. 257–281.
34. J. A. Hogan and J. D. Lakey, Sampling and oversampling in shift-invariant and multiresolution spaces II: Estimation and aliasing, in preparation.
35. N. Jacobson, *Basic Algebra*, W. H. Freeman, New York, 1985.
36. A. J .E .M. Janssen, The Zak transform: A signal transform for sampled time-continuous signals, *Philips J. Res.*, **43** (1998), pp. 23–69.
37. A. J. E. M. Janssen, The Zak transform and sampling theorems for wavelet subspaces, *IEEE Trans. Signal Proc.*, **41** (1993), pp. 3360–3364.
38. A. Papoulis, Generalized sampling expansion, *IEEE Trans. Circuits and Systems*, **24** (1977), pp. 652–654.
39. O. Rioul, Simple regularity criteria for subdivision schemes, *SIAM J. Math. Anal.*, **23** (1992), pp. 1544–1576.
40. D. Seidner and M. Feder, Noise amplification of periodic nonuniform sampling, *IEEE Trans. Signal Proc.*, **48** (2000), pp. 275–277.
41. W. Sun and X. Zhou, Reconstruction of functions in spline subspaces from local averages, *Proc. Amer. Math. Soc.*, **131** (2003), pp. 2561–2571.
42. M. Unser and J. Zerubia, A generalized sampling theory without bandlimiting constraints, *IEEE Trans. Circuits and Systems II*, **45** (1998), pp. 959–969.
43. P. P. Vaidyanathan, Sampling theorems for non bandlimited signals: theoretical impact and practical applications in: Proc. SampTA, Orlando, 2001, pp. 17–26.
44. M. Vrhel and A. Aldroubi, Sampling procedures in function spaces and asymptotic equivalence with Shannon's sampling theory, *Numer. Funct. Anal. Optim.*, **15** (1994), pp. 1–21.
45. G. Walter, A sampling theorem for wavelet subspaces, *IEEE Trans. Inform. Theory*, **38** (1992), pp. 881–884.
46. X.-G. Xia and Z. Zhang, On sampling theorem, wavelets, and wavelet transforms, *IEEE Trans. Signal Proc.*, **41** (1993), pp. 3524–3535.
47. J. L. Yen, On the nonuniform sampling of bandwidth-limited signals, *IRE Trans. Circuit Theory* **3** (1956), 251–257.

13

Sampling on Unions of Shifted Lattices in One Dimension

Bjarte Rom[1] and David Walnut[2]

[1] Department of Mathematical Sciences, The Norwegian University of Science and Technology, 7491 Trondheim, Norway bjarter@math.ntnu.no

[2] Department of Mathematical Sciences, George Mason University, Fairfax, Virginia 22030, USA dwalnut@gmu.edu

Summary. We give a complete solution to the problem of sampling and interpolation of functions in $PW_\sigma(\mathbf{R})$ on finite unions of shifted lattices in $\mathbf{R}$ of the form $\Lambda_j = \frac{1}{2\sigma_j}\mathbf{Z} + \alpha_j$, $j = 1, \ldots, m$ where $\sigma_j > 0$, $\alpha_j \in \mathbf{R}$, and $\sum_j \sigma_j = \sigma$. At points where more than one lattice intersect, we sample the function and its derivatives. None of the results or techniques employed is new, but a systematic and elementary treatment of this situation does not seem to exist in the literature. Sampling on unions of shifted lattices includes classical sampling, bunched or periodic sampling, and sampling with derivatives. Such sampling sets arise in deconvolution, tomography, and in the theory of functions bandlimited to convex regions in the plane.

13.1 Introduction

In the Classical Sampling Theorem, samples of a bandlimited function are taken on a sequence of evenly spaced points on the real line, in other words, on a lattice in $\mathbf{R}$. Specifically, if $f \in PW_\sigma(\mathbf{R})$, the space of functions $f \in L^2(\mathbf{R})$ whose Fourier transform vanishes outside the interval $[-\sigma, \sigma]$, then

$$f(x) = \sum_n f\left(\frac{n}{2\sigma}\right) \frac{\sin(2\pi x\sigma - \pi n)}{2\pi x\sigma - \pi n},$$

where the series converges uniformly and in $L^2(\mathbf{R})$. The sampling theorem has been generalized for example to higher dimensions, and to nonuniform sampling sets (see for example [36], [3], [4], [16], [31]). A substantial literature exists regarding further generalizations of the theorem (see [30], [42], [43], [20], [21] [24]).

A particular generalization that has substantial applications has been to consider sets of the form $\cup_{i=1}^m \{n/2\sigma + \alpha_i\}_{n\in\mathbf{Z}}$ where $\alpha_i \in \mathbf{R}$ [41], [34, Chap. 6], or in d dimensions on sets of the form $\cup_{i=1}^m \{Wn + \alpha_i\}_{n\in\mathbf{Z}^d}$ where W is a nonsingular $d \times d$ matrix, and $\alpha_i \in \mathbf{R}^d$ [14], [9], [23]. These sampling sets are unions of shifts of a single lattice and this type of sampling is referred to as periodic or bunched sampling.

One can also consider sampling on shifts of different lattices, that is, on sets of the form $\cup_{i=1}^{m}\{n/2r_i + \alpha_i\}_{n\in\mathbf{Z}}$ in one dimension or more generally on sets $\cup_{i=1}^{m}\{W_i n + \alpha_i\}_{n\in\mathbf{Z}^d}$, where $\{W_i\}_{i=1}^{m}$ is a collection of nonsingular $d \times d$ matrices. Sampling on such sets arises naturally in a number of applications in signal processing and remote sensing [5], [6], [7], [8], [40], in interlaced sampling schemes in tomography [12], [15], in recovery of functions from their local averages [39], [19], and also in the theory of sampling and interpolation of functions bandlimited to polygonal regions in the plane [29].

In most cases, the problem of unique determination and recovery of a bandlimited function from its samples on a union of lattices is dealt with in an ad hoc manner and there seem to be very few papers that deal with this case specifically and in a systematic way. However, techniques for completely solving this problem exist in the literature and are quite elementary. They involve some complex analysis, the theory of Hilbert spaces of entire functions, and some notions about polynomial interpolation in the complex plane. Our goal in this chapter is to present these results in a systematic and elementary way.

Specifically, we consider the case of sampling on a finite union of arbitrary shifted lattices on $\mathbf{R}$. Our setting throughout the chapter is the following. Let positive numbers $\sigma_1 \leq \sigma_2 \leq \cdots \leq \sigma_m$ and arbitrary real numbers $\alpha_1, \alpha_2, \ldots, \alpha_m$ be given. Let $\sigma = \sum_{i=1}^{m} \sigma_i$ and consider the shifted lattices $\Lambda_i = \frac{1}{2\sigma_i}\mathbf{Z} + \alpha_i$. Let $\Lambda = \cup_{i=1}^{m} \Lambda_i$. We note several elementary observations about the set Λ.

(1) A point $\lambda \in \Lambda$ can sit in more than one but no more than m shifted lattices Λ_j. Indeed there is nothing stopping us from taking two (σ_j, α_j) to be the same and hence having two of the shifted lattices Λ_j coincide.

(2) If the numbers σ_j are pairwise rationally related, that is, if σ_i/σ_j is rational for every i and j, then no matter what α_j are chosen the set Λ is *periodic* in the sense that there is a finite subset $F \subseteq \Lambda$ and a number Σ such that

$$\Lambda = \bigcup_{n\in\mathbf{Z}} (F + n\Sigma).$$

In this case, the set Λ is *separated*, that is,

$$\inf\{|\lambda - \mu| : \lambda, \mu \in \Lambda, \lambda \neq \mu\} > 0.$$

(3) If not all of the numbers σ_j are rationally related, then Λ is not periodic and not separated. Given $\varepsilon > 0$, it is possible, by going sufficiently far away from the origin, to find pairs of points in Λ that do not coincide but which are within ε of each other. In fact, one can find "bunches" of diameter ε containing at least two and no more than m points.

(4) Whether the σ_j are all rationally related or not, it is always possible to choose shifts α_j in such a way that each of the points of Λ is contained in exactly one shifted lattice Λ_j.

We will associate to each $\lambda \in \Lambda$ a number p_λ, called the *multiplicity* of λ, which is the number of lattices Λ_j to which λ belongs. When we talk about the samples of a bandlimited function f on Λ we will mean the values $\{f^{(j)}(\lambda)\}$ where $\lambda \in \Lambda$ and for each λ, $0 \leq j < p_\lambda$. We are interested in the question of whether or not a function bandlimited to the interval $[-\sigma, \sigma]$ can be recovered from its samples on Λ. By standard duality arguments, this question is equivalent to the question of whether a function in $L^2[-\sigma, \sigma]$ is completely determined by its inner products with the complex exponentials $\{t^j\, e^{2\pi i \lambda t} : \lambda \in \Lambda,\, 0 \leq j < p_\lambda\}$. We will address this question by showing the following.

(a) A function $f \in PW_\sigma$ whose samples on Λ vanish must be identically zero. This is equivalent to the statement that the collection $\{t^j\, e^{2\pi i \lambda t}\}$ is complete (Definition 13.10) in $L^2[-\sigma, \sigma]$.

(b) There exist functions $\{g_{\lambda,j}\}$ that interpolate on Λ, that is, each $g_{\lambda,j}$ is in PW_σ and if $\mu \in \Lambda$ and $0 \leq k < p_\mu$, then $g_{\lambda,j}^{(k)}(\mu) = 1$ if $(k, \mu) = (j, \lambda)$ and 0 otherwise. This is equivalent to the statement that the collection $\{t^j\, e^{2\pi i \lambda t}\}$ is minimal (Definition 13.10) in $L^2[-\sigma, \sigma]$.

(c) For $f \in PW_\sigma$ we can write

$$f(x) = \sum_{\lambda \in \Lambda} \sum_{j=0}^{p_\lambda - 1} f^{(j)}(\lambda)\, g_{\lambda,j}(x),$$

where the sum converges unconditionally in PW_σ if Λ is separated, and by means of a certain block summation procedure (Section 13.2.3) if Λ is not separated. This is equivalent to the statement that the collection $\{t^j\, e^{2\pi i \lambda t}\}$ is a Riesz basis if Λ is separated, and generates a Riesz basis from subspaces (Definition 13.21) if Λ is not separated.

In proving these results a central role will be played by entire functions of exponential type whose zero sets coincide with the set on which we are sampling. One reason for this is that in this case such a function is simply a product of sine functions and hence very easy to work with. This point of view is classical and appeared in [33] and [28]. The interested reader is invited to consult [27], and [22] for more information on these beautiful and powerful ideas. Since the set Λ of interest here is contained in the real line and has such a simple structure, the full power of these techniques will not really be evident in this exposition. Indeed the only theorems from complex analysis that we will need appear in any first year graduate course. Specifically, we will need Liouville's Theorem (e.g., [11, Thm. 3.4]), the Maximum Modulus

Theorem (e.g., [11, Thm. 1.2, Exercise 1, p. 129]), and the Residue Theorem (e.g., [11, Thm. 2.2], [1, Chap. 4, Sec. 5]).

The chapter is organized as follows. In Section 13.2, we collect the basic results from complex analysis, Hilbert space theory, and polynomial interpolation that will be used later on. The more standard results will not be proved, but an effort has been made to make the chapter as self-contained as possible. This section also includes some basic results on bases in Hilbert spaces. In Section 13.3, we will demonstrate parts (a) and (b) above, and in particular give explicit formulas for the interpolating functions by using some of the formulas and ideas from polynomial interpolation theory. In Section 13.4, we show part (c).

13.2 Basic Results and Definitions

13.2.1 The Hardy Space H^2

Definition 13.1. *The Hardy space H^2 is the space of all functions analytic in the upper half plane of $\mathbf{C}$, $\{z \in \mathbf{C} : \Im(z) > 0\}$, such that*

$$\|f\|_{H^2} = \sup_{y>0} \left(\int_{-\infty}^{\infty} |f(x+iy)|^2 \, dx \right)^{1/2} < \infty.$$

The following lemma contains the basic facts about H^2 that will be used in this chapter. More details can be found in [13, Chap. 11].

Lemma 13.2. (a) *A function $f(z)$ analytic in the upper half plane is in H^2 if and only if there is a function $\varphi \in L^2[0,\infty)$ such that*

$$f(z) = \int_0^{\infty} e^{2\pi i z t} \, \varphi(t) \, dt.$$

(b) $\|f\|_{H^2} = \left(\int_{-\infty}^{\infty} |f(x)|^2 \, dx \right)^{1/2} = \left(\int_0^{\infty} |\varphi(t)|^2 \, dt \right)^{1/2}$ *where $f(x)$ is the boundary function $f(x) = \lim_{y \to 0^+} f(x+iy)$ a.e. H^2 is a Hilbert space with respect to the inner product induced by this norm.*

(c) *If $f \in H^2$, then for every $z \in \mathbf{C}$ with $\Im(z) > 0$,*

$$f(z) = \int_{-\infty}^{\infty} \frac{f(t)}{t-z} \, dt.$$

(d) *Let $\{\lambda_n\}_{n \in \mathbf{Z}} \subseteq \mathbf{R}$ be a separated sequence, and let $k \in \mathbf{Z}$, $k \geq 0$, and $h > 0$ be given. Then there is a constant $C = C(h,k)$ such that for all $f \in H^2$,*

$$\left(\sum_n |f^{(k)}(\lambda_n + ih)|^2 \right)^{1/2} \leq C \, \|f\|_{H^2}.$$

Proof. The proofs of (a), (b), (c), and (d) (when $k = 0$) can be found for example in [42, pp. 145–146] and [13, pp. 194–196]. We will prove (d) for $k \in \mathbf{N}$. Note that if $h_0 > 0$, then by part (a),

$$f(x + ih_0) = \int_0^\infty e^{2\pi i(x+ih_0)t}\, \varphi(t)\, dt = \int_0^\infty e^{2\pi ixt}\, e^{-2\pi h_0 t}\, \varphi(t)\, dt$$

and differentiation under the integral sign gives

$$f^{(k)}(x + ih_0) = \int_0^\infty e^{2\pi ixt}\, (2\pi it)^k\, e^{-2\pi h_0 t}\, \varphi(t)\, dt.$$

Since $t^k\, e^{-2\pi h_0 t}$ is bounded on $[0, \infty)$, $f^{(k)}(\cdot + ih_0) \in H^2$.

Assuming $0 < h_0 < h$ and applying the result of (b) and (d) with $k = 0$, there is a $C > 0$ such that for all $f \in H^2$,

$$\begin{aligned}\Big(\sum_n |f^{(k)}(\lambda_n + ih)|^2\Big)^{1/2} &= \Big(\sum_n |f^{(k)}(\lambda_n + ih_0 + i(h - h_0))|^2\Big)^{1/2} \\ &\le C\left(\int_{-\infty}^\infty |f^{(k)}(x + ih_0)|^2\, dx\right)^{1/2} \\ &= C\left(\int_0^\infty |(2\pi it)^k\, e^{-2\pi h_0 t}\, \varphi(t)|^2\, dt\right)^{1/2} \\ &\le C\,(2\pi)^k\, \|t^k e^{-2\pi h_0 t}\|_\infty \left(\int_0^\infty |\varphi(t)|^2\, dt\right)^{1/2} \\ &= C(h_0, k)\, \|f\|_{H^2}. \quad \square\end{aligned}$$

13.2.2 The Paley–Wiener Space PW_σ

Definition 13.3. *Given $\sigma > 0$, the* Paley–Wiener space, $PW_\sigma(\mathbf{R})$, *is the space of all functions $f \in L^2(\mathbf{R})$ such that*

$$f(x) = \int_{-\sigma}^{\sigma} e^{2\pi ixt}\, \varphi(t)\, dt$$

for some $\varphi \in L^2[-\sigma, \sigma]$. $PW_\sigma(\mathbf{R})$ is a Hilbert space under the norm

$$\|f\|_{PW_\sigma} = \|\varphi\|_{L^2[-\sigma,\sigma]}.$$

Theorem 13.4 (Paley–Wiener Theorem). *An entire function $f(z)$ is the extension to $\mathbf{C}$ of a function $f \in PW_\sigma$ if and only if* (a) *f is square-integrable on the real line, and* (b) *there is a constant $C > 0$ such that for all $z \in \mathbf{C}$, $|f(z)| \le C\, e^{2\pi\sigma|\Im z|}$. In this case, we may write for $z \in \mathbf{C}$,*

$$f(z) = \int_{-\sigma}^{\sigma} e^{2\pi izt}\,\varphi(t)\,dt$$

for some $\varphi \in L^2[-\sigma, \sigma]$.

Lemma 13.5 ([42, p. 92, Problem 1]). *If* $f \in PW_\sigma(\mathbf{R})$, *then for every* $y \in \mathbf{R}$, *the function* $f(\cdot + iy)$ *is in* $PW_\sigma(\mathbf{R})$ *and there is a number* $C = C(y)$ *such that*

$$\frac{1}{C}\,\|f\|_{PW_\sigma} \le \left(\int_{-\infty}^{\infty} |f(x+iy)|^2\,dx\right)^{1/2} \le C\,\|f\|_{PW_\sigma}.$$

Theorem 13.6 (Bernstein's Inequality). *Suppose that* $f \in PW_\sigma(\mathbf{R})$. *Then* f' *exists, is in* $PW_\sigma(\mathbf{R})$, *and*

$$\|f'\|_{PW_\sigma} \le 2\pi\sigma\,\|f\|_{PW_\sigma}.$$

Proof. Since $f(z) = \int_{-\sigma}^{\sigma} e^{2\pi izt}\,\varphi(t)\,dt$, differentiating under the integral sign gives

$$f'(z) = 2\pi i \int_{-\sigma}^{\sigma} e^{2\pi izt}\,t\,\varphi(t)\,dt$$

so that

$$\begin{aligned}\|f'\|_{PW_\sigma} &= 2\pi\,\|t\,\varphi(t)\|_{L^2[-\sigma,\sigma]} \\ &\le 2\pi\sigma\,\|\varphi\|_{L^2[-\sigma,\sigma]} \\ &= 2\pi\sigma\,\|f\|_{PW_\sigma}. \quad \square\end{aligned}$$

Theorem 13.7 (Plancherel–Pólya Theorem [42, Thm. 17, p. 82]**).** *Let* $\{\lambda_n\}_{n\in\mathbf{Z}} \subseteq \mathbf{R}$ *be a separated sequence. Then there is a constant* C *such that for all* $f \in PW_\sigma$,

$$\left(\sum_n |f(\lambda_n)|^2\right)^{1/2} \le C\,\|f\|_{PW_\sigma}.$$

The notion of a sine-type function was introduced by B. Y. Levin [25] advancing a method which appeared in [33] and [28] in which collections of complex exponentials $\{e^{i\lambda_n x}\}_{n\in\mathbf{Z}}$ are studied by examining functions of exponential type whose zero sets coincide exactly with the set $\{\lambda_n\}$. Such functions are called *generating functions* for the collection $\{e^{i\lambda_n x}\}$. Fundamental results appear in [25], [27, Lecture 22–23], and [35]. Some of the main results of those papers are reproduced below as Theorems 13.22–13.24.

Definition 13.8 (Sine-Type Function). *An entire function* $F(z)$ *is said to be of* sine type r *provided that*

(a) *for some* $C > 0$, $|F(z)| \le C\,e^{2\pi r|\Im z|}$ *for all* $z \in \mathbf{C}$,

(b) *if* Γ *is the zero set of* $F(z)$, *then* $\sup_{\gamma\in\Gamma}\{|\Im\gamma|\} = H < \infty$, *and*

(c) *for every* $\varepsilon > 0$ *there is a positive constant* m_ε *such that whenever* $\operatorname{dist}(z, \Gamma) > \varepsilon$,

$$m_\varepsilon\, e^{2\pi r|\Im(z)|} \leq |F(z)|.$$

Example 13.9. (a) The function $\sin(\pi z)$ is a function of sine type 1/2. First, it is clear that $\sin(\pi z)$ is entire and satisfies (a) and (b) of Definition 13.8. To verify (c) of Definition 13.8, note that with $z = x + iy$,

$$|\sin(\pi z)|^2 = \frac{1}{2}\,(\cosh(2\pi y) - \cos(2\pi x)),$$

and that

$$e^{-2\pi|y|}\,\cosh(2\pi y) = \frac{1}{2}\,(1 + e^{-4\pi|y|}).$$

Hence

$$\begin{aligned} e^{-2\pi|y|}\,|\sin(\pi z)|^2 &= \frac{1}{2}\,(e^{-2\pi|y|}\,\cosh(2\pi y) - e^{-2\pi|y|}\,\cos(2\pi x)) \\ &= \frac{1}{4}\,(1 + e^{-4\pi|y|} - 2\,e^{2\pi|y|}\,\cos(2\pi x)) \\ &= \frac{1}{4}\,(1 + e^{-4\pi|y|} - 2\,e^{-2\pi|y|}) + \frac{1}{2}\,e^{-2\pi|y|}\,(1 - \cos(2\pi x)) \\ &= \frac{1}{4}\,(1 - e^{-2\pi|y|})^2 + \frac{1}{2}\,e^{-2\pi|y|}\,(1 - \cos(2\pi x)). \end{aligned}$$

Now suppose that $z = x + iy$ is outside the union of boxes of sidelength $0 < \delta < 1/4$ centered at $n \in \mathbf{Z}$. If $|y| \geq \delta/2$, then $(1 - e^{-2\pi|y|})^2 \geq (1 - e^{-\pi\delta})^2 > 0$. If $|y| < \delta/2$, then necessarily $|x - n| \geq \delta/2$ for all $n \in \mathbf{Z}$. Consequently, $e^{-2\pi|y|}(1 - \cos(2\pi x)) \geq e^{-\pi\delta}(1 - \cos(\pi\delta)) > 0$ and (c) follows.

Note also that the same argument applies to the function $\sin(\pi z + z_0)$ for any fixed $z_0 \in \mathbf{C}$.

(b) Let $r_1, r_2, \ldots, r_m$ be given positive numbers such that $\sum_{i=1}^m r_i = R$, and let $s_1, s_2, \ldots, s_m$ be arbitrary real numbers. Then the function $S(z) = \prod_{i=1}^m \sin(2\pi r_i(z - s_i))$ is a function of sine type R. It is clear that $S(z)$ is entire and satisfies (a) and (b) of Definition 13.8. By part (a) of this example, each term in the product defining $S(z)$ satisfies for each $\varepsilon > 0$, $|\sin(2\pi r_i(z - s_i))| \geq m_{\varepsilon,i}\, e^{2\pi r_i|y|}$ for some $m_{\varepsilon,i} > 0$ whenever $\operatorname{dist}(z, \mathbf{Z}/2r_i) > \varepsilon$. From this, (c) of Definition 13.8 follows with $m_\varepsilon = \prod_{i=1}^m m_{\varepsilon,i}$.

13.2.3 Bases in Hilbert Space

In this section we collect some basic definitions and results on bases in Hilbert spaces. The results are general but what we have in view are bases of exponentials of the form $\{t^j\, e^{2\pi i\lambda t}\}$ in Hilbert spaces $L^2(I)$ where $I \subseteq \mathbf{R}$ is an interval. The results relate to the primary goal of this chapter insofar as results about completeness and basis properties of the collection

$\{t^j e^{2\pi i\lambda t} : \lambda \in \Lambda, 0 \le j < p_\lambda\}$ in $L^2[-\sigma,\sigma]$ have equivalent expression in terms of sampling and interpolation of functions in PW_σ. Most of the results in this section are adapted from [42]. We also recommend [18, Chap. 5]

Definition 13.10. *A sequence of vectors $\{x_n\}_{n=1}^\infty$ is* complete *in a Hilbert space H if* span$(\{x_n\})$ *is dense in H. Equivalently, $\{x_n\}_{n=1}^\infty$ is complete in H if for any $x \in H$, $\langle x, x_n\rangle = 0$ for all n implies $x = 0$.*

A sequence $\{y_n\}_{n=1}^\infty$ is biorthogonal *to $\{x_n\}_{n=1}^\infty$ if $\langle x_n, y_m\rangle = \delta_{n,m}$. A sequence $\{x_n\}_{n=1}^\infty$ is* minimal *if it possesses a biorthogonal sequence. A sequence that is both minimal and complete is called* exact.

Definition 13.11. *A sequence of vectors $\{x_n\}_{n=1}^\infty$ in a Hilbert space H is a* basis *or a* Schauder basis *for H if to each vector $x \in H$ there is a unique sequence of scalars $\{a_n(x)\}_{n=1}^\infty$ such that*

$$x = \sum_{n=1}^{\infty} a_n(x)\, x_n,$$

where the sum converges in H.

Theorem 13.12 ([42, Thm. 3, p. 19]). *If $\{x_n\}_{n=1}^\infty$ is a basis for a Hilbert space H, then for each n, the mapping $x \mapsto a_n(x)$ is a continuous linear functional.*

It is obvious that a basis for H is complete in H. By the Riesz representation theorem, if $\{x_n\}_{n=1}^\infty$ is a basis in H, there is a sequence $\{y_n\}_{n=1}^\infty$ in H such that for each $x \in H$ and each n, $a_n(x) = \langle x, y_n\rangle$. It is easy to see that $\{y_n\}_{n=1}^\infty$ is biorthogonal to $\{x_n\}_{n=1}^\infty$. Moreover, it can be shown that $\{y_n\}_{n=1}^\infty$ is also a basis for H ([42, Thm. 5, p. 22]). We also have the following useful result.

Theorem 13.13 ([42, Thm. 3, p. 19]). *Let $\{x_n\}$ be a basis for a Hilbert space H with $\{y_n\}$ its biorthogonal sequence. Then there is a constant $M > 0$ such that for all n,*

$$1 \le \|x_n\|\,\|y_n\| \le M.$$

Definition 13.14. *A sequence $\{x_n\}_{n=1}^\infty$ in a Hilbert space H is a* Bessel sequence *if there is a constant B such that for all $x \in H$,*

$$\sum_{n=1}^{\infty} |\langle x, x_n\rangle|^2 \le B\,\|x\|^2. \tag{13.1}$$

Theorem 13.15. *Let $\{x_n\}_{n=1}^\infty$ be a sequence in a Hilbert space H. Then $\{x_n\}$ is a Bessel sequence in H if and only if there is a number C such that for all finite sequences of numbers $\{c_n\}$,*

$$\left\|\sum_n c_n\, x_n\right\| \le C\left(\sum_n |c_n|^2\right)^{1/2}.$$

Proof. ($\Longleftarrow$) Given $x \in H$ and a finite sequence $\{c_n\}$ with $\sum |c_n|^2 = 1$,

$$\begin{aligned}
\left|\sum_n \langle x, x_n\rangle \overline{c_n}\right| &= \left|\left\langle x, \sum_n c_n x_n \right\rangle\right| \\
&\leq \|x\| \left\|\sum_n c_n x_n\right\| \\
&\leq C \|x\| \left(\sum_n |c_n|^2\right)^{1/2} \\
&= C \|x\|.
\end{aligned}$$

Taking the supremum of the left side over all such sequences gives

$$\left(\sum_n |\langle x, x_n\rangle|^2\right)^{1/2} \leq C \|x\|$$

and $\{x_n\}$ is a Bessel sequence.

($\Longrightarrow$) Now suppose that $\{x_n\}$ is a Bessel sequence with Bessel bound B. For any $y \in H$ with $\|y\| = 1$, and for any finite sequence of numbers $\{c_n\}$,

$$\begin{aligned}
\left|\left\langle \sum_n c_n x_n, y \right\rangle\right| &= \left|\sum_n c_n \langle x_n, y\rangle\right| \\
&\leq \left(\sum_n |c_n|^2\right)^{1/2} \left(\sum_n |\langle y, x_n\rangle|^2\right)^{1/2} \\
&\leq B \|y\| \left(\sum_n |c_n|^2\right)^{1/2} \\
&= B \left(\sum_n |c_n|^2\right)^{1/2}.
\end{aligned}$$

Taking the supremum of the left side over all such y gives the result. □

Definition 13.16. *An exact sequence* $\{x_n\}_{n=1}^{\infty}$ *is a* Riesz basis *if there are constants* A, $B > 0$ *such that for all* $x \in H$,

$$A \|x\|^2 \leq \sum_{n=1}^{\infty} |\langle x, x_n\rangle|^2 \leq B \|x\|^2.$$

There are many equivalent characterizations of Riesz bases (see e.g., [42, Thm. 9, p. 27]). We now give two characterizations (Theorem 13.17 and the remark following) that will be used later on.

Theorem 13.17. *An exact sequence* $\{x_n\}_{n=1}^{\infty}$ *is a Riesz basis for a Hilbert space* H *if and only if it and its biorthogonal sequence are Bessel sequences.*

Proof. ($\Longrightarrow$) Suppose that the exact sequence $\{x_n\}$ is a Riesz basis. Then by Definition 13.16 it is also a Bessel sequence. Let $\{y_n\}$ be biorthogonal to $\{x_n\}$, and let $\{c_n\}$ be a finite sequence of numbers. Then

$$\begin{aligned}
\left\|\sum_n c_n\, y_n\right\| &\leq \frac{1}{A^{1/2}}\left(\sum_m\left|\left\langle\sum_n c_n\, y_n,\, x_m\right\rangle\right|^2\right)^{1/2} \\
&= \frac{1}{A^{1/2}}\left(\sum_m\left|\sum_n c_n\,\langle y_n, x_m\rangle\right|^2\right)^{1/2} \\
&= \frac{1}{A^{1/2}}\left(\sum_m |c_m|^2\right)^{1/2}.
\end{aligned}$$

Hence by Theorem 13.15, $\{y_n\}$ is a Bessel sequence.

($\Longleftarrow$) Suppose that the exact sequence $\{x_n\}$ is a Bessel sequence. In order to show that it is a Riesz basis, it is sufficient to show that it also has a lower bound A as in Definition 13.16. Let $\{y_n\}$ be biorthogonal to $\{x_n\}$ and assume that it is a Bessel sequence with bound $A > 0$. Let $x \in H$ and let $y \in \operatorname{span}\{x_n\}$ with $\|y\| = 1$. Then $y = \sum_n \langle y, y_n\rangle\, x_n$ and

$$\begin{aligned}
|\langle x, y\rangle| &= \left|\left\langle x,\, \sum_n \langle y, y_n\rangle\, x_n\right\rangle\right| \\
&\leq \sum_n |\langle x, x_n\rangle|\,|\langle y, y_n\rangle| \\
&\leq \left(\sum_n |\langle x, x_n\rangle|^2\right)^{1/2}\left(\sum_n |\langle y, y_n\rangle|^2\right)^{1/2} \\
&\leq \left(\sum_n |\langle x, x_n\rangle|^2\right)^{1/2} A^{1/2}\|y\| \\
&= A^{1/2}\left(\sum_n |\langle x, x_n\rangle|^2\right)^{1/2}.
\end{aligned}$$

Since $\operatorname{span}\{x_n\}$ is dense in H, taking the supremum of the left side over all such y gives

$$\frac{1}{A^{1/2}}\,\|x\| \leq \left(\sum_n |\langle x, x_n\rangle|^2\right)^{1/2}.$$

Hence $\{x_n\}$ is a Riesz basis. □

If $\{x_n\}_{n=1}^{\infty}$ is a Riesz basis, it is also a basis and has the property that for every $x \in H$,

$$x = \sum_n \langle x, y_n\rangle\, x_n = \sum_n \langle x, x_n\rangle\, y_n,$$

where the sequence $\{y_n\}$ is biorthogonal to $\{x_n\}$ and where the sum converges unconditionally, that is, regardless of the order in which the terms are summed. In fact, a Schauder basis $\{x_n\}$ whose elements satisfy $0 < \inf_n \|x_n\| \leq \sup_n \|x_n\| < \infty$ and for which the expansion sum of every $x \in H$ converges unconditionally must be a Riesz basis ([18, Chap. VI, Sec. 2, Thm. 2.2]). Such a collection is called a *bounded unconditional basis.* We will prove some basic results on unconditional convergence below.

Definition 13.18. *A series* $\sum_{n=1}^{\infty} v_n$ *in a Hilbert space* H converges unconditionally *if for every permutation* ρ *of* $\mathbf{N}$, $\sum_{n=1}^{\infty} v_{\rho(n)}$ *converges.*

Lemma 13.19. *The following are equivalent.*

(a) $\sum_{n=1}^{\infty} v_n$ *converges unconditionally.*

(b) *For every* $\varepsilon > 0$ *there is a finite set* $F \subseteq \mathbf{N}$ *such that for every finite set* $G \subseteq \mathbf{N}$, $\|\sum_{n\in G\setminus F} v_n\| < \varepsilon$.

Proof. (a) $\Longrightarrow$ (b). Suppose that (b) does not hold. We will find a permutation ρ of $\mathbf{N}$ such that the series $\sum_{n=1}^{\infty} v_{\rho(n)}$ is not Cauchy. Since (b) fails, there is an $\varepsilon > 0$ such that for all finite sets F there is a finite set G such that $\|\sum_{n\in G\setminus F} v_n\| \geq \varepsilon$. We can therefore find sequences of finite sets F_k and G_k such that $F_k = \{1, \ldots, N_k\}$ for some sequence N_k increasing to infinity with k, $F_k \subset G_k \subseteq F_{k+1}$, and $\|\sum_{n\in G_k\setminus F_k} v_n\| \geq \varepsilon$. Define the permutation ρ of $\mathbf{N}$ as follows: First enumerate F_1, then $G_1 \setminus F_1$, then $F_2 \setminus G_1$, etc. Given any $N \in \mathbf{N}$, then for all k sufficiently large that $N_k > N$,

$$\left\| \sum_{n=N_k+1}^{|G_k|} v_{\rho(n)} \right\| = \left\| \sum_{n\in G_k\setminus F_k} v_n \right\| \geq \varepsilon.$$

(b) $\Longrightarrow$ (a). Let ρ be a permutation of $\mathbf{N}$ and let $\varepsilon > 0$ be given. By (b) there is a finite set $F \subseteq \mathbf{N}$ such that for all finite sets $G \subseteq \mathbf{N}$, $\|\sum_{n\in G\setminus F} v_n\| < \varepsilon$. Let N_0 be so large that

$$F \subseteq \{\rho(1), \rho(2), \ldots, \rho(N_0)\},$$

let $N, M \geq N_0$, and let $G = \{\rho(N), \rho(N+1), \ldots, \rho(M)\}$. Then G is finite and $G \cap F = \emptyset$. Therefore

$$\left\| \sum_{n=N}^{M} v_{\rho(n)} \right\| = \left\| \sum_{n\in G} v_n \right\| = \left\| \sum_{n\in G\setminus F} v_n \right\| < \varepsilon. \quad \square$$

Note that the above characterization is also valid in a Banach space. The definition of unconditional convergence allowed for the possibility that different permutations of the series could converge to different limits. The next lemma shows that all permutations of such a series converge to the same limit.

Lemma 13.20. *Suppose that $\sum_{n=1}^{\infty} v_n$ converges unconditionally. Then there is a $v \in H$ such that for every perturbation ρ of $\mathbf{N}$,*

$$v = \sum_{n=1}^{\infty} v_{\rho(n)}.$$

Proof. Let $v = \sum_{n=1}^{\infty} v_n$ and let $\varepsilon > 0$. Let N_0 be so large that if $N \geq N_0$, then $\|v - \sum_{n=1}^{N} v_n\| < \varepsilon/2$, let $F \subseteq \mathbf{N}$ be a finite set such that for every finite set $G \subseteq \mathbf{N}$, $\|\sum_{n \in G \setminus F} v_n\| < \varepsilon/2$, and let $N_1 \geq N_0$ and $N_2 \geq N_1$ be such that

$$F \subseteq \{1, \ldots, N_1\} \subseteq \{\rho(1), \rho(2), \ldots, \rho(N_2)\}.$$

Now if $N \geq N_2$, and with

$$G = \{\rho(1), \rho(2), \ldots, \rho(N_2)\} \setminus \{1, \ldots, N_1\}$$

so that $G = G \setminus F$,

$$\left\| v - \sum_{n=1}^{N} v_{\rho(n)} \right\| \leq \left\| v - \sum_{n=1}^{N_1} v_n \right\| + \left\| \sum_{n \in G \setminus F} v_n \right\| < \frac{\varepsilon}{2} + \frac{\varepsilon}{2} = \varepsilon. \qquad \square$$

Since our sampling set Λ need not be separated, the collection $\{t^j\, e^{2\pi i \lambda t} : \lambda \in \Lambda, 0 \leq j < p_\lambda\}$ need not be a Riesz basis or a Schauder basis for $L^2[-\sigma, \sigma]$ (see [42, Cor. 2, p. 151] and Theorem 13.39). Hence in order to discuss the recovery of a function in PW_σ from its samples on Λ, we require the more general notion of a Riesz basis referred to in [32] as a *Riesz basis from subspaces* (see also [2, Sec. I.1.4] and [18, Chap. VI, Sec. 5]). This type of basis is also related to the notion of a *polynomial basis* (see [37, Sec. 20.1]).

Definition 13.21. *Let H be a Hilbert space and $\mathcal{H} = \{H_n\}_{n=1}^{\infty}$ a collection of subspaces of H. $\mathcal{H}$ is a* Riesz basis from subspaces *(RBS) for H if the following hold.*

(a) *There exist bounded linear operators $\mathcal{P}_n$ on H such that $\mathcal{P}_n$ restricted to H_k is zero if $n \neq k$ and the identity if $n = k$.*

(b) *For every $x \in H$,*

$$x = \sum_{n=1}^{\infty} \mathcal{P}_n\, x$$

unconditionally in H.

Here we make some remarks concerning the relationship between Riesz bases and Riesz bases from subspaces. Suppose that $\{x_n\}$ is a Riesz basis with biorthogonal system $\{y_n\}$. Define H_n to be the one-dimensional subspace of H spanned by x_n. Clearly span$\{H_n\}$ is dense in H and the projection operators $\mathcal{P}_n$ satisfying Definition 13.21(a) are given by $\mathcal{P}_n x = \langle x, y_n \rangle\, x_n$. Since for each

x, $x = \sum_n \langle x, y_n \rangle\, x_n = \sum_n \mathcal{P}_n x$ unconditionally, then $\{H_n\}$ is an RBS. Note that no mention is made of the *boundedness* of $\{x_n\}$ and $\{y_n\}$, and that if we for example replaced x_n by nx_n and y_n by $\frac{1}{n} y_n$, then we could define H_n and $\mathcal{P}_n$ exactly as before and still have an RBS, although $\{nx_n\}$ is clearly not a Riesz basis. In other words, an RBS requires only the unconditional convergence of the sum of the *projection operators* $\mathcal{P}_n$ to the identity in the strong operator topology (that is, pointwise for each $x \in H$).

On the other hand, suppose that each subspace H_n in an RBS is finite dimensional and that for each n, $\{x_{n,k}\}_{k=1}^{\dim(H_n)}$ is a basis for H_n with dual basis $\{y_{n,k}\}_{k=1}^{\dim(H_n)}$. Then

$$\mathcal{P}_n\, x = \sum_{k=1}^{\dim(H_n)} \langle x, y_{n,k} \rangle\, x_{n,k}$$

and for all $x \in H$ we have the expansion

$$x = \sum_{n=1}^{\infty} \sum_{k=1}^{\dim(H_n)} \langle x, y_{n,k} \rangle\, x_{n,k},$$

where the sum over n can be ordered in any way we choose. By renumbering the sequence $\{x_{n,k}\}$ as

$$\{x_\ell\}_{\ell=1}^{\infty} = \{x_{1,1},\, x_{1,2},\, \dots,\, x_{1,\dim(H_1)},\, x_{2,1},\, \dots,\, x_{2,\dim(H_2)},\, x_{3,1},\, \dots\}$$

and similarly renumbering $\{y_{n,k}\}$ as $\{y_\ell\}$, it does not necessarily follow that $\{x_\ell\}$ is a basis for H. This is because the sum

$$x = \sum_{\ell=1}^{\infty} \langle x, y_\ell \rangle\, x_\ell$$

need not converge in H. The problem is that the angle between pairs of elements in $\{x_{n,k}\}_{k=1}^{\dim H_n}$ may become arbitrarily small as n goes to infinity. This will in particular force the norms $\|y_{n,k}\|$ to become arbitrarily large. This is indeed precisely what happens when the set Λ is not separated. However, as long as the angle between the *subspaces* $\{H_n\}_{n=1}^{\infty}$ does not become arbitrarily small, the collection $\{x_\ell\}$ can be seen to correspond to an RBS. Of course, it is always possible to choose bases $\{\widetilde{x}_{n,k}\}_{k=1}^{\dim H_n}$ for the spaces H_n for which this does not happen (for example, by choosing orthonormal bases). In this case, the corresponding sequence $\{\widetilde{x}_\ell\}$ will be a Riesz basis for H.

One final observation about Riesz bases from subspaces: Let $\ell_1 = \dim H_1$, $\ell_2 = \ell_1 + \dim H_2$, and in general $\ell_k = \ell_{k-1} + \dim H_k$ for $k = 1, 2, \dots$. Then we do have the expansion

$$x = \lim_{k\to\infty} \sum_{\ell=1}^{\ell_k} \langle x, y_\ell \rangle\, x_\ell$$

for all $x \in H$. In this case, $\{x_\ell\}$ is an example of a *generalized basis* [25] or a *basis with braces.* The notion is classical and originated in the context of spectral projections and eigenfunction expansions. The main result of [25] is the following.

Theorem 13.22. *Let $F(z)$ be a function of sine type r with zero set $\{\gamma_n\}_{n=1}^\infty$ and suppose that each γ_n has multiplicity p_n. Then $\{t^j\, e^{2\pi i \gamma_n t}\}_{n=1, 0\le j<p_n}^\infty$ is a generalized basis for $L^2[-r,r]$.*

This means that there is a sequence $\{n_k\}$ such that for all $\varphi \in L^2[-r,r]$ there exist unique scalars $a_{n,j}$ such that

$$\varphi = \lim_{k\to\infty} \sum_{n=1}^{n_k} \sum_{j=0}^{p_n-1} a_{n,j}\, t^j\, e^{2\pi i \gamma_n t}.$$

Equivalently, this means that for all $f \in PW_r$,

$$f = \lim_{k\to\infty} \sum_{n=1}^{n_k} \sum_{j=0}^{p_n-1} f^{(j)}(\gamma_n)\, g_{\gamma_n,j},$$

where the $g_{\gamma_n,j}$ are interpolating functions satisfying $g_{\gamma_n,j}^{(k)}(\gamma_\ell) = 1$ if $(n,j) = (\ell,k)$ and 0 otherwise. Explicit formulas for the $g_{\gamma_n,j}$ equivalent to those given by Corollary 13.34, Section 13.3.2 can be found in [25].

Sine-type functions play a significant role in Levin's approach to finding Riesz bases of exponentials for the spaces $L^2(I)$ where $I \subseteq \mathbf{R}$ is an interval. In particular, the following theorem holds [27, Lecture 23].

Theorem 13.23. *Let $F(z)$ be a function of sine type r with simple zeros given by $\{\gamma_n\}_{n=1}^\infty$ and assume that the zeros are separated. Then $\{e^{2\pi i \gamma_n t}\}_{n=1}^\infty$ is a Riesz basis for $L^2[-r,r]$.*

The following characterization of Riesz bases of exponentials is due to Pavlov [35].

Theorem 13.24. *The collection $\{e^{2\pi i \gamma_n t}\}_{n=1}^\infty$ where $\{\gamma_n\}_{n=1}^\infty$ are the zeros of an entire function $F(z)$ of exponential type is a Riesz basis for $L^2[-r,r]$ if and only if*

(a) $\sup_n |\Im \gamma_n| < \infty$,

(b) $|F(z)| \le C\, e^{2\pi r |\Im z|}$ *for all* $z \in \mathbf{C}$, *and*

(c) *the function* $W = |F|^2$ *satisfies Muckenhoupt's condition*

$$\sup \left\{ \left(\frac{1}{|I|} \int_I W \right) \left(\frac{1}{|I|} \int_I W^{-1} \right) \right\} < \infty,$$

where the supremum is taken over all finite intervals $I \subseteq \mathbf{R}$.

13.2.4 Polynomial Interpolation in C

In this section we collect some basic results on polynomial interpolation in $\mathbf{C}$. These results will be used in the definition of the interpolating functions $g_{\lambda,j}$ on Λ in Section 13.3.2 and in the definition of the projection operators $\mathcal{P}_n$ in Section 13.4.2. The results in this section are adapted from [10] and [17].

Interpolation on Distinct Points

Suppose that we are given $n+1$ distinct points in $\mathbf{C}$, $z_0, z_1, \ldots, z_n$, and a function $f(z)$ defined on $\mathbf{C}$. We seek a polynomial of degree at most n that interpolates f at the points z_k, that is, $p_n(z_k) = f(z_k)$ for $0 \leq k \leq n$.

It is possible to write an explicit formula for $p_n(z)$ called the *Lagrange form*. Define the function $\psi_n(z)$ by

$$\psi_n(z) = (z - z_0)\,(z - z_1)\,\cdots\,(z - z_n).$$

Then $\psi_n(z)$ is a degree $n+1$ polynomial that vanishes at the points z_k. It is easy to see that

$$\psi_n'(z_k) = (z_k - z_0)\,\cdots\,(z_k - z_{k-1})\,(z_k - z_{k+1})\,\cdots\,(z_k - z_n)$$

and that if we define the degree n polynomials $l_k(z)$ by

$$l_k(z) = \frac{\psi_n(z)}{(z - z_k)\,\psi_n'(z_k)},$$

then the l_k satisfy

$$l_k(z_j) = \begin{cases} 1, & \text{if } j = k, \\ 0, & \text{if } j \neq k, \end{cases}$$

where we have defined $l_k(z_k)$ as $\lim_{z\to z_k} l_k(x)$. Hence

$$p_n(z) = \sum_{k=0}^{n} f(z_k)\,l_k(z) \tag{13.2}$$

solves the interpolation problem. If $q_n(z)$ is another solution to the problem, then $p_n(z) - q_n(z)$ is a polynomial of degree n vanishing at the $n+1$ points $z_0, z_1, \ldots, z_n$ and so vanishes identically. Thus $p_n(z)$ is the unique solution to the interpolation problem.

Definition 13.25 (Divided Differences). *Let $f(z)$ be a function on $\mathbf{C}$, and let $z_0, z_1, \ldots, z_k$ be $k+1$ distinct points in $\mathbf{C}$. Then the kth-order* Newton divided difference *of f, denoted $f[z_0, \ldots, z_k]$, is defined to be the coefficient of z^k in the polynomial p_k defined by* (13.2).

The Newton divided difference has the following properties.

Theorem 13.26. (a) *$f[z_0, z_1, \ldots, z_k]$ is unchanged under a permutation of the points z_j.*

(b) *The polynomial p_n defined by (13.2) can be written as*

$$p_n(z) = \sum_{k=0}^{n} f[z_0, \ldots, z_k] \prod_{j=0}^{k-1} (z - z_j).$$

This formula is referred to as the Newton form *of the polynomial p_n.*

(c) $f[z_0, z_1, \ldots, z_k] = \dfrac{f[z_1, \ldots, z_k] - f[z_0, \ldots, z_{k-1}]}{z_k - z_0}$.

Proof. (a) follows from the observation that the polynomial p_n, and hence its leading term, is independent of the order of the points z_j.

To see (b), note that we can write

$$p_n(z) = p_{n-1}(z) + f[z_0, \ldots, z_n] \prod_{j=0}^{n-1} (z - z_j).$$

Now $p_{n-1}(z)$ has degree $n-1$ since by definition the coefficients of z^n in $p_n(z)$ and in $f[z_0, \ldots, z_n] \prod_{j=0}^{n-1}(z - z_j)$ match. Next note that $p_{n-1}(z_k) = p_n(z_k) = f(z_k)$ for $0 \leq k \leq n-1$ so that p_{n-1} uniquely solves the interpolation problem for f at the points $z_0, \ldots, z_{n-1}$. Hence we can write

$$p_{n-1}(z) = p_{n-2}(z) + f[z_0, \ldots, z_{n-1}] \prod_{j=0}^{n-2} (z - z_j)$$

and assert that p_{n-2} is the unique degree $n-2$ polynomial solving the interpolation problem at $z_0, \ldots, z_{n-2}$. Proceeding in this fashion, we arrive at the result.

To see (c), note that, as above, we can write

$$p_n(z) = p_{n-1}(z) + f[z_0, \ldots, z_n] \prod_{j=0}^{n-1} (z - z_j),$$

where p_{n-1} is the unique degree $n-1$ polynomial interpolating f at the points $z_0, \ldots, z_{n-1}$, and that we can also write

$$p_n(z) = q_{n-1}(z) + f[z_0, \ldots, z_n] \prod_{j=1}^{n} (z - z_j),$$

where q_{n-1} is the unique degree $n-1$ polynomial interpolating f at $z_1, \ldots, z_n$. Setting the two expressions equal and rearranging gives

$$\frac{q_{n-1}(z) - p_{n-1}(z)}{z_n - z_0} = f[z_0, \ldots, z_n] \prod_{j=1}^{n-1} (z - z_j).$$

Equating the leading term of each side gives the result. □

Interpolation with Multiple Points

Now we consider the interpolation problem when the points in the sequence $z_0, z_1, \ldots, z_n$ may occur more than once. In this case, the interpolation problem is the following. Suppose that no point in the sequence $z_0, z_1, \ldots, z_n$ occurs more than $k+1$ times, and that our function f is analytic in an open neighborhood of the points z_j. We seek a polynomial $p_n(z)$ of degree no more than n such that $p^{(j)}(z) = f^{(j)}(z)$ for $0 \le j \le r-1$ at each point z which occurs r times in the sequence $z_0, z_1, \ldots, z_n$. This problem is referred to as *Hermite* or *osculatory interpolation.*

It turns out that the Newton divided difference formalism is already set up to accommodate this situation. We start with the observation that a unique solution to the Hermite interpolation problem for the points $z_0, \ldots, z_n$ exists. In fact, we can write down the analogues of the Lagrange interpolating functions for this case. To do this, we will reindex the interpolation points as follows. Consider now the sequence $z_0, \ldots, z_q$ of distinct points and suppose that the point z_j occurs μ_j times in the original sequence. For $k = 0, \ldots, q$, define the function

$$\psi_n^k(z) = \frac{\psi_n(z)}{(z - z_k)^{\mu_k}}$$

and for $0 \le r < \mu_k$ define

$$\ell_k^r(z) = \psi_n^k(z)\, Q_r(z)\, P_r(z),$$

where

$$Q_r(z) = \sum_{\ell=0}^{\mu_k - 1 - r} \frac{1}{\ell!}\, (1/\psi_n^k)^{(\ell)}(z_k)\, (z - z_k)^{\ell}$$

and

$$P_r(z) = \frac{1}{r!}\, (z - z_k)^r.$$

Then for $0 \le s < \mu_j$,

$$(\ell_k^r)^{(s)}(z_j) = \begin{cases} 1, & \text{if } (j,s) = (k,r), \\ 0, & \text{otherwise.} \end{cases} \tag{13.3}$$

In order to see this, note that if $j \ne k$, then clearly $\psi_n^k(z_j) = 0$ so that (13.3) holds in this case. Let $j = k$. Since $P_r(z)$ has a zero of order r at $z = z_k$, then (13.3) holds for $0 \le s < r$. If $0 \le r \le s < \mu_k$, then

$$\begin{aligned} (\ell_k^r)^{(s)}(z_k) &= \sum_{j=0}^{s} \binom{s}{j} (\psi_n^k \cdot Q_r)^{(j)}(z_k)\, P_r^{(s-j)}(z_k) \\ &= \binom{s}{s-r} (\psi_n^k \cdot Q_r)^{(s-r)}(z_k) \end{aligned}$$

$$= \binom{s}{s-r} \sum_{j=0}^{s-r} \binom{s-r}{j} (\psi_n^k)^{(j)}(z_k)\, Q_r^{(s-r-j)}(z_k)$$

$$= \binom{s}{s-r} \sum_{j=0}^{s-r} \binom{s-r}{j} (\psi_n^k)^{(j)}(z_k) \left(\frac{1}{\psi_n^k}\right)^{(s-r-j)}(z_k)$$

$$= \binom{s}{s-r} \left(\psi_n^k \cdot \frac{1}{\psi_n^k}\right)^{(s-r)}(z_k)$$

$$= \binom{s}{s-r} (1)^{(s-r)}(z_k)$$

$$= \begin{cases} 1, & \text{if } s = r, \\ 0, & \text{if } r < s < \mu_k. \end{cases}$$

Hence

$$p_n(z) = \sum_{k=0}^{q} \sum_{r=0}^{\mu_j - 1} f^{(r)}(z_k)\, \ell_k^r(z)$$

is the unique solution to the Hermite interpolation problem.

As before, we define the nth-order Newton divided difference as the coefficient of z^n in p_n. The divided differences so defined satisfy essentially the same properties as the divided differences defined for distinct points. The proof of the following theorem is the same as that of Theorem 13.26.

Theorem 13.27. *The Newton divided differences satisfy the following properties.*

(a) $f[z_0, z_1, \ldots, z_k]$ *is unchanged under a permutation of the points* z_j.

(b) *The polynomial* p_n *solving the Hermite interpolation problem can be written*

$$p_n(z) = \sum_{k=0}^{n} f[z_0, \ldots, z_k] \prod_{j=0}^{k-1} (z - z_j).$$

(c) *If* $z_k \neq z_0$ *then*

$$f[z_0, z_1, \ldots, z_k] = \frac{f[z_1, \ldots, z_k] - f[z_0, \ldots, z_{k-1}]}{z_k - z_0}.$$

Divided differences for arbitrary sets of points may also be defined by taking the limit of divided differences for sets of distinct points. In order to show this, we will use an alternate representation of the divided difference for distinct points known as the Hermite–Gennochi formula. The proof of this formula is not reproduced here but amounts to showing that the expression satisfies Theorem 13.26(c) (see [17, Sec. 1.4.2]).

Theorem 13.28 (Hermite–Gennochi formula). *Suppose that f has k continuous derivatives on an open set containing the convex hull of the points z_j. Then*

$$f[z_0, z_1, \ldots, z_k] = \int_0^1 \int_0^{t_1} \cdots \int_0^{t_{k-1}} f^{(k)}(z_1 + (z_2 - z_1)\, t_1 + (z_3 - z_2)\, t_2 + \cdots + (z_0 - z_k)\, t_k) \, dt_k \, dt_{k-1} \cdots dt_1.$$

Corollary 13.29. *Suppose that f satisfies the hypotheses of Theorem 13.28. Then there is a point ξ in the convex hull of the points $\{z_0, \ldots, z_k\}$ such that*

$$f[z_0, z_1, \ldots, z_k] = \frac{1}{k!}\, f^{(k)}(\xi).$$

Proof. Note that the argument of $f^{(k)}$ in the integrand in Theorem 13.28 can be written

$$\begin{aligned} z_1 &+ (z_2 - z_1)\, t_1 + (z_3 - z_2)\, t_2 + \cdots + (z_0 - z_k)\, t_k \\ &= (1 - t_1)\, z_1 + (t_1 - t_2)\, z_2 + \cdots + (t_{k-1} - t_k)\, z_k + t_k\, z_0 \\ &= \lambda_0\, z_0 + \lambda_1\, z_1 + \lambda_2\, z_2 + \cdots + \lambda_k\, z_k, \end{aligned}$$

where $\sum_{j=0}^{k} \lambda_k = 1$. Hence the integral in Theorem 13.28 is taken over the convex hull of the points z_j. In light of the fact that

$$\int_0^1 \int_0^{t_1} \cdots \int_0^{t_{k-1}} dt_k \, dt_{k-1} \cdots dt_1 = \frac{1}{k!},$$

the result follows from the integral version of the Mean Value Theorem. □

We can now prove the following theorem.

Theorem 13.30. *Let $z_0, \ldots, z_n$ be a sequence of points, not necessarily distinct, such that no point in the sequence occurs more than $k + 1$ times and assume that f is analytic in a neighborhood of the convex hull of the points z_j. Then the following hold.*

(a) *If $z_0 = z_1 = \cdots = z_n = \xi$ then*

$$f[z_0, \ldots, z_n] = \frac{f^{(n)}(\xi)}{n!}.$$

(b) *If for each $r \in \mathbf{N}$, $z_0^{(r)}, \ldots, z_n^{(r)}$ are distinct points in $\mathbf{C}$ and if for each j, $\lim_{r\to\infty} z_j^{(r)} = z_j$, then*

$$\lim_{r\to\infty} f[z_0^{(r)}, \ldots, z_n^{(r)}] = f[z_0, \ldots, z_n].$$

Proof. (a) follows from the definition of the divided difference and the observation that if $z_0 = z_1 = \cdots = z_n = \xi$ then $p_n(z)$ is simply the Taylor polynomial of f at ξ.

The proof of (b) is by induction on n. If $n = 0$ then the result is obvious. Now suppose $n > 0$ and assume first that $z_0 = z_1 = \cdots = z_n = \xi$. By Corollary 13.29, there is a point $\xi^{(r)}$ in the convex hull of the points $z_k^{(r)}$ satisfying

$$f[z_0^{(r)}, z_1^{(r)}, \ldots, z_n^{(r)}] = \frac{1}{n!} f^{(n)}(\xi^{(r)}).$$

Since $\lim_r z_j^{(r)} = z_j$ for each j, $\lim_r \xi_j^{(r)} = \xi$ and the result follows.

Now suppose that at least two of the points z_j are distinct. Since the ordering of the points in the divided difference doesn't matter, we may take them to be z_0 and z_n. In this case, by the induction hypothesis and Theorem 13.27(c),

$$\begin{aligned}\lim_r f[z_0^{(r)}, z_1^{(r)}, \ldots, z_n^{(r)}] &= \lim_r \frac{f[z_0^{(r)}, \ldots, z_{n-1}^{(r)}] - f[z_1^{(r)}, \ldots, z_n^{(r)}]}{z_n^{(r)} - z_0^{(r)}} \\ &= \frac{f[z_1, \ldots, z_n] - f[z_0, \ldots, z_{n-1}]}{z_n - z_0} \\ &= f[z_0, z_1, \ldots, z_n]. \quad \square\end{aligned}$$

For functions in H^2, we have a particularly convenient integral representation of the divided difference.

Theorem 13.31. *Let $f \in H^2$ and let $z_0, \ldots, z_n$ be a sequence of points, not necessarily distinct, such that $\Im(z_j) > 0$ for each j. Then*

$$f[z_0, z_1, \ldots, z_n] = \int_{-\infty}^{\infty} \frac{f(t)}{(z_0 - t)(z_1 - t) \cdots (z_n - t)}\, dt.$$

Proof. Assume first that all of the z_k are distinct, and proceed by induction. If $n = 0$ then the result is a restatement of Lemma 13.2(b). For $n > 0$ note that

$$\begin{aligned}&\frac{1}{(z_0 - t)(z_1 - t) \cdots (z_{n-1} - t)} - \frac{1}{(z_1 - t)(z_2 - t) \cdots (z_n - t)} \\ &\qquad = \frac{z_n - z_0}{(z_0 - t)(z_1 - t) \cdots (z_{n-1} - t)(z_n - t)}.\end{aligned}$$

Hence by the induction hypothesis,

$$\begin{aligned}
\int_{-\infty}^{\infty} & \frac{f(t)}{(z_0 - t)(z_1 - t) \cdots (z_n - t)}\, dt \\
&= \frac{1}{z_n - z_0} \int_{-\infty}^{\infty} \Bigg(\frac{f(t)}{(z_0 - t)(z_1 - t) \cdots (z_{n-1} - t)} \\
&\qquad\qquad - \frac{f(t)}{(z_1 - t)(z_2 - t) \cdots (z_n - t)} \Bigg)\, dt \\
&= \frac{f[z_1, \ldots, z_n] - f[z_0, \ldots, z_{n-1}]}{z_n - z_0} \\
&= f[z_0, z_1, \ldots, z_n].
\end{aligned}$$

The case in which not all the z_j are distinct is handled by applying Theorem 13.30 and by observing that since $f \in H^2$, the Dominated Convergence Theorem allows for taking the limit under the integral sign. □

13.3 Exactness of Samples on a Union of Lattices

The goal of this section is to show that the collection $\{t^j\, e^{2\pi i\lambda t} : \lambda \in \Lambda, 0 \leq j < p_\lambda\}$ is exact in $L^2[-\sigma, \sigma]$. This is done by first showing in Section 13.3.1 that any function $f \in PW_\sigma$ is completely determined by its samples on the union of shifted lattices Λ and then by constructing explicitly in Section 13.3.2 the system $\{g_{\lambda,j} : \lambda \in \Lambda,\ 0 \leq j < p_\lambda\}$ of interpolating functions for the set Λ. This explicit form will be useful in showing that the collection forms a Riesz basis or corresponds to an RBS in Section 13.4.

13.3.1 Completeness

That any function in PW_σ is completely determined by its samples on Λ is equivalent to the completeness of the collection $\{t^j\, e^{2\pi i\lambda t} : \lambda \in \Lambda, 0 \leq j < p_\lambda\}$ in $L^2[-\sigma, \sigma]$. Necessary and sufficient conditions for completeness of such collections have been given in [33, Sec. 27] and in [28] (see [26, App. III, Sec. 1]) and relate to the existence or nonexistence of a nonzero entire function of exponential type vanishing precisely on Λ. The proof we present here is due to Levin [26], [27].

Theorem 13.32. *If $f \in PW_\sigma$ vanishes on Λ, then $f \equiv 0$.*

Proof. Let $S(z) = \prod_{i=1}^{m} \sin(2\pi\sigma_i(z - \alpha_i))$. Note that S vanishes precisely on Λ and that by Example 13.9(b), S is a function of sine type σ. Consider the function

$$\Phi(z) = \frac{f(z)}{S(z)}.$$

Since both f and S are entire functions and since f vanishes on Λ to the same order as S, the function $\Phi(z)$ is also entire. By the Paley–Wiener theorem (Therorem 13.4) $|f(z)| \le C\, e^{2\pi\sigma|\Im(z)|}$. Given $\varepsilon > 0$, define the set C_ε by

$$C_\varepsilon = \bigcup_{\lambda\in\Lambda} \{z : |z-\lambda| < \varepsilon\}.$$

It is easy to see that if $\varepsilon > 0$ is small enough, C_ε consists of a countable union of bounded open sets in $\mathbf{C}$ and that if $z \notin C_\varepsilon$ then $\mathrm{dist}(z,\Lambda) > \varepsilon$, and on the boundary of each bounded component of C_ε, the function $\Phi(z)$ is bounded above by C/m_ε (where the constant m_ε comes from Definition 13.8(c)). The Maximum Modulus Theorem allows us to extend this estimate to the interior of each component. Hence $\Phi(z)$ is a bounded entire function and so constant by Liouville's Theorem.

To see that in fact $\Phi(z)$ vanishes identically, let $h > H$ (where the constant H appears in Definition 13.8(b)) and consider the function $\Phi(x+ih)$, $x \in \mathbf{R}$. By Definition 13.8(c) and Lemma 13.5, it follows that $\Phi(x+ih)$ is in $L^2(\mathbf{R})$. Since it is also identically constant, that constant must be zero. □

13.3.2 Minimality: The Interpolating Functions

Theorem 13.33. *Let Ω be a finite subset of Λ. Define the function*

$$\Delta_\Omega(z) = \frac{\prod_{\lambda\in\Omega}(z-\lambda)^{p_\lambda}}{S(z)},$$

where

$$S(z) = \prod_{i=1}^{m} \sin(2\pi\sigma_i(z-\alpha_i)).$$

Let $f(z)$ be analytic in a neighborhood of Ω, and finally let $P_\Omega(z)$ be the polynomial of lowest degree interpolating the function $f\Delta_\Omega$ at each point $\lambda \in \Omega$ up to multiplicity p_λ. Define

$$g_\Omega(z) = \frac{S(z)}{\prod_{\lambda\in\Omega}(z-\lambda)^{p_\lambda}}\, P_\Omega(z)$$

Then $g_\Omega(z)$ is in PW_σ and satisfies

(a) *for all $\lambda \in \Lambda\setminus\Omega$, $g_\Omega^{(j)}(\lambda) = 0$ for all $0 \le j < p_\lambda$, and*

(b) *for all $\lambda \in \Omega$, $g_\Omega^{(j)}(\lambda) = f^{(j)}(\lambda)$ for all $0 \le j < p_\lambda$.*

Proof. It is clear that $g_\Omega(z)$ has the right exponential growth and that since the degree of $P_\Omega(z)$ is no more than $(\sum_{\lambda\in\Omega} p_\lambda) - 1$, $g_\Omega(z)$ is in $L^2(\mathbf{R})$. Hence $g_\Omega \in PW_\sigma$.

To see (a), note that $S(z)$ vanishes on Λ and that the function

$$\frac{1}{\prod_{\lambda\in\Omega}(z-\lambda)^{p_\lambda}}\,P_\Omega(z)$$

has no poles on the set $\Lambda\setminus\Omega$.

Finally, to see (b), let $\lambda\in\Omega$ and let $0\le j<p_\lambda$. Then using the fact that by definition $P_\Omega^{(j-k)}(\lambda)=(f\Delta_\Omega)^{(j-k)}(\lambda)$,

$$\begin{aligned}
g_\Omega^{(j)}(\lambda) &= \sum_{k=0}^{j}\binom{j}{k}\frac{d^k}{dz^k}\left(\frac{S(z)}{\prod_{\mu\in\Omega}(z-\mu)^{p_\mu}}\right)\Bigg|_{z=\lambda} P_\Omega^{(j-k)}(\lambda)\\
&= \sum_{k=0}^{j}\binom{j}{k}\frac{d^k}{dz^k}\left(\frac{S(z)}{\prod_{\mu\in\Omega}(z-\mu)^{p_\mu}}\right)\Bigg|_{z=\lambda} (f\Delta_\Omega)^{(j-k)}(\lambda)\\
&= \frac{d^j}{dz^j}\left(\frac{S(z)}{\prod_{\mu\in\Omega}(z-\mu)^{p_\mu}}(f\Delta_\Omega)(z)\right)\Bigg|_{z=\lambda}\\
&= f^{(j)}(\lambda). \quad \square
\end{aligned}$$

From Theorem 13.33 it is easy to specify the interpolating functions corresponding to the set Λ (cf. [25]). We will denote $\Delta_{\{\lambda\}}$ by Δ_λ.

Corollary 13.34. *Given $\lambda\in\Lambda$ and $0\le j<p_\lambda$, the function $g_{\lambda,j}\in PW_\sigma$ given by*

$$g_{\lambda,j}(z)=\sum_{k=j}^{p_\lambda-1}\binom{k}{j}\Delta_\lambda^{(k-j)}(\lambda)\,\frac{S(z)}{(z-\lambda)^{p_\lambda-k}} \tag{13.4}$$

satisfies $g_{\lambda,j}^{(\ell)}(\mu)=1$ if $(\mu,\ell)=(\lambda,j)$ and 0 otherwise.

Proof. Let $f(z)=(1/j!)\,(z-\lambda)^j$. Then for $0\le k<p_\lambda$, $f^{(k)}(\lambda)=1$ if $k=j$ and 0 otherwise. Let $\Omega=\{\lambda\}$, define the polynomial P_λ by

$$\begin{aligned}
P_\lambda(z) &= \sum_{k=0}^{p_\lambda-1}\frac{1}{k!}(f\,\Delta_\lambda)^{(k)}(\lambda)\,(z-\lambda)^k\\
&= \sum_{k=0}^{p_\lambda-1}\frac{1}{k!}\sum_{\ell=0}^{k}\binom{k}{\ell}f^{(\ell)}(\lambda)\,\Delta_\lambda^{(k-\ell)}(\lambda)\,(z-\lambda)^k\\
&= \sum_{k=j}^{p_\lambda-1}\frac{1}{k!}\binom{k}{j}\Delta_\lambda^{(k-j)}(\lambda)\,(z-\lambda)^k,
\end{aligned}$$

and let

$$g_{\lambda,j}(z)=\frac{S(z)}{(z-\lambda)^{p_\lambda}}\,P_\lambda(z).$$

The result now follows from Theorem 13.33. $\square$

13.4 Reconstruction of $f \in PW_\sigma$ from its Samples on Λ

Throughout this section, $S(z)$ will denote the function

$$S(z) = \prod_{i=1}^{m} \sin(2\pi\sigma_i(z - \alpha_i)).$$

13.4.1 The Case in Which Λ is Separated

Our goal is to show that $\{g_{\lambda,j}\}$ is a Riesz basis. By Theorem 13.17 it is sufficient to show that both it and its biorthogonal system are Bessel sequences. The latter is accomplished by the following theorem.

Theorem 13.35. *There is a constant C such that for all $f \in PW_\sigma$,*

$$\sum_{\lambda\in\Lambda} \sum_{j=0}^{p_\lambda - 1} |f^{(j)}(\lambda)|^2 \le C\, \|f\|^2_{PW_\sigma}. \tag{13.5}$$

Proof. Note that by the definition of the set Λ, $\sup_{\lambda\in\Lambda} p_\lambda \le m$ where m is the number of shifted lattices in Λ. Consequently, we can rearrange the sum in (13.5) as

$$\sum_{\lambda\in\Lambda} \sum_{j=0}^{p_\lambda - 1} |f^{(j)}(\lambda)|^2 = \sum_{k=1}^{m} \sum_{\{\lambda : p_\lambda \le k\}} |f^{(k)}(\lambda)|^2.$$

Since for each k, the set $\{\lambda \in \Lambda : p_\lambda \le k\}$ is real and separated, it follows from the Plancherel–Pólya Theorem (Theorem 13.7) and Bernstein's Inequality (Theorem 13.6) that

$$\sum_{\{\lambda : p_\lambda \le k\}} |f^{(k)}(\lambda)|^2 \le C\, \|f^{(k)}\|^2_{PW_\sigma} \le C_k \|f\|^2_{PW_\sigma},$$

and the result follows with $C = \sum_k C_k$. □

The argument that $\{g_{\lambda,j}\}$ is a Bessel sequence uses the theory of H^2 spaces. Note that by (13.4), each function $g_{\lambda,j}$ is a finite linear combination of the functions $S(z)/(z-\lambda)^n$, $1 \le n \le m$ where the coefficients in these linear combinations involve the numbers $\Delta_\lambda^{(\ell)}(\lambda)$. The next lemma shows that these numbers are bounded by a constant independent of λ and $0 \le \ell < m$.

Lemma 13.36. $\sup\{|\Delta_\lambda^{(\ell)}(\lambda)| : \lambda \in \Lambda,\ 0 \le \ell < m\} = C < \infty.$

Proof. The proof is by induction on ℓ. Let $S_\lambda(z) = S(z)/(z-\lambda)^{p_\lambda}$ and note that for every ℓ, $S_\lambda^{(\ell)}(\lambda)$ is bounded uniformly in λ. Also note that, since λ is at least some fixed positive distance away from any other zero of $S(z)$, $\Delta_\lambda(\lambda)$ is bounded uniformly in λ. Hence the conclusion holds with $\ell = 0$. For $\ell > 0$, note that

$$\begin{aligned}0 &= (S_\lambda\, \Delta_\lambda)^{(\ell)}(\lambda)\\ &= \sum_{k=0}^{\ell} \binom{\ell}{k} S_\lambda^{(k)}(\lambda)\, \Delta_\lambda^{(\ell-k)}(\lambda)\\ &= S_\lambda(\lambda)\, \Delta_\lambda^{(\ell)}(\lambda) + \sum_{k=1}^{\ell} \binom{\ell}{k} S_\lambda^{(k)}(\lambda)\, \Delta_\lambda^{(\ell-k)}(\lambda).\end{aligned}$$

Hence

$$|\Delta_\lambda^{(\ell)}(\lambda)| \le |\Delta_\lambda(\lambda)| \sum_{k=1}^{\ell} \binom{\ell}{k} |S_\lambda^{(k)}(\lambda)|\, |\Delta_\lambda^{(\ell-k)}(\lambda)|,$$

and the result follows for ℓ from the induction hypothesis. □

Lemma 13.37. *The sequence*

$$\left\{ \frac{S(z)}{(z-\lambda)^n} \right\}_{\lambda\in\Lambda, 1\le n\le p_\lambda}$$

is a Bessel sequence in PW_σ.

Proof. We will use the characterization of Bessel sequences given in Theorem 13.15. It will suffice to show that there is a constant $C > 0$ such that for all finite sequences $\{c_{\lambda,n}\}$,

$$\left\| \sum_{\lambda,n} c_{\lambda,n} \frac{S(z)}{(z-\lambda)^n} \right\|_{PW_\sigma} \le C \left(\sum_{\lambda,n} |c_{\lambda,n}|^2 \right)^{1/2}.$$

To that end, let $\{c_{\lambda,n}\}$ be such a sequence.

Note that given $h > 0$, the function $S(t+ih)$ never vanishes and is bounded on $\mathbf{R}$. Also note that the function

$$S(z+ih)^{-1} \sum_{\lambda,n} c_{\lambda,n} \frac{S(z+ih)}{(z+ih-\lambda)^n} = \sum_{\lambda,n} \frac{c_{\lambda,n}}{(z+ih-\lambda)^n}$$

is in H^2. Therefore, in light of Lemma 13.5, and with C representing a constant independent of $\{c_{\lambda,n}\}$ which may change from line to line,

$$\begin{aligned}\left\| \sum_{\lambda,n} c_{\lambda,n} \frac{S(z)}{(z-\lambda)^n} \right\|_{PW_\sigma} &\le C \left(\int_{-\infty}^{\infty} \left| \sum_{\lambda,n} c_{\lambda,n} \frac{S(t+ih)}{(t+ih-\lambda)^n} \right|^2 dt \right)^{1/2}\\ &= C \left(\int_{-\infty}^{\infty} |S(t+ih)|^2 \left| \sum_{\lambda,n} \frac{c_{\lambda,n}}{(t+ih-\lambda)^n} \right|^2 dt \right)^{1/2}\\ &\le C\, \|S(t+ih)\|_\infty \left(\int_{-\infty}^{\infty} \left| \sum_{\lambda,n} \frac{c_{\lambda,n}}{(t+ih-\lambda)^n} \right|^2 dt \right)^{1/2}\\ &= C \left\| \sum_{\lambda,n} \frac{c_{\lambda,n}}{(t+ih-\lambda)^n} \right\|_{H^2}.\end{aligned}$$

Hence it suffices to find a constant $C > 0$ independent of $\{c_{\lambda,n}\}$ such that

$$\Big\|\sum_{\lambda,n} \frac{c_{\lambda,n}}{(t+ih-\lambda)^n}\Big\|_{H^2} \le C\Big(\sum_{\lambda,n} |c_{\lambda,n}|^2\Big)^{1/2}.$$

To see this, take $\varphi \in H^2$ with $\|\varphi\|_{H^2} = 1$. Then

$$\begin{aligned}
\Big|\Big\langle \sum_{\lambda,n} \frac{c_{\lambda,n}}{(t+ih-\lambda)^n}, \varphi(t)\Big\rangle\Big| &= \Big|\sum_{\lambda,n} c_{\lambda,n} \int_{-\infty}^{\infty} \frac{\overline{\varphi(t)}}{(t-\overline{(\lambda+ih)})^n}\, dt\Big| \\
&= \Big|\sum_{\lambda,n} c_{\lambda,n} \overline{\int_{-\infty}^{\infty} \frac{\varphi(t)}{(t-(\lambda+ih))^n}\, dt}\Big| \\
&= \Big|\sum_{\lambda,n} c_{\lambda,n}\, \overline{\varphi^{(n)}(\lambda+ih)}\Big| \\
&\le \Big(\sum_{\lambda,n} |c_{\lambda,n}|^2\Big)^{1/2} \Big(\sum_{\lambda,n} |\varphi^{(n)}(\lambda+ih)|^2\Big)^{1/2}.
\end{aligned}$$

Note that as in the proof of Theorem 13.35, in light of Lemma 13.2(c), and using the fact that Λ is separated, we can write

$$\begin{aligned}
\sum_{\lambda,n} |\varphi^{(n)}(\lambda+ih)|^2 &= \sum_{n=1}^{m} \sum_{\{\lambda : n \le p_\lambda\}} |\varphi^{(n)}(\lambda+ih)|^2 \\
&= \sum_{n=1}^{m} C(h,n)\, \|\varphi\|_{H^2} \\
&= C\, \|\varphi\|_{H^2},
\end{aligned}$$

and the result follows. □

Theorem 13.38. *The collection $\{g_{\lambda,j} : \lambda \in \Lambda,\ 0 \le j < p_\lambda\}$ is a Bessel sequence in PW_σ.*

Proof. Let $\{c_{\lambda,j}\}$ be a finite sequence and consider the sum

$$\sum_{\lambda\in\Lambda} \sum_{j=0}^{p_\lambda-1} c_{\lambda,j}\, g_{\lambda,j}.$$

Fixing $\lambda \in \Lambda$, and using (13.4), we can write

$$\begin{aligned}
\sum_{j=0}^{p_\lambda-1} c_{\lambda,j}\, g_{\lambda,j}(z) &= \sum_{j=0}^{p_\lambda-1} c_{\lambda,j} \sum_{k=j}^{p_\lambda-1} \binom{k}{j} \Delta_\lambda^{(k-j)}(\lambda)\, \frac{S(z)}{(z-\lambda)^{p_\lambda-k}} \\
&= \sum_{j=0}^{p_\lambda-1} c_{\lambda,j} \sum_{n=1}^{p_\lambda-j} \binom{p_\lambda-n}{j} \Delta_\lambda^{(p_\lambda-n-j)}(\lambda)\, \frac{S(z)}{(z-\lambda)^n}
\end{aligned}$$

$$= \sum_{n=1}^{p_\lambda} \left(\sum_{j=0}^{p_\lambda - n} c_{\lambda,j} \binom{p_\lambda - n}{j} \Delta_\lambda^{(p_\lambda - n - j)}(\lambda) \right) \frac{S(z)}{(z-\lambda)^n}$$

$$\equiv \sum_{n=1}^{p_\lambda} C_{\lambda,n} \frac{S(z)}{(z-\lambda)^n}.$$

Now, by Lemma 13.37,

$$\left\| \sum_{\lambda \in \Lambda} \sum_{j=0}^{p_\lambda - 1} c_{\lambda,j}\, g_{\lambda,j} \right\|_{PW_\sigma} = \left\| \sum_{\lambda \in \Lambda} \sum_{n=1}^{p_\lambda} C_{\lambda,n} \frac{S(z)}{(z-\lambda)^n} \right\|_{PW_\sigma}$$
$$\leq C \left(\sum_{\lambda \in \Lambda} \sum_{n=1}^{p_\lambda} |C_{\lambda,n}|^2 \right)^{1/2}.$$

Now, by the Cauchy–Schwarz inequality and Lemma 13.36,

$$|C_{\lambda,n}|^2 = \left| \sum_{j=0}^{p_\lambda - n} c_{\lambda,j} \binom{p_\lambda - n}{j} \Delta_\lambda^{(p_\lambda - n - j)}(\lambda) \right|^2 \leq C \sum_{j=0}^{p_\lambda - n} |c_{\lambda,j}|^2$$

for some C independent of λ, n, and j. Finally note that

$$\sum_{\lambda \in \Lambda} \sum_{n=1}^{p_\lambda} |C_{\lambda,n}|^2 \leq C \sum_{\lambda \in \Lambda} \sum_{n=1}^{p_\lambda} \sum_{j=0}^{p_\lambda - n} |c_{\lambda,j}|^2$$
$$= C \sum_{\lambda \in \Lambda} \sum_{j=0}^{p_\lambda - 1} \sum_{n=1}^{p_\lambda - j} |c_{\lambda,j}|^2$$
$$\leq C\, p_\lambda \sum_{\lambda \in \Lambda} \sum_{j=0}^{p_\lambda - 1} |c_{\lambda,j}|^2,$$

and the result follows since $p_\lambda \leq m$ for all $\lambda \in \Lambda$. □

13.4.2 The Case in Which Λ is Not Separated

In this case, the collection $\{g_{\lambda,j}\}$ cannot be a Schauder basis. In fact the following holds.[3]

Theorem 13.39. *Suppose that $\{x_n\}$ is a Schauder basis for a Hilbert space H with the property that $\inf_n \|x_n\| > 0$. Then there is an $\varepsilon > 0$ such that $\|x_n - x_m\| \geq \varepsilon$ for all n, m.*

Proof. Let $\{y_n\}$ be the sequence biorthogonal to $\{x_n\}$. By Theorem 13.13, there is a constant $M > 0$ such that for all n

[3]Thanks to R. M. Young for pointing this out to the second author.

$$1 \leq \|x_n\| \, \|y_n\| \leq M.$$

Consequently, $\sup_n \|y_n\| = B < \infty$ and we can write

$$1 = \langle x_n - x_m, y_n \rangle \leq \|y_n\| \, \|x_n - x_m\| \leq B \, \|x_n - x_m\|,$$

and it follows that $\|x_n - x_m\| \geq 1/B$ for all n, m. □

Corollary 13.40. *If $\{t^j \, e^{2\pi i \lambda t} : \lambda \in \Lambda, 0 \leq j < p_\lambda\}$ is a Schauder basis for $L^2[-\sigma, \sigma]$ then $\inf_{\lambda, \mu \in \Lambda, \lambda \neq \mu} |\lambda - \mu| > 0$.*

Proof. Note that for each $\lambda \in \Lambda$, the element $e^{2\pi i \lambda t}$ is included in the basis. By Theorem 13.39, there is an $\varepsilon > 0$ such that for all λ, $\mu \in \Lambda$, with $\lambda \neq \mu$,

$$\|e^{2\pi i \lambda t} - e^{2\pi i \mu t}\|_{L^2[-\sigma,\sigma]} \geq \varepsilon.$$

Next observe that as long as, say $|\mu - \lambda| \leq 1$,

$$\begin{aligned}
\|e^{2\pi i \lambda t} - e^{2\pi i \mu t}\| &= \|e^{2\pi i \lambda t}(1 - e^{2\pi i (\mu - \lambda) t})\| \\
&\leq \|1 - e^{2\pi i (\mu - \lambda) t}\| \\
&= \left\| \sum_{n=1}^{\infty} \frac{(2\pi i)^n}{n!} \, t^n \, (\mu - \lambda)^n \right\| \\
&\leq \sum_{n=1}^{\infty} \frac{(2\pi)^n}{n!} |\mu - \lambda|^n \, \|t^n\| \\
&\leq (2\sigma)^{1/2} \sum_{n=1}^{\infty} \frac{(2\pi\sigma)^n}{n!} |\mu - \lambda|^n \\
&= (2\sigma)^{1/2} \, (e^{2\pi\sigma|\mu - \lambda|} - 1) \\
&\leq (2\sigma)^{1/2} \, e^{2\pi\sigma} \, |\mu - \lambda|,
\end{aligned}$$

and the result follows. □

Since $\{g_{\lambda,j}\}$ is a basis for PW_σ if and only if $\{t^j \, e^{2\pi i \lambda t}\}$ is a basis for $L^2[-\sigma, \sigma]$, we see that if Λ is not separated, then $\{g_{\lambda,j}\}$ cannot be a basis for PW_σ.

The way around this difficulty is to partition the set Λ into "bunches" with the property that (1) each bunch consists of no more than m points in Λ, (2) the diameter of each bunch is no greater than some constant $\varepsilon > 0$, and (3) the separation between adjacent bunches is greater than some constant $\varepsilon' > 0$. To each bunch will be assigned a subspace given as the span of the interpolating functions corresponding to the points in the bunch. We will then show that the resulting sequence of subspaces forms a Riesz basis from subspaces for PW_σ. Specifically, we make the following definition.

Definition 13.41. *Given any discrete set $\Gamma \subseteq \mathbf{R}$, and ε, $\varepsilon' > 0$, we say that a subset $\Omega \subseteq \Gamma$ is an $\varepsilon, \varepsilon'$-*block *provided that*

(a) *if* λ, $\mu \in \Omega$ *then* $|\lambda - \mu| < \varepsilon$, *and*

(b) $\mathrm{dist}(\Omega, \Gamma \setminus \Omega) > \varepsilon'$.

Lemma 13.42. *For all $\varepsilon > 0$ sufficiently small, Λ can be written as a disjoint union of $\varepsilon, \varepsilon/m$-blocks. Moreover, each block can be chosen to contain no more than m points.*

Proof. Fix $\varepsilon < 1/(4\sigma_m)$ and write $\mathbf{R}$ as a disjoint union of intervals of length $\varepsilon' = \varepsilon/m$ by defining $I_k = [k\,\varepsilon', (k+1)\,\varepsilon')$. Now define the sequence k_n as follows. Let k_1 be the smallest index $k \geq 1$ with the property that $I_{k_1} \cap \Lambda$ is empty and define recursively k_n to be the smallest index strictly larger than k_{n-1} such that $I_{k_n - 1} \cap \Lambda$ is nonempty while $I_{k_n} \cap \Lambda$ is empty. Then let $\Omega_n = \cup_{k=k_n}^{k_{n+1}} (I_k \cap \Lambda)$. For $n \leq -1$, define $\Omega_n = -\Omega_{-n}$ and define $\Omega_0 = \{0\}$. It is obvious that $\Lambda = \cup_{n \in \mathbf{Z}} \Omega_n$ and that (a) and (b) are satisfied with $\varepsilon' = \varepsilon/m$.

The sets Ω_n satisfy $\#(\Omega_n) \leq m$ for all $n \in \mathbf{Z} \setminus \{0\}$. To see this, note first that for each $n \geq 1$, $k_{n+1} - k_n \leq m + 1$. If this were not the case, then there would be $m+1$ contiguous values of k for which $I_k \cap \Lambda$ would be nonempty. Since each interval I_k has length $\varepsilon' = \varepsilon/m$, this means that there would be an interval of length $(m+1)\varepsilon/m \leq 2\varepsilon$ containing at least $m+1$ points of Λ. This would imply that there was a j with $1 \leq j \leq m$ such that at least two of these points were contained in Λ_j. Hence the distance between these points would be at least $1/2\sigma_m$. But $2\varepsilon < 1/2\sigma_m$, a contradiction. Now, since $k_{n+1} - k_n \leq m+1$, Ω_n is contained in an interval of length ε since there can be no more than m contiguous values of k with $I_k \cap \Lambda$ nonempty. Since no more than m points of Λ can sit in any such interval, the claim is proved. It is now clear that each Ω_n is an ε-block with $\varepsilon' = \varepsilon/m$. $\square$

Now suppose that a $0 < \varepsilon < 1/(4\sigma_m)$, as described in Lemma 13.42, is fixed and let $\mathcal{A}$ denote the collection of such ε-blocks partitioning Λ. We make the following definition.

Definition 13.43. *For each $\Omega \in \mathcal{A}$, define the subspace $H_\Omega \subseteq PW_\sigma$ by*

$$H_\Omega = \mathrm{span}\{g_{\lambda,j} : \lambda \in \Omega,\ 0 \leq j < p_\lambda\},$$

and the projector $\mathcal{P}_\Omega$ by

$$\mathcal{P}_\Omega f = \sum_{\lambda \in \Omega} \sum_{j=0}^{p_\lambda - 1} f^{(j)}(\lambda)\, g_{\lambda,j}.$$

Our goal is to show that the collection $\{H_\Omega\}_{\Omega \in \mathcal{A}}$ is a Riesz basis from subspaces for PW_σ. We will verify this directly. First of all, it is clear from the definition of $\mathcal{P}_\Omega$ and of the interpolating functions $g_{\lambda,j}$ that part (a) of Definition 13.21 holds. The proof that part (b) of Definition 13.21 holds will rely on Lemmas 13.45 and 13.46 below and will consist in arguing that (1) for each $f \in PW_\sigma$, the sum $\sum_\Omega \mathcal{P}_\Omega f$ converges unconditionally, and that (2) it must converge to f.

Before getting to that, we make some reductions based on the observation that if we take a finite partition of $\mathcal{A}$ then we may argue separately on each element of the partition. To that end, note that to each $\Omega \in \mathcal{A}$ there is associated a subset Γ_Ω of $\{1, 2, \ldots, m\}$ given by $j \in \Gamma_\Omega$ if and only if $\Omega \cap \Lambda_j \neq \emptyset$. Now we partition the set $\mathcal{A}$ as follows. Given $\Gamma \subseteq \{1, 2, \ldots, m\}$ define the set $\mathcal{A}_\Gamma$ by $\mathcal{A}_\Gamma = \{\Omega \in \mathcal{A} : \Gamma_\Omega = \Gamma\}$. Clearly, the collection $\{\mathcal{A}_\Gamma : \Gamma \subseteq \{1, \ldots, m\}\}$ partitions $\mathcal{A}$ into finitely many disjoint subsets.

Now fix a subset Γ and an $\Omega \in \mathcal{A}_\Gamma$. By renumbering the indices if necessary, we can assume without loss of generality that $\Gamma = \{1, \ldots, p\}$ for some $p \leq m$. We enumerate Ω as a sequence in which points appear as many times as their multiplicity. Specifically, we can take $\Omega = \{\lambda_1, \ldots, \lambda_p\}$ where $\lambda_\ell \in \Lambda_\ell$ for each ℓ. In this case, according to Theorem 13.33,

$$\begin{aligned}
\mathcal{P}_\Omega f(z) &= \frac{S(z)}{\prod_{k=1}^{p}(z-\lambda_k)}\, P_\Omega(z) \\
&= \frac{S(z)}{\prod_{k=1}^{p}(z-\lambda_k)} \sum_{j=1}^{p} (f\,\Delta_\Omega)[\lambda_1, \ldots, \lambda_j] \prod_{\ell=0}^{j-1}(z-\lambda_\ell) \\
&= \sum_{j=1}^{p} (f\,\Delta_\Omega)[\lambda_1, \ldots, \lambda_j]\, \frac{S(z)}{\prod_{k=j}^{p}(z-\lambda_k)}.
\end{aligned}$$

The proof of the following lemma is similar to that of Lemma 13.36.

Lemma 13.44. *For each $\Omega \in \mathcal{A}$, let*

$$I_\Omega = [\min\{\lambda \in \Omega\}, \max\{\lambda \in \Omega\}].$$

Then

$$\sup_{\Omega \in \mathcal{A}} \sup_{1 \leq j \leq m} \sup_{x \in I_\Omega} |\Delta_\Omega^{(j)}(x)| = M < \infty.$$

Lemma 13.45. *Let $\Gamma = \{1, \ldots, p\}$ for some $p \leq m$, and fix $1 \leq j \leq p$. Then there is a constant $C > 0$ such that for all $f \in PW_\sigma$ and all finite subsets $\mathcal{B} \subseteq \mathcal{A}_\Gamma$,*

$$\left(\sum_{\Omega \in \mathcal{B}} |(f\,\Delta_\Omega)[\lambda_1, \ldots, \lambda_j]|^2 \right)^{1/2} \leq C\, \|f\|_{PW_\sigma}.$$

Proof. Corollary 13.29 says that there is a point $\xi_\Omega^j \in I_\Omega$ such that

$$\begin{aligned}
|(f\,\Delta_\Omega)[\lambda_1, \ldots, \lambda_j]| &= \frac{1}{(j-1)!}\,(f\,\Delta_\Omega)^{(j-1)}(\xi_\Omega^j) \\
&= \frac{1}{(j-1)!} \sum_{k=0}^{j-1} \binom{j-1}{k} f^{(k)}(\xi_\Omega^j)\, \Delta_\Omega^{(j-1-k)}(\xi_\Omega^j).
\end{aligned}$$

Now,

$$\Big(\sum_{\Omega\in\mathcal{B}} |(f\,\Delta_\Omega)[\lambda_1, \dots, \lambda_j]|^2\Big)^{1/2}$$

$$= \frac{1}{(j-1)!}\Big(\sum_{\Omega\in\mathcal{B}}\Big|\sum_{k=0}^{j-1}\binom{j-1}{k} f^{(k)}(\xi_\Omega^j)\,\Delta_\Omega^{(j-1-k)}(\xi_\Omega^j)\Big|^2\Big)^{1/2}$$

$$\le \frac{1}{(j-1)!}\sum_{k=0}^{j-1}\binom{j-1}{k}\Big(\sum_{\Omega\in\mathcal{B}} |f^{(k)}(\xi_\Omega^j)|^2\,|\Delta_\Omega^{(j-1-k)}(\xi_\Omega^j)|^2\Big)^{1/2}$$

$$\le M\,\frac{1}{(j-1)!}\sum_{k=0}^{j-1}\binom{j-1}{k}\Big(\sum_{\Omega\in\mathcal{B}} |f^{(k)}(\xi_\Omega^j)|^2\Big)^{1/2}$$

$$\le M\,\frac{1}{(j-1)!}\sum_{k=0}^{j-1}\binom{j-1}{k} C_k\,\|f^{(k)}\|_{PW_\sigma}$$

$$\le C\,\|f\|_{PW_\sigma},$$

where the constant M comes from Lemma 13.44, the constant C_k from the Plancherel–Pólya inequality (Theorem 13.7), and where the last line is a consequence of Bernstein's inequality (Theorem 13.6). □

Lemma 13.46. *Let $\Gamma = \{1, \dots, p\}$ for some $p \le m$, and fix $1 \le j \le p$. For each $\Omega \in \mathcal{A}_\Gamma$, $\Omega = \{\lambda_1, \dots, \lambda_p\}$, define the function $S_\Omega(z)$ by*

$$S_\Omega(z) = \frac{S(z)}{\prod_{k=j}^p (z-\lambda_k)}.$$

Then the collection $\{S_\Omega\}_{\Omega\in\mathcal{A}_\Gamma}$ is a Bessel sequence in PW_σ.

Proof. We will show that there is a constant $C > 0$ such that for all finite sequences $\{c_\Omega\}_{\Omega\in\mathcal{A}_\Gamma}$,

$$\Big\|\sum_\Omega c_\Omega\, S_\Omega\Big\|_{PW_\sigma} \le C\Big(\sum_\Omega |c_\Omega|^2\Big)^{1/2}.$$

By repeating the first part of the proof of Lemma 13.37, we have that for some $h > 0$ and some constant C,

$$\Big\|\sum_\Omega c_\Omega\, S_\Omega\Big\|_{PW_\sigma} \le C\,\Big\|\sum_\Omega \frac{c_\Omega}{\prod_{k=j}^p (t+ih-\lambda_k)}\Big\|_{H^2}.$$

To complete the proof, let $\varphi \in H^2$ with $\|\varphi\|_{H^2} = 1$, and let $z_k = \lambda_k + ih$. Then,

$$\begin{aligned}
&\left|\left\langle \sum_{\Omega} \frac{c_\Omega}{\prod_{k=j}^{p}(t+ih-\lambda_k)},\, \varphi(t)\right\rangle\right| \\
&\qquad = \left|\sum_{\Omega} c_\Omega \int_{-\infty}^{\infty} \frac{\overline{\varphi(t)}}{\prod_{k=j}^{p}(t-\overline{z_k})}\, dt\right| \\
&\qquad = \left|\sum_{\Omega} c_\Omega \overline{\int_{-\infty}^{\infty} \frac{\varphi(t)}{\prod_{k=j}^{p}(t-z_k)}\, dt}\right| \\
&\qquad = \left|\sum_{\Omega} c_\Omega \overline{\varphi[z_j,\, \dots,\, z_p]}\right| \\
&\qquad \le C\left(\sum_{\Omega} |c_\Omega|^2\right)^{1/2} \left(\sum_{\Omega} |\varphi[z_j,\, \dots,\, z_p]|^2\right)^{1/2}.
\end{aligned}$$

By Corollary 13.29, there is a point $\xi_\Omega^j \in I_\Omega + ih$ such that

$$\varphi[z_j,\, \dots,\, z_p] = \frac{1}{(p-j)!}\, \varphi^{(p-j)}(\xi_\Omega^j).$$

Note that since the intervals I_Ω are separated, so are the points ξ_Ω^j. Hence, in light of Lemma 13.2(d), we can write

$$\sum_{\Omega} |\varphi[z_j,\, \dots,\, z_p]|^2 = \frac{1}{(p-j)!} \sum_{\Omega} |\varphi^{(p-j)}(\xi_\Omega^j)|^2 \le C\, \|\varphi\|_{H^2}.$$

The conclusion of the theorem now follows. □

In order to see that $\sum_\Omega \mathcal{P}_\Omega f$ converges unconditionally for each f, note that by Lemma 13.46, there is a constant C such that for all finite subsets $\mathcal{B}$ of $\mathcal{A}_\Gamma$,

$$\begin{aligned}
\left\|\sum_{\Omega\in\mathcal{B}} \mathcal{P}_\Omega f\right\|_{PW_\sigma} &= \left\|\sum_{\Omega\in\mathcal{B}} \sum_{j=1}^{p} (f\,\Delta_\Omega)[\lambda_1,\, \dots,\, \lambda_j]\, \frac{S(z)}{\prod_{k=j}^{p}(z-\lambda_k)}\right\|_{PW_\sigma} \\
&\le C\left(\sum_{j=1}^{p} \sum_{\Omega\in\mathcal{B}} |(f\,\Delta_\Omega)[\lambda_1,\, \dots,\, \lambda_j]|^2\right)^{1/2}.
\end{aligned}$$

Moreover Lemma 13.45 implies that the sum

$$\sum_{j=1}^{p} \sum_{\Omega} |(f\,\Delta_\Omega)[\lambda_1,\, \dots,\, \lambda_j]|^2$$

converges absolutely and hence unconditionally. Therefore the above inequality allows us to establish a Cauchy criterion as in Lemma 13.19(b) on the series $\sum_\Omega \mathcal{P}_\Omega f$ which in turn implies that the series converges unconditionally to some function $g \in PW_\sigma$. By the definition of $\mathcal{P}_\Omega$, g agrees with f on Λ, and so by Theorem 13.32, $g = f$.

Acknowledgments

The second author is grateful to his advisor, colleague, and friend John Benedetto, master expositor, for teaching him many things including the importance of expository writing in mathematics.

The authors also wish to express their deep appreciation to Yura Lyubarskii, whose detailed and thoughtful comments greatly enhanced this manuscript. We wish to thank him especially for exposing us to the techniques used in this chapter and for providing us with many useful and hard-to-find references.

References

1. L. Ahlfors, *Complex Analysis* (Third Edition), McGraw-Hill, New York, 1979.
2. S. A. Avdonin and S. A. Ivanov, *Families of Exponentials* (Translated from the Russian and revised by the authors), Cambridge University Press, Cambridge, 1995.
3. J. J. Benedetto, Irregular sampling and frames, in: *Wavelets: A Tutorial in Theory and Applications*, C. K. Chui, ed., Academic Press, New York, 1992, pp. 445–507.
4. J. J. Benedetto and W. Heller, Irregular sampling and the theory of frames, I, *Mat. Note*, **10** (suppl. 1) (1990), pp. 103–125.
5. C. A. Berenstein, S. Casey, and E. V. Patrick, Systems of convolution equations, deconvolution, and wavelet analysis, U. Maryland Systems Research Center whitepaper (1990).
6. C. A. Berenstein and E. V. Patrick, Exact deconvolution for multiple convolution operators—an overview, plus performance characterizations for imaging sensors, Proc. IEEE, **78** (1990), pp. 723–734.
7. C. A. Berenstein, A. Yger, and B. A. Taylor, Sur quelques formules explicites de deconvolution, *J. Optics (Paris)*, **14** (1983), pp. 75–82.
8. S. D. Casey and D. F. Walnut, Systems of convolution equations, deconvolution, Shannon sampling and the Gabor and wavelet transform, SIAM Review, **36** (1994), pp. 537–577.
9. K. F. Cheung, A multidimensional extension of Papoulis' generalized sampling expansion with the application in minimum density sampling, in: *Advanced Topics in Shannon Sampling and Interpolation Theory*, R. J. Marks II, Springer-Verlag, New York, 1993, pp. 85–119.
10. S. D. Conte and C. de Boor, *Elementary Numerical Analysis: An Algorithmic Approach* (Third Edition), McGraw-Hill, New York, 1980.
11. J. B. Conway, *Functions of One Complex Variable* (Second Edition), Springer-Verlag, New York, 1978.
12. L. Desbat, Efficient sampling on coarse grids in tomography, *Inverse Problems*, **9** (1993), pp. 251–269.
13. P. L. Duren, *Theory of H^p Spaces*, Academic Press, New York, 1970.
14. A. Faridani, A generalized sampling theorem for locally compact Abelian groups, *Math. Comp.*, **63** (1994), pp. 307–327.

15. A. Faridani, An application of a multidimensional sampling theorem to computed tomography, in: *Integral geometry and tomography* (Arcata, CA, 1989), E. Grinberg and E. T. Quinto, eds., Contemp. Math., Vol. 113, American Mathematical Society, Providence, RI, 1990, pp. 65–80.
16. H. G. Feichtinger and K. Gröchenig, Theory and practice of irregular sampling, in: *Wavelets: Mathematics and Applications*, J. J. Benedetto and M. W. Frazier, M., eds., CRC Press, Boca Raton, FL, 1993, pp. 305–363.
17. A. O. Gel'fond, *Calculus of Finite Differences* (Translated from the Russian), Hindustan Publishing Corp., Delhi, India, 1971.
18. I. C. Gohberg and M. G. Krein, *Introduction to the Theory of Linear Nonselfadjoint Operators*, Americal Mathematical Society, Providence, RI, 1969.
19. K. Gröchenig, C. Heil, and D. Walnut, Nonperiodic sampling and the local three squares theorem, *Ark. Mat.*, **38** (2000), pp. 77–92.
20. J. R. Higgins, Five short stories about the cardinal series, *Bull. Amer. Math. Soc. (N.S.)*, **12** (1985), pp. 45–89.
21. J. R. Higgins, *Sampling Theory in Fourier and Signal Analysis*, Oxford University Press, Oxford, 1996.
22. S. V. Hruščev, N. K. Nikol'skiĭ, and B. S. Pavlov, Unconditional basees of exponentials and of reproducing kernels, in: *Complex Analysis and Spectral Theory*, V. P. Havin and N. K. Nikol'skiĭ, eds., Lecture Notes in Math., Vol. 864, Springer, Berlin, 1981, pp. 214–335.
23. S. H. Izen, Generalized sampling expansion on lattices, *IEEE Trans. Signal Processing*, **53** (2005), pp. 1949–1963.
24. A. Jerri, The Shannon sampling theorem—its various extensions and applications: A tutoring review, *Proc. IEEE*, **65** (1977), pp. 1565–1596.
25. B. Y. Levin, On bases of exponential functions in L^2, *Zap. Mekh.-Mat. Fak. i Khar'kov. Mat. Obshch.*, **27** (1961), pp. 39–48 (Russian).
26. B. Ja. Levin, *Distribution of Zeros of Entire Functions* (Translated from the Russian, Revised Edition), Translations of Mathematical Monographs, Vol. 5, American Mathematical Society, Providence, RI, 1980.
27. B. Ya. Levin, *Lectures on Entire Functions* (Translated from the Russian), Translations of Mathematical Monographs, Vol. 150, American Mathematical Society, Providence, RI, 1996.
28. N. Levinson, *Gap and Density Theorems*, American Mathematical Society Colloquium Publications, Vol. 26, American Mathematical Society, New York, 1940.
29. Y. I. Lyubarskii and A. Rashkovskii, Complete interpolating sequences for Fourier transforms supported by convex symmetric polygons, *Ark. Mat.*, **38** (2000), pp. 139–170.
30. R. J. Marks II, *Introduction to Shannon Sampling and Interpolation Theory*, Springer-Verlag, New York, 1991.
31. F. Marvasti, Nonuniform sampling, in: *Advanced Topics in Shannon Sampling and Interpolation Theory*, R. J. Marks II, Springer-Verlag, New York, 1993, pp. 121–156.
32. N. K. Nikol'skiĭ, *Treatise on the Shift Operator* (Translated from the Russian), Grundlehren der Mathematischen Wissenschaften, Vol. 263, Springer-Verlag, Berlin, 1986.
33. R. E. A. C. Paley and N. Wiener, *Fourier Transforms in the Complex Domain*, American Mathematical Society Colloquium Publications, Vol. 19, American Mathematical Society, New York, 1934.

34. A. Papoulis, *Signal Analysis*, McGraw-Hill, New York, 1977.
35. B. S. Pavlov, Basicity of an exponential system and Muckenhoupt's condition, *Soviet Math. Doklady*, **20** (1979), pp. 655–659.
36. D. P. Petersen and D. Middleton, Sampling and reconstruction of wave-number-limited functions in N-dimensional Euclidean spaces, *Inform. Control*, **5** (1962), pp. 279–323.
37. I. Singer, *Bases in Banach Spaces, I*, Springer-Verlag, New York, 1970.
38. D. Walnut, Nonperiodic sampling of bandlimited functions on unions of rectangular lattices, *J. Fourier Anal. Appl.*, **2** (1996), pp. 435–452.
39. D. Walnut, Solutions to deconvolution equations using nonperiodic sampling, *J. Fourier Anal. Appl.*, **4** (1998), pp. 669–709.
40. D. F. Walnut, Nonperiodic sampling, conditional convergence and the local three squares theorem, in: Proc. SampTA'99 (1999 International Workshop on Sampling Theory and Applications, Loen, Norway, August 11-14, 1999), 1999, pp. 25–30.
41. J. L. Yen, On non-uniform sampling of bandwidth-limited signals, IRE Trans. Circuit Theory, **3** (1956), pp. 251–257.
42. R. M. Young, *An Introduction to Nonharmonic Fourier Series* (Revised First Edition), Academic Press, San Diego, 2001.
43. A. Zayed, ed., *Advances in Shannon's Sampling Theory*, CRC Press, Boca Raton, FL, 1994.

14

Learning the Right Model from the Data

Akram Aldroubi[1], Carlos Cabrelli[2], and Ursula Molter[2]

[1] Department of Mathematics, Vanderbilt University, Nashville, TN 37240, USA
aldroubi@math.vanderbilt.edu

[2] Departamento de Matemática, Facultad de Ciencias Exactas y Naturales, Universidad de Buenos Aires, Ciudad Universitaria, Pabellón I, 1428 Capital Federal, Argentina, and CONICET, Argentina
cabrelli@dm.uba.ar, umolter@dm.uba.ar

Summary. In this chapter we discuss the problem of finding the shift-invariant space model that best fits a given class of observed data $\mathcal{F}$. If the data is known to belong to a fixed—but unknown—shift-invariant space $V(\Phi)$ generated by a vector function Φ, then we can probe the data $\mathcal{F}$ to find out whether the data is sufficiently rich for determining the shift-invariant space. If it is determined that the data is not sufficient to find the underlying shift-invariant space V, then we need to acquire more data. If we cannot acquire more data, then instead we can determine a shift-invariant subspace $S \subset V$ whose elements are generated by the data. For the case where the observed data is corrupted by noise, or the data does not belong to a shift-invariant space $V(\Phi)$, then we can determine a space $V(\Phi)$ that fits the data in some optimal way. This latter case is more realistic and can be useful in applications, e.g., finding a shift-invariant space with a small number of generators that describes the class of chest X-rays.

To John, whose mathematics and humanity have inspired us.

14.1 Introduction

In many signal and image processing applications, images and signals are assumed to belong to some shift-invariant space of the form:

$$V(\Phi) := \left\{ f = \sum_{i=1}^{n} \sum_{j \in \mathbb{Z}^d} \alpha_i(j)\phi_i(\cdot + j) : \alpha_i \in l^2(\mathbb{Z}^d),\ i = 1, \dots, n \right\}, \qquad (14.1)$$

where $\Phi = [\phi_1, \phi_2, \dots, \phi_n]^t$ is a column vector whose elements ϕ_i are functions in $L^2(\mathbb{R}^d)$. These functions are *a set of generators* for the space $V = V(\Phi)$. For example, if $n = 1$, $d = 1$, and $\phi(x) = \text{sinc}(x)$, then the underlying space is the space of band-limited functions (often used in communications) [4], [5], [6].

However, in most applications, the shift-invariant space chosen to describe the underlying class of signals is not derived from experimental data—for example most signal processing applications assume "band-limitedness" of the signal, which has theoretical advantages, but generally does not necessarily reflect the underlying class of signals accurately. Thus, in order to derive the appropriate signal model for a class of signals, we consider the following two types of problems:

(I) Given a class of signals belonging to a certain fixed—but unknown—shift-invariant space V, the problem is whether it is possible to determine the space V from a set of m experimental data $\mathcal{F} = \{f_1, f_2, \dots, f_m\}$, where f_i are observed functions (signals) belonging to $V(\Phi)$.

(II) Given a large set of experimental data $\mathcal{F} = \{f_1, f_2, \dots, f_m\}$, where f_i are observed functions (signals) that are not necessarily from a shift-invariant space with a small fixed number of generators, we wish to determine some small space V that models the signals in "some" best way.

For Problem I to be meaningful, we must have some a priori assumption about our signal space V. In particular, we assume that V is a shift-invariant space that can be generated by a set of exactly n generators, $\Phi = [\phi_1, \phi_2, \dots, \phi_n]^t$, such that $\{\phi_i(\cdot - k) : k \in \mathbb{Z}, i = 1, \dots, n\}$ forms a Riesz basis for $V(\Phi)$. If a finite set $\mathcal{F}$ of signals is sufficient to determine $V(\Phi)$, then $\mathcal{F}$ is called a *determining set for* $V(\Phi)$. The goal is to see if we can perform operations on the observations $\mathcal{F} = \{f_1, f_2, \dots, f_m\}$ to deduce whether they are sufficient to determine the unknown shift-invariant space $V(\Phi)$, and if so, use them to find some set of generators Ψ for $V(\Phi)$, i.e., find some $\Psi = [\psi_1, \psi_2, \dots, \psi_n]^t$ such that $V(\Psi) = V(\Phi)$. If the observations are not sufficient to determine $V(\Phi)$, then we need to obtain more observations until a determining set is found.

This then becomes a learning problem: If the data is insufficient to determine the model, then the set $S(\mathcal{F}) = \text{closure}_{L^2}\big(\text{span}\{f_i(\cdot - k) : i = 1, \dots, m, k \in \mathbb{Z}^d\}\big)$ is a proper shift-invariant subspace of V. Thus the data determines some "smaller" shift-invariant space. The acquisition of new data will allow us to "learn" more about the right model, i.e., with the new information we can obtain a more complete description of the space.

In practice however, the a priori hypothesis that the class of signals belongs to a shift-invariant space with a known number of generators may not be satisfied. For example, the class of functions from which the data is drawn may not be a shift-invariant space. Another example is when the shift-invariant space hypothesis is correct but the assumptions about the number of generators is wrong. A third example is when the a priori hypothesis is correct but the data is corrupted by noise. For these three more realistic cases, we must consider Problem II.

Similarly to Problem I, we must impose some a priori conditions on the space V. In particular, we will search for the optimal space V among those spaces that are generated by exactly n generators. Consider the class $\mathcal{V}$ of all the shift-invariant spaces that are generated by some set of generators $\Phi = [\phi_1, \phi_2, \ldots, \phi_n]^t$, $\phi_i \in L^2(\mathbb{R}^d)$, with the property that $\{\phi_i(\cdot - k) : k \in \mathbb{Z}^d, i = 1, \ldots, n\}$ is a Riesz basis for $V(\Phi)$. The problem is then to find a space $V \in \mathcal{V}$ such that

$$V = \underset{V \in \mathcal{V}}{\operatorname{argmin}} \sum_{i=1}^{m} w_i \, \|f_i - P_V f_i\|^2, \tag{14.2}$$

where w_i are positive weights and where P_V is the orthogonal projection on V. The weights w_i can be chosen to normalize or to reflect our confidence about the data. For example we can choose $w_i = \|f_i\|^{-2}$ to place the data on a sphere or we can choose a small weight w_i for a given f_i if—due to noise or other factors—our confidence about the accuracy of f_i is low. The goal is to use the observations $\mathcal{F} = \{f_1, f_2, \ldots, f_m\}$ to find some set of generators $\Psi = [\psi_1, \psi_2, \ldots, \psi_n]^t$ that generates the optimal space $V = V(\Psi)$ in (14.2).

14.2 Notation and Preliminaries

Throughout this chapter, we assume that the unknown space V is a Riesz shift-invariant space, i.e., a shift-invariant space that has a set of generators $\Phi = [\phi_1, \ldots, \phi_n]^t$ such that $\{\phi_i(x - k) : i = 1, \ldots, n, \; k \in \mathbb{Z}^d\}$ forms a Riesz basis for V. That is, there exist $0 < A \leq B$ such that for all $f \in V(\Phi)$,

$$A\|f\|^2 \leq \sum_{i=1}^{n} \sum_{k \in \mathbb{Z}^d} |\langle \phi_i(\cdot - k), f\rangle|^2 \leq B\|f\|^2.$$

This Riesz basis assumption can be restated in the Fourier domain using the Grammian matrix of Φ. Specifically, the Grammian G_Θ of a vector function $\Theta = [\theta_1, \ldots, \theta_n]^t$ is defined by

$$G_\Theta(\omega) = \sum_{k \in \mathbb{Z}^d} \widehat{\Theta}(\omega + k) \, \widehat{\Theta}^*(\omega + k)$$

where $\widehat{\Theta}(\omega) := \int_{\mathbb{R}^d} \Theta(x) \, e^{-2\pi i \omega x} \, dx$, and $\widehat{\Theta}^*$ is the adjoint of $\widehat{\Theta}$. With this definition, it is well known that Φ induces a Riesz basis of the space $V = V(\Phi)$ defined by (14.1) if and only if there exist two positive constants $A > 0$ and $B > 0$ such that

$$AI \leq G_\Phi(\omega) \leq BI, \quad \text{a.e. } \omega, \tag{14.3}$$

where I is the $n \times n$ identity matrix (see, e.g., [1], [8], [9]). The set $B = \{\phi_i(x-k) : i = 1, \ldots, n, \; k \in \mathbb{Z}^d\}$ forms an orthonormal basis if and only if $A = B = 1$ in (14.3). Throughout the chapter we assume that $\Phi = [\phi_1, \ldots, \phi_n]^t$ satisfies (14.3).

We use $\mathcal{F}$ to indicate a set of functions and F to denote the vector-valued function whose components are the elements of $\mathcal{F}$ in some fixed order.

14.3 Problem I

A complete account of the results considered in this section, with proofs, is contained in [3].

Our main goal is to find necessary and sufficient conditions on subsets $\mathcal{F} = \{f_1, \dots, f_m\}$ of $V(\Phi)$ such that any $g \in V$ can be recovered from $\mathcal{F}$ as defined precisely next. A set $\mathcal{F}$ with such a property will be called a *determining set* for $V(\Phi)$. Specifically we have the following definition.

Definition 14.1. *The set $\mathcal{F} = \{f_1, f_2, \dots, f_m\} \subset V(\Phi)$ is said to be a* determining set *for $V(\Phi)$ if any $g \in V(\Phi)$ can be written as*

$$\widehat{g}(\omega) = \widehat{\alpha}_1(\omega)\, \widehat{f}_1(\omega) + \widehat{\alpha}_2(\omega)\, \widehat{f}_2(\omega) + \cdots + \widehat{\alpha}_m(\omega)\, \widehat{f}_m(\omega),$$

where $\widehat{\alpha}_1, \dots, \widehat{\alpha}_m$ are some 1-periodic measurable functions. In addition, if $\mathcal{F}$ is a determining set of $V(\Phi)$, then we will say that $V(\Phi)$ is determined by *$\mathcal{F}$.*

Remark 14.2. (i) The integer translates of the functions in the set $\mathcal{F} = \{f_1, \dots, f_m\}$ need not form a Riesz basis for V. In fact, series of the form

$$\sum_{i=1}^{m} \sum_{k \in \mathbb{Z}^d} c_i(k) f_i(x - k)$$

need not even be convergent for all $c_i \in l^2$.

(ii) An equivalent definition of a determining set is the following (e.g., see [8, Thm. 1.7]): a set $\mathcal{F}$ is a determining set for $V(\Phi)$ if and only if $V(\Phi) \subset \text{closure}_{L_2}\big(\text{span}\{f_i(x-k) : f_i \in \mathcal{F}\}\big)$.

It is not surprising that if V has a Riesz basis of n generators, then the cardinality m of a determining set $\mathcal{F}$ must be larger or equal to n. This result is stated in the following proposition.

Proposition 14.3. *Let V be a shift-invariant space generated by some Riesz basis $\{\phi_i(x-k) : i = 1, \dots, n,\ k \in \mathbb{Z}^d\}$, where $\Phi = [\phi_1, \dots, \phi_n]^t$ is a vector of functions in V. If $\mathcal{F}$ is a determining set for V, then card$(\mathcal{F}) \geq n$.*

Because of the proposition above, we will only consider sets $\mathcal{F}$ of cardinality m larger than or equal to the number n of the generators for V. Given such a set $\mathcal{F}$ there are $L = \binom{m}{n}$ subsets $\mathcal{F}_\ell \subset \mathcal{F}$ of size n. For each such subset $\mathcal{F}_\ell$ of size n, we define the set

$$A_\ell = \{\omega : \det G_{F_\ell}(\omega) \neq 0\}, \qquad 1 \leq \ell \leq L, \tag{14.4}$$

where G_{F_ℓ} is the $n \times n$ Grammian matrix for the vector F_ℓ, and we "disjointize" the sets A_ℓ by introducing the sets $\{B_\ell\}_{\ell=1}^{L}$ defined by

$$B_1 := A_1, \quad B_\ell := A_\ell - \bigcup_{j=1}^{\ell-1} A_j, \quad \ell = 2, \dots, L.$$

Below, we state (and give a reduced version of the proof of) a theorem from [3] that solves Problem I. The result characterizes determining sets, and produces an orthonormal basis for a shift-invariant space V when it is determined by the data.

Theorem 14.4 ([3]). *A set $\mathcal{F} = \{f_1, \dots, f_m\} \subset V(\Phi)$ is a determining set for $V(\Phi)$ if and only if the set $\bigcup_{\ell=1}^{L} A_\ell$ has Lebesgue measure one.*

Moreover, if $\mathcal{F}$ is a determining set for $V(\Phi)$, then the vector function

$$\widehat{\Psi}(\omega) := G_{F_1}^{-\frac{1}{2}}(\omega)\, \widehat{F_1}(\omega)\, \chi_{B_1}(\omega) + \cdots + G_{F_L}^{-\frac{1}{2}}(\omega)\, \widehat{F_L}(\omega)\, \chi_{B_L}(\omega) \tag{14.5}$$

generates an orthonormal basis $\{\psi_i(x-k) : i = 1, \dots, n,\ k \in \mathbb{Z}^d\}$ of $V(\Phi)$.

Proof (Sketch). Since $\mathcal{F}_\ell \subset V(\Phi)$ and $\mathrm{card}(\mathcal{F}_\ell) = n$, we can write $\widehat{F_\ell} = \widehat{C_{F_\ell}}\widehat{\Phi}$ for some $n \times n$ square matrix $\widehat{C_{F_\ell}}$ with $L^2([0,1]^d)$ entries, and we have

$$\begin{aligned} G_{F_\ell}(\omega) &= \sum_k (\widehat{C_{F_\ell}}(\omega+k)\, \widehat{\Phi}(\omega+k))\, (\widehat{C_{F_\ell}}(\omega+k)\, \widehat{\Phi}(\omega+k))^* \\ &= \sum_k \widehat{C_{F_\ell}}(\omega+k)\, \widehat{\Phi}(\omega+k)\, \widehat{\Phi}^*(\omega+k)\, \widehat{C_{F_\ell}}^*(\omega+k) \\ &= \widehat{C_{F_\ell}}(\omega)\, G_\Phi(\omega)\, \widehat{C_{F_\ell}}^*(\omega), \end{aligned}$$

since $\widehat{C_{F_\ell}}(\omega)$ is 1-periodic.

Moreover, since Φ induces a Riesz basis, it follows that G_Φ is positive definite. It is also true that $\widehat{C_{F_\ell}}(\omega)$ is non-singular for a.e. $\omega \in B_\ell$. Thus, G_{F_ℓ} is self-adjoint and positive definite on B_ℓ.

Therefore, if we define Ψ as in (14.5), then it can be seen that the set $\{\psi_i(x-k) : i = 1, \dots, n,\ k \in \mathbb{Z}^d\}$ forms an orthonormal basis for $V(\Phi)$.

For the converse see [3]. □

Remark 14.5. (i) Theorem 14.4 provides a method for checking whether and when a set of functions generates a fixed (yet unknown) shift-invariant space generated by some unknown Φ of known size n. Since other than the value n, the only requirement is that the set of functions must belong to the same (unknown) shift-invariant space, we can apply the theorem to a set of *observed* functions (the data) if we know that they are all from some shift-invariant space V. We can either determine the space, or conclude that we do not have enough data to do so and need to acquire more data. If we cannot acquire more data, we can still determine the space $S(\mathcal{F}) = \mathrm{closure}_{L^2}(\mathrm{span}\{f_i(\cdot - k) : i = 1, \dots, m, k \in \mathbb{Z}\})$, which is a subspace of the unknown space V. However, the subspace $S(\mathcal{F})$ is not necessarily generated by a Riesz basis.

(ii) The functions of the orthonormal basis constructed in Theorem 14.4 are in L^2 but not in $L^1 \cap L^2$ in general. Further investigation is needed for the construction of better-localized bases.

14.4 Problem II

The intuition—or idea—behind Problem II is that one has a large amount of data (for example the data base of all chest X-rays during the last 10 years). The space

$$\mathcal{S}(\mathcal{F}) = \text{closure}_{L_2}\left(\text{span}\{f_i(x-k) : f_i \in \mathcal{F}\}\right)$$

generated by our set of experimental data contains all the data as possible signals, but it is too large to be an appropriate model for use in applications. A space with a "small" number of generators is more suitable, since if the space is chosen correctly, it would reduce noise, and would give a computationally manageable model for a given application. Since in general the data does not belong to a shift-invariant space with n generators (n small), the goal is to find—among all possible shift-invariant spaces with n generators—the one that fits the data optimally.

Accordingly, in this section we do not assume that $\mathcal{F} = \{f_1, \dots, f_m\}$ belongs to a space V with exactly n generators.

Let us consider the function

$$r(\omega) = \text{rank } G_{\mathcal{F}}(\omega),$$

where $G_{\mathcal{F}}(\omega)$ is the Grammian matrix at ω. Let $r_{\min}$ and $r_{\max}$ denote the minimum and the maximum value that $r(\omega)$ can attain in $[0,1]^d$, i.e.,

$$r_{\min} = \min_{\omega \in [0,1]^d} r(\omega) \quad \text{and} \quad r_{\max} = \max_{\omega \in [0,1]^d} r(\omega).$$

Clearly, if $r_{\max}$ is already small, the problem is not interesting. So we will assume that $r_{\min} \geq n$, where n is the number of generators for the space V that we are seeking to model the observed data $\mathcal{F}$. This hypothesis is not strictly necessary for our results, but we will impose it for simplicity.

Consider as before the class $\mathcal{V}$ of all the shift-invariant spaces that are generated by some set of n generators Φ with the property that $\{\phi_i(\cdot - k) : k \in \mathbb{Z}^d,\ i = 1, \dots, n\}$ is an orthogonal basis for $V(\Phi)$. Note that the assumption of orthogonality does not change the class $\mathcal{V}$ considered in the introduction. Let $w = (w_1, \dots, w_m)$ be a vector of weights, (i.e., $w_i \in \mathbb{R}$, $w_i > 0$).

Our goal is, given $\mathcal{F}$, to find a space $V \in \mathcal{V}$ such that V minimizes the least square error

$$E(\mathcal{F}, w, n) = \sum_{i=1}^{m} w_i \left\| f_i - P_V f_i \right\|^2,$$

where P_V is the orthogonal projection onto V. This problem can be viewed as a nonlinear infinite-dimensional constrained minimization problem. It is remarkable that it has a constructive solution, as shown in [2]. This problem may also be viewed in the framework of the recent learning theory developed in [7] and estimates of "model fit" in terms of noise and approximation space may be derived (see the next section).

The first question that arises is if such a space exists at all. In the case that it exists, in order to be useful for applications, it will be important to have a way to construct the generators of the space and to estimate the error $E(\mathcal{F}, w, n)$.

Surprisingly, in [2] the following theorem is proved.

Theorem 14.6. *With the previous notation, let n be given, assume that $n \leq r_{\min}$, and let w be a vector of weights, $w = (w_1, \ldots, w_m)$. Then there exists a space $V \in \mathcal{V}$ such that*

$$\sum_{i=1}^{m} w_i \left\| f_i - P_V f_i \right\|^2 \leq \sum_{i=1}^{m} w_i \left\| f_i - P_{V'} f_i \right\|^2, \quad \forall V' \in \mathcal{V}. \qquad (14.6)$$

Proof (Sketch). The proof is quite technical and therefore not suitable for this chapter (see [2]); however, it is constructive. We will skip the details and try to give an idea of the construction of the space V.

We consider the space $\mathcal{S}(\mathcal{F})$ and look at the Grammian matrix $G_{\mathcal{F}}$. Since $r_{\min} \geq n$, we always have at least n non-zero eigenvalues. For $i = 1, \ldots, n$, consider $\hat{g}_i(\omega) = v_1^i \hat{f}_1 + \cdots + v_m^i \hat{f}_m(\omega)$ where $v^i \in \mathbb{C}^m$ are some choice of eigenvectors associated to the n largest eigenvalues of $G_{\mathcal{F}}(\omega)$. If this choice can be made in such a way that the resulting functions are linearly independent functions in $\mathcal{S}(\mathcal{F})$, then the space generated by these n functions will be the space V we are looking for.

Note that it is not immediate to see that the functions obtained in this way belong to L^2 (or are even measurable functions!). However, after solving this technical part ([2]), one sees that if $r_{\min}$ is greater than or equal to n, we can *always* solve Problem II. □

We will call a space $V \in \mathcal{V}$ satisfying (14.6) an *optimal space* (for the data $\mathcal{F}$). Moreover, it can be seen that the space V is (under minor assumptions) unique.

In view of the preceding construction, we can now state two consequences of the previous theorem that are relevant for this chapter.

Theorem 14.7. *Let $V \in \mathcal{V}$ be an optimal space. Then $V \subset \mathcal{S}(\mathcal{F})$.*

This shows that every optimal space should be contained in the space $\mathcal{S}(\mathcal{F})$ spanned by the data.

Further, we have the following estimate for the error.

Theorem 14.8. *Let again $V \in \mathcal{V}$ be an optimal space, $w = (w_1, \dots, w_m)$ a vector of weights, and $\lambda_1(\omega) \geq \lambda_2(\omega) \geq \cdots \geq \lambda_{r_{\max}}(\omega)$ the eigenvalues of $G_{\mathcal{F}}$ at ω. Then*

$$E(\mathcal{F}, w, n) = \sum_{i=1}^{m} w_i \|f_i - P_V f_i\|^2 = \sum_{i=n+1}^{r_{\max}} w_i \int_{[0,1]^d} \lambda_i(\omega) d\omega.$$

Remark 14.9. Obviously, if $n = m$ then the error between the model and the observation is null. However, by plotting the error in Theorem 14.8 in terms of the number of generators, an optimal number n may be derived if the behavior of the error in terms of n shows a horizontal asymptote.

14.5 Problem II as a Learning Problem

Problem II has an interpretation as a learning problem as defined in [7].

Consider a class of signals or images (e.g., electroencephalograms or MRI images). This class of signals belongs to some unknown space that we can assume to be a shift-invariant space $\mathcal{T} \subset L^2(\mathbb{R}^d)$. The space $\mathcal{T}$ (*the target space*) is often very large. For processing, analysis, and manipulation of the data it is necessary to restrict the model to a smaller class of spaces with enough structure. For example, shift-invariant spaces that can be generated by a Riesz basis are appropriate, and are often used in many signal processing applications.

Therefore, we fix a positive integer n, and consider the class $\mathcal{V}$ (*the hypothesis class*) as before. We want to learn about the space $\mathcal{T}$ from some sample elements. Assume that we have m sample signals, say $\mathcal{F} = \{f_1, \dots, f_m\}$ (*the training set*). Using Theorem 14.6, we see that from our data set $\mathcal{F}$ we can obtain some space $V_{\mathcal{F}} \in \mathcal{V}$ that best fits our data.

However, a realistic assumption should consider that our samples are noisy. Therefore, they may not belong to the space $\mathcal{T}$. This means that the space $V_{\mathcal{F}}$ will, in general, be different from the space $V_{\tilde{\mathcal{F}}}$ that we would have found from signals that are not corrupted by noise.

The noisy data introduces an error. This error can be quantified using some distance between the subspaces $V_{\mathcal{F}}$ and $V_{\tilde{\mathcal{F}}}$. (We can, for example, consider the distance between the orthogonal projections in some operator norm.) This error is usually called the *sample error* in learning theory.

There is another error (*the approximation error*) due to the fact that our family of spaces is constrained to have only n generators.

Estimation of these errors in terms of the number of samples and the number of generators is an ongoing research by the authors.

Acknowledgments

The authors were partially supported by grants from the NSF (DMS-0103104, DMS-0139740), CIES (CIES-87426189), the University of Buenos Aires (UBA-CyT X058 and 108), and the ANPCyT (BID 1201/OC-AR PICT 15033).

References

1. A. Aldroubi, Oblique projections in atomic spaces, *Proc. Amer. Math. Soc.*, **124** (1996), pp. 2051–2060.
2. A. Aldroubi, C. A. Cabrelli, D. Hardin, and U. M. Molter, Optimal shift-invariant spaces and their Parseval frame generators, preprint (2006).
3. A. Aldroubi, C. A. Cabrelli, D. Hardin, U. Molter, and A. Rodado, Determining sets of shift invariant spaces, in: *Wavelets and their Applications* (Chennai, January 2002), M. Krishna, R. Radha, and S. Thangavelu, eds., Allied Publishers, New Delhi (2003), pp. 1–8.
4. A. Aldroubi and K. Gröchenig, Non-uniform sampling and reconstruction in shift-invariant spaces, *SIAM Review*, **43** (2001), pp. 585–620.
5. J. J. Benedetto and P. J. S. G. Ferreira, eds., *Modern Sampling Theory: Mathematics and Applications*, Birkhäuser, Boston, 2001.
6. J. J. Benedetto and A. I. Zayed, eds., *Sampling, Wavelets, and Tomography*, Birkhäuser, Boston, 2004.
7. F. Cucker and S. Smale, On the mathematical foundations of learning, *Bull. Amer. Math. Soc. (N.S.)*, **39** (2002), pp. 1–49.
8. C. de Boor, R. De Vore, and A. Ron, The structure of finitely generated shift-invariant subspaces of $L_2(R^d)$, *J. Funct. Anal.*, **119** (1994), pp. 37–78.
9. T. N. T. Goodman, S. L. Lee, and W. S. Tang, Wavelets in wandering subspaces, *Trans. Amer. Math. Soc.*, **338** (1993), pp. 639–654.

15

Redundancy in the Frequency Domain

Lawrence Baggett

Department of Mathematics, University of Colorado, Boulder, CO 80309, USA
baggett@euclid.colorado.edu

Summary. A description of the fine structure of a refinable, shift-invariant subspace of $L^2(\mathbb{R})$ is presented. This fine structure is exhibited through the existence of a canonical frame of functions in such a space, and a related notion of frequency content in these frame elements uniquely determines a multiplicity function that quantifies a redundancy of the frequencies. The refinability of the subspace can then be described by a pair of matrices of periodic functions that satisfy a set of equations, related to the multiplicity function, which play the role of high-dimensional filter equations.

15.1 Introduction

The very idea of redundancy is traditionally abhorred in mathematics. We usually search for the minimal set of hypotheses, the shortest possible proof, the sharpest possible constant in an inequality, and so on. However, we have learned from engineers that sacrificing some efficiency by allowing a bit of redundancy can often serve a useful purpose. For example, in communication problems, where a fundamental task is to interpret a transmitted signal as accurately as possible, the sender could repeat the transmission two or three times, allowing the receiver to use this redundant information to make a more intelligent decision about what the true message was. In the pure mathematics of Banach spaces, and even more dramatically in Hilbert spaces, the concept of a frame $\{f_n\}$, as a generalization of a basis, has provided very rich mathematical questions and answers. The very definition of a frame suggests a notion of redundancy, for it hints that the collection $\{f_n\}$ is not an independent set; i.e., some of the f_n's can be ignored without loss of information. Among the specific Hilbert spaces to which frame theory has successfully been applied is the Hilbert space $L^2(\mathbb{R})$. Our aim here is to describe a more intrinsic notion of redundancy or multiplicity in that space, a notion that is more clearly evident in the frequency domain, the Fourier transform side. More on this later.

The study of frames is widespread nowadays, and the reader will find ample discussion of this concept in many of the chapters in this very volume.

In addition, a detailed presentation is given in [5]. We give here only the small part of frame theory required for the goals of this chapter, and our intent is for this development, except for some basic results on Hilbert spaces and unitary operators, to be self-contained. In the end, we present a detailed kind of fine structure of a refinable, shift-invariant subspace of $L^2(\mathbb{R})$.

Definition 15.1. *A* frame *for a (separable) Hilbert space H is a countable set $\{f_n\}_{n\in I}$ of vectors for which there exist positive constants A and B such that*

$$A\|g\|^2 \leq \sum_{n\in I} |\langle g, f_n\rangle|^2 \leq B\|g\|^2 \tag{15.1}$$

for every $g \in H$. The constants A and B are called lower and upper frame bounds, *the first inequality is called the* lower frame inequality, *and the second is called the* upper frame inequality.

If the two frame bounds can be taken to be equal, i.e., $A = B$, then the frame is called a *tight frame*, and if $A = B = 1$, then the frame is called a *Parseval frame*, for in this case the defining inequalities coincide with the classical Parseval Equality for orthonormal bases:

$$\sum_{n\in I} |\langle g, f_n\rangle|^2 = \|g\|^2.$$

Clearly, if $\{f_n\}$ is a frame for a Hilbert space H, and if U is a unitary operator from H onto another Hilbert space H', then the collection $\{U(f_n)\}$ is a frame for H' having the same bounds as the frame $\{f_n\}$. This follows because a unitary operator preserves norms and inner products.

Frequently we will want to consider a collection $\{f_n\}$ of vectors in a Hilbert space H, and we will know in advance that the span of the f_n's is not dense in H. In this case, the question will be whether the f_n's form a frame for their closed linear span. When they do, we call the collection $\{f_n\}$ a *frame sequence.*

Perhaps the simplest example of a frame is the following.

Example 15.2. Let $\{\eta_n\}$ be an orthonormal basis for H. We construct a sequence $\{f_n\}$ by repeating each vector η_n two times:

$$f_1 = \eta_1,\ f_2 = \eta_1,\ f_3 = \eta_2,\ f_4 = \eta_2,\ \ldots.$$

At this point in our discussion, we could think of redundancy as some kind of multiplicity; i.e., information is being repeated, but there are other kinds of redundancy. For instance, let $\{\eta_n\}$ and $\{\zeta_n\}$ be two orthonormal bases for H. Define vectors f_n by setting $f_{2n} = \eta_n/\sqrt{2}$ and $f_{2n+1} = \zeta_n/\sqrt{2}$. In this case,

$$\sum_n |\langle g, f_n\rangle|^2 = \frac{1}{2}\sum_n |\langle g, \eta_n\rangle|^2 + \frac{1}{2}\sum_n |\langle g, \zeta_n\rangle|^2 = \|g\|^2,$$

so that both constants A and B can be taken to be equal to 1, whence these f_n's form a Parseval frame. Clearly, these f_n's don't have to be orthogonal; they don't even have to be linearly independent.

An obvious generalization of the construction above shows that the union of any finite set of orthonormal bases, appropriately normalized, forms a Parseval frame. For instance, if d is any positive integer, then the functions $\{\sqrt{d}^{-1} e^{2\pi i(n/d)x}\}_{n\in\mathbb{Z}}$ form a Parseval frame for $L^2([0,1))$.

In fact, a naive person might have guessed that the Parseval Equality holds only for orthonormal sets, but, as the preceding example shows, this is far from true. Moreover, although every specific Hilbert space has infinitely many orthonormal bases, none of them may be "natural" choices for a given space. On the other hand, as we will see, all shift-invariant subspaces of $L^2(\mathbb{R})$ have natural Parseval frames.

An especially important fact is this. If $\{f_n\}$ is any Parseval frame, then each element $g \in H$ can be recovered (reconstructed), just as in the case of an orthonormal basis, from its inner products with the elements of the frame:

$$g = \sum_{n\in I} \langle g, f_n \rangle f_n.$$

Before proving this important fact about Parseval frames, we introduce the concept of the analysis map associated to a frame.

Definition 15.3. *Let $\{f_n\}_{n\in I}$ be a frame for a Hilbert space H. Define a map T from H into the set of functions on the index set I by*

$$T(g)_n = [T(g)](n) = \langle g, f_n \rangle.$$

This map T is called the analysis map *associated to the frame $\{f_n\}$.*

It is immediate from the upper frame inequality that the functions $T(g)$ in the range of T are square-summable functions on I, i.e., elements of $l^2(I)$. It is also clear that T is a linear transformation of H into $l^2(I)$, and T is a bounded linear transformation whose operator norm is $\leq \sqrt{B}$. The lower frame inequality implies that the range R of T is a (possibly proper) closed subspace of the Hilbert space $l^2(I)$, and in fact that inequality implies that T is 1-1, and the inverse map from R to H is a bounded linear transformation whose norm is $\leq 1/\sqrt{A}$.

Now we can prove the important fact mentioned above about recovering the elements of H from their inner products with a Parseval frame.

Theorem 15.4. *Let $\{f_n\}$ be a Parseval frame for a Hilbert space H. Then, for each $g \in H$, we have the following* reconstruction formula:

$$g = \sum_{n\in I} \langle g, f_n \rangle f_n. \tag{15.2}$$

Proof. If T is the analysis map from H into $l^2(I)$, then, because $\{f_n\}$ is a Parseval frame, T is an isometry. Of course this implies that T is a unitary map from H onto its range R, and so if $T^* : l^2 \to H$ denotes the adjoint operator of T, then the restriction of T^* to the range R of T is the inverse of T.

Now, if e^k is the element of $l^2(I)$ given by $e^k_m = \delta_{k,m}$, then we claim that $T^*(e^k) = f_k$. Indeed,

$$\begin{aligned}
[T(T^*(e^k))]_n &= \langle T^*(e^k), f_n\rangle \\
&= \langle e^k, T(f_n)\rangle \\
&= \sum_m e^k_m \, \overline{[T(f_n)]_m} \\
&= \sum_m \delta_{k,m} \, \overline{\langle f_n, f_m\rangle} \\
&= \langle f_k, f_n\rangle \\
&= [T(f_k)]_n,
\end{aligned}$$

implying, since T is 1-1, that $T^*(e^k) = f_k$. Next, since T^* is continuous, and the e^n's form an orthonormal basis for $l^2(I)$, we have that for any element $\{a_n\} \in l^2(I)$,

$$T^*(\{a_n\}) = T^*\left(\sum_n a_n e^n\right) = \sum_n a_n f_n.$$

Finally, if g is in H, then

$$g = T^{-1}(T(g)) = T^*(T(g)) = T^*(\{[T(g)]_n\}) = T^*(\{\langle g, f_n\rangle\}) = \sum_n \langle g, f_n\rangle f_n,$$

as claimed. □

Remark 15.5. Rules for complete disclosure require us to point out that the reconstruction equation (15.2) does not imply that the element g is uniquely represented as an infinite linear combination of the frame elements. In fact, this kind of unique representation is almost never the case. The Parseval frame in Example 15.2 demonstrates the possible nonuniqueness of representations.

Another property satisfied by Parseval frames, and by now an anticipated one, is this.

Theorem 15.6. *Suppose that $\{f_n\}$ is a Parseval frame for a Hilbert space H, and assume that f_k is an element of the frame that is not orthogonal to every other element f_n of the frame. That is, there exists at least one $n \neq k$ for which f_k is not orthogonal to f_n. Then every element $g \in H$ can be written in the form*

$$g = \sum_{n \neq k} c_n f_n;$$

i.e., g is recoverable from the frame with f_k removed.

Proof. Notice first that, because

$$\|f_k\|^2 = \sum_n |\langle f_k, f_n \rangle|^2,$$

and $\langle f_k, f_n \rangle$ is not 0 for at least one $n \neq k$, then we have

$$\|f_k\|^2 > |\langle f_k, f_k \rangle|^2 = \|f_k\|^4,$$

and this implies that $\|f_k\| < 1$. Now, by Theorem 15.4,

$$f_k = \sum_n \langle f_k, f_n \rangle f_n = \langle f_k, f_k \rangle f_k + \sum_{n \neq k} \langle f_k, f_n \rangle f_n,$$

and we have

$$f_k = \sum_{n \neq k} \frac{\langle f_k, f_n \rangle}{1 - \|f_k\|^2} f_n.$$

So, for any $g \in H$, we have

$$\begin{aligned} g &= \sum_n \langle g, f_n \rangle f_n \\ &= \langle g, f_k \rangle f_k + \sum_{n \neq k} \langle g, f_n \rangle f_n \\ &= \sum_{n \neq k} \left(\frac{\langle g, f_k \rangle \langle f_k, f_n \rangle}{1 - \|f_k\|^2} + \langle g, f_n \rangle \right) f_n, \end{aligned}$$

proving the theorem. □

Theorem 15.6 is a special case of a more general result proved by Duffin and Schaeffer (the original "framers"), which asserts that if a frame is not a basis then some element of the frame can be removed and still leave a frame [6].

Conceptually, we imagine that the elements f_n of a Parseval frame are basic building blocks, and the redundancy discussed in the preceding theorem we interpret as a reflection of some kind of overlap or multiplicity among them. This kind of redundancy, in the context of the Hilbert space $L^2(\mathbb{R})$, we will call *time-side redundancy*. As mentioned earlier, we will investigate in this chapter a kind of frequency-side redundancy, but we postpone giving a precise definition of this notion until later.

There are many examples of Parseval frames, the most important of which for our purposes is exhibited in the following example.

Example 15.7. Let E be a Borel subset of the torus $[0,1)$ and let $H = L^2(E)$. Define functions $\{f_n\}_{-\infty}^{\infty}$ in H by $f_n(\omega) = e^{2\pi in\omega}\chi_E(\omega)$. Then the collection $\{f_n\}$ forms a Parseval frame for H. Indeed, if $g \in H = L^2(E)$, then, using the ordinary Parseval Equality for Fourier series, we have

$$\begin{aligned}
\|g\|^2 &= \int_E |g(\omega)|^2\, d\omega \\
&= \int_0^1 |g(\omega)\chi_E(\omega)|^2\, d\omega \\
&= \sum_{n=-\infty}^{\infty} \left| \int_0^1 g(\omega)\, \chi_E(\omega)\, e^{-2\pi in\omega}\, d\omega \right|^2 \\
&= \sum_{n=-\infty}^{\infty} \left| \int_E g(\omega)\, \overline{f_n(\omega)}\, d\omega \right|^2 \\
&= \sum_{n=-\infty}^{\infty} |\langle g, f_n\rangle|^2,
\end{aligned}$$

proving that these f_n's form a Parseval frame for $H = L^2(E)$.

Of course, the Hilbert space $L^2(E)$ has an orthonormal basis, but there is no obvious or natural choice of one. On the other hand, these exponential functions are quite natural, and they do form a Parseval frame. Hence, this very example provides a convincing explanation of the value of Parseval frames.

Although we concentrate in this chapter on the Hilbert space $H = L^2(\mathbb{R})$, all the arguments extend directly to $\mathbb{R}^d$. We use the following formula for the Fourier transform:

$$\widehat{f}(\omega) = \int_{\mathbb{R}} f(x)\, e^{-2\pi i\omega x}\, dx.$$

15.2 Shift-Invariant Subspaces and Cyclic Frames

For each real number x define the *translation operator* τ_x on $L^2(\mathbb{R})$ by $[\tau_x(f)](t) = f(t+x)$. Ordinarily, we call a closed subspace V of $L^2(\mathbb{R})$ *shift-invariant* if it is closed under all integer translations. That is, for each $f \in V$ and each integer n, we require that $\tau_n(f)$ also be in V. Here, we also consider generalized shift-invariant subspaces, i.e., subspaces that are invariant under shifts by multiples of some positive number r, i.e., f in V implies that $\tau_{nr}(f)$ is also in V for all integers n. When we need to be explicit, we will say that such a subspace is *invariant under shifts by multiples of* r. Shift-invariant subspaces have been studied extensively by many authors, and much of what we present in this section is well known to many. However, we believe our perspective is novel and instructive, since it is based on intuition gleaned from unitary representation theory.

One special kind of frame for a generalized shift-invariant subspace of $L^2(\mathbb{R})$ is one that is built from the translates $\{\tau_{nr}(f)\}$ of a single function f by integral multiples of r.

Definition 15.8. *Let r be a positive number, and let f be a nonzero element of $L^2(\mathbb{R})$. The collection $\{f_{nr}\} \equiv \{\tau_{nr}(f)\}$ of translates of f is called a* cyclic collection, *and f is called a* cyclic vector *for the subspace V that is the closed linear span of the f_{nr}'s.*

Ordinarily, these functions $\{\tau_{nr}(f)\}$ will not be orthogonal, but, unlike what we saw in Example 15.2, they are always linearly independent. Indeed, arguing by contradiction, if $\sum_{j=-N}^{N} c_j f_{jr} = 0$, then, by taking the Fourier transform of both sides, we obtain

$$\sum_{j=-N}^{N} c_j \, e^{2\pi i j r \omega} \, \widehat{f}(\omega) = p(\omega) \, \widehat{f}(\omega) = 0,$$

where p is the periodic, trigonometric polynomial on the interval $[0, 1/r)$ given by $p(\omega) = \sum_{j=-N}^{N} c_j e^{2\pi i j r \omega}$. Since such a nontrivial trigonometric polynomial is nonzero almost everywhere, this implies that $\widehat{f}$ is 0 a.e., contradicting the assumption that f is nonzero. An immediate consequence of this observation, and one we will need later, is that the map that sends the element $\sum_{j=-N}^{N} c_j f_{jr}$ to the trigonometric polynomial $\sum_{j=-N}^{N} c_j e^{2\pi i j r \omega}$ is a linear isomorphism from the finite linear span of the set of all translates f_{jr} of f onto the space of all periodic trigonometric polynomials on $[0, 1/r)$.

An interesting and old question is what conditions on f will guarantee that such a cyclic collection $\{f_{nr}\}$ forms a frame for its closed linear span. The answer, discovered by several mathematicians, including our own hero Gianni Benedetto [4], is summarized in the following.

Theorem 15.9. *Let f be a nonzero element of $L^2(\mathbb{R})$, let $r > 0$, and set ρ_f^r equal to the periodic function, with period $1/r$, given by*

$$\rho_f^r(\omega) = \sum_{z=-\infty}^{\infty} \left| \widehat{f}\Big(\omega + z\frac{1}{r}\Big) \right|^2. \tag{15.3}$$

Let E_f^r denote the set of ω's in $[0, 1/r)$ for which $\rho_f^r(\omega) > 0$. Finally, let V be the Hilbert space that is the closure of the linear span of the f_{nr}'s. Then the cyclic collection $\{f_{nr}\}$ is a frame for V, with frame bounds A and B, if and only if

$$Ar \leq \rho_f^r(\omega) \leq Br$$

for almost all $\omega \in E_f^r$.

The cyclic collection is a Parseval frame for V if and only if $\rho_f^r = r\chi_{E_f^r}$, and it is an orthonormal basis for V if and only if $\rho_f^r \equiv r$ on $[0, 1/r)$.

Proof. Let g be in the finite linear span of the f_{nr}'s. Then, $g = \sum_{j=-N}^{N} c_{jr} f_j$, and as above, $\widehat{g} = p\widehat{f}$ for some (periodic) trigonometric polynomial p on the interval $[0, 1/r)$. So,

$$\begin{aligned}
\|g\|^2 = \|\widehat{g}\|^2 &= \int_{\mathbb{R}} |p(\omega)|^2 \, |\widehat{f}(\omega)|^2 \, d\omega \\
&= \sum_{z=-\infty}^{\infty} \int_{z\frac{1}{r}}^{(z+1)\frac{1}{r}} |p(\omega)|^2 \, |\widehat{f}(\omega)|^2 \, d\omega \\
&= \sum_{z=-\infty}^{\infty} \int_0^{\frac{1}{r}} |p(\omega)|^2 \left| \widehat{f}\left(\omega + z\frac{1}{r}\right) \right|^2 d\omega \\
&= \int_0^{\frac{1}{r}} |p(\omega)|^2 \, \rho_f^r(\omega) \, d\omega. \qquad (15.4)
\end{aligned}$$

And, making use of the ordinary Parseval Equality on $L^2([0, 1/r))$, we obtain

$$\begin{aligned}
\sum_{n=-\infty}^{\infty} |\langle g, f_{nr} \rangle|^2 &= \sum_{n=-\infty}^{\infty} |\langle \widehat{g}, \widehat{f_{nr}} \rangle|^2 \\
&= \sum_{n=-\infty}^{\infty} \left| \int_{\mathbb{R}} p(\omega) \, \widehat{f}(\omega) \, \overline{\widehat{f}(\omega) \, e^{2\pi i n r \omega}} \, d\omega \right|^2 \\
&= \sum_{n=-\infty}^{\infty} \left| \int_0^{\frac{1}{r}} p(\omega) \, \rho_f^r(\omega) \, e^{-2\pi i n r \omega} \, d\omega \right|^2 \\
&= \frac{1}{r} \int_0^{\frac{1}{r}} |p(\omega)|^2 \, {\rho_f^r}^2(\omega) \, d\omega. \qquad (15.5)
\end{aligned}$$

Hence, the cyclic collection $\{f_{nr}\}$ is a frame, with frame bounds A and B, if and only if

$$A \int_0^{\frac{1}{r}} |p(\omega)|^2 \, \rho_f^r(\omega) \, d\omega \leq \frac{1}{r} \int_0^{\frac{1}{r}} |p(\omega)|^2 \, {\rho_f^r}^2(\omega) \, d\omega \leq B \int_0^{\frac{1}{r}} |p(\omega)|^2 \, \rho_f^r(\omega) \, d\omega$$

for every trigonometric polynomial p. This is equivalent to

$$A\rho_f^r(\omega) \leq \frac{1}{r} {\rho_f^r}^2(\omega) \leq B\rho_f^r(\omega)$$

for almost all ω, from which the theorem follows. □

Remark 15.10. Many different functions f determine the same function ρ^r and set E^r. Nevertheless, these two bits of data are important attributes of a function f and the subspace V. In particular, the set E^r gives some crude information about the frequency content of both the function f and the subspace V. We will return to these ideas later when we discuss redundancy in the frequency domain.

There is a useful alternative way to describe the function ρ_f^r of the preceding theorem.

Theorem 15.11. *Let f be as in the preceding theorem. Then*

$$\rho_f^r(\omega) = \sum_{n=-\infty}^{\infty} \sqrt{r}\,\langle f, \tau_{nr}(f)\rangle\,\sqrt{r}\,e^{2\pi i n r \omega}. \tag{15.6}$$

That is, the nth Fourier coefficient c_n of the periodic function ρ_f^r on the interval $[0, 1/r)$ is given by

$$c_n(\rho_f^r) = \int_0^{\frac{1}{r}} \rho_f^r(\omega)\sqrt{r}\,e^{-2\pi i n r \omega}\,d\omega = \sqrt{r}\,\langle f, \tau_{nr}(f)\rangle.$$

Proof. We calculate

$$\begin{aligned}
c_n(\rho_f^r) &= \int_0^{\frac{1}{r}} \rho_f^r(\omega)\,\sqrt{r}\,e^{-2\pi i n r \omega}\,d\omega \\
&= \int_0^{\frac{1}{r}} \sqrt{r}\,e^{-2\pi i n r \omega} \sum_{z=-\infty}^{\infty} \left|\widehat{f}\left(\omega + z\frac{1}{r}\right)\right|^2 d\omega \\
&= \int_{-\infty}^{\infty} \sqrt{r}\,e^{-2\pi i n r x}\,|\widehat{f}(x)|^2\,dx \\
&= \int_{-\infty}^{\infty} \sqrt{r}\,e^{-2\pi i n r x}\,\widehat{f}(x)\,\overline{\widehat{f}(x)}\,dx \\
&= \sqrt{r}\int_{-\infty}^{\infty} \widehat{f}(x)\,\overline{\widehat{f_{nr}}(x)}\,dx \\
&= \sqrt{r}\int_{-\infty}^{\infty} f(t)\,\overline{f_{nr}(t)}\,dt \\
&= \sqrt{r}\,\langle f, f_{nr}\rangle,
\end{aligned}$$

as claimed. □

Most investigations of shift-invariant subspaces have dealt with translation by integers, i.e., when $r = 1$. It is instructive to examine Theorem 15.9 in the integer translation case for functions ϕ_α of the form $\phi_\alpha = \sqrt{\alpha}^{-1}\chi_{[0,\alpha)}$, where α is a positive number. The corresponding periodic function $\rho_{\phi_\alpha} \equiv \rho_{\phi_\alpha}^1$ is given by

$$\rho_{\phi_\alpha}(\omega) = \sum_{z=-\infty}^{\infty} \frac{|\sin(\alpha\pi(\omega + z))|^2}{\alpha\pi^2(\omega + z)^2}.$$

By the Weierstrass M-test, $\rho_{\phi_\alpha}(\omega)$ is continuous in both variables α and ω, $\alpha > 0$ and $0 \le \omega < 1$. For $\alpha \le 1$, the integer translates of ϕ_α are clearly orthonormal, and so by the theorem above $\rho_{\phi_\alpha}(\omega) \equiv 1$. (Note that a direct verification of this equality is by no means a triviality.) If α is an integer $p > 1$, then we see directly that $\rho_{\phi_\alpha}(1/p) = 0$. Since ρ_{ϕ_α} is continuous as a function of ω, we then see that ρ_{ϕ_α} is not bounded below by any positive number A on the set E_{ϕ_α}, whence the integer translates of ϕ_α do not form a frame for their linear span if α is an integer > 1. Finally, if $\alpha > 1$, and α is not equal to an integer, then $\rho_{\phi_\alpha}(\omega)$ is clearly positive except possibly when $\omega = q/\alpha$ for some integer q. But even then ρ_{ϕ_α} is positive because

$$\frac{|\sin(\alpha\pi(\frac{q}{\alpha}+1))|^2}{\alpha\pi^2(\frac{q}{\alpha}+1)^2} = \frac{|\sin(\alpha\pi)|^2}{\alpha\pi^2(\frac{q}{\alpha}+1)^2} > 0.$$

So, again, because ρ_{ϕ_α} is continuous as a function of ω, it must be bounded below on the entire interval $[0,1)$, whence the integer translates of ϕ_α do form a frame for their linear span in this case.

Now, suppose f is a nonzero element of $L^2(\mathbb{R})$, and write V for the closed linear span of the integer translates of f. Although the integer translates of f may not even form a frame for their closed linear span, we do have the following fundamental structural result for V.

Theorem 15.12. *Let f be a nonzero element of $L^2(\mathbb{R})$, let V denote the closure of the linear span of the integer translates of f, and let $\rho_f \equiv \rho_f^1$ and $E_f \equiv E_f^1$ be as in Theorem 15.9. Then there exists an element $\eta \in V$ whose integer translates $\{\eta_n\}_{-\infty}^{\infty}$ form a Parseval frame for V. Moreover, $E_\eta = E_f$, and there exists a unitary map J from V onto $L^2(E_\eta)$ for which the intertwining condition*

$$[J(\tau_n(g))](\omega) = e^{2\pi i n\omega}\,[J(g)](\omega) \tag{15.7}$$

holds for every $g \in V$, every integer n, and almost every $\omega \in [0,1)$.

Proof. Define a linear transformation J on the finite linear span V' of the functions f_n by

$$[J(g)](\omega) = \left[J\Big(\sum_{j=-N}^{N} c_j f_j\Big)\right](\omega) = \sqrt{\rho_f(\omega)}\sum_{j=-N}^{N} c_j e^{2\pi i j\omega},$$

and note that J is an isometry with respect to the L^2 norms on V' and $L^2([0,1))$, because

$$\begin{aligned}
\int_0^1 |[J(g)](\omega)|^2\,d\omega &= \int_0^1 \left|\left[J\Big(\sum_{j=-N}^{N} c_j f_j\Big)\right](\omega)\right|^2 d\omega \\
&= \int_0^1 \rho_f(\omega)\left|\sum_{j=-N}^{N} c_j e^{2\pi i j\omega}\right|^2 d\omega \\
&= \int_0^1 \sum_{z=-\infty}^{\infty} |\widehat{f}(\omega+z)|^2 \left|\sum_{j=-N}^{N} c_j e^{2\pi i j\omega}\right|^2 d\omega
\end{aligned}$$

$$= \int_{-\infty}^{\infty} \left| \widehat{f}(x) \sum_{j=-N}^{N} c_j e^{2\pi i j x} \right|^2 dx$$
$$= \int_{-\infty}^{\infty} \left| \sum_{j=-N}^{N} c_j \widehat{f}_j(x) \right|^2 dx$$
$$= \int_{-\infty}^{\infty} |\widehat{g}(x)|^2 \, dx = \|\widehat{g}\|^2 = \|g\|^2.$$

In fact, we can see that for every $g \in V'$, $|J(g)|^2$ is exactly the function ρ_g, because

$$\begin{aligned}
|[J(g)](\omega)|^2 &= \left| \left[J\Big(\sum_{j=-N}^{N} c_j f_j \Big) \right] (\omega) \right|^2 \\
&= \rho_f(\omega) \left| \sum_{j=-N}^{N} c_j e^{2\pi i j \omega} \right|^2 \\
&= \sum_{z=-\infty}^{\infty} |\widehat{f}(\omega + z)|^2 \left| \sum_{j=-N}^{N} c_j e^{2\pi i j \omega} \right|^2 \\
&= \sum_{z=-\infty}^{\infty} \left| \sum_{j=-N}^{N} c_j e^{2\pi i j \omega} \, \widehat{f}(\omega + z) \right|^2 \\
&= \sum_{z=-\infty}^{\infty} |\widehat{g}(\omega + z)|^2 = \rho_g(\omega).
\end{aligned}$$

Moreover, the intertwining condition (15.7) holds for each $g = \sum_{j=N}^{N} c_j f_j \in V'$, because

$$\begin{aligned}
[J(\tau_n(g))](\omega) &= \left[J\Big(\Big[\sum_{j=-N}^{N} c_j f_j \Big]_n \Big) \right] (\omega) \\
&= \left[J\Big(\sum_{j=-N}^{N} c_j f_{j+n} \Big) \right] (\omega) \\
&= \sum_{j=-N}^{N} c_j e^{2\pi i (j+n)\omega} \, \rho_f(\omega) \\
&= e^{2\pi i n \omega} \sum_{j=-N}^{N} c_j e^{2\pi i j \omega} \, \rho_f(\omega) \\
&= e^{2\pi i n \omega} \left[J\Big(\sum_{j=-N}^{N} c_j f_j \Big) \right] (\omega) \\
&= e^{2\pi i n \omega} \, [J(g)](\omega),
\end{aligned}$$

as desired.

Since J is an isometry, it clearly has a unique extension (also called J) to an isometry (unitary operator) from the closure V of the linear span V' of the f_n's onto the space $L^2(E_f)$. Further, it follows that Equation (15.7) holds for all $g \in V$, and also that $|J(g)|^2 = \rho_g$ for all $g \in V$.

Now, let η be the unique element of V for which $J(\eta) = \chi_{E_f}$. Then, because the functions $\{e^{2\pi i n\omega}\chi_{E_f}\}$ form a Parseval frame for $L^2(E_f)$ (see Example 15.7), and because J is a unitary operator, it follows that $\{\eta_n\}$ must be a Parseval frame for V. Finally, since $\rho_\eta = |J(\eta)|^2 = \chi_{E_f}$, we see that the two sets E_η and E_f coincide. □

Remark 15.13. The intertwining condition in Equation (15.7) clearly resembles the intertwining condition satisfied by the Fourier transform. However, one should be careful and notice that the intertwining condition of Equation (15.7) is quite different from the one for the Fourier transform. Namely, the intertwining condition satisfied by the Fourier transform holds for any translation operator τ_x, $\widehat{\tau_x(f)}(\omega) = e^{2\pi i x\omega}\widehat{f}(\omega)$, while Equation (15.7) only holds for integer translation operators. The map J is **not** the Fourier transform.

Remark 15.14. It follows directly from Equation (15.7) that for every Borel subset $F \subseteq E_\eta$, the subspace $V_F = J^{-1}(L^2(F))$ is a subspace that is invariant under integer shifts, and the intertwining condition continues to hold for the restriction of J to V_F. In fact, again using the ideas in Example 15.7, the integer translates of the element $J^{-1}(\chi_F)$ form a Parseval frame for V_F. We will need this observation later on. Note that such a subspace V_F need not be invariant under other shifts, e.g., τ_r for $r \neq 1$.

There is the natural analog to Theorem 15.12 for translates of a function by multiples of another positive number $r \neq 1$, and we will need this generalization later on. Its proof is completely analogous to the preceding one, with appropriate use of Theorem 15.9.

Theorem 15.15. *Let f be a nonzero element of $L^2(\mathbb{R})$, let r be a positive number, and let V denote the closure of the linear span of the translates f_{nr} of f. Let ρ_f^r and E_f^r be as in Theorem 15.9. Then there exists an element $\eta \in V$ whose translates η_{nr} form a Parseval frame for V. Moreover, $E_\eta^r = E_f^r$, and there exists a unitary map J from V onto $L^2(E_\eta^r)$ for which the intertwining condition*

$$[J(\tau_{nr}(g))](\omega) = e^{2\pi i n r\omega}\,[J(g)](\omega) \tag{15.8}$$

holds for every $g \in V$, every integer n, and almost every $\omega \in [0, 1/r)$.

The next theorem is just an induction argument applied to the two preceding ones. We state it for integer translations and leave the general version to the reader. We call a generalized shift-invariant subspace V *finitely generated* if there exist finitely many elements $f_1, \ldots, f_M$ in V whose translates by multiples of r together span a dense subspace of V.

Theorem 15.16. *Let V be a finitely generated shift-invariant (by integers) subspace of $L^2(\mathbb{R})$. Then there exists a finite collection $\eta_1, \ldots, \eta_M$ of elements of V such that the collection of integer translates of the η_k's together form a Parseval frame for V. In addition, if $k \neq k'$, then every integer translate of η_k is orthogonal to every integer translate of $\eta_{k'}$.*

Finally, there exists a unitary map $J : V \mapsto \bigoplus_{k=1}^{M} L^2(E_{\eta_k})$ for which the intertwining condition

$$[J(\tau_n(g))]_k(\omega) = e^{2\pi i n\omega}\,[J(g)]_k(\omega) \tag{15.9}$$

holds for every $g \in V$, every integer n, every k, and almost every $\omega \in [0,1)$.

Proof. Let $f_1, \ldots, f_M$ be a finite collection of elements of V whose integer translates together span a dense subspace of V. We argue by induction on the integer M. Notice that the case for $M = 1$ is precisely Theorem 15.12. Let W be the closure of the span of the integer translates of the functions $f_1, \ldots, f_{M-1}$, and by the inductive hypothesis let $\eta_1, \ldots, \eta_{M-1}$ be elements of W that satisfy the requirements of the theorem. Next, let V' be the projection of the subspace V into the orthogonal complement of W. (Of course, this subspace might be $\{0\}$, in which case we are done.) Because W is invariant under integer translations, it follows that the projection operator p onto the orthogonal complement of W commutes with all integer translations, and hence V' contains a cyclic vector, namely $p(f_M)$. Using Theorem 15.12, let η_M be an element of V' whose integer translates form a Parseval frame for V'. Clearly, the elements $\eta_1, \ldots, \eta_M$ satisfy the first two requirements of the theorem. Moreover, if we write V_k for the closed linear span of the integer translates of η_k, then the V_k's are pairwise orthogonal, and there exist unitary operators $J_k : V_k \to L^2(E_{\eta_k})$ satisfying the intertwining condition in Equation (15.7). It is immediate that the desired unitary operator J mapping $V = \bigoplus V_k$ onto $\bigoplus L^2(E_{\eta_k})$ can be built from these J_k's, and the theorem is proved. □

Remark 15.17. Evidently the sets E_{η_k} provide frequency information about the subspace V, and a potentially important function is the sum m of the indicator functions of these sets, which we will refer to as the multiplicity function:

$$m = \sum_{k=1}^{M} \chi_{E_{\eta_k}}.$$

Where sets overlap, we will be observing some "multiplicity" or "redundancy" in the frequency domain. Before making additional comments about this multiplicity function and its implications, we must show that it is uniquely determined by the subspace, i.e., not dependent on the choice of the elements $\eta_1, \ldots, \eta_M$.

The next theorem gives our main structure result for finitely generated, shift-invariant subspaces of $L^2(\mathbb{R})$, and it is essentially the Spectral Multiplicity Theorem for a collection of commuting unitary operators (see [7] or [8]).

Although the preceding theorem tells us that any such subspace is equivalent to a direct sum of L^2 spaces, it gives no restrictions on the sets E_{η_k}. This next result organizes these sets in a consistent, and as it happens, a unique way.

Theorem 15.18. *Let V be a finitely generated, shift-invariant subspace of $L^2(\mathbb{R})$. Then there exist unique, nontrivial, Borel subsets $S_1 \supseteq S_2 \supseteq \cdots \supset S_c$ of $[0,1)$, and a unitary operator J from V onto the direct sum $\bigoplus_{j=1}^c L^2(S_j)$ for which the intertwining condition*

$$[J(\tau_n(g))]_k(\omega) = e^{2\pi i n \omega}\,[J(g)]_k(\omega)$$

holds for every $g \in V$, every integer n, every $1 \leq k \leq c$, and almost every $\omega \in [0,1)$.

Proof. Let $\eta_1, \ldots, \eta_M$ and J be as in Theorem 15.16. Again, we will argue by induction on M. Note that the case for $M = 1$ is nothing more than Theorem 15.12. For simplicity of notation, write E_k for the set E_{η_k}. Now define disjoint sets F_k by $F_1 = E_1$, $F_2 = E_2 \setminus F_1$, $F_3 = E_3 \setminus (F_1 \cup F_2), \ldots$. Note that each F_k is a subset of the corresponding E_k. Set $S_1 = \cup_{k=1}^M F_k$, and observe that

$$\begin{aligned}
V \equiv J(V) &= \bigoplus_{k=1}^M L^2(E_k) \\
&= \bigoplus_{k=1}^M L^2(F_k \cup (E_k \setminus F_k)) \\
&= \bigoplus_{k=1}^M (L^2(F_k) \oplus L^2(E_k \setminus F_k)) \\
&= \bigoplus_{k=1}^M L^2(F_k) \oplus \bigoplus_{k=2}^M L^2(E_k \setminus F_k) \\
&= L^2(\cup_{k=1}^M F_k) \oplus \bigoplus_{k=2}^M L^2(E_k \setminus F_k) \\
&= L^2(S_1) \oplus \bigoplus_{k=2}^M L^2(E_k \setminus F_k) \\
&\equiv V_1 \oplus W_1,
\end{aligned}$$

where $V_1 = J^{-1}(L^2(S_1))$, and $W_1 = J^{-1}(\bigoplus_{k=2}^M L^2(E_k \setminus F_k))$. Setting η_k' equal to $J^{-1}(\chi_{E_k \setminus F_k})$ for each $2 \leq k \leq M$, we may apply the inductive hypothesis to the subspace W_1 to produce nontrivial, Borel sets $S_2 \supseteq S_3 \supseteq \cdots \supseteq S_c$ satisfying the conditions of the theorem. Clearly, $S_1 \supseteq S_2$, so that these sets $S_1, S_2, \ldots, S_c$ satisfy the nested requirement of the theorem, and the intertwining condition is basically a consequence of Remark 15.14.

We omit the proof of the uniqueness of the S_k's, except to hint at its proof by remarking that if U is a unitary operator from a direct sum of j copies of $L^2(E)$ into a direct sum of j' copies of $L^2(E')$, which commutes with multiplication by all the exponential functions $e^{2\pi in\omega}$, then E must be a subset of E' and j must be $\leq j'$. With this observation, and a bit of notation, one can construct a proof of the uniqueness. Okay, I lied about this chapter being self-contained. Mi scusi. □

A useful corollary to Theorem 15.18 is the following.

Corollary 15.19. *Let V be a finitely generated, shift-invariant subspace of $L^2(\mathbb{R})$, and let $S_1, \ldots, S_c$ be as in Theorem 15.18. Let $\eta_1, \ldots, \eta_M$ be any elements of V satisfying the requirements of Theorem 15.16. Then*

$$\sum_{i=1}^{c} \chi_{S_i} = \sum_{k=1}^{M} \chi_{E_{\eta_k}}. \tag{15.10}$$

Proof. Since the η_k's in the proof of Theorem 15.18 were arbitrary, we need only verify Equation (15.10) for those particular η_k's. Again we argue by induction on M. It follows directly from the constructions in the proof above that $\chi_{S_1} = \sum_{k=1}^{M} \chi_{F_k}$. So, using the inductive hypothesis on the space W_1 and the elements $\eta'_2, \ldots, \eta'_M$, we have

$$\begin{aligned}
\sum_{i=1}^{c} \chi_{S_i} &= \chi_{S_1} + \sum_{i=2}^{c} \chi_{S_i} \\
&= \sum_{k=1}^{M} \chi_{F_k} + \sum_{k=2}^{M} \chi_{\eta'_k} \\
&= \sum_{k=1}^{M} \chi_{F_k} + \sum_{k=2}^{M} \chi_{E_k \setminus F_k} \\
&= \chi_{F_1} + \sum_{k=2}^{M} (\chi_{F_k} + \chi_{E_k \setminus F_k}) \\
&= \chi_{E_1} + \sum_{k=2}^{M} \chi_{E_k} \\
&= \sum_{k=1}^{M} \chi_{E_k},
\end{aligned}$$

proving the corollary. □

As usual, a result analogous to the preceding theorem holds for subspaces that are invariant under shifts by multiples of another positive number $r \neq 1$, but we will not need it here.

Definition 15.20. *Let V be a closed, shift-invariant subspace of $L^2(\mathbb{R})$ and assume that V is finitely generated. Let $S_1, \ldots, S_c$ and J be as in Theorem 15.18. Define the* multiplicity function m *associated to V by*

$$m = \sum_{j=1}^{c} \chi_{S_j}.$$

Define generalized scaling functions $\phi_1, \ldots, \phi_c$ *in V by $\phi_j = J^{-1}(\chi_j)$, where χ_j is the element of $\bigoplus_{k=1}^{c} L^2(S_k)$ whose jth component is χ_{S_j} and all of whose other components are 0.*

Remark 15.21. If V is a shift-invariant subspace, and J is a unitary map from V onto a direct sum $\bigoplus_{i=1}^{c} L^2(E_i)$ that satisfies Equation 15.9, then the multiplicity function m associated to V equals the sum of the χ_{E_i}'s. We simply define $\eta_i = J^{-1}(E_i)$ as we did in the proof to the corollary above. One simple consequence of this is the fact that if a shift-invariant subspace V is the direct sum of two shift-invariant subspaces V_1 and V_2, then the multiplicity function for V is the sum of the two multiplicity functions for V_1 and V_2. Just express each subspace V_i in terms of a direct sum of L^2 spaces, so that $V_1 \oplus V_2$ is a direct sum of two direct sums of L^2 spaces.

Now we are able to describe precisely what we mean by redundancy or multiplicity in the frequency domain. The generalized scaling functions $\phi_1, \ldots, \phi_c$ together with their integer translates form a Parseval frame for V, and as such these generalized scaling functions represent a kind of redundancy, or multiplicity, in the time domain. In fact, the set of translates of each individual function ϕ_i is in all likelihood a redundant set, as suggested in the introduction. That there are several generalized scaling functions suggests even more redundancy, but in fact this extra redundancy is subtle. It can be truly more redundancy, no extra redundancy, for instance when all the $\tau_n(\phi_i)$'s are orthonormal, and more interestingly, it can be a kind of mixture. The multiplicity function m, on the other hand, indicates a repetition, or multiplicity, in the frequencies contained in the elements of V. This multiplicity of frequency is more subtle than the corresponding multiplicity in time, since the multiplicity of a frequency ω varies, depending on which sets S_j that ω belongs to. Indeed, we will see in the next theorem that V decomposes as a direct sum of subspaces V_k, and each of these subspaces has its own distinct uniform multiplicity in the frequency domain.

Let V, m, and the S_i's be as in the preceding definition. Define sets $\{T_k\}$ in $[0,1)$ by $T_k = \{\omega : m(\omega) = k\}$. Because the sets S_j are unique, it follows that the multiplicity function m, and therefore the sets T_k, are also uniquely determined by the subspace V. Also, it is evident that the T_k's are pairwise disjoint. Here is an alternative version of Theorem 15.18.

Theorem 15.22. *Let V be a closed subspace of $L^2(\mathbb{R})$ that is invariant under the translation operators τ_n, and suppose that V is finitely generated. Let*

$T_1, \dots, T_c$ be the subsets defined above. Write $L^2(T_k, \mathbb{C}^k)$ for the Hilbert space of all square-integrable $\mathbb{C}^k$-valued functions on T_k. Then there exists a unitary map J from V onto the direct sum $\bigoplus_{k=1}^{c} L^2(T_k, \mathbb{C}^k)$ for which the intertwining condition

$$[J(\tau_n(g))]_k(\omega) = e^{2\pi i n\omega}\,[J(g)]_k(\omega)$$

holds for every $g \in V$, every integer n, every $1 \le k \le c$, and almost every $\omega \in [0,1)$.

Proof. This is just a bit of set theory. Note first that

$$\bigoplus_{j=1}^{n} L^2(S) \equiv L^2(S, \mathbb{C}^n).$$

Second, because $S_j = \{\omega : m(\omega) \ge j\}$, and $T_k = \{\omega : m(\omega) = k\}$, we see that $S_j = \cup_{k=j}^{c} T_k$. Then,

$$\begin{aligned}
V &\equiv \bigoplus_{j=1}^{c} L^2(S_j) \\
&= \bigoplus_{j=1}^{c} L^2\left(\bigcup_{k=j}^{c} T_k\right) \\
&= \bigoplus_{j=1}^{c} \bigoplus_{k=j}^{c} L^2(T_k) \\
&= \bigoplus_{k=1}^{c} \bigoplus_{j=1}^{k} L^2(T_k) \\
&\equiv \bigoplus_{k=1}^{c} L^2(T_k, \mathbb{C}^k).
\end{aligned}$$

Again, the intertwining condition holds because of Remark 15.14. □

15.3 Frequency Redundancy in a Refinable Subspace

Finally, let us apply the analysis we have developed for frequency redundancy in shift-invariant subspaces of $L^2(\mathbb{R})$ to a subspace that is also refinable. We will discover an intricate fine structure for such subspaces, and this will bring us to the brink of wavelet and multiresolution analysis theory.

Definition 15.23. *Let d be a positive integer > 1, and define the* dilation operator *δ on $L^2(\mathbb{R})$ by $[\delta(f)](t) = \sqrt{d}\, f(dt)$. A subspace $V \subseteq L^2(\mathbb{R})$ is called* refinable for dilation by *d if $V \subseteq \delta(V)$.*

For the remainder of this chapter, we will refer to a subspace being refinable, omitting the explicit reference to the value of the dilation d. It is immediate that the operator δ is unitary, so that the definition of refinable can be rephrased as $\delta^{-1}(V) \subseteq V$. It is then evident that V is refinable if, whenever a function $f(x)$ is an element of V, so is the function $f(x/d)$.

It was, at least for me, a remarkable discovery ([3] and [1]) that the multiplicity function m associated to a refinable, finitely generated, shift-invariant subspace must satisfy a certain functional equation called the *consistency equation*, given below. However, a preliminary observation concerns the interaction between the dilation operator δ and the translation operators τ_x. The relation is easy to check, and it is

$$\delta^{-1}\tau_x\delta = \tau_{dx}. \tag{15.11}$$

Let V be a finitely generated, refinable, shift-invariant subspace of $L^2(\mathbb{R})$. We see directly from Equation (15.11) that the subspace $\delta(V)$ is invariant under all shifts $\tau_{n/d}$, i.e., by all multiples of $1/d$. For, if $f = \delta(g) \in \delta(V)$, then we have

$$\tau_{\frac{n}{d}}(f) = \tau_{\frac{n}{d}}(\delta(g)) = \delta(\delta^{-1}(\tau_{\frac{n}{d}}(\delta(g)))) = \delta(\tau_n(g)) \in \delta(V).$$

It then is clear that $\delta(V)$ is itself shift-invariant, i.e., by integers. Let W be the orthogonal complement of V in $\delta(V)$. Then, because the τ_n's are unitary, W also is shift-invariant.

Theorem 15.24. *If m denotes the multiplicity function associated to the refinable, shift-invariant subspace V, and $\widetilde{m}$ denotes the multiplicity function associated to the shift-invariant subspace W, then, for almost all $\omega \in [0,1)$, we have*

$$m(\omega) + \widetilde{m}(\omega) = \sum_{l=0}^{d-1} m\Big(\frac{\omega + l}{d}\Big). \tag{15.12}$$

Proof. As seen above, the subspace $\delta(V)$ is shift-invariant by integers, and we denote its associated multiplicity function by m'. The proof of this theorem will follow by computing m' in two different ways.

First of all, it follows from Remark 15.21 that

$$m'(\omega) = m(\omega) + \widetilde{m}(\omega), \tag{15.13}$$

and this is the first way of computing $m'(\omega)$.

As in Definition 15.20, let $\phi_1, \ldots, \phi_c$ be generalized scaling functions for V, and write Y_i for the closed linear span of the translates $\tau_n(\phi_i)$. Then $V = \bigoplus_{i=1}^{c} Y_i$, whence

$$\delta(V) = \bigoplus_{i=1}^{c} \delta(Y_i).$$

Further, since the translates $\{\tau_n(\phi_i)\}$ form a Parseval frame for Y_i, and δ is a unitary operator, it follows from Equation (15.11) that the translates $\{\tau_{\frac{n}{d}}(\delta(\phi_i))\}$ form a Parseval frame for $\delta(Y_i)$.

Next, we find a relation between the two functions ρ_{ϕ_i} and $\rho^{\frac{1}{d}}_{\delta(\phi_i)}$ of Theorem 15.9, and we do this using Theorem 15.11:

$$\begin{aligned}
c_n(\rho^{\frac{1}{d}}_{\delta(\phi_i)}) &= \frac{1}{\sqrt{d}} \left\langle \delta(\phi_i), \tau_{\frac{n}{d}}(\delta(\phi_i)) \right\rangle \\
&= \frac{1}{\sqrt{d}} \left\langle \phi_i, \delta^{-1}(\tau_{\frac{n}{d}}(\delta(\phi_i))) \right\rangle \\
&= \frac{1}{\sqrt{d}} \left\langle \phi_i, \tau_n(\phi_i) \right\rangle \\
&= \frac{1}{\sqrt{d}} c_n(\rho_{\phi_i}).
\end{aligned}$$

Therefore,

$$\begin{aligned}
\rho^{\frac{1}{d}}_{\delta(\phi_i)}(\omega) &= \sum_{n=-\infty}^{\infty} c_n(\rho^{\frac{1}{d}}_{\delta(\phi_i)}) \frac{1}{\sqrt{d}} e^{2\pi i \frac{n}{d}\omega} \\
&= \sum_{n=-\infty}^{\infty} \frac{1}{\sqrt{d}} c_n(\rho_{\phi_i}) \frac{1}{\sqrt{d}} e^{2\pi i n \frac{\omega}{d}} \\
&= \frac{1}{d} \rho_{\phi_i}\left(\frac{\omega}{d}\right) \\
&= \frac{1}{d} \chi_{S_i}\left(\frac{\omega}{d}\right) \\
&= \frac{1}{d} \chi_{dS_i}(\omega).
\end{aligned}$$

Hence, according to Theorem 15.9, the subset $E^{1/d}_{\delta(\phi_i)}$ of the interval $[0,d)$ coincides with dS_i. Hence, applying Theorem 15.15 applied to the subspace $\delta(Y_i)$, we let J_i be a unitary operator from $\delta(Y_i)$ onto $L^2(dS_i)$ satisfying Equation (15.8):

$$[J_i(\tau_{\frac{n}{d}}(g))](\omega) = e^{2\pi i \frac{n}{d}\omega} [J_i(g)](\omega)$$

holds for every $g \in \delta(Y_i)$, every integer n, and almost every $\omega \in [0,d)$. Combining these J_i's into a single operator J from $\delta(V)$ to $\bigoplus_{i=1}^{c} L^2(dS_i)$, we have the following direct sum decomposition of $\delta(V)$:

$$\begin{aligned}
\delta(V) = \delta\left(\bigoplus_{i=1}^{c} Y_i\right) &= \bigoplus_{i=1}^{c} \delta(Y_i) \\
&\equiv \bigoplus_{i=1}^{c} L^2(dS_i)
\end{aligned}$$

$$= \bigoplus_{i=1}^{c} \bigoplus_{l=0}^{d-1} L^2(dS_i \cap [l, l+1))$$

$$= \bigoplus_{i=1}^{c} \bigoplus_{l=0}^{d-1} L^2((dS_i \cap [l, l+1)) - l).$$

Setting $\eta_{i,l}$ equal to $J_i^{-1}(\chi_{(dS_i \cap [l,l+1))-l})$, and using Equation (15.10), we have

$$\begin{aligned}
m'(\omega) &= \sum_{i=1}^{c} \sum_{l=0}^{d-1} \chi_{\eta_{i,l}}(\omega) \\
&= \sum_{i=1}^{c} \sum_{l=0}^{d-1} \chi_{(dS_i \cap [l,l+1))-l}(\omega) \\
&= \sum_{i=1}^{c} \sum_{l=0}^{d-1} \chi_{dS_i \cap [l,l+1)}(\omega + l) \\
&= \sum_{l=0}^{d-1} \sum_{i=1}^{c} \chi_{dS_i}(\omega + l) \\
&= \sum_{l=0}^{d-1} \sum_{i=1}^{c} \chi_{S_i}\Big(\frac{\omega + l}{d}\Big) \\
&= \sum_{l=0}^{d-1} m\Big(\frac{\omega + l}{d}\Big),
\end{aligned} \tag{15.14}$$

and this gives the second way of computing m', and completes the proof of the theorem. □

Remark 15.25. Note that the consistency equation implies that $\widetilde{m}$ is finite a.e. because m is. In other words, the subspace W is finitely generated.

Again, let V be a finitely generated, shift-invariant, refinable subspace of $L^2(\mathbb{R})$, and let $S_1, \ldots, S_c$ and J be as in Theorem 15.18. Let $\{\phi_i\} = \{J^{-1}(\chi_i)\}$ be a standard set of generalized scaling functions in V. Then, because V is refinable, we must have that $\delta^{-1}(\phi_i)$ is also an element of V. Therefore, there must exist a $c \times c$ matrix $H = [h_{i,j}]$ of periodic functions satisfying

$$J(\delta^{-1}(\phi_i)) = \bigoplus_{j=1}^{c} h_{i,j}.$$

Similarly, since W is finitely generated and shift-invariant, Theorem 15.18 implies that there exist Borel sets $\widetilde{S}_1 \supseteq \cdots \supseteq \widetilde{S}_{\widetilde{c}}$ of $[0,1)$, and a unitary map $\widetilde{J}$ from W onto $\bigoplus_{k=1}^{\widetilde{c}} L^2(\widetilde{S}_k)$ that satisfies the intertwining equation (15.9). As in Definition 15.20, define generalized scaling functions $\psi_1, \ldots, \psi_{\widetilde{c}}$ in W

by $\psi_k = \widetilde{J}^{-1}(\chi_{\widetilde{S}_k})$. Then, because $\delta^{-1}(\psi_k)$ must belong to V, it follows that there exists a $\widetilde{c} \times c$ matrix $G = [g_{k,j}]$ of periodic functions for which

$$J(\delta^{-1}(\psi_k)) = \bigoplus_{j=1}^{c} g_{k,j}.$$

Another remarkable fact is that the matrices H and G must satisfy their own functional equations related to the multiplicity functions m and $\widetilde{m}$.

Theorem 15.26. *Let m, $\widetilde{m}$, H, and G be as in the preceding paragraphs. Then, for almost all $\omega \in [0,1)$, we have the following three* generalized filter equations:

$$\sum_{j=1}^{c}\sum_{l=0}^{d-1} h_{i,j}\Big(\omega + \frac{l}{d}\Big)\,\overline{h_{i',j}\Big(\omega + \frac{l}{d}\Big)} = d\,\delta_{i,i'}\,\chi_{S_i}(d\omega), \tag{15.15}$$

$$\sum_{j=1}^{c}\sum_{l=0}^{d-1} g_{k,j}\Big(\omega + \frac{l}{d}\Big)\,\overline{g_{k',j}\Big(\omega + \frac{l}{d}\Big)} = d\,\delta_{k,k'}\,\chi_{\widetilde{S}_k}(d\omega), \tag{15.16}$$

$$\sum_{j=1}^{c}\sum_{l=0}^{d-1} h_{i,j}\Big(\omega + \frac{l}{d}\Big)\,\overline{g_{k,j}\Big(\omega + \frac{l}{d}\Big)} = 0, \tag{15.17}$$

for all $1 \le i \le c$ and $1 \le k \le \widetilde{c}$.

Proof. We prove the first equation, noting that totally analogous arguments would prove the second and third equations. For each pair (i, i'), define a function $F_{i,i'}$ on $[0,1)$ by

$$F_{i,i'}(\omega) = \sum_{j=1}^{c}\sum_{l=0}^{d-1} h_{i,j}\Big(\omega + \frac{l}{d}\Big)\,\overline{h_{i',j}\Big(\omega + \frac{l}{d}\Big)}.$$

Of course, the first equation will follow if we can satisfactorily compute these functions. Clearly, $F_{i,i'}$ is periodic with period $1/d$, and so can be expressed as a Fourier series in terms of exponentials $e^{2\pi i n d\omega}$. Suppose first that $i \neq i'$. Computing the Fourier coefficients $c_{nd}(F_{i,i'})$ for $F_{i,i'}$ we obtain

$$\begin{aligned} c_{nd}(F_{i,i'}) &= \sqrt{d}\int_0^{\frac{1}{d}} F_{i,i'}(\omega)\, e^{-2\pi i n d\omega}\, d\omega \\ &= \sqrt{d}\int_0^{\frac{1}{d}} \sum_{j=1}^{c}\sum_{l=0}^{d-1} h_{i,j}\Big(\omega + \frac{l}{d}\Big)\,\overline{h_{i',j}\Big(\omega + \frac{l}{d}\Big)}\, e^{-2\pi i n d\omega + \frac{l}{d}}\, d\omega \\ &= \sqrt{d}\int_0^{1} \sum_{j=1}^{c} h_{i,j}(\omega)\,\overline{e^{2\pi i n d\omega}\, h_{i',j}(\omega)}\, d\omega \end{aligned}$$

$$= \sqrt{d}\sum_{j=1}^{c}\int_0^1 h_{i,j}(\omega)\,\overline{e^{2\pi ind\omega}\,h_{i',j}(\omega)}\,d\omega$$
$$= \sqrt{d}\,\big\langle J(\delta^{-1}(\phi_i)),\, J(\tau_{nd}(\delta^{-1}(\phi_{i'})))\big\rangle$$
$$= \sqrt{d}\,\big\langle \delta^{-1}(\phi_i),\, \tau_{nd}(\delta^{-1}(\phi_{i'}))\big\rangle$$
$$= \sqrt{d}\,\big\langle \phi_i,\, \tau_n(\phi_{i'})\big\rangle$$
$$= 0.$$

On the other hand, if $i = i'$, then we have

$$\begin{aligned} c_{nd}(F_{i,i}) &= \sqrt{d}\,\big\langle \phi_i,\, \tau_n(\phi_i)\big\rangle \\ &= \sqrt{d}\,\big\langle J(\phi_i),\, J(\tau_n(\phi_i))\big\rangle \\ &= \sqrt{d}\int_0^1 \chi_{S_i}(\omega)\,\overline{e^{2\pi in\omega}\,\chi_{S_i}(\omega)}\,d\omega \\ &= \sqrt{d}\int_0^1 \chi_{S_i}(\omega)\,e^{-2\pi in\omega}\,d\omega \\ &= \sqrt{d}\,c_n(\chi_{S_i}). \end{aligned}$$

Therefore,

$$F_{i,i}(\omega) = \sum_{n=-\infty}^{\infty} \sqrt{d}\;c_{nd}(F_{i,i})e^{2\pi ind\omega} = \sum_{n=-\infty}^{\infty} d\,c_n(\chi_{S_i})\,e^{2\pi in(d\omega)} = d\,\chi_{S_i}(d\omega),$$

which completes the proof of Equation (15.15). Changing the indices i to k, and making the same computations, yields Equation (15.16), and replacing one of the h's with g, and making similar computations, gives Equation (15.17). □

To conclude, what we have presented here is a rather complex structure, from a frequency-side perspective, of a refinable, finitely generated, shift-invariant subspace V of $L^2(\mathbb{R})$. We have called the three equations in the last theorem generalized filter equations, because they are generalizations of classical filter equations in wavelet theory. In fact, the original work of Mallat and Meyer ([9], [10]) on wavelets was largely concerned with building wavelets out of such classical filters. A general version of that classical method has recently been presented in [2], and interested readers of this chapter are encouraged to move on to that work.

We mention finally that there is yet a third remarkable functional equation that is satisfied by the functions $h_{i,j}$ and $g_{k,j}$. It is derived by combining Equation (15.12) and the three filter equations of the last theorem. We do not include the proof here, but here is the equation:

$$\sum_{i=1}^{c} h_{i,j}\left(\frac{\omega+l}{d}\right)\overline{h_{i,j'}\left(\frac{\omega+l'}{d}\right)} + \sum_{k=1}^{\widetilde{c}} g_{k,j}\left(\frac{\omega+l}{d}\right)\overline{g_{k,j'}\left(\frac{\omega+l'}{d}\right)}$$
$$= d\,\delta_{j,j'}\,\delta_{l,l'}\,\chi_{S_j}\left(\frac{\omega+l}{d}\right).$$

Acknowledgments

Many thanks to the National Science Foundation, who has supported this research over the years, and most recently with the grant NSF-0139366. I gratefully appreciate the powerful insights of my many collaborators on the subject of generalized multiresolution analyses, insights that are fundamental to the ideas in this chapter. Very special thanks go to the mathematicians Jennifer Courter, Palle Jorgensen, Herbert Medina, Kathy Merrill, and Judith Packer. Thanks, too, to Christopher Heil for inviting me to contribute an chapter to this volume. And thanks definitely to the man himself, JJB, for having the personal magnetism and the mathematical talent and spirit to have brought us all together.

References

1. L. Baggett, J. Courter, and K. Merrill, The construction of wavelets from generalized conjugate mirror filters in $L^2(\mathbb{R}^n)$, *Appl. Comput. Harmon. Anal.*, **13** (2002), pp. 201–233.
2. L. Baggett, P. Jorgensen, K. Merrill, and J. Packer, Construction of Parseval wavelets from redundant filter systems, *J. Math. Phys.*, **46** (2005), 28 pp.
3. L. Baggett, H. Medina, and K. Merrill, Generalized multiresolution analyses and a construction procedure for all wavelet sets in $\mathbb{R}^n$, *J. Fourier Anal. Appl.*, **5** (1999), pp. 563–573.
4. J.J. Benedetto and S. Li, The theory of multiresolution analysis frames and applications to filter banks, *Appl. Comput. Harmon. Anal.*, **5** (1998), pp. 389–427.
5. O. Christensen, *An Introduction to Frames and Riesz Bases*, Birkhäuser, Boston, 2003.
6. R. J. Duffin and A. C. Schaeffer, A class of nonharmonic Fourier series, *Trans. Amer. Math. Soc.*, **72** (1952), pp. 341–366.
7. P. Halmos, *Introduction to Hilbert Space and the Theory of Spectral Multiplicity*, Chelsea Publishing Co., New York, 1951.
8. H. Helson, *The Spectral Theorem*, Lecture Notes in Mathematics, Vol. 1227, Springer-Verlag, New York, 1986.
9. S. Mallat, Multiresolution approximations and wavelet orthonormal bases of $L^2(\mathbb{R})$, *Trans. Amer. Math. Soc.*, **315** (1989), pp. 69–87.
10. Y. Meyer, *Wavelets and Operators*, Cambridge University Press, Cambridge, 1992.

16

Density Results for Frames of Exponentials

Peter G. Casazza[1], Ole Christensen[2], Shidong Li[3], and Alexander Lindner[4]

[1] Department of Mathematics, University of Missouri–Columbia, MO 65211, USA
pete@math.missouri.edu

[2] Department of Mathematics, Technical University of Denmark, Building 303, 2800 Kgs. Lyngby, Denmark
Ole.Christensen@mat.dtu.dk

[3] Department of Mathematics, San Francisco State University, San Francisco, CA 94132, USA
shidong@sfsu.edu

[4] Center for Mathematical Sciences, Technical University of Munich, D-85747 Garching, Germany
lindner@ma.tum.de

Summary. For a separated sequence $\Lambda = \{\lambda_k\}_{k\in\mathbb{Z}}$ of real numbers there is a close link between the lower and upper densities $D^-(\Lambda), D^+(\Lambda)$ and the frame properties of the exponentials $\{e^{i\lambda_k x}\}_{k\in\mathbb{Z}}$: in fact, $\{e^{i\lambda_k x}\}_{k\in\mathbb{Z}}$ is a frame for its closed linear span in $L^2(-\gamma,\gamma)$ for any $\gamma \in (0, \pi D^-(\Lambda)) \cup (\pi D^+(\Lambda), \infty)$. We consider a classical example presented already by Levinson [11] with $D^-(\Lambda) = D^+(\Lambda) = 1$; in this case, the frame property is guaranteed for all $\gamma \in (0,\infty) \setminus \{\pi\}$. We prove that the frame property actually breaks down for $\gamma = \pi$. Motivated by this example, it is natural to ask whether the frame property can break down on an interval if $D^-(\Lambda) \neq D^+(\Lambda)$. The answer is yes: We present an example of a family Λ with $D^-(\Lambda) \neq D^+(\Lambda)$ for which $\{e^{i\lambda_k x}\}_{k\in\mathbb{Z}}$ has no frame property in $L^2(-\gamma,\gamma)$ for any $\gamma \in (\pi D^-(\Lambda), \pi D^+(\Lambda))$.

Dedicated to Professor John Benedetto.

16.1 Introduction

While frames nowadays are a recognized tool in many branches of harmonic analysis and signal processing, it is interesting to remember that Duffin and Schaeffer [7] actually introduced the concept in the context of systems of complex exponentials. Much of the study is rooted in the study of sampling theories tracing back to Paley–Wiener, Levinson, Plancherel–Polya and Boas; a complete treatment is given by John Benedetto in his paper *Irregular Sampling and Frames* [3]. That paper also contains original work on irregular sampling using Fourier frames, as well as new properties of Fourier frames. In particular, density issues for Fourier frames and distinctions as well as interconnections among *uniform density*, *natural density*, and the *lower and upper Beurling densities* $D^-(\Lambda)$ and $D^+(\Lambda)$ of a sequence $\Lambda \equiv \{\lambda_k\}_{k\in\mathbb{Z}} \subset \mathbb{R}$ are discussed in detail in Sections 7 and 9 of [3].

The frame properties for systems of exponentials $\{e^{i\lambda_k x}\}_{k\in\mathbb{Z}}$ are closely related to density issues concerning the sequence $\{\lambda_k\}_{k\in\mathbb{Z}}$, as revealed by, e.g., [9], [10], [13]. A combination of well-known results shows that for a separated sequence $\{\lambda_k\}_{k\in\mathbb{Z}} \subset \mathbb{R}$, the exponentials $\{e^{i\lambda_k x}\}_{k\in\mathbb{Z}}$ form a frame sequence in $L^2(-\gamma,\gamma)$ for all $\gamma \in (0, \pi D^-(\Lambda)) \cup (\pi D^+(\Lambda), \infty)$. On the other hand, it is known that there might be no frame property for the limit cases $\gamma = \pi D^-(\Lambda), \gamma = \pi D^+(\Lambda)$. This appears, e.g., in a classical example presented by Levinson [11] which we will consider in Example 16.11: in that example, $D^-(\Lambda) = D^+(\Lambda) = 1$, and the exponentials $\{e^{i\lambda_k x}\}_{k\in\mathbb{Z}}$ form a frame sequence in $L^2(-\gamma,\gamma)$ exactly for $\gamma \in (0,\infty) \setminus \{\pi\}$. The above considerations leave an interesting gap on the interval $(\pi D^-(\Lambda), \pi D^+(\Lambda))$. In particular, it is natural to ask whether there are exponentials $\{e^{i\lambda_k x}\}_{k\in\mathbb{Z}}$ with $D^-(\Lambda) \neq D^+(\Lambda)$ and having no frame property in the gap $(\pi D^-(\Lambda), \pi D^+(\Lambda))$. The answer turns out to be yes; the proof of Theorem 16.12 will provide a concrete example.

This chapter is organized as follows. Section 16.2 concerns the general frame terminology and definitions. Section 16.3 summarizes known results of Seip and Beurling, in which it is shown that exponentials $\{e^{i\lambda_k x}\}_{k\in\mathbb{Z}}$ form a frame sequence in $L^2(-\gamma,\gamma)$ for all $\gamma \notin [\pi D^-(\Lambda), \pi D^+(\Lambda)]$. A necessary condition for $\{e^{i\lambda_k x}\}_{k\in\mathbb{Z}}$ being a frame for $L^2(-\gamma,\gamma)$, due to Landau [10], is phrased as a no-go theorem as a preparation for applications in later sections. In Section 16.4, a lemma by Jaffard is presented and discussed; this is the foundation for the construction of the aforementioned example in Section 16.6. Section 16.5 analyzes the limit case of the frame version of the classical Kadec's 1/4-Theorem and discusses the example presented by Levinson. Then in Section 16.6, we give an explicit example with no frame property for $\gamma \in (\pi D^-(\Lambda), \pi D^+(\Lambda))$.

16.2 General Frames

We will formulate the basic concepts in somewhat larger generality than needed in the present chapter. Thus, in this section we consider a separable Hilbert space $\mathcal{H}$ with the inner product $\langle\cdot,\cdot\rangle$ linear in the first entry. In later sections we will mainly consider $\mathcal{H} = L^2(-\gamma,\gamma)$ for some $\gamma \in (0,\infty)$.

We begin with some definitions.

Definition 16.1. *Let $\{f_k\}_{k=1}^\infty$ be a sequence in $\mathcal{H}$. We say that*

(i) *$\{f_k\}_{k=1}^\infty$ is a frame for $\mathcal{H}$ if there exist constants $A, B > 0$ such that*

$$A\,||f||^2 \le \sum_{k=1}^\infty |\langle f, f_k\rangle|^2 \le B\,||f||^2, \ \forall f \in \mathcal{H}. \tag{16.1}$$

(ii) $\{f_k\}_{k=1}^\infty$ *is a frame sequence if there exist constants* $A, B > 0$ *such that*

$$A\ ||f||^2 \leq \sum_{k=1}^{\infty} |\langle f, f_k \rangle|^2 \leq B\ ||f||^2, \ \forall f \in \overline{\text{span}}\{f_k\}_{k=1}^\infty.$$

(iii) $\{f_k\}_{k=1}^\infty$ *is a Riesz basis for* $\mathcal{H}$ *if* $\overline{\text{span}}\{f_k\}_{k=1}^\infty = \mathcal{H}$ *and there exist constants* $A, B > 0$ *such that*

$$A \sum |c_k|^2 \leq \left\| \sum c_k f_k \right\|^2 \leq B \sum |c_k|^2 \tag{16.2}$$

for all finite scalar sequences $\{c_k\}$*.*

(iv) $\{f_k\}_{k=1}^\infty$ *is a Riesz sequence if there exist constants* $A, B > 0$ *such that*

$$A \sum |c_k|^2 \leq \left\| \sum c_k f_k \right\|^2 \leq B \sum |c_k|^2$$

for all finite scalar sequences $\{c_k\}$*.*

The frame definition goes back to the paper [7] by Duffin and Schaeffer. More recent treatments can be found in the books [14], [6], or [4].

Any numbers $A, B > 0$ which can be used in (16.1) will be called *frame bounds.* If $\{f_k\}_{k=1}^\infty$ is a frame for $\mathcal{H}$, there exists a dual frame $\{g_k\}_{i=1}^\infty$ such that

$$f = \sum_{i=1}^{\infty} \langle f, g_k \rangle f_k = \sum_{i=1}^{\infty} \langle f, f_k \rangle g_k, \ \forall f \in \mathcal{H}. \tag{16.3}$$

The series in (16.3) converges unconditionally; for this reason, we can index the frame elements in an arbitrary way. In particular, we can apply the general frame results discussed in this section to frames of exponentials, which are usually indexed by $\mathbb{Z}$.

A Riesz basis is a frame, so a representation of the type (16.3) is also available if $\{f_k\}_{k=1}^\infty$ is a Riesz basis. Furthermore, the possible values of A, B in (16.2) coincide with the frame bounds. Riesz bases are characterized as the class of frames which are ω-independent, see [8]: that is, a frame $\{f_k\}_{k=1}^\infty$ is a Riesz basis if and only if $\sum_{k=1}^\infty c_k f_k = 0$ implies that $c_k = 0$ for all $k \in \mathbb{N}$. Thus, Riesz bases are the frames which are at the same time bases. This means that a frame which is not a Riesz basis is *redundant:* it is possible to remove elements without destroying the frame property. However, in general, it is not the case that *arbitrary* elements can be removed, as shown in the following example.

Example 16.2. If $\{e_k\}_{k=1}^\infty$ is an orthonormal basis for $\mathcal{H}$, then

$$\{e_1, e_2, e_2, e_3, e_3, e_4, e_4, \dots\}$$

is a frame for $\mathcal{H}$. If any e_k with $k \geq 2$ is removed, the remaining family is still a frame for $\mathcal{H}$. However, if e_1 is removed, the remaining family is merely a frame sequence.

We note in passing that the above example is typical: the removal of a single element from a frame might leave an incomplete family, which cannot be a frame for the same space. However, the remaining family will always be a frame sequence. These considerations of course generalize to the removal of a *finite* number of elements, but *not* to removal of arbitrary collections; for more details on removing infinite subsets, see [2].

Note that a frame sequence also leads to representations of the type (16.3)—but only for $f \in \overline{\text{span}}\{f_k\}_{k=1}^{\infty}$.

16.3 Frames of Exponentials

In this section, we consider the frame properties of a sequence of complex exponentials $\{e^{i\lambda_k x}\}_{k\in\mathbb{Z}}$, where $\Lambda = \{\lambda_k\}_{k\in\mathbb{Z}}$ is a sequence of real numbers. Before we discuss the frame properties, we introduce some central concepts related to the sequence Λ.

We say that Λ is *separated* if there exists $\delta > 0$ such that $|\lambda_k - \lambda_l| \geq \delta$ for all $k \neq l$. If Λ is a finite union of separated sets, we say that Λ is *relatively separated.* It can be proved that $\{e^{i\lambda_k x}\}_{k\in\mathbb{Z}}$ satisfies the upper frame condition if and only if Λ is relatively separated.

In this chapter, we concentrate on separated sequences Λ. Given $r > 0$, let $n^-(r)$ (resp. $n^+(r)$) denote the minimal (resp. maximal) number of elements from Λ to be found in an interval of length r. The lower (resp. upper) density of Λ is defined by

$$D^-(\Lambda) = \lim_{r\to\infty} \frac{n^-(r)}{r} \quad \text{resp. } D^+(\Lambda) = \lim_{r\to\infty} \frac{n^+(r)}{r}.$$

The sufficiency parts of Theorems 2.1 and 2.2 in [13] can be formulated as follows.

Theorem 16.3. *Let $\Lambda = \{\lambda_k\}_{k\in\mathbb{Z}}$ be a separated sequence of real numbers. Then the following holds:*

(a) *$\{e^{i\lambda_k x}\}_{k\in\mathbb{Z}}$ forms a frame for $L^2(-\gamma,\gamma)$ for any $\gamma < \pi D^-(\Lambda)$.*
(b) *$\{e^{i\lambda_k x}\}_{k\in\mathbb{Z}}$ forms a Riesz sequence in $L^2(-\gamma,\gamma)$ for any $\gamma > \pi D^+(\Lambda)$.*

As an illustration of this result, we encourage the reader to consider the family $\{e^{ikx}\}_{k\in\mathbb{Z}}$: it is an orthonormal basis for $L^2(-\pi,\pi)$, a frame for $L^2(-\gamma,\gamma)$ for any $\gamma \in (0,\pi)$, and a (non-complete) Riesz sequence in $L^2(-\gamma,\gamma)$ for $\gamma > \pi$. This corresponds to the fact that $D^-(\mathbb{Z}) = D^+(\mathbb{Z}) = 1$. This example is considered in detail in [8] in this volume.

In terms of frame sequences Theorem 16.3 gives the following.

Corollary 16.4. *Let Λ be a separated sequence of real numbers. Then $\{e^{i\lambda_k x}\}_{k\in\mathbb{Z}}$ is a frame sequence in $L^2(-\gamma,\gamma)$ whenever*

$$\gamma \in (0, \pi D^-(\Lambda)) \cup (\pi D^+(\Lambda), \infty).$$

Corollary 16.4 serves as motivation for our new results in Section 16.5 and Section 16.6. In Section 16.5, we consider an example with $D^-(\Lambda) = D^+(\Lambda) = 1$, where it turns out that the frame property holds in $L^2(-\gamma,\gamma)$ for any $\gamma \neq \pi$. In Section 16.6, we prove that if $D^-(\Lambda) \neq D^+(\Lambda)$, then $\{e^{i\lambda_k x}\}_{k\in\mathbb{Z}}$ might not have any frame property when considered in $L^2(-\gamma,\gamma), \gamma \in (\pi D^-(\Lambda), \pi D^+(\Lambda))$.

We need a deep result due to Landau [10] (see Ortega-Cerda and Seip [12, pp. 791, 792] for a discussion of this result).

Theorem 16.5. *A separated family of complex exponentials $\{e^{i\lambda x}\}_{\lambda\in\Lambda}$ is not a frame for $L^2(-\gamma,\gamma)$ if $\pi D^-(\Lambda) < \gamma$.*

In [13], Seip considers removal of elements from a frame of exponentials. In particular, he proves that it always is possible to remove elements in a way such that the remaining family is still a frame, but now corresponding to a separated family.

Lemma 16.6. *Assume that Λ is relatively separated and that $\{e^{i\lambda_k x}\}_{k\in\mathbb{Z}}$ is a frame for $L^2(-\gamma,\gamma)$ for some $\gamma > 0$. Then there exists a separated subfamily $\Lambda' \subseteq \Lambda$ such that $\{e^{i\lambda x}\}_{\lambda\in\Lambda'}$ is a frame for $L^2(-\gamma,\gamma)$.*

16.4 Jaffard's Lemma

We need a lemma by Jaffard, [9, Lemma 4, p. 344]. Jaffard states the lemma as follows: *Suppose a sequence of functions $\{e_k\}_{k\in\mathbb{Z}}$ is a frame for $L^2(I)$ for an interval I. Then $\{e_k\}_{k\neq 0}$ is a frame on each interval $I' \subset I$ such that $|I'| < |I|$.*

Unfortunately, the lemma is false in the stated generality. Before we illustrate this with an example, let us examine the mistake which appears in the first two lines of the proof. There the following is stated: *The $\{e_k\}_{k\in\mathbb{Z}}$ are a frame of $L^2(I')$. Then, either $\{e_k\}_{k\neq 0}$ is a frame of $L^2(I')$, and we have nothing to prove, or the $\{e_k\}_{k\in\mathbb{Z}}$ are a Riesz basis of $L^2(I')$.* This is not true. If $\{e_k\}_{k\in\mathbb{Z}}$ is a redundant frame for $L^2(I)$ and if e_0 is not in the closed linear span of $\{e_k\}_{k\neq 0}$ in $L^2(I')$ then $\{e_k\}_{k\neq 0}$ is not a frame for $L^2(I')$ but $\{e_k\}_{k\in\mathbb{Z}}$ is still a redundant frame for its span and not a Riesz basis. For example, let $\{e_k\}_{k\in\mathbb{Z}}$ be an orthonormal basis for $L^2(0,1)$ with supp $e_0 \subset [0,1/2]$. Since $e_0 \perp e_k$ for all $k \neq 0$ on $L^2(0,1)$ and $e_0(t) = 0$, for all $1/2 \leq t \leq 1$, it follows that $e_0 \perp e_k$ for all $k \neq 0$ on $L^2(0,1/2)$. Hence, $\{e_k\}_{k\neq 0}$ is not a frame for $L^2(0,1/2)$.

However, a very special case of this lemma is true, and, fortunately, it is all that we need (as, in fact, did Jaffard).

Lemma 16.7. *Let $\{e^{i\lambda x}\}_{\lambda\in\Lambda}$ be a frame for $L^2(-R,R)$ and let $J\subset\Lambda$ be a finite set. Then $\{e^{i\lambda x}\}_{\lambda\in\Lambda-J}$ is a frame for $L^2(-A,A)$ for all $0<A<R$.*

Jaffard's proof works to prove Lemma 16.7 because of a result of Seip [13, Lemma 3.15, p. 142] which makes the first two lines of Jaffard's proof correct in this special case. We state Seip's lemma below.

Lemma 16.8. *If $\{e^{i\lambda x}\}_{\lambda\in\Lambda}$ is a frame for $L^2(-R,R)$, then either $\{e^{i\lambda x}\}_{\lambda\in\Lambda}$ is a Riesz basis for $L^2(-R,R)$ or $\{e^{i\lambda x}\}_{\lambda\in\Lambda-\{\lambda_0\}}$ is a frame for $L^2(-R,R)$ for every $\lambda_0\in\Lambda$.*

However, for completeness we now give an independent proof of Lemma 16.7 that was communicated to us by D. Speegle.

Proof of Lemma 16.7. We will rely on Theorem 16.5. First, we assume that Λ is separated. By Theorem 16.5 we know that $D^-(\Lambda)\geq\frac{R}{\pi}$. It is clear that D^- is unchanged if we delete a finite number of elements. So for all $0<A<R$,

$$D^-(\Lambda-J)=D^-(\Lambda)\geq\frac{R}{\pi}>\frac{A}{\pi}.$$

So by Theorem 16.3 we have that $\{e^{i\lambda x}\}_{\lambda\in\Lambda-J}$ is a frame for $L^2(-A,A)$.

If Λ is not separated, Lemma 16.6 shows that there is a $\Lambda^{'}\subset\Lambda$, with $\Lambda^{'}$ separated so that $\{e^{i\lambda x}\}_{\lambda\in\Lambda'}$ is also a frame for $L^2(-R,R)$. Now, $\{e^{i\lambda x}\}_{\lambda\in\Lambda'-J}$ is a frame for $L^2(-A,A)$ for all $A<R$ and hence so is $\{e^{i\lambda x}\}_{\lambda\in\Lambda-J}$. □

16.5 The Limit Case of Kadec's 1/4-Theorem

The classical Kadec's 1/4-Theorem concerns perturbations of the orthonormal basis $\{e^{ikx}\}_{k\in\mathbb{Z}}$ for $L^2(-\pi,\pi)$: it states that if $\{\lambda_k\}_{k\in\mathbb{Z}}$ is a sequence of real numbers and $\sup_{k\in\mathbb{Z}}|\lambda_k-k|<1/4$, then $\{e^{i\lambda_k x}\}_{k\in\mathbb{Z}}$ is a Riesz basis for $L^2(-\pi,\pi)$. It is well known that the result is sharp, in the sense that $\{e^{i\lambda_k x}\}_{k\in\mathbb{Z}}$ might not be a Riesz basis for $L^2(-\pi,\pi)$ if $\sup_{k\in\mathbb{Z}}|\lambda_k-k|=1/4$ (see Example 16.11 below). In Proposition 16.10 we shall sharpen this result, using the following extension of Kadec's theorem to frames, which was proved independently by Balan [1] and Christensen [5].

Theorem 16.9. *Let $\{\lambda_k\}_{k\in\mathbb{Z}}$ and $\{\mu_k\}_{k\in\mathbb{Z}}$ be real sequences. Assume that $\{e^{i\lambda_k x}\}_{k\in\mathbb{Z}}$ is a frame for $L^2(-\pi,\pi)$ with bounds A,B, and that there exists a constant $L<1/4$ such that*

$$|\mu_k-\lambda_k|\leq L \text{ and } 1-\cos\pi L+\sin\pi L<\frac{A}{B}.$$

Then $\{e^{i\mu_k x}\}_{k\in\mathbb{Z}}$ is a frame for $L^2(-\pi,\pi)$ with bounds

$$A\left(1-\frac{B}{A}(1-\cos\pi L+\sin\pi L)\right)^2,\quad B(2-\cos\pi L+\sin\pi L)^2.$$

With the help of Theorem 16.9 we can now prove the following.

Proposition 16.10. *Let $\{\lambda_k\}_{k\in\mathbb{Z}}$ be a sequence of real numbers such that*

$$\sup_{k\in\mathbb{Z}} |\lambda_k - k| = 1/4.$$

Then either $\{e^{i\lambda_k x}\}_{k\in\mathbb{Z}}$ is a Riesz basis for $L^2(-\pi,\pi)$ or it is not a frame for $L^2(-\pi,\pi)$.

Proof. For $t \in [0,1]$, define

$$\lambda_k(t) := k + t(\lambda_k - k).$$

Then

$$\sup_{k\in\mathbb{Z}} |\lambda_k(t) - k| = t \cdot \sup_{k\in\mathbb{Z}} |\lambda_k - k| = \frac{t}{4}.$$

Furthermore, $\lambda_k(1) = \lambda_k$ for all $k \in \mathbb{Z}$. Now, suppose that $\{e^{i\lambda_k x}\}_{k\in\mathbb{Z}}$ is a frame for $L^2(-\pi,\pi)$, with lower bound A, say. By Theorem 16.9, there exists a $t_0 \in [0,1)$ such that $\{e^{i\lambda_k(t)x}\}_{k\in\mathbb{Z}}$ is a frame for $L^2(-\pi,\pi)$ with lower bound $A/2$ for any $t \in [t_0,1)$. But, by Kadec's 1/4-Theorem, $\{e^{i\lambda_k(t)x}\}_{k\in\mathbb{Z}}$ is a Riesz basis for $L^2(-\pi,\pi)$ for all $t \in [t_0,1)$, too; since Riesz bounds and frame bounds coincide, it is thus a Riesz basis with lower bound $A/2$. Thus, we have for $t \in [t_0,1)$:

$$\frac{A}{2}\sum_{k=-N}^{N} |c_k|^2 \le \left\| \sum_{k=-N}^{N} c_k e^{i\lambda_k(t)(\cdot)} \right\|^2_{L^2(-\pi,\pi)} \qquad \forall\, N \in \mathbb{N}, c_{-N},\ldots,c_N \in \mathbb{C}.$$

Taking the limit $t \to 1$, we obtain

$$\frac{A}{2}\sum_{k=-N}^{N} |c_k|^2 \le \left\| \sum_{k=-N}^{N} c_k e^{i\lambda_k(\cdot)} \right\|^2_{L^2(-\pi,\pi)} \qquad \forall\, N \in \mathbb{N}, c_{-N},\ldots,c_N \in \mathbb{C}.$$

Thus $\{e^{i\lambda_k x}\}_{k\in\mathbb{Z}}$ satisfies the lower Riesz sequence condition in $L^2(-\pi,\pi)$. Since it is also a frame by assumption, it is complete and satisfies the upper condition, too. Thus, it is a Riesz basis for $L^2(-\pi,\pi)$. □

We will now reconsider the classical example presented by Levinson [11].

Example 16.11. Consider the sequence

$$\lambda_k := \begin{cases} k - 1/4 & \text{if } k > 0 \\ k + 1/4 & \text{if } k < 0 \\ 0 & \text{if } k = 0. \end{cases}$$

It is clear that $D^-(\Lambda) = D^+(\Lambda) = 1$; thus, by Corollary 16.4, $\{e^{i\lambda_k x}\}_{k\in\mathbb{Z}}$ is a frame sequence in $L^2(-\gamma,\gamma)$ when $\gamma \in (0,\pi)\cup(\pi,\infty)$. It is also known [14] that $\{e^{i\lambda_k x}\}_{k\in\mathbb{Z}}$ is complete in $L^2(-\pi,\pi)$ but not a Riesz basis for $L^2(-\pi,\pi)$. Thus, by Proposition 16.10 we conclude that $\{e^{i\lambda_k x}\}_{k\in\mathbb{Z}}$ is not a frame sequence in $L^2(-\pi,\pi)$.

16.6 Exponentials $\{e^{i\lambda_k x}\}_{k\in\mathbb{Z}}$ with No Frame Property in $L^2(-\gamma,\gamma)$ for $\gamma \in (\pi D^-(\Lambda), \pi D^+(\Lambda))$

Recall that Corollary 16.4 does not provide us with any conclusion for $\gamma \in [\pi D^-(\Lambda), \pi D^+(\Lambda)]$; and for $\gamma = \pi D^-(\Lambda)$ and $\gamma = \pi D^+(\Lambda)$, Example 16.11 shows that no frame property might be available. If $D^-(\Lambda) < D^+(\Lambda)$ we are thus missing information on a whole interval. Our purpose is now to show that Corollary 16.4 is optimal in the sense that for any $a, b \in (0,\infty)$, $a < b$, we can construct sequences $\Lambda \subset \mathbb{R}$ with $a = D^-(\Lambda)$, $b = D^+(\Lambda)$, for which $\{e^{i\lambda_k x}\}_{k\in\mathbb{Z}}$ has no frame property for *any* $\gamma \in (\pi a, \pi b)$. This is stronger than Example 16.11, where we had $D^-(\Lambda) = D^+(\Lambda)$.

Theorem 16.12. *For any $0 < a < b$, there are real numbers $\Lambda = \{\lambda_k\}_{k\in\mathbb{Z}}$ satisfying*

(1) $D^-(\Lambda) = a$,
(2) $D^+(\Lambda) = b$,
(3) $\{\frac{1}{\sqrt{2\pi b}} e^{i\lambda_k x}\}_{k\in\mathbb{Z}}$ *is a subsequence of an orthonormal basis for* $L^2(-\pi b, \pi b)$,
(4) $\{e^{i\lambda_k x}\}_{k\in\mathbb{Z}}$ *spans* $L^2(-\gamma,\gamma)$ *for every* $0 < \gamma < \pi b$.

It follows that

(5) $\{e^{i\lambda_k x}\}_{k\in\mathbb{Z}}$ *is a frame for* $L^2(-\gamma,\gamma)$ *for all* $0 < \gamma \le \pi a$,
(6) $\{e^{i\lambda_k x}\}_{k\in\mathbb{Z}}$ *is not a frame sequence in* $L^2(-\gamma,\gamma)$ *for* $\pi a < \gamma < \pi b$,
(7) $\{e^{i\lambda_k x}\}_{k\in\mathbb{Z}}$ *is a Riesz sequence in* $L^2(-\gamma,\gamma)$ *for all* $\pi b \le \gamma$.

Proof. To simplify the notation, we will do the case $a = 1$, $b = 2$. The general case follows immediately from here by a change of variables. We let

$$\{f_k\}_{k\in J} = \{e^{ikx}, e^{i(k+1/2)x}\}_{k\in\mathbb{Z}}.$$

Our purpose is to exhibit a subfamily $\{f_k\}_{k\in\Lambda}$, $\Lambda \subset J$, which has the required properties.

Now, $\{f_k\}_{k\in J}$ is an orthogonal basis for $L^2(-2\pi, 2\pi)$. The idea of the construction is to carefully delete (by induction) a family of subsets of J, $J_1 \subset J_2 \subset \cdots$, so that $\Lambda = J - \cup_{n=1}^{\infty} J_n$ has the required properties. The difficult part is to maintain property (4). First, let $J_1 = \{1 + 1/2\}$ and $\alpha = 1 + 1/2$. By Lemma 16.7, there is a sequence $\{a_k^{\alpha,1}\} \in \ell^2$ so that

$$e^{i\alpha x} = \sum_{k\in J-J_1} a_k^{\alpha,1} f_k \in L^2(-2\pi+\pi, 2\pi-\pi).$$

Choose a natural number $N_1 > 1 + \frac{1}{2}$ so that

$$\left\| \sum_{|k|\ge N_1} a_k^{\alpha,1} f_k \right\|_{L^2(-2\pi+\pi,2\pi-\pi)} \le 1.$$

Let

$$J_2 = \{1 + 1/2, N_1 + 1/2, N_1 + 1 + 1/2\} = J_1 \cup \{N_1 + 1/2, N_1 + 1 + 1/2\}.$$

Since $\{f_k\}_{k \in J - J_2}$ is a frame for $L^2(-2\pi + \frac{\pi}{2}, 2\pi - \frac{\pi}{2})$ by Lemma 16.7, for each $\alpha \in J_2$ there is a sequence of scalars $\{a_k^{\alpha,2}\} \in \ell^2$ so that

$$e^{i\alpha x} = \sum_{k \in J - J_2} a_k^{\alpha,2} f_k \in L^2(-2\pi + \frac{\pi}{2}, 2\pi - \frac{\pi}{2}).$$

Choose a natural number $N_2 > N_1 + 1 + \frac{1}{2}$ so that for each $\alpha \in J_2$ we have

$$\left\| \sum_{|k| \geq N_2} a_k^{\alpha,2} f_k \right\|_{L^2(-2\pi + \frac{\pi}{2}, 2\pi - \frac{\pi}{2})} \leq 1/2.$$

Now by induction, for each n we can choose natural numbers $\{N_j\}_{j=1}^n$ with $N_{j-1} + j - 1 + \frac{1}{2} < N_j$ and sets $J_1 \subset J_2 \subset \cdots \subset J_n$ with

$$J_n = \cup_{j=1}^{n-1} J_j \cup \{N_{n-1} + 1/2, N_{n-1} + 1 + 1/2, \ldots, N_{n-1} + n - 1 + 1/2\}$$

satisfying

(i) $\{f_k\}_{k \in J - J_n}$ is a frame for $L^2(-2\pi + \pi/n, 2\pi - \pi/n)$,

(ii) for every $\alpha \in J_n$ there is a sequence $\{a_k^{\alpha,n}\} \in \ell^2$ so that

$$e^{i\alpha x} = \sum_{k \in J - J_n} a_k^{\alpha,n} f_k \in L^2(-2\pi + \pi/n, 2\pi - \pi/n),$$

and

(iii)
$$\left\| \sum_{|k| \geq N_n} a_k^{\alpha,n} f_k \right\|_{L^2(-2\pi + \pi/n, 2\pi - \pi/n)} \leq \frac{1}{n}.$$

Now, let $\Lambda = J - \cup_{n=1}^{\infty} J_n$; we claim that $\{f_k\}_{k \in \Lambda}$ has the required properties.

For (1), by the definition of the sets J_n,

$$\begin{aligned}
&\frac{|\Lambda \cap [N_{n-1} + \frac{1}{2}, N_{n-1} + n - 1 + \frac{1}{2}]|}{(N_{n-1} + n - 1 + \frac{1}{2}) - (N_{n-1} + \frac{1}{2})} \\
&= \frac{|\{N_{n-1} + \frac{1}{2}, N_{n-1} + 1 + \frac{1}{2}, \ldots, N_{n-1} + n - 1 + \frac{1}{2}\}|}{(N_{n-1} + n - 1 + \frac{1}{2}) - (N_{n-1} + \frac{1}{2})} \\
&= \frac{n}{n-1}.
\end{aligned}$$

So $D^-(\Lambda) \leq 1$. But, since Λ contains $\mathbb{Z}$ it follows that $D^-(\Lambda) \geq 1$.

For (2), for any $N \in \mathbb{N}$ we have

$$\frac{|\Lambda \cap [-2N,-N]|}{-N-(-2N)} = \frac{|\{-2N,-2N+\frac{1}{2},-2N+1,\ldots,-N\}|}{N} = \frac{2N+1}{N}.$$

So $2 \leq D^+(\Lambda) \leq D^+(J) = 2$.

(3) This is obvious.

(4) Fix $0 < \gamma < 2\pi$ and choose $M > 0$ such that $\gamma < 2\pi - \frac{\pi}{M}$. Fix $j \in \mathbb{N}$ and choose any $\alpha \in J_j$. For all $n \geq \max\{j, M\}$, we have

$$\begin{aligned} &\left\| e^{i\alpha x} - \sum_{k \in J - J_n, |k| \leq N_n} a_k^{\alpha,n} f_k \right\|_{L^2(-\gamma,\gamma)} \\ &\leq \left\| e^{i\alpha x} - \sum_{k \in J - J_n, |k| \leq N_n} a_k^{\alpha,n} f_k \right\|_{L^2(-2\pi+\pi/n, 2\pi-\pi/n)} \\ &\leq \frac{1}{n}. \end{aligned}$$

In the rest of the argument for (4) we consequently consider the vectors f_k as elements in the vector space $L^2(-\gamma,\gamma)$. Since $J_1 \subset J_2 \subset \cdots$ and

$$\max J_n \leq N_n < \min(J_{n+1} - J_n),$$

it follows that

$$\sum_{k \in J - J_n, |k| \leq N_n} a_k^{\alpha,n} f_k \in \text{span}\, \{f_k\}_{k \in \Lambda}.$$

Hence,

$$e^{i\alpha x} \in \overline{\text{span}}\, \{f_k\}_{k \in \Lambda}.$$

That is, for all j,

$$f_j \in \overline{\text{span}}\, \{f_k\}_{k \in \Lambda}.$$

Since $\{f_k\}_{k \in J}$ is a frame for $L^2(-\gamma,\gamma)$, we have (4).

(5) For $0 < \gamma < \pi a$, this follows from Theorem 16.3(a) and the fact that $D^-(\Lambda) = a = 1$. For $\gamma = \pi a$, we note that $\{f_k\}_{k \in \Lambda}$ contains $\{e^{ikx}\}_{k \in \mathbb{Z}}$, which is an orthonormal basis for $L^2(-\pi,\pi)$. Therefore, $\{f_k\}_{k \in \Lambda}$ is a frame for $L^2(-\pi,\pi)$.

(6) Since the closed linear span of $\{f_k\}_{k \in \Lambda}$ equals $L^2(-\gamma,\gamma)$ by (4), if it were a frame sequence then it would be a frame, contradicting Theorem 16.5.

(7) For $\gamma > \pi b$, this is a consequence of Theorem 16.3(b). For $\gamma = \pi b$, we note that $\{f_k\}_{k \in \Lambda}$ is a subset of $\{f_k\}_{k \in J}$, which is an orthogonal basis for $L^2(-2\pi, 2\pi)$; this implies that $\{f_k\}_{k \in \Lambda}$ is a Riesz sequence in $L^2(-2\pi, 2\pi)$.

This completes the proof of the theorem. □

Acknowledgments

The first author was supported by NSF DMS 0405376.

The authors would like to thank Darrin Speegle for useful discussions concerning Jaffard's lemma.

References

1. R. Balan, Stability theorems for Fourier frames and wavelet Riesz bases, *J. Fourier Anal. Appl.*, **3** (1997), pp. 499–504.
2. R. Balan, P. G. Casazza, C. Heil, and Z. Landau, Deficits and excesses of frames, *Adv. Comp. Math.*, 18 (2002), pp. 93–116.
3. J. Benedetto, Irregular sampling and frames, in: *Wavelets: A Tutorial in Theory and Applications*, C. K. Chui, ed., Academic Press, Boston, pp. 445–507.
4. O. Christensen, *An Introduction to Frames and Riesz Bases*, Birkhäuser, Boston, 2003.
5. O. Christensen, Perturbations of frames and applications to Gabor frames, in: *Gabor Analysis and Algorithms, Theory and Applications*, H. G. Feichtinger and T. Strohmer, eds., Birkhäuser, Boston, 1997, pp. 193–209.
6. I. Daubechies, *Ten Lectures on Wavelets*, SIAM, Philadelphia, 1992.
7. R. J. Duffin and A. C. Schaeffer, A class of nonharmonic Fourier series, *Trans. Amer. Math. Soc.*, **72** (1952), pp. 341–366.
8. C. Heil, Linear independence of finite Gabor systems, Chapter 9, this volume (2006).
9. S. Jaffard, A density criterion for frames of complex exponentials. *Michigan Math. J.*, **38** (1991), pp. 339–348.
10. H. J. Landau, Necessary density conditions for sampling and interpolation of certain entire functions, *Acta Math.*, **117** (1967), pp. 37–52.
11. N. Levinson, On non-harmonic Fourier series, *Ann. of Math. (2)*, **37** (1936), pp. 919–936.
12. J. Ortega-Cerda and K. Seip, Fourier frames, *Ann. of Math. (2)*, **155** (2002), pp. 789–806.
13. K. Seip, On the connection between exponential bases and certain related sequences in $L^2(-\pi, \pi)$, *J. Funct. Anal.*, **130** (1995), pp. 131–160.
14. R. M. Young, *An Introduction to Nonharmonic Fourier Series*, Revised First Edition, Academic Press, San Diego, 2001.

Index